Handbook of Research on Green Engineering Techniques for Modern Manufacturing

M. Uthayakumar
Kalasalingam University, India

S. Aravind Raj
Vellore Institute of Technology, India

Tae Jo Ko
Yeungnam University, South Korea

S. Thirumalai Kumaran
Kalasalingam University, India

J. Paulo Davim
University of Aveiro, Portugal

A volume in the Advances in Mechatronics and Mechanical Engineering (AMME) Book Series

IGI Global
DISSEMINATOR OF KNOWLEDGE

Published in the United States of America by
IGI Global
Engineering Science Reference (an imprint of IGI Global)
701 E. Chocolate Avenue
Hershey PA, USA 17033
Tel: 717-533-8845
Fax: 717-533-8661
E-mail: cust@igi-global.com
Web site: http://www.igi-global.com

Library of Congress Cataloging-in-Publication Data

Names: Uthayakumar, M., 1975- editor.
Title: Handbook of research on green engineering techniques for modern manufacturing / M. Uthayakumar, S. Aravind Raj, Tae Jo Ko, S. Thirumalai Kumaran, and J. Paulo Davim, editors.
Description: Hershey, PA : Engineering Science Reference, [2019]
Identifiers: LCCN 2017054764| ISBN 9781522554455 (h/c) | ISBN 9781522554462 (eISBN)
Subjects: LCSH: Manufacturing processes--Environmental aspects. | Green products. | Sustainable engineering.
Classification: LCC TS155.7 .H355 2019 | DDC 628--dc23 LC record available at https://lccn.loc.gov/2017054764

This book is published in the IGI Global book series Advances in Mechatronics and Mechanical Engineering (AMME) (ISSN: 2328-8205; eISSN: 2328-823X)

British Cataloguing in Publication Data
A Cataloguing in Publication record for this book is available from the British Library.

For electronic access to this publication, please contact: eresources@igi-global.com.

Advances in Mechatronics and Mechanical Engineering (AMME) Book Series

J. Paulo Davim
University of Aveiro, Portugal

ISSN:2328-8205
EISSN:2328-823X

Mission

With its aid in the creation of smartphones, cars, medical imaging devices, and manufacturing tools, the mechatronics engineering field is in high demand. Mechatronics aims to combine the principles of mechanical, computer, and electrical engineering together to bridge the gap of communication between the different disciplines.

The **Advances in Mechatronics and Mechanical Engineering (AMME) Book Series** provides innovative research and practical developments in the field of mechatronics and mechanical engineering. This series covers a wide variety of application areas in electrical engineering, mechanical engineering, computer and software engineering; essential for academics, practitioners, researchers, and industry leaders.

Coverage

- Nanomaterials and nanomanufacturing
- Tribology and surface engineering
- Micro and nanomechanics
- Design and Manufacture
- Autonomous Systems
- Mechanisms and machines
- Computational Mechanics
- Biologically Inspired Robotics
- Intelligent Sensing
- Sustainable and green manufacturing

IGI Global is currently accepting manuscripts for publication within this series. To submit a proposal for a volume in this series, please contact our Acquisition Editors at Acquisitions@igi-global.com or visit: http://www.igi-global.com/publish/.

The Advances in Mechatronics and Mechanical Engineering (AMME) Book Series (ISSN 2328-8205) is published by IGI Global, 701 E. Chocolate Avenue, Hershey, PA 17033-1240, USA, www.igi-global.com. This series is composed of titles available for purchase individually; each title is edited to be contextually exclusive from any other title within the series. For pricing and ordering information please visit http://www.igi-global.com/book-series/advances-mechatronics-mechanical-engineering/73808. Postmaster: Send all address changes to above address.

Titles in this Series

For a list of additional titles in this series, please visit: www.igi-global.com/book-series

Recent Advancements in Airborne Radar Signal Processing Emerging Research and Opportunities
Amir Almslmany (Independent Researcher, Egypt)
Engineering Science Reference • copyright 2018 • 190pp • H/C (ISBN: 9781522554363) • US $145.00 (our price)

Stochastic Methods for Estimation and Problem Solving in Engineering
Seifedine Kadry (Beirut Arab University, Lebanon)
Engineering Science Reference • copyright 2018 • 275pp • H/C (ISBN: 9781522550457) • US $205.00 (our price)

Design and Optimization of Mechanical Engineering Products
K. Kumar (Birla Institute of Technology, India) and J. Paulo Davim (University of Aveiro, Portugal)
Engineering Science Reference • copyright 2018 • 347pp • H/C (ISBN: 9781522534013) • US $235.00 (our price)

Managerial Approaches Toward Queuing Systems and Simulations
Salvador Hernandez-Gonzalez (Tecnológico Nacional de México en Celaya, Mexico) and Manuel Dario Hernandez Ripalda (Tecnológico Nacional de México en Celaya, Mexico)
Engineering Science Reference • copyright 2018 • 311pp • H/C (ISBN: 9781522552642) • US $195.00 (our price)

Socio-Technical Decision Support in Air Navigation Systems Emerging Research and Opportunities
Tetiana Shmelova (National Aviation University, Ukraine) Yuliya Sikirda (National Aviation University, Ukraine) Nina Rizun (Gdansk University of Technology, Poland) Abdel-Badeeh M. Salem (Ain Shams University, Egypt) and Yury N. Kovalyov (National Aviation University, Ukraine)
Engineering Science Reference • copyright 2018 • 305pp • H/C (ISBN: 9781522531081) • US $175.00 (our price)

Soft Computing Techniques and Applications in Mechanical Engineering
Mangey Ram (Graphic Era University (Deemed), India) and J. Paulo Davim (University of Aveiro, Portugal)
Engineering Science Reference • copyright 2018 • 336pp • H/C (ISBN: 9781522530350) • US $215.00 (our price)

Advanced Numerical Simulations in Mechanical Engineering
Ashwani Kumar (Government of Uttar Pradesh, India) Pravin P. Patil (Graphic Era University, India) and Yogesh Kr. Prajapati (National Institute of Technology Uttarakhand, India)
Engineering Science Reference • copyright 2018 • 242pp • H/C (ISBN: 9781522537229) • US $165.00 (our price)

IGI Global
DISSEMINATOR OF KNOWLEDGE

701 East Chocolate Avenue, Hershey, PA 17033, USA
Tel: 717-533-8845 x100 • Fax: 717-533-8661
E-Mail: cust@igi-global.com • www.igi-global.com

List of Reviewers

V. Anbumalar, *Velammal College of Engineering and Technology, India*
S. Appavu, *K. L. N. College of Engineering, India*
S. Bathrinath, *Kalasalingam University, India*
K. Jayakrishna, *VIT University, India*
Shouxun Ji, *Brunel University London, UK*
M. Kathiresan, *Thiagarajar College of Engineering, India*
M. Adam Khan, *Kalasalingam University, India*
Rendi Kurniawan, *Yeungnam University, South Korea*
Magdalena Niemczewska-Wojcik, *Cracow University of Technology, Poland*
Chander Prakash, *Lovely Professional University, India*
B. Vijay Ramnath, *Sri Sairam Engineering College, India*
N. S. Reddy, *Gyeongsang National University, South Korea*
M. Saravanamohan, *Kumaraguru College of Technology, India*
Sunpreet Singh, *Lovely Professional University, India*
Adam Slota, *Cracow University of Technology, Poland*
M. T. H. Sultan, *Universiti Putra Malaysia, Malaysia*
Temel Varol, *Karadeniz Technical University, Turkey*
Yusri Yusof, *UTHM, Malaysia*

List of Contributors

Table of Contents

Detailed Table of Contents

The demand for the orthopedic and dental implants has increased sharply in last decade due to physical traumas and age-related deficiencies. The material used for orthopedic and dental implants should be biocompatible to ensure the adaptability of the implant in the human body. The mechanical stability of implants is dependent on mechanical properties and surface characteristics essential to ensure corrosion and wear resistance. The requirement of mechanical properties also differs substantially from load-bearing to non-load-bearing implants. There are many problems arising due to lack of sufficient biocompatibility, like infection, poor osseointegration, and excessive foreign body response. Fatigue failure, stress shielding, and bone resorption are some major problems associated with lack of mechanical stability. Numerous conventional materials, coatings, and nanomaterials have been used to enhance the implant stability.

The chapter proposes an integrated manufacturing system consisting of three main components: digital prototyping, physical prototyping, and lost core technology. The integrated system combines the beneficial aspects of computer-aided design, computer-aided engineering, rapid prototyping, and rapid tooling. The proposed integrated system is an attempt to compress the product development time while saving cost.

The system can be efficient in designing of mold, parts with complex ducts and cavities, and carrying out design analysis through optimization and simulations. The system is therefore an attempt to minimize the waste of material that occurs in the development of a product and is therefore an efficient green technology for the manufacturing industries.

Conventional investment casting (IC) has suffered from numerous limitations such as rigidity of the process, longer production cycles, higher tooling cost, and waste during different manufacturing stages. With the invent of additive manufacturing (AM) technologies, it is now possible to overcome the aforesaid issues along with additional benefits in terms of comparatively better quality characteristics of the resulting castings. The collaboration of AM and IC provided numerous avenues, specifically in biomedical, aerospace, and automobile sectors. AM technologies supported the IC process both in direct and indirect ways where these systems can be used for both job and mass production applications, respectively. In the chapter, the author will try to discuss the assistance of AM process to IC in detail. Each and every step to be followed will be supported with the practical findings, either by the contributing author or published somewhere else. Moreover, some of the case studies will be discussed in detail to highlight the practical importance of the duo.

The objective behind the development of miniaturization or small specimen test technology is to reduce the cost and quantity of material involved during the characterization of the material. The idea of the development of miniaturization took attention when the nuclear industry starts developing as these materials are very costly and it is not economically feasible to waste large amount of these materials for the sole purpose of testing. The second factor which promotes the miniaturization is that the working of machine is not affected while at the same time its material is being tested. At present, the idea of miniaturization is being applied to other materials also. The miniaturization of standards for metals has been done successfully in the past. For composites, not much work has been done. In the chapter, the specimen size effects on tensile properties of glass fiber composite have been identified by varying the length and width simultaneously and have established a relationship between the ASTM standard specimen and the small size specimen.

Robots are electromechanical systems that need mechatronic approach before manufacturing to reduce the development cost. In this chapter, the modelling of the 8 degrees of freedom (DOF) SCARA robot with a multiple gripper using SolidWorks CAD software and the dynamic study with the aid of MATLAB/SimMechanics is presented. The SCARA with multiple gripper is used for pick and place operation in manufacturing industries. The SolidWorks CAD model of SCARA with multiple grippers is converted into SimMechanics block diagram by exporting the 3D CAD model to the MATLAB/SimMechanics second generation technology environment. The motion sensing capability of the SimMechanics is used for determining the dynamic parameters of the manipulators. The SimMechanics block diagrams and the results of the dynamic study presented in this chapter infer that the structure of the robot can be changed to get the required dynamic parameters.

Section 2
Optimization Techniques and Material

Chapter 6
PageRank Algorithm-Based Recommender System Using Uniformly Average Rating Matrix........... 99
Bathrinath S., Kalasalingam University, India
Saranyadevi S., Kalasalingam University, India
Thirumalai Kumaran S., Kalasalingam University, India
Saravanasankar S., Kalasalingam University, India

Applications of web data mining is the prediction of user behavior with respect to items. Recommender systems are being applied in knowledge discovery techniques to the problem of making decisions on personalized recommendation of information. Traditional CF approaches involve the amount of effort increases with number of users. Hence, new recommender systems need to be developed to process high quality recommendations for large-scale networks. In this chapter, a model for UAR matrix construction method for item rank calculations, a Page Rank-based item ranking approach are proposed. The analysis of various techniques for computing item-item similarities to identify relationship between the selected items and to produce a qualified recommendation for users to acquire the items as their wish. As a result, the new item rank-based approaches improve the quality of recommendation outcome. Results show that the proposed UAR method outperforms than the existing method. The same method is applied for the large real-time rating dataset like Movie Lens.

Chapter 7
Performance of PM Linear Generator Under Various Ferromagnetic Materials for Wave Energy Conversion 113
Izzeldin Idris Abdalla Yagoube, Universiti Teknologi PETRONAS, Malaysia
Taib Ibrahim, Universiti Teknologi PETRONAS, Malaysia
Nursyarizal Mohd Nor, Universiti Teknologi PETRONAS, Malaysia
Perumal Nallagownden, Universiti Teknologi PETRONAS, Malaysia

This chapter examines the influence of the various ferromagnetic materials on the performance of a single-phase tubular permanent-magnet linear generator (TPMLG) for wave energy conversion. Four ferromagnetic materials were considered in this study. They are non-oriented electrical steel, Permalloy (Ni-Fe-Mn), Accucore, and Somaloy 700. The generator equipped with a tubular stator carries a single coil and employs a quasi-Halbach magnetized moving-magnet translator. Therefore, in order to obtain an accurate performance analysis, the nonlinear time-stepping finite-element analysis (FEA) technique has

Section 3
Industrial Engineering and Management

Key performance indicators (KPIs) are a critical tool to support activities and results' monitoring in any industrial organization. The published literature and the available approaches on KPIs focus on the business and administrative level, being computed with information retrieved at the shop-floor level. Despite that, there is a scarcity of structured and comprehensive approaches to support the generation of KPIs to be used at the shop-floor level (the few existent approaches are empiric-based). In this chapter, a methodology to support the selection and organization of KPIs at the shop-floor level is proposed. Departing from the Hoshin Kanri strategy deployment, it identifies the levels of decision and control in the company regarding the production activities and derives the most adequate KPIs for each level based on universal questions about "what performance to assess." The build-up of visual management boards for each level is also proposed.

Reverse logistics stands out as a rapidly gaining concept due to its contribution to both the environment and the economy. There are many problems with reverse logistics. The decision of recovery options is a fundamental issue that serves many purposes. Choosing the right recovery option will also provide the environmental and economic contribution to maximize the benefits. For this purpose, many solution approaches have been produced for different objectives, which are based on the selection of better recovery options. Since solution approaches are directly interacting with problem models and objectives, it is important to determine an appropriate approach to achieve better results. Until now, many different approaches have been implemented, and results are shared. This chapter systematically examines these solution approaches and reveals the achievements in the literature in order to provide directions for future studies.

Section 4
Green Manufacturing and Sustainable Engineering Concepts

The eco-efficiency assessment is a powerful metric to introduce two components of sustainability assessment in the industrial companies' decisions making: the concurrent consideration of economic and environmental performance. The application of the eco-efficiency concept and of the normative documents is not an easy task, mainly because there are myriad environmental related indicator to consider and acquire. This barrier is higher in the realm of plastic injection molding, where each mold is unique, requiring a recurrent effort of data retrieving for such one-of-a-kind molds. To overcome this barrier, an integrated framework to support the eco-efficiency calculation on a life cycle perspective for a specific type of products, injection molds, is proposed in this chapter. It retrieves a small but representative selected set of eco-efficiency performance indicators. A tool was developed to apply the proposed framework and the results of its application to four real industrial case studies is discussed.

The present industry scenario focuses on green manufacturing, in terms of effective reuse and recycling of the industrial wastes generated in enormous amount while preparing the product. The wastes also act as a threat to the society by causing various kinds of pollution. Therefore, the proper safe disposal of the same is a very critical factor. Most of the industries struggled with the enormous disposal of these wastes and finding ways for reuse and disposal. In this chapter, one such way of reuse of these wastes for making composite product is explored. Industrial wastes such as flyash and ricehusk used as fillers of varying weight percentages, 6%, 8%, 30%, 40%, and 50%, wt%, respectively, are reinforced with matrix. The prepared composites were subjected to flexural studies to know the load withstand ability. Results show that the incorporation of both fly ash and rice husk industrial wastes as filler into the polymer matrix increases the flexural strength. In addition, a low-cost product with high strength and good performance is obtained by adding this waste.

Growth of nature is an additive process that gives sustainable existence to the structures developed; on the other hand, traditional manufacturing techniques can be wasteful as they are subtractive. Additive manufacturing produces almost nil waste and accordingly preserves raw materials resulting in cost reduction for the procurement of the same. It will also cut down on the carbon emissions that are usually generated from industrial manufacturing. Additive printed objects are lighter as well, making them more efficient, especially when used in the automobile and aerospace industry. Further, the intrinsic characteristics and the promising merits of additive manufacturing process are expected to provide a solution to improve the sustainability of the process. This chapter comprehensively reports on various additive manufacturing processes and their sustainable applications for green technology. The state of the art, opportunities, and future, related to sustainable applications of additive manufacturing have been presented at length.

In this chapter, low elastic modulus porous Mg-Zn-Mn-(Si, HA) alloy was fabricated by mechanical alloying and spark plasma sintering technique. The microstructure, topography, elemental, and chemical composition of the as-sintered bio-composite were characterized by optical microscope, FE-SEM, EDS, and XRD technique. The mechanical properties such as hardness and elastic modulus were determined by nanoindentation technique. The as-sintered bio-composites show low ductility due to the presence of Si, Ca, and Zn elements. The presence of Mg matrix was observed as primary grain and the presence of coarse Mg2Si, Zn, and CaMg as a secondary grain boundary. EDS spectrum and XRD pattern confirms the formation of intermetallic biocompatible phases in the sintered compact, which is beneficial to form apatite and improved the bioactivity of the alloy for osseointegration. The lowest elastic modulus of 28 GPa was measured. Moreover, the as-sintered bio-composites has high corrosion resistance and corrosion rate of the Mg was decreased by the addition of HA and Si element.

The increasing competition among the manufacturing organizations and stringent government regulation forces the manufacturing organizations to implement sustainability principles in manufacturing. Sustainability focuses on material, product development, and manufacturing process orientations. End of life (EoL) disposal of the product is very much important in the modern scenario. The remanufacturing is a vital strategy for attaining sustainability in manufacturing. The assessment of remanufacturability of products needs to be done during the design stage so as to provide the manufacturers the guidelines for sustainable product development. In this context, this chapter presents the insights on remanufacturability index assessment for a typical automotive product. The practical implications of the study are also being discussed.

Increasing legislative concerns and rapidly transforming technologies pressurizes the global competitive landscape to deploy smart, safe, and sustainable green manufacturing. This chapter scrutinizes organizational sustainability of the automobile components manufacturing organizations located in the state of Tamil Nadu, India using hierarchy cluster analysis towards setting up a benchmark on sustainability of organizations. Along with the triple bottom line (TBL) of sustainable development, the organizational responsibility and government legislation in achieving sustainability were selected as the five major governing variables during the conduct of this case study. As a result, 25 automotive components manufacturing organizations chosen from for this study were classified into three clusters, confirming a particular organization as the most suitable one for the conduct of green manufacturing sustainability studies. According to the distinctiveness of the assorted clusters, suggestions were also proposed for improving the organizational sustainability further.

Foreword

Sustainable manufacturing is the need of the hour today. Environmental benign manufacturing technologies are the requirements of today's scenario. Academic, industry and research organizations are now interested in implementing green engineering techniques. This book covers a selection of such important green manufacturing techniques for modern manufacturing. This volume has four sections with 17 focused chapters. The newer techniques are covered in a precise way.

Multidisciplinary approach is to be realized to solve the real-world tasks. Section 1 deals with biomedical manufacturing, 3D printing, advanced additive manufacturing and necessity on miniaturization. Section 2 covers the advanced optimization techniques. Page rank algorithm-based recommender system is elucidated. Particle swarm optimization is explained to optimize the process parameters in advanced machining process. Advanced machining methods are explained especially for difficult to machine materials. Performance evaluation of various erromagnetic materials in wave energy conversion is illustrated. New lanthanum based composite machining is discussed in a detailed way. Section 3 deals with the industrial engineering and management related aspects. Methodology for shop floor key performance indicators (KPI) is discussed. Solution approaches for reverse logistics is reviewed. Section 4 deals with sustainable engineering concepts. Sustainability associated with plastic parts is discussed. An integrated tool for industrial use is demonstrated. Adding value to waste by way of utilizing waste to manufacture product is immense helpful in alleviating problems in waste disposal. Demonstration of making polymeric composite from industrial waste is carried out in this chapter. Green manufacturing strategies, additive manufacturing towards green manufacturing, manufacturability index related to automotive industry are giving information on newer trends in manufacturing.

This book covers articles in multidiscipline such as bio technology, additive manufacturing, productivity improvement, optimization techniques, utilization of waste to make product etc. The readers will find interesting to see all the important and relevant topics in a book with a connected way. Certainly, this book introduces the green manufacturing to readers and it encourages the readers towards the implementation.

S. Aravindan
Indian Institute of Technology Delhi, India

Preface

INTRODUCTION

Global scenario and contemporary manufacturing necessitates the importance towards "Green Manufacturing" (GM), since its related to all manufacturing processes and its related activities. The main aim of GM is to reduce the emissions, pollutants and other by-products which is harmful to the environment. Comparing with older manufacturing methods and practices recent techniques such as additive manufacturing and sustainable manufacturing focuses about environmental friendly fabrication. Also green industrial concepts such as Green Supply Chain Management (GSCM), reverse logistics and other related techniques to practice sustainable utilization of resources. The ideas and concepts related to GM is prospered globally in past decade and gradually implementation of newer ideas was practiced by various researchers, laboratories and practitioners across the world. This book focuses emerging researchers, scientist and students to acquire recent advancements in green engineering techniques for modern manufacturing.

ORGANIZATION OF THE BOOK

This book is organized with 17 chapters by various researchers, faculty and practitioners across the world who are currently working on various phases of GM.

Handbook of Research on Green Engineering Techniques for Modern Manufacturing is organized in four different sections

Section 1: "Recent Trends in Manufacturing" focuses about recent manufacturing concepts related to GM and its impact compared to conventional manufacturing processes.

Section 2: "Optimization Techniques and Material" is devoted to newer optimization techniques, mathematical models, problem solving techniques which can be used for sustainable practices and measurement.

Section 3: "Industrial Engineering and Management" presents current industrial engineering and managerial concepts to know extent of green practices in a firm.

Section 4: "Green Manufacturing and Sustainable Engineering Concepts" contributes the sustainability and sustainable practices in manufacturing firms

Section 1

Chapter 1 presents the biomechanical properties of orthopaedic and dental implants as a review. This gives an extensive idea for researchers who works in similar field by exploring various research articles and its implications for biomedical components to be implanted in human body. This article also gives its view on biocompatibility, like infection, poor osseointegration and excessive foreign body response of materials.

Chapter 2 focuses about the zero waste in manufacturing using additive manufacturing techniques. The authors discussed about various processes such as stereolithography, fused deposition modelling, polyjet, selective laser sintering, electron beam melting. It is evident that above mentioned processes was useful for making complex geometric products with zero wastages since its additive manufacturing techniques.

Chapter 3 presents about recent advancements in investment casting Investment casting is a special casting process which is otherwise known as precision investment casting. The authors discussed about the various limitation in conventional investment casting process such as such as rigidity of the process, longer production cycles, higher tooling cost and wastages during different manufacturing stages. The conventional process was slightly modified with inclusion of additive manufacturing techniques to overcome the limitations.

Chapter 4 focuses about miniaturization of test specimen since standard specimen preparation leads to excessive cost in fabrication as well as testing. The authors explored this concept based on nuclear engineering testing where very small specimen was used for testing which is giving efficient and cost effective results. Since miniaturization of metal component testing was done earlier, this work focuses in glass fibre testing and relationship between results with ASTM standard specimen.

Chapter 5 discusses about the simulation of SCARA robot with eight degrees of freedom. The SCARA robot with multiple grippers was used in this study and simulation was carried in SimMechanics environment. The simulation results shows design and structural changes can be done with respect to the work environment.

Section 2

Chapter 6 study about application of web data mining to explore and compute the consumer behaviour using PageRank algorithm. Decision making problem can be solved using this technique. Page Rank-based item ranking approach are proposed to handle large data sets. The analysis was done to compute relationship between the selected items and to produce a qualified recommendation. This method provides improved outcome compared to previous methods

Chapter 7 presents performance of PM linear generator under various ferromagnetic materials for wave energy conversion, four ferromagnetic materials were considered in this study non-oriented electrical steel, Permalloy (Ni-Fe-Mn), Accucore and Somaloy 700. To achieve accurate results non linear time stepping finite element analysis was used. The results shows positive strengths of newer of materials such as ease of manufacture, highest force capability, lower cost and minimum eddy-current loss.

Chapter 8 discusses about the optimization of process parameters for Electro-Chemical Machining (ECM) which is most commonly used non-conventional machining process. In this study ECM process of EN 18 material was carried out using particle swamp optimization technique. The following parameters were considered for optimization Electrolyte concentration (%), Voltage (V), Feed rate (mm/min) and Inter-electrode gap (mm).

Chapter 9 discusses about the performance study of $LaPO_4$-Y_2O_3 composite fabricated by Sol-Gel process using Abrasive Water Machining (AWM). This study focus to improve the material removal rate with reduced kerf angle and surface roughness, by considering the input parameters to achieve output. Jet pressure plays a vital role in AWM and it contributes major part in material removal. Microscopic examination was carried to verify the plastic deformation and cracks in specimen.

Section 3

Chapter 10 presents the Key Performance Indicators (KPIs) in shop floor. Various research article was publishes about the importance of KPI in manufacturing firms but the implementation level of KPIs in shop floor is not discussed well. This article focused to collect data sets from shop floor and interpret the importance of data on business and administrative level.

Chapter 11 discusses about reverse logistics and its importance on environment and economy. Implementation of reverse logistics concepts in a firm has various practical issues. In this work authors have explored best recovery option for providing environmental and economic benefits. Several solution approaches were used to find better recovery options. This study examine the validity of solution approaches and its achievements in reverse logistics.

Section 4

Chapter 12 present the sustainability analysis of plastic parts production. Conventional plastic production involves various steps and this study explores plastic injection moulding. In conventional process moulds are unique for individual applications. To overcome all other issues the authors have developed an integrated framework to improve the eco-efficiency of the process as well as life cycle of particular process was analysed.

Chapter 13 discuss about the effective utilization of industrial wastages for preparing polymer matrix composites. This chapter gives a new perspective in the field of recycling ofwastage collected from industry and reuse of same as raw material for composite preparation. Fly ash and rice husk were used in various composition as filler material and the specimen was subjected to flexural load testing. The results shows increase in strength with addition of filler materials.

Chapter 14 presents about application of green concepts and sustainability through additive manufacturing. Compared to conventional manufacturing process additive manufacturing provides better solution with ease in manufacturing with less emission. This plays a vital role in sustainability and green concepts. The merits of additive manufacturing was discussed with state of art manufacturing and future perspectives of it.

Chapter 15 explores about the fabrication of biodegradable low elastic porous Mg-Zn-Mn-Si-HA alloy using spark plasma sintering process. The mechanical properties of the component was analysed. The microstructure, topography, elemental and chemical composition of bio-composite were characterized by optical microscope, FE-SEM, EDS and XRD technique.

Chapter 16 discusses about the remanufacturability index of an automotive component. This study focuses about effective utilization of automotive component based on life cycle analysis and end of life concepts. Remanufacturability of products needs to be done during the design stage which provide the manufacturers, the necessary guidelines for sustainable product development and the selected component was used for remanufacturing after its end of life.

Chapter 17 presents about application of cluster analysis to select competent automotive firms to conduct the sustainability studies. Twenty-five automotive firms were selected for this study further grouped into three different clusters based on computation. The cluster was formed based on data analysis and responses provided by the case organizations. Most suitable firm was selected for sustainability study. Further suggestions were provided for rest of firms to improve their level.

Section 1
Recent Trends in Manufacturing

Chapter 1
Biomechanical Properties of Orthopedic and Dental Implants:
A Comprehensive Review

Manjeet Kumar
Panjab University, India

Rajesh Kumar
Panjab University, India

Sandeep Kumar
Guru Jambheshwar University of Science and Technology, India

Chander Prakash
Lovely Professional University, India

ABSTRACT

The demand for the orthopedic and dental implants has increased sharply in last decade due to physical traumas and age-related deficiencies. The material used for orthopedic and dental implants should be biocompatible to ensure the adaptability of the implant in the human body. The mechanical stability of implants is dependent on mechanical properties and surface characteristics essential to ensure corrosion and wear resistance. The requirement of mechanical properties also differs substantially from load-bearing to non-load-bearing implants. There are many problems arising due to lack of sufficient biocompatibility, like infection, poor osseointegration, and excessive foreign body response. Fatigue failure, stress shielding, and bone resorption are some major problems associated with lack of mechanical stability. Numerous conventional materials, coatings, and nanomaterials have been used to enhance the implant stability.

DOI: 10.4018/978-1-5225-5445-5.ch001

INTRODUCTION

With increasing physical trauma, inherent structural defects and age related deformities, it is necessary to develop quantitative and qualitative enhanced orthopaedic implants (OI) and dental implants (DI). In 2016, Global OI's market was valued at \$47,261 million and it is expected to rise to \$74,796 million by 2023, registering compound annual growth rate (CAGR) of 6.8% during the forecast period 2017 - 2023 (TMR, 2016). The global dental implants market is projected to expand at a modest CAGR of 6.9% between 2017 and 2025 (Chandra, 2014). OIs are further divided into load bearing and non-load bearing implants. Load bearing OIs includes articulating joint replacement like hip, knee, shoulder and finger replacements and other prosthesis. While non load bearing includes various structural elements like pins, rods and plates which supports the damaged orthopedic parts when they heals them-self properly (Alivu et al.,2018). Historically, there have been two different types of DI: (1) endosteal and (2) subperiosteal. Endosteal refers to an implant that is "in the bone," and subperiosteal refers to an implant that rests on top of the jawbone under the gum tissue. Subperiosteal implants are no longer in use today because of their poor long-term results in comparison to endosteal dental implants.

Metals, ceramic and polymers have been used as bio-material in implants. In metals, titanium based alloys are leading material due to their unique properties. Other metals that have been used are steel and cobalt - chromium based alloys. In ceramics, calcium phosphate based composites are widely used due to their proximity with bone apatite. Hydroxyapatite ($Ca_{10}(PO_4)6(OH)2$) is most commonly used calcium phosphate bioactive ceramic. Other ceramics which have been used are alumina (Al_2O_3) and zirconia (ZrO_2) which fall in category of bioinert ceramics. Polymers have been used not only in OI and DI but also in tissue engineering and drug delivery due to their unique properties. In polymers, mainly ultra high molecular weight Poly ethylene (UHMWPE), Polytetrafluorethyelene (PTFE), Polymethyl methacrylate (PMMA), Polylactide (PLA), Polyglycolide (PGA) and Polyetheretherketon (PEEK) have been used. With recent advances in material science, various nanomaterials have been developed which enhanced the capabilities of implants. Nanomaterials have structural compatibility with hard tissues like bone. Nanomaterials based implants are qualitatively better because of enhancement in essential properties.

The prerequisite properties of materials for OI and DI are characterized as mechanical, biological and other surface properties. The performance of implants highly depends on these properties. It is necessary to focus the research on these prerequisite properties to develop future materials or enhance the capabilities of existing materials. This chapter discuss the necessary prerequisite properties of materials for OI and DI.

PREREQUISITE PROPERTIES

Orthopedic and dental structure of living bodies is like a mechanical system imbedded in a biological environment. The biomaterial which is intended to use in implantation should have desired mechanical and biological properties. Implant's material and design is most dominant factor which decides both the short term and the long term performance of the implant (Prakash et al.,2016). With increase in human life expectancy, the better quality and long life spanned OIs and DIs are needed (Narayan, 2012). The human body's internal environment is very austere having an oxygenated saline solution with salt content of about 0.9% at pH 7.4 and temperature of 37°C (Saro & Sidhu, 2012). This environment accelerates the fatigue failure, corrosion and wears rates several times that cause decrements in the implant life.

Corrosion and wear cause many particle borne diseases and infection that may put risk on patient's life (Prakash et al.,2015). So, properties of material are not only important from functional point of view but also from patient's life point of view. The implant's material properties can be classified as mechanical properties, biological properties, and corrosion resistance as shown in figure 1. These properties are interrelated and affect each other.

An ideal material for implants should have sufficient level of all three types of properties. For example, if a material is highly biocompatible and bioactive but have poor mechanical properties, then it cannot be used in load bearing OIs.

Biological Properties

The biological properties of implant's material are essential to ensure the compatibility of the implant with the surrounding environment in human body. The absence of biological properties may trigger sever toxicity, infection, acute inflammation, and sever foreign body reactions. Ultimately, this may cause morbidity. The major biological property is biocompatibility of materials. Bioactivity is subpart of biocompatibility which allows the tissue growth on materials' surface (Prakash et al.,2017). In fact, no material is accepted by our biological system completely as our tissues and immune system treat all materials as foreign element. But the materials which are termed as bio-materials are more biocompatible than other materials, which have been proved by *in vivo and in vitro tests.*

Biocompatibility

The material for implantation should not be toxic, compatible or have harmony with surrounding tissues and allow the growth of tissues on its surfaces. The material itself or any degraded products from it should not cause cell death, intense inflammation, suppress immunity or other harmful effects on cellular or tissue functions. The living tissues have ability to grow and renew itself with time due to the

Figure 1. Classification of prerequisite properties of OI's and DI's material

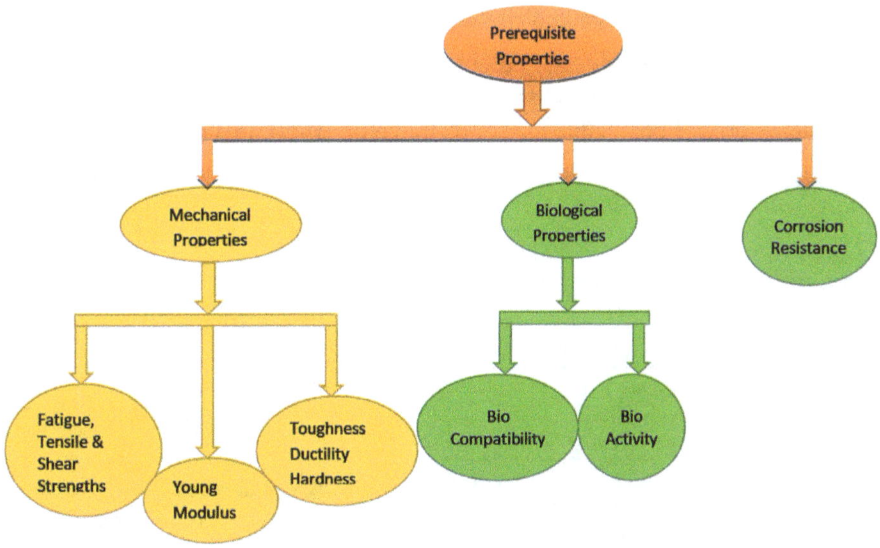

metabolism activities of human body. The Implant's material should allow the growth of surrounding tissues to ensure Osseointegration. The implant surface should be osseoconductive and osseoinductive. Toxicity is further divided into cytotoxicty, genotoxicity and carcinogenicity. Cytotoxicty causes cell death while genotoxicity damages the DNA. Carcinogenicity causes the cancer in surrounding cells. Toxicity effects from bulk as well as degraded material on host body and cells & tissues attachment on surface are checked by *in vitro* and *in vivo* tests to observe the biocompatibility of material usually. The biocompatible material is divided into three categories, biotolerant, bioinert, and bioactive. Bio tolerant materials do not release nontoxic substances but these lead to encapsulation within connective tissues. No harmful elements are released by bioinert materials but nil or very less tissue growth is allowed on their surface (Bauer & Schmuki, 2013; Wintermantel, 2002). Bioactive materials allow tissue growth on surface. Bioactivity will be discussed in next section. Biocompatibility is a surface phenomenon as only surface of material is interacted with living tissues. Different types of biocompatibilty tests are shown in table 1.

Bioactivity

Bioactivity of an implant is shown by adhesion of osteoblast and cohesion of fibroblast (Balasundarm, 2007). The cell adhesion is controlled by adsorption of proteins on implant surface (Schakenraad, 1996). Some proteins such as fibronectin and victronectin in biological fluid convey the adhesion, osteoblast differentiation, and growth of worthy cells on implant surface. Further studies found that protein adsorption depends upon surface features like pore size and roughness (topography), surface composition and surface energy of implants (Durrieu, 2004). These factors affect both concentration and type of protein adsorption. Explicit amino acid sequences in adsorbed protein adheres the osteoblast preferentially (Balasundram, 2007; Porte-Durrieu, 2004). These specific amino acid sequences are mainly arginine-glycine-aspartic (RGD) and heparin-sulphate (Balasundaram & Sato, 2006; Balasundaram & Webster, 2006). The protein adsorption and cellular adhesion mechanism is shown in figure 2. The researchers have paid less attention in understanding of cellular recognition to initially absorbed protein on biomaterial surfaces and it is one of key reasons for short life span (On average less than 15 years) of OIs (Balasundaram & Webster, 2006). Because if the material's surface is not designed to provoke the absorption of protein initially, the osteoblast adhesion and fibroblast cohesion on implant surface will be very low or negligible. Ultimately, the bone growth on implant's surface will be poor and this will lead to failure of OI. Mainly three types of cell adhesions are observed on implant surface depending upon the distance from surface (Durrieu, 2004). Focal adhesion (at distance of 10-20 nm) is strongest followed by 'close contact adhesion' (at distance of 30-50 nm) and extra cellular matrix contacts or fibrillar adhesion (at

Table 1. Biocompatibility and Bioactivity tests

	***In Vitro* Tests**	***In Vivo* Tests**
Bio-compatibility tests	• Cell Culture test • Haemolysis test • Platelet adhesion test	• Micro-CT analysis • Histology and histomorphometric test
Bio-activity tests	• Stimulated Body Fluid test (SBF) • Water Contact angle on surface (Hydrophilicity)	• Micro-CT analysis • Histology and histomorphometric test

Figure 2. Mechanism of Protein adsorption and osteoblast adhesion

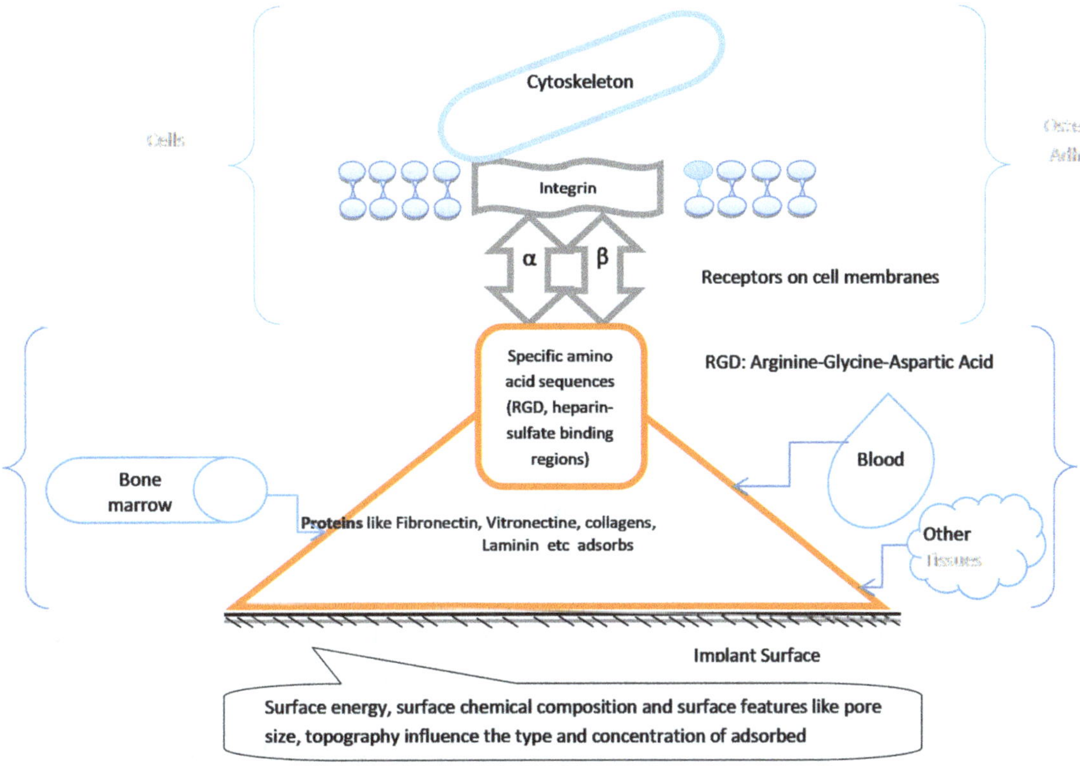

distance greater than 100 nm). Focal adhesion is desirable and symbol of better bioactivity of implant surface. Tests for bioactivity *in vitro* and *in vivo* are given in table 1.

The implant surfaces have been classified on basis of different surface parameters such as roughness texture and orientation of irregularities (Wennerberg, 2010; Chaturvedi, 2009; Singh, 2015; Prakash et al.,2017). Wennerberg et al.(2010) have classified surfaces according to roughness as minimal rough, intermediate rough and rough as shown in figure 3. The additive treatments on implants such as coatings make them concave textured while subtractive treatments such as itching make them convex textured. The similar or directional topography is based on orientation of irregularities. The adsorption of proteins is also affected by surface energy of implant (Misch, 1998). The surface energy or surface tension is determined by wettability or hydrophillicity of surface (Singh, 2015).

Corrosion Resistance

As mentioned earlier, the human body's environment is very harsh with oxygenated saline solution which favours and accelerates the corrosion process. The human blood plasma contains chlorine ions which further makes highly aggressive environment. Most metallic materials undergo chemical and electro-chemical dissolution in this environment (Karamachimudali, 2003). The reactions in aqueous solution which are responsible for corrosion in metals are electrochemical in nature. The body temperature of 37 °C accelerates these electrochemical reactions (Burstein, 2005). In this environment, metal cations are formed on metal surface. Instead of formation of oxide layer on surface, metal ions permanently

Figure 3. Classification of Implant surfaces
(Wennerberg, A., 2010, Singh, M., 2015)

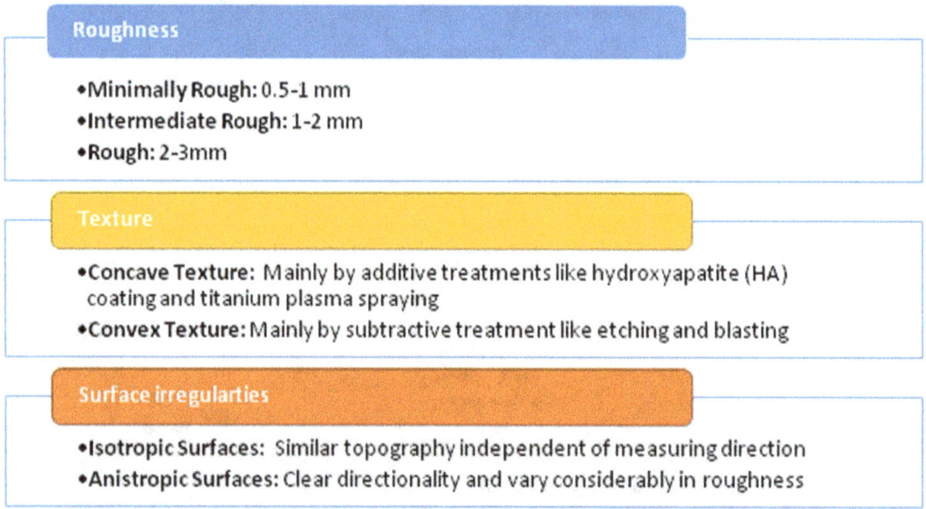

solvatized and thus metal starts dissolving. The dissolution results in loss of metallic ions in this saline solution. The corrosion and wear debris have fine particles, these interact with surrounding tissues and cause chronic inflammation and may result in osteolysis (Waerhaug, 1956). The release of elements can cause allergic as well as toxic reactions in adjacent tissues (Gettleman, 1977). These reactions may force to perform another surgery for cutting out the implant and in worst case may lead to death of patient. Corrosion gnarled the surface of implant and which leads to abrasion and discolouration of adjacent tissues (Ashman, 1971).

Corrosion resistance is not only important from biological point of view but also important from mechanical point of view. Corrosion makes implant weaker and causes breakage. Many mechanical phenomenon such as fatigue strength, wear resistance and stress shielding are influenced due to corrosion (Ramakrishna, 2001; Prakash et al.,2018). The wear and corrosion reported to be major cause for particle borne diseases, loosening and premature failure of hip and knee implants (Ramakrishna, S., 2001). Types of corrosion, which are pertinent to the currently used alloys, (shown in figure 4) are pitting, crevice, galvanic, inter-granular, stress-corrosion cracking, corrosion fatigue, and fretting corrosion (Ramakrishna, 2001; Adya 2005).

Mechanical Properties

Human body is a living machine and orthopaedic and dental structure is like a mechanical system embedded in biological environment. The paramount motive of implants is to provide mechanical stability, movements, and avoid relative micro-motion at bone-implant surface. The mechanical properties of implant materials are essential for proper alignment and retaining of function of bone during physiologic loading of bones and joints. Carter et al. (1998) revealed that the mechanical stability helps in biological aspect of bone healing by decreasing unnecessary shear stress. While, Liu et al. (2008) and kienapfe et al. (1999) observed that minimization of micro motion at bone-implant interface would promote the

Figure 4. Classification of corrosions in Orthopedic Implants
(Adya, N., 2005)

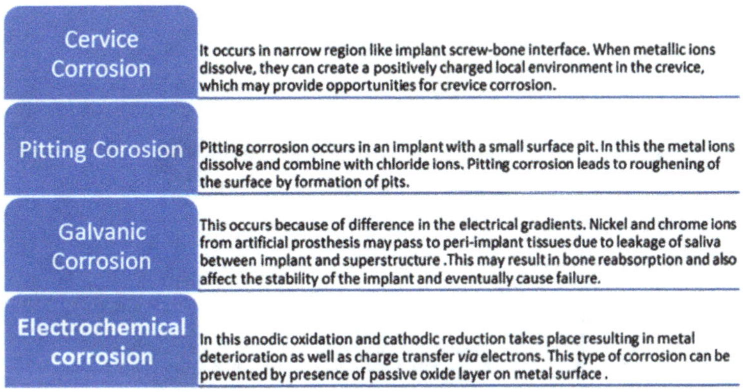

bone formation and remodeling. Thus, mechanical and biological aspects of implants and bone healing are closely inter-related and mechanical properties also help to achieve the biological goals.

The Young's Modulus (YM) of implant material should be comparable to that of bone to avoid stress shielding effect (Niinomi, 2011). The YM of cortical bone ranges from 10-30 Gpa (Bauer & Schmuki, 2013), but on an average, it is considered as 18 Gpa. Wintermantel et al. (2002) observed that if YM of implant material is much larger than that of cortical bone, the load distribution would not be uniform and this may result into stress shielding. As more load is to be borne by implants, therefore stresses are shielded for bone. According to wolf law, if load on bone is decreased compare to normal loading, bone starts losing mass and bone morphology changed. Stress shielding effect is very dangerous and can lead to the loosening of device. Figure 5 shows the stress-strain curves of biomaterials, while figure 6 compares the YM of various biomaterials with hard tissues.

The mechanical faliure of cyclic load bearing implants is mainly resulted from poor fatigue strength. The implant should be capable to withstand very large number of load cycles during service without failure for long time (Kienapfel, 1999). Fatigue strength is essential property for load bearing orthopedic as well as dental implants (Prakash et al.,2016). The high value of tensile stress and fluctuations in loading also causes fatigue failure along with numerous cycles of loading (Niinomi, 2003; Teoh, 2000). An implant material should also have high tensile strength, compressive strength as well as shear strength to prevent fractures and to improve functional stability. Ductility improves the formability and machinability of metal implant material and ultimately decreases the manufacturing cost. Hardness is directly related to wear as increase in hardness, increases the wear resistance of implant. High toughness is required to prevent mechanical failure or fracture of implant. The strength, ductility, hardness, and wear resistance of material is affected by grain size and shape. Smaller grain size leads to greater hardness, strength, wear resistance but poor ductility and vice versa (Ratner, 2004). Metallurgical methods like heat treatments, structure refining and work hardening are applied to improve the mechanical properties. Presences of porosity, cavities or channels in material also play a very important role and allow controlled ingrowth of tissue and blood vessels (vascularization) into the material. Therefore it improves bioactivity and biocompatibility of OIs but weakens the mechanical properties of materials and fails the implants at much lower level of load.

Figure 5. Stress stain behaviour of bio materials

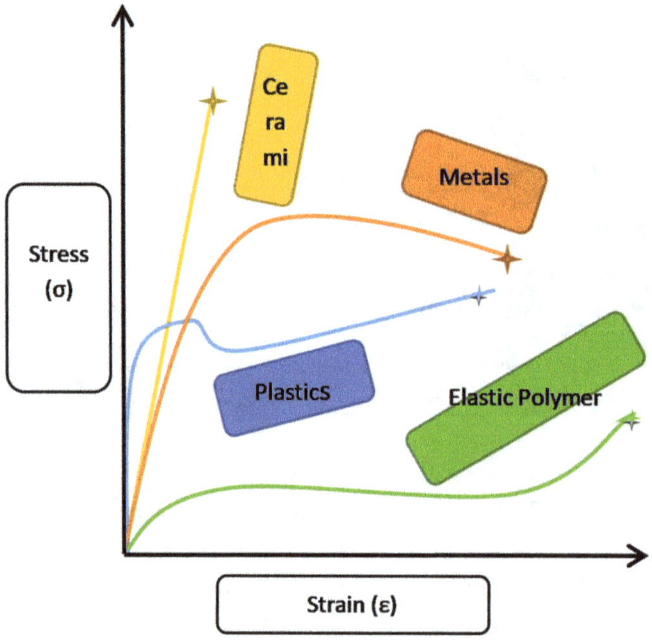

Figure 6. Relative values of Young Modulus (YM)

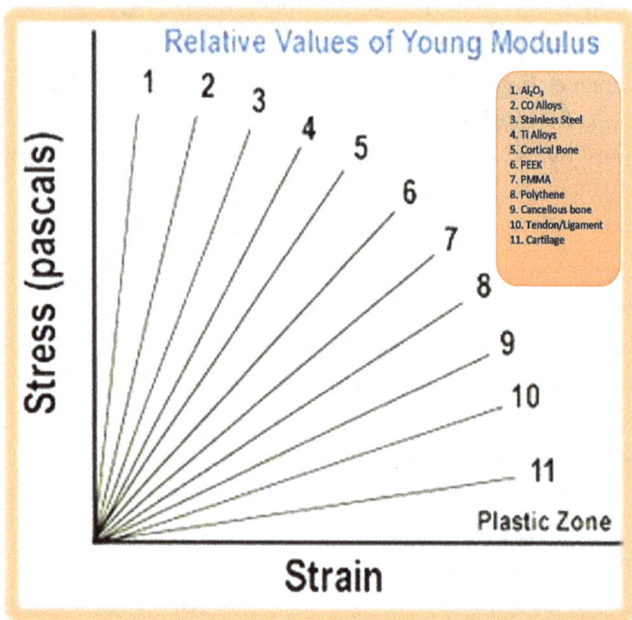

FUTURE RESEARCH DIRECTIONS

Improvement in biological and mechanical properties of material will definitely produce qualitative implants. The future research is focused on nanotechnology and nanomaterials for OI and DI is blooming area of research. Nanotopography increases the surface energy which further increases the protein adsorption and osteoblast adhesion (McHale, 2003). The nanoscale roughness also enhances the osteoblast adhesion, proliferation and calcium deposition on implant surface (Price, 2003). The nano sized grains improve the toughness, ductility and corrosion resistance of material (Yan, 2012). More research is required which can substantially make understand the mechanism behind the improvement in properties of nanomaterials so that life and performance of implants can be improved. Therefore, nanomaterials can be used for other functionalities such as sensing and detection of bone diseases, infection resistance, target drug delivery, inhibiting cancer cells etc. These functions can be attached with future orthopedic and dental implants to improve the functionalities and performance. This can help not only in implantation but also in diseases control and other part of bio medical science.

CONCLUSION

Orthopedic and dental implants are essential need of today's health sector and expected to grow fast in coming years. Both quantitative and qualitative improvement is needed to achieve the futuristic goal. The performance of implants is largely dependent on the properties of material. Orthopedic and dental implant are bio-mechanical device whose performance depends upon biological and mechanical properties. The material should be biocompatible to ensure intoxicity and cohesion with local tissues and osseointegration. The material should be bioactive to ensure the growth of tissues on its surface and to become integral part of body. Human body's internal environment is saline and therefore it provokes corrosion in implant's material. Corrosion and wear debris causes allergic reaction and infection in local tissues. So, the material should be corrosion and wear resistant. OI and DI are primarily a mechanical device which supports the loads and movements of body. The young's modulus of material should be comparable to that of bone to avoid stress shielding effect and loosening of implant. High fatigue strength is necessary for long life of load bearing implants. Hardness improves wear resistance and ductility ease the manufacturing of implants. Other mechanical properties such as higher tensile and shear strength are higher desirable to ensure the normal functioning and long life of implants. The future alloys and composites for orthopedic and dental implantation purpose should be design by considering all these properties to ensure betterment of implantation. Nanomaterials have better mechanical as well as biological properties compared to conventional materials. These also provide other functionalties which can help to achieve the goal of ideal implantation. These can be used as future materials for OI and DI.

REFERENCES

Adya, N., Alam, M., Ravindranath, T., Mubeen, A., & Saluja, B. (2005). Corrosion in titanium dental implants: Literature review. *Journal of Indian Prosthodontic Society*, *5*(3), 126–131. doi:10.4103/0972-4052.17104

Aliyu, A. A. A., Abdul-Rani, A. M., Ginta, T. L., Prakash, C., Axinte, E., Razak, M. A., & Ali, S. (2017). A review of additive mixed-electric discharge machining: Current status and future perspectives for surface modification of biomedical implants. *Advances in Materials Science and Engineering*, *2017*, 1–23. doi:10.1155/2017/8723239

Ashman, A. (1971). Acrylic resin tooth implant: A progress report. *The Journal of Prosthetic Dentistry*, *25*(3), 342–347. doi:10.1016/0022-3913(71)90197-1 PMID:5276857

Balasundaram, G., Sato, M., & Webster, T. J. (2006). Using hydroxyapatite nanoparticles and decreased crystallinity to promote osteoblast adhesion similar to functionalizing with RGD. *Biomaterials*, *27*(14), 2798–2805. doi:10.1016/j.biomaterials.2005.12.008 PMID:16430957

Balasundaram, G., & Webster, T. J. (2006). A perspective on nanophase materials for orthopedic implant applications. *Journal of Materials Chemistry*, *16*(38), 3737–3745. doi:10.1039/b604966b

Balasundarm, G., & Webster, T. J. (2007). An overview of Nano-Polymers for Orthopaedic Applications. *Macromolecular Bioscience Journal*, *7*(5), 635–642. doi:10.1002/mabi.200600270

Bauer, S., Schmuki, P., von der Mark, K., & Park, J. (2013, April). Engineering biocompatible implant surfaces part 1: Materials and surfaces. *Progress in Materials Science*, *58*(3), 261–326. doi:10.1016/j.pmatsci.2012.09.001

Burstein, G. T., Liu, C., & Souto, R. (2005). The effect of temperature on the nucleation of corrosion pits on titanium in Ringer's physiological solution. *Biomaterials*, *26*(3), 245–256. doi:10.1016/j.biomaterials.2004.02.023 PMID:15262467

Carter, D. R., & Beaupre, G. S. (1998). Mechanobiology of skeletal regeneration. *Clinical Orthopaedics and Related Research*, *335*, 541–555. PMID:9917625

Chandra, G. (2014). *Bio-implants Market by Product Types- Global Opportunity Analysis and Industry Forecast, 2013 – 2020*. Retrieved from https://www.alliedmarketresearch.com/bio-implants-market

Chaturvedi, T. P. (2009). An overview of the corrosion aspect of dental implants (titanium and its alloys). *Indian Journal of Dental Research*, *20*(1), 91–98. doi:10.4103/0970-9290.49068 PMID:19336868

Durrieu, M. C., Pallu, S., Guillemot, F., Bareille, R., Amédée, J., Baquey, C., ... Dard, M. (2004). Grafting RGD containing peptides onto hydroxyapatite to promote osteoblastic cells adhesion. *Journal of Materials Science. Materials in Medicine*, *15*(7), 779–786. doi:10.1023/B:JMSM.0000032818.09569.d9 PMID:15446238

Gettleman, L., Nathanson, D., & Myerson, R. L. (1977). Effect of rapid curing procedures on polymer implant materials. *The Journal of Prosthetic Dentistry*, *37*(1), 74–82. doi:10.1016/0022-3913(77)90195-0 PMID:264323

Kamachimudali, U., & Sridhar, T. (2003). Corrosion of bio implants. *Sadhana Academy Proceedings in Engineering Sciences*, *28*, 601–637.

Kienapfel, H., Sprey, C., Wilke, A., & Griss, P. (1999). Implant fixation by bone ingrowth. *The Journal of Arthroplasty*, *14*(3), 355–368. doi:10.1016/S0883-5403(99)90063-3 PMID:10220191

Liu, X., & Niebur, G. L. (2008). Bone ingrowth into a porous coated implant predicted by mechano-regulatory tissue differentiation algorithm. *Biomechanics and Modeling in Mechanobiology*, *7*(4), 335–344. doi:10.100710237-007-0100-3 PMID:17701434

Long, M., & Rack, H. J. (1998). Titanium alloys in total joint replacement – a materials science perspective. *Biomaterials*, *19*(18), 1621–1639. doi:10.1016/S0142-9612(97)00146-4 PMID:9839998

McHale, G., Shirtcliffe, N. J., Aqil, S., Perry, C. C., & Newton, M. I. (2004). Topography driven spreading. *Physical Review Letters*, *93*(3), 36–102. doi:10.1103/PhysRevLett.93.036102 PMID:15323838

Misch, C. E. (1998). *Contemporary implant dentistry*. St. Louis, MO: Mosby Elsevier.

Narayan, R. (Ed.). (2012). Materials for medical devices: ASM Handbook, Vol 23. ASM International.

Niinomi, M. (2003). Fatigue performance and cyto-toxicity of low rigidity titanium alloy, Ti–29Nb–13Ta–4.6Zr. *Biomaterials*, *24*(16), 2673–2683. doi:10.1016/S0142-9612(03)00069-3 PMID:12711513

Niinomi, M., & Nakai, M. (2011). Titanium-Based Biomaterials for Preventing Stress Shielding between Implant Devices and Bone. *International Journal of Biomaterials*, *2011*, 1–10. doi:10.1155/2011/836587 PMID:21765831

Porte-Durrieu, M. C., Guillemot, F., Pallu, S., Labrugère, C., Brouillaud, B., Bareille, R., ... Baquey, C. (2004). Cyclo-(DfKRG) peptide grafting onto Ti-6Al-4V: Physical characterization and interest towards human osteoprogenitor cells adhesion. *Biomaterials*, *25*(19), 4837–4846. doi:10.1016/j.biomaterials.2003.11.037 PMID:15120531

Prakash, C. (2016). Powder Mixed Electric Discharge Machining an Innovative Surface Modification Technique to Enhance Fatigue Performance and Bioactivity of β-Ti Implant for Orthopaedics Application. *Journal of Computing and Information Science in Engineering*, *14*, 1–9.

Prakash, C., & … . (2016). Electric discharge machining- a potential choice for surface modification of metallic implants for orthopedics applications: A review. Proceedings of the Institution of Mechanical Engineers, Part B. *Journal of Engineering*, *230*, 231–253.

Prakash, C., Kansal, H. K., Pabla, B. S., & Puri, S. (2015). Processing and characterization of novel biomimetic nanoporous bioceramic surface on β-Ti implant by powder mixed electric discharge machining. *Journal of Materials Engineering and Performance*, *24*(9), 3622–3633. doi:10.100711665-015-1619-6

Prakash, C., Kansal, H. K., Pabla, B. S., & Puri, S. (2016). Multi-objective optimization of powder mixed electric discharge machining parameters for fabrication of biocompatible layer on β-Ti alloy using NSGA-II coupled with Taguchi based response surface methodology. *Journal of Mechanical Science and Technology.*, *30*(9), 4195–4204. doi:10.100712206-016-0831-0

Prakash, C., Kansal, H. K., Pabla, B. S., & Puri, S. (2017). Experimental Investigations in Powder Mixed Electrical Discharge Machining of Ti-35Nb-7Ta-5Zr β-Ti Alloy. *Materials and Manufacturing Processes*, *32*(3), 274–285. doi:10.1080/10426914.2016.1198018

Prakash, C., Singh, S., Pabla, B. S., & Uddin, M. S. (2018). Synthesis, characterization, corrosion and bioactivity investigation of nano-HA coating deposited on biodegradable Mg-Zn-Mn alloy. *Surface and Coatings Technology*, *346*, 9–18. doi:10.1016/j.surfcoat.2018.04.035

Prakash, C., & Uddin, M.S. (2017). Surface modification of β-phase Ti implant by hydroaxyapatite mixed electric discharge machining to enhance the corrosion resistance and in-vitro bioactivity. *Surface and Coatings Technology, 236*(Part A), 134-145.

Price, R. L., Waid, M. C., Haberstroh, K. M., & Webster, T. J. (2003). Selective bone cell adhesion on formulations containing carbon nanofibers. *Biomaterials*, *24*(11), 1877–1887. doi:10.1016/S0142-9612(02)00609-9 PMID:12615478

Ramakrishna, S., Mayer, J. E., Wintermantel, E., & Leong, K. W. (2001). Wintermantel and K.W. Leong, "Biomedical applications of polymer composite materials: A review. *Composites Science and Technology*, *61*(9), 1189–1224. doi:10.1016/S0266-3538(00)00241-4

Ratner, B. D., & Hoffman, A. S. (2004). *Biomaterials science*. Amsterdam: Elsevier.

Saro, T. P. S., & Sidhu, H. S. (2012). Characterization and in vitro corrosion investigation of thermal sprayed hydroxyapatite and hydroxyapatite-titania coating on TI alloy. *Metallurigical and Materials Transaction A: Physical Metalurgy and Material Science, 43*, 4365-4376. 10.100711661-012-1175-8

Schakenraad, J. M. (1996). Cells: Their surfaces and interactions with materials. In B. D. Ratner (Ed.), *Biomaterial Science: An Introduction to Materials in Medicine* (pp. 141–147). San Diego, CA: Academic Press.

Singh, M., & Singh, Y. (2015). Implant biomaterials: A comprehensive review. *World Journal of Clinical Cases*, *3*(1), 52–57. doi:10.12998/wjcc.v3.i1.52 PMID:25610850

Teoh, S. H. (2000). Fatigue of biomaterials: A review. *International Journal of Fatigue*, *22*(10), 825–837. doi:10.1016/S0142-1123(00)00052-9

Transparency Market Research. (2016). *Global Dental Implants Market: Rising Incidence of Periodontal Diseases Stokes Growth.* Retrieved from https://www.transparencymarketresearch.com/pressrelease/dental-implants-market.htm

Waerhaug, J. & Zander, H.A. (1956). Implantation of acrylic roots in tooth sockets. *Oral Surgery, Oral Medicine, Oral Pathology, Oral Radiology, and Endodontology*, *9*, 46–54.

Wennerberg, A., & Albrektsson, T. (2010). On implant surfaces: A review of current knowledge and opinions. *International Journal of Oral & Maxillofacial Implants*, *25*, 63–74. PMID:20209188

Wintermantel, E., & Suk-Woo, H. (2002). *Medizintechnik mit biokompatiblen Werkstoffen und Verfahren*. Berlin, Germany: Springer.

Yan, F., Liu, G., Tao, N. R., & Lu, K. (2012). Strength and ductility of 316L austenitic stainless steel strengthened by nano-scale twin bundles. *Acta Materialia*, *60*(3), 1059–1071. doi:10.1016/j.actamat.2011.11.009

ADDITIONAL READING

Chennell, P., & Feschet-Chassot, E. (2013). In vitro evaluation of TiO2 nanotubes as cefuroxime carriers on orthopaedic implants for the prevention of periprosthetic joint infections. *International Journal of Pharmaceutical Technology and Biotechnology*, *455*(1–2), 298–305. PMID:23892151

Kaneesh, K. (2013). Evaluation of implant success: A review of past and present concepts. *Journal of Pharmacy & Bioallied Sciences*, *5*(5), S117–S119. doi:10.4103/0975-7406.113310 PMID:23946563

Refojo, M. F. (1996). *Application of Materials in Medice and Dentistry: Ophthalmologic Applications*. San Diego: Academic Press.

Temenoff, S., & Mikos Antonios, G. (2008). *Biomaterials: the intersection of biology and materials science., Upper Saddle*. Pearson/Prentice Hall.

Yang, L. (2015). *Nanotechnology-Enhanced Orthopedic Materials: Fabrications, Applications and Future Trends*. Cambridge: Woodhead Publishing. doi:10.1016/B978-0-85709-844-3.00009-4

Chapter 2
Integrated Manufacturing System for Complex Geometries:
Towards Zero Waste in Additive Manufacturing

Divya Zindani
National Institute of Technology Silchar, India

Kaushik Kumar
Birla Institute of Technology, India

ABSTRACT

The chapter proposes an integrated manufacturing system consisting of three main components: digital prototyping, physical prototyping, and lost core technology. The integrated system combines the beneficial aspects of computer-aided design, computer-aided engineering, rapid prototyping, and rapid tooling. The proposed integrated system is an attempt to compress the product development time while saving cost. The system can be efficient in designing of mold, parts with complex ducts and cavities, and carrying out design analysis through optimization and simulations. The system is therefore an attempt to minimize the waste of material that occurs in the development of a product and is therefore an efficient green technology for the manufacturing industries.

INTRODUCTION

The escalating competition worldwide and the globalization have resulted in immense pressure on different manufacturing units to reduce the production time and cost of the product while meeting the quality requirements. The time-to-market the products have been now been reduced to weeks and therefore the new product must be made quickly and cheaply meeting the market requirements. One of the costly and time consuming phases in product development is the manufacturing of moulds for both the

DOI: 10.4018/978-1-5225-5445-5.ch002

development of prototype part and production component manufacturing. The conventional machining approach to produce mould entails long lead times and costs. And therefore the sequential approach of conventional machining doesn't meet the requirement of rapid product development. Thus it has become imperative on the part of manufacturing units to research on to innovate new technologies that can aid in rapid product development.

Additive manufacturing has emerged as potential technology that has the ability to reduce the production development time. Rapid prototyping (RP) was the first of a kind process for creating a 3D object through layer by layer technique and computer-aided design (CAD). The RP process was developed in 1980's for production of models and prototypes and therefore help design engineers to create what they had in mind. The major advantages of the RP processes are the reduction in cost and time as well as the possibility to create any complex shape (Ashley, 1991). The versatile RP processes have been applied by medical doctors, scientists, professors, artists and market researchers (Noorani, 2006; Flowers & Moniz, 2002; Chua et al., 1998) for creation and analysis of models for various studies. For instance doctors create a model for a body part for which the operation is to be performed. They use the model to better plan the procedure. The prototypes made from rapid prototyping (RP) can be used for different evaluations such as visual inspection, ergonomic evaluations etc., thereby leading to comprehensive design analysis earlier in the product development cycle. Since its inception, RP has been used in different industries where both time and precision are of paramount importance. RP is however not used for large scale production.

RP is now evolving towards rapid tooling (RT) which has the capability to be used for commercial scale production. Therefore RT is used effectively for making of moulds for commercial purposes. Further, there are products that are required to be provided with complex cavities, bypass or even to completely close hollow bodies. There are other products in which case the assembly is to be avoided such as in the case of pneumatic and hydraulic parts. For such cases one of the evolving technologies is the Lost Core or Soluble Fusible metal core technology.

Industries are now looking for a feasible option to use the advantages of both the RT and RP. The present chapter proposes an integrated manufacturing system to produce complex shape products that have complex internal cavities.

RAPID PROTOTYPING PROCESSES

The different rapid prototyping processes can be categorized into: powder based, solid based and liquid based.

Liquid Based Processes

Stereolithography (SL)

The SL process was developed by 3D systems, Inc. and is one of the most widely used process of rapid prototyping. SL is a liquid based process wherein the ultraviolet light makes contact with the resin which leads to solidification or curing of photosensitive polymer. The SL process incepts with a 3D CAD model and is translated to STL file. In the STL format, the 3D CAD model is cut into slices and the information of each layer is stored. A platform supports the piece and other overhanging structures. The UV laser then solidifies the selective locations of the resin. The platform supporting the piece is lowered on completion of a layer and the excess material is then drained on completion of the entire product. The drained material then can be reused (Noorani, 2006; Cooper, 2001; Kruth, 1991). Microstereolithography is the improvised version ofstereolithography supporting for higher resolution.

Certain errors are introduced in the final product from SL process. One such error is that of overcuring which occurs mainly on the overhung parts. The overcuring may be attributed to the absence of fusion with the bottom layer. Scanned line shape is another error which is introduced as a result of scanning process. The layer thickness is variable because of the high viscosity of the resin and therefore introduces an error in the border position control. Lack of surface finish is another error of the SL process (Kim et al., 2010).

SL offers to build part with different materials and the process is called multiple material SL (Nagy & Matyasi, 2003).

Fused Deposition Modeling (FDM)

In FDM process a thin filament plastic is fed to a machine. The machine consists of print heads that melts the plastic filament and extrude it. The materials used in the FDM process are acrylonitrile butadiene styrene (ABS), polycarbonate (PC), polyphenylsulfone (PPSF), PC-ISO and PC-ABS blends. The FDM process has numerous advantages such as no requirement of chemical post-processing, less expensive machines and no resins to cure (Noorani, 2006; Cooper, 2011). The major disadvantage is the relatively low resolution on the z axis in comparison to other processes, requirement of finishing processes and relatively high amount of time required to build a part. Some models however allow for two modes of operation: dense mode and sparse mode, which aids in saving of production time but at the cost of mechanical properties (Morvan et al., 2005). Figure 1 delineates the FDM process outline.

Polyjet

The polyjet is a liquid based rapid prototyping process that uses inkjet to produce physical models. A photopolymer is deposited by the x and y axes moving inkjet head. The deposited photopolymer is then cured by the ultraviolet lamps after the deposition of each layer is completed. The overhanging parts are supported by using a gel-type polymer which is then removed after the part is completed. The polyjet process can produce parts of multiple colors (Wholers, 2010; Singh, 2011; Petrovic et al., 2011). Further, the polyjet process produces parts with very high resolution. However, the parts produced by the polyjet process are relatively weaker than that produced by other RP processes.

Powder Based Processes

Selective Laser Sintering (SLS)

SLS is a 3D printing process in which a carbon dioxide laser beam is used to fuse powders. The chamber is heated to the melting point of the material. The carbon dioxide laser beam fuses the powder, lying on a bed, at specific position in accordance with the design. The bed is supported with a piston which lowers on completion of a layer by the amount equivalent to that of the layer thickness. A wide range of materials can be used with the SLS process such as plastics, metals, polymers and combination of

Figure 1. Process layout for fused deposition modeling
(Make Parts Fast, 2018)

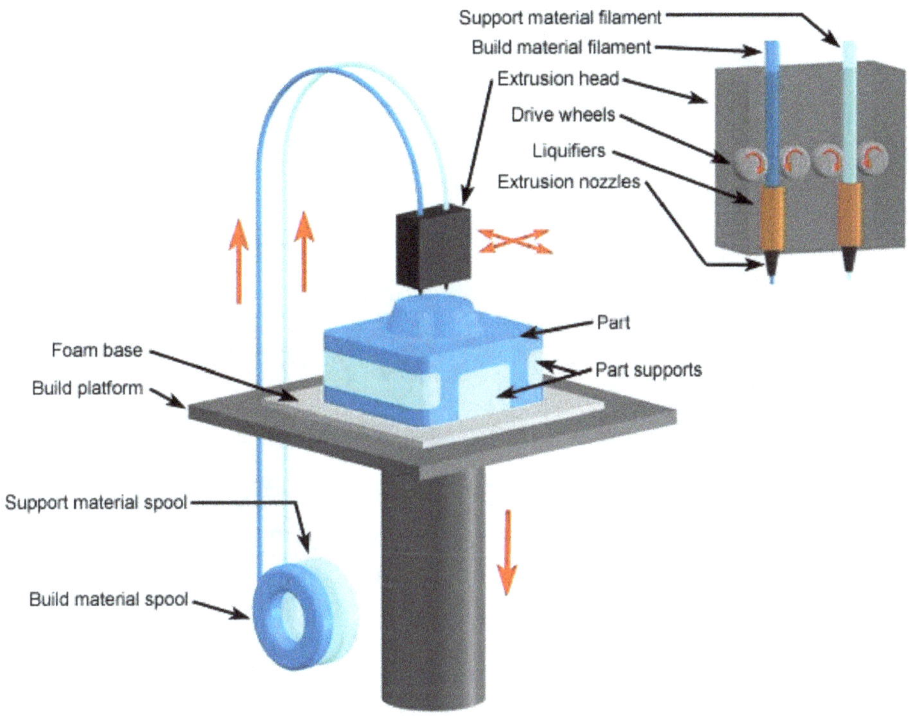

metals, metals and ceramics, metals and polymers (Halloran, 2011; Hwa-Hsing et al., 2011; Salmoria et al., 2011). Another advantage of the SLS process is the recycling of the unused powder. The SLS process however suffers from a number of disadvantages such as the accuracy of the part is determined by the size of the material particles, requirement of inert gas to avoid the oxidation and to ensure that the process is carried out at constant temperature. Process outline of SLS is depicted in Figure 2.

Electron Beam Melting (EBM)

EBM process is similar to the SLS process. A very high voltage electron beam melts powder. The use of electron beam is only the differentiating factor between the EBM and the SLS processes. The voltage of electron beam ranges between 30 to 60 kV. The EBM process takes place in a vacuum chamber. The main advantage of the EBM process production of parts with a wide variety of pre-alloyed metals. Further, the process has find applicability in outer space (Murr et al., 2012; Semetay, 2007). The process is mainly used for producing metal parts which is one of the disadvantages of the EBM process.

Prometal

This powder-based process is used for the fabrication of tools and dies. The process uses a liquid binder to bind the steel powder. The powder is located on a piston supported bed that lowers when each layer is completed. The excess powder is removed when the required thickness of the layer is achieved. No

Figure 2. Process layout for selective laser sintering
(Make Parts Fast, 2018)

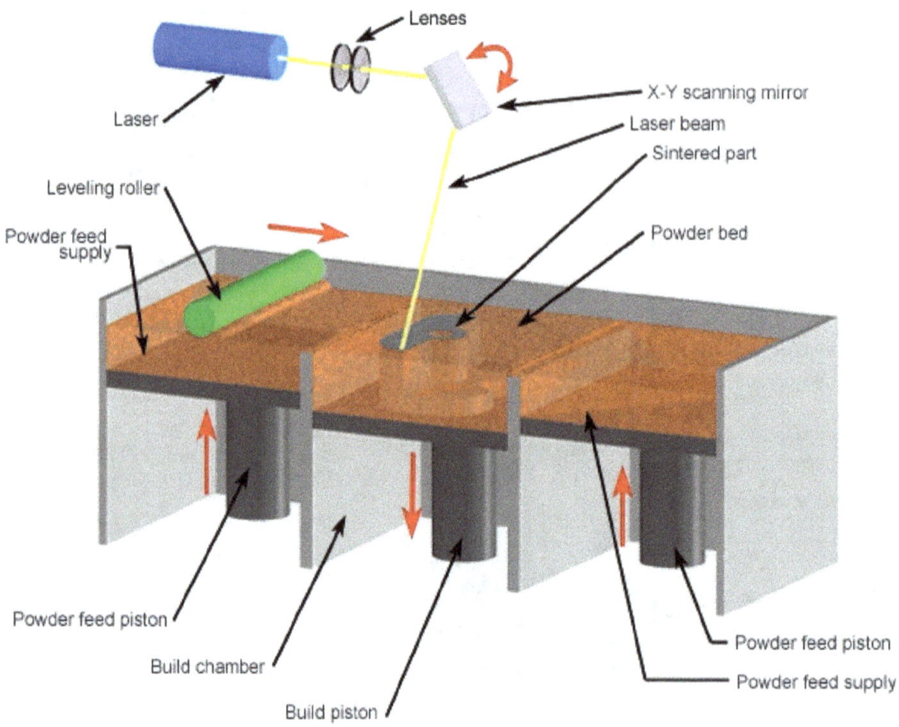

postprocesssing is required while building a mold. However, finishing operations such as infiltration and sintering are required if a functional part is being built (Cooper, 2001; Kruth, 1991).

Laser Engineering Net Shaping (LENS)

A high powered laser beam is used to melt the metal powder. The metal powder is injected at a specific location. The molten metal material solidifies on cooling. The LENS process takes place in a chamber with argon atmosphere. A wide range of materials and their combinations can be used with the LENS process. For instance materials such as stainless steel, titanium-6, nickel based alloys, aluminium-4 vanadium, copper alloys etc can be used with the LENS process. The LENS process is also suitable for repairing of parts that are impossible or expensive to be repaired by other processes. One of the major disadvantages of the LENS process is the presence of residual stress due to uneven cooling and heating process.

3DP

The data from the CAD drawing is printed with the aid of water-based liquid binder supplied in a jet onto a starch based powder. The starch-based powder is supported on a powder bed and glues to each other with the aid of liquid based blinder. The 3DP process is similar to that of inkjet printing process that is used for two-dimensional printing. The 3DP process is capable of handling a high variety of polymers (Halloran, 2011; Cooper, 2001).

Solid Based Processes

Laminated Object Manufacturing (LOM)

Helisys of Torrance, CA developed the LOM process. The LOM process combines both the subtractive and additive techniques to build a part. The material is fed in the form of sheet that is advanced over a platform with the help of sheet feed mechanism. A heated roller is used to bond the sheets together with the use of heat energy. The outline of the part is then cut with the help of laser. The platform is then lowered by a depth equal to the thickness of the sheet and then the new layer of sheet advances over the previously built layer. The main advantages of the LOM process are the low cost, no deformation, and requirement of no postprocessing and the possibility to build large parts. The disadvantages of the LOM process are wastage of material, difficulty to produce complex cavities and low surface definition. The LOM process can be used for creating models of papers, composite and metals (Noorani, 2006; Cooper, 2001; Vaupotic, 2006). Process outline of laminated object manufacturing is shown in Figure 3.

LOST CORE TECHNOLOGY (LCT)

The Lost core technology requires the core of the part with complex cavities. A low melting point alloy is used for casting the component's interior. A composite or plastic material is used to overlay the casted component. The overlay may be cured in some cases. The low melting alloy is allowed to melt out with the overlay still in place. The melting temperature doesn't leads to any distortion or damage to the fabricated part. The steps involved in making the core consist of the following:

Figure 3. Process layout for laminated object manufacturing
(Make Parts Fast, 2018)

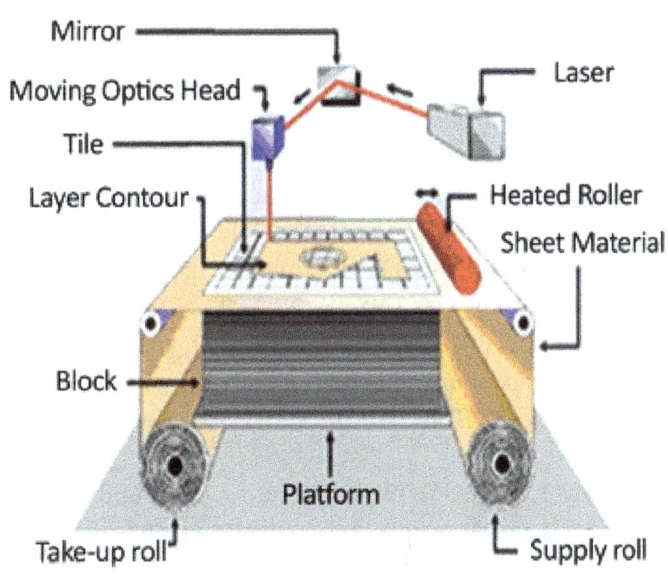

- Casting of the core.
- Inserting the core into mold.
- Performance of the casting process.
- Ejecting the mold and the core.
- Core and mold separation by melting

The alloys used for making of cores have their melting point ranging 50°C to 300°C. it should be ensured that the melting temperature of the alloy is consistent with that of the working temperature of the over molding. The mechanical properties of the alloy are to be given due consideration in cases where the core is heavy or is required to withstand high pressures. The amount of energy consumed by melting is another factor to be given due consideration (Polifke et al., 1998). A wide range of materials can be used for making cores such as wax, Tin-Bismuth alloy, Zn-Al eutectic mixture etc. The lost core technology is suitable for producing complex internal ducts and cavities.

THE INTEGRATED MANUFACTURING SYSTEM

A product development system should consider all aspects related to the product such as product design, its manufacturability and recycling. All these aspects should be considered at the early stage of design cycle so that design engineers can easily make changes during the iterative process of product development. The integrated manufacturing system for producing geometries with complex cavities and ducts consists of three main phases: Digital prototype, Physical prototype and Rapid tooling (lost core technology process). The integrated manufacturing system is shown in the Figure 4 below.

The digital prototyping phase begins with the 3D model using the computer aided design software (CAD). The dimensional as well as the aesthetical aspects of the product is fixed at this stage. The mold and its other components are also designed using the CAD software such as Pro-Engineer, CATIA, SOLIDWORKS, Unigraphics etc. For an already existing component reverse engineering (RE) process can be used for creating the CAD model. The RE process consists of three main steps: digitization of the existing part, extraction of part features and the 3D CAD model. Variety of contact and non-contact digitizers can be used for digitization of the part. Segmentation of the digitized data accomplishes the task of feature extraction. Fitting of the digitized surfaces to the set of data points will accomplish the 3D CAD modeling of the part under consideration. Part digitization can be accomplished with a wide range of systems such as coordinate measuring machine (CMM), laser triangular scanner (LTS), computer tomography (CT), cross-sectional imaging measurement etc.

The optimized product is the primary requirement of the manufacturer. To guide through the process of optimization of the product design, a number of computer aided engineering (CAE) systems can be used. Few examples of CAE systems are ANSYS, AUTOFORM, DYNAFORM etc. the 3D CAD model can be transferred to the CAE systems for the optimization analysis. CAE environment can help in predicting the filling pattern, temperature and pressure for the mold. The CAE systems aids in optimizing the part geometry, and the core. The CAE model will help in reduction of shrinkage, warpage and residual stresses.

The digital prototype is now converted to physical prototype using the different rapid prototyping techniques. The 3D CAD model is converted to STL file format which is typically an interfacing system between the 3D CAD model and the RP systems. The RP process creates the physical prototype layer

Figure 4. Process layout for integrated manufacturing system

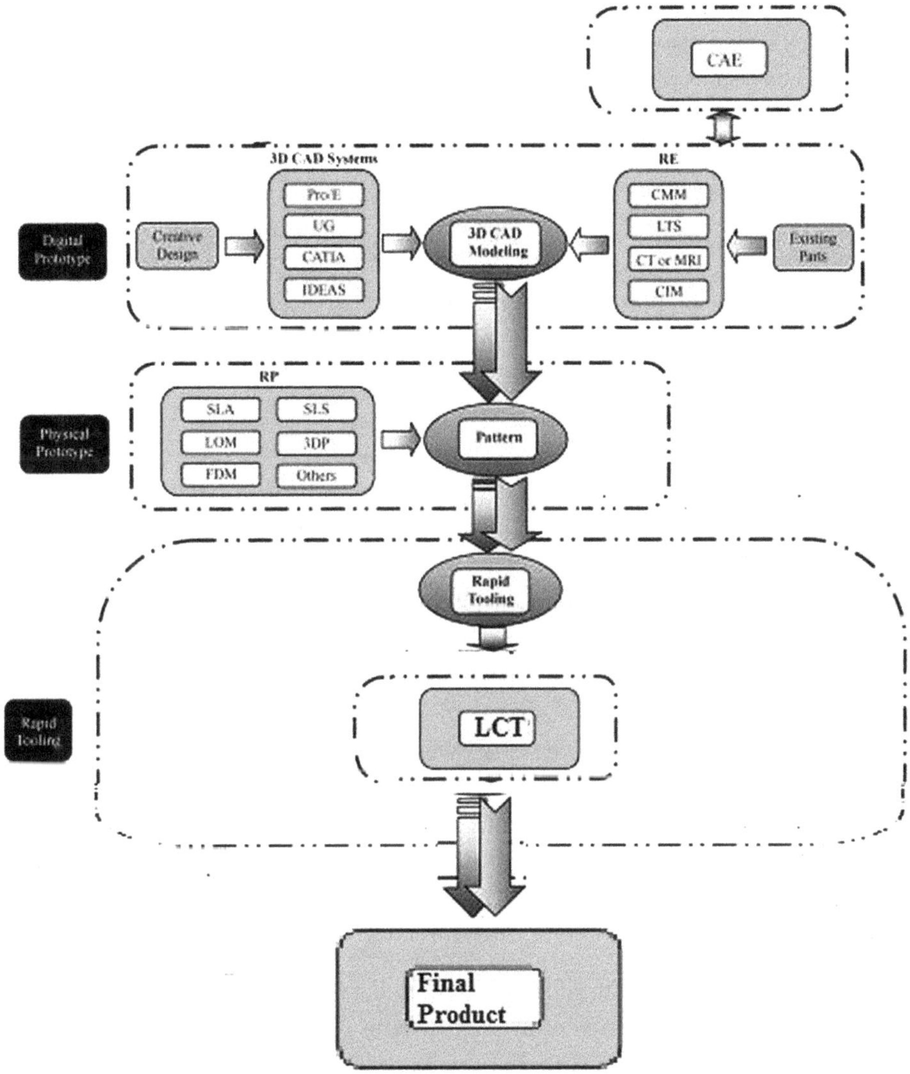

by layer from the STL format of the 3D CAD model. The RP techniques save a lot of time to market the product with improved quality and reduced cost. Over the years, industries have used a wide range of RP techniques such as SL, SLS, FDM, LOM, 3DP etc.

Next the mold is prepared with the physical prototype of the part. The physical prototype is now master pattern for preparing the mold. The master pattern is suspended in a mold frame equipped with the gates and runners. The master pattern is set up on the parting line of the mold casting frame. The casting frame will then be filled with silicon-rubber resin which is properly mixed using computer-controlled equipment. The resin fills the casting frame surrounding the master pattern. The casting is then moved to a heating chamber for curing. The master pattern is then removed from the silicone mold after the hardening is accomplished. The gates and runners are cut off to obtain the final mold cavity.

For the parts with complex cavities and ducts, the lost core technique is then followed. Lost core technology is one of the rapid tooling process. The core prepared from the RP process is used for the fabrication of complex shaped cavities and ducts. The core is prepared using a low melting alloy such as tin-bismuth alloy. The core is placed inside the already prepared silicon mold. The mold is now filled with the desired resin. The mold is then moved to the heating chamber and the casting is removed from the mold after the completion of the hardening process. The obtained casting has core embedded in it. Therefore to remove the core, the casting is then placed in the oil bath. The temperature of the oil bath is similar to the melting point of the core material. The core material melts out leaving behind the complex cavity.

The desired part with the complex cavity is finally obtained after sequential procedural steps.

CONCLUSION

The chapter presents an integrated manufacturing system that optimizes the product design while minimizing the wastage of the material. The system comprises of three different modules: Digital prototyping, Physical prototyping and Rapid tooling. 3D CAD systems are used to establish the digital prototypes and the CAE systems are used to optimize product design. The RP techniques then are used to obtain the master pattern and the mold. The RT in the form of lost core technology is used to produce the complex shaped mold cavity. Thus the proposed system can reap benefits in number of ways such as saving of time, cost and material. The system can thus be used by different manufacturing industries seeking to develop cost effective product with enhanced quality and less time to market the product.

REFERENCES

Ashley, S. (1991). Rapid prototyping systems. *Mechanical Engineering (New York, N.Y.)*, *113*(4), 34.

Chua, C. K., Chou, S. M., Lin, S. C., Eu, K. H., & Lew, K. F. (1998). Rapid prototyping assisted surgery planning. *International Journal of Advanced Manufacturing Technology*, *14*(9), 624–630. doi:10.1007/BF01192281

Cooper, K. (2001). *Rapid prototyping technology: selection and application*. CRC Press. doi:10.1201/9780203910795

Core Tech System Inc. (n.d.). *Engineering Talk From Core Tech Systems*. Core Tech Systems.

Flowers, J., & Moniz, M. (2002). Rapid prototyping in technology education. *Technology Teacher*, *62*(3), 7.

Halloran, J. W., Tomeckova, V., Gentry, S., Das, S., Cilino, P., Yuan, D., ... Alabi, T. R. (2011). Photopolymerization of powder suspensions for shaping ceramics. *Journal of the European Ceramic Society*, *31*(14), 2613–2619. doi:10.1016/j.jeurceramsoc.2010.12.003

Hwa-Hsing, T., Ming-Lu, C., & Hsiao-Chuan, Y. (2011). Slurry based selective laser sintering of polymer-coated ceramic powders to fabricate high strength alumina parts. *Journal of the European Ceramic Society*, *31*(8), 1383–1388. doi:10.1016/j.jeurceramsoc.2011.02.020

Kim, H., Choi, J. W., & Wicker, R. (2010). Scheduling and process planning for multiple material stereolithography. *Rapid Prototyping Journal*, *16*(4), 232–240. doi:10.1108/13552541011049243

Kruth, J. P. (1991). Material incress manufacturing by rapid prototyping techniques. *CIRP Annals-Manufacturing Technology*, *40*(2), 603–614. doi:10.1016/S0007-8506(07)61136-6

Lost Core Injection Moulding. (n.d.). Copyright. Retrieved from http://LostCoreInjectionMoulding.htm

Make Parts Fast. (2018). Retrieved from www.makepartsfast.com

MCP. (n.d.). Retrieved from http://www.mcp-group.de

Morvan, S., Hochsmann, R., & Sakamoto, M. (2005). ProMetal RCT (TM) process for fabrication of complex sand molds and sand cores. *Rapid Prototyping*, *11*(2), 1–7.

Murr, L. E., Gaytan, S. M., Ramirez, D. A., Martinez, E., Hernandez, J., Amato, K. N., ... Wicker, R. B. (2012). Metal fabrication by additive manufacturing using laser and electron beam melting technologies. *Journal of Materials Science and Technology*, *28*(1), 1–14. doi:10.1016/S1005-0302(12)60016-4

Noorani, R. (2006). *Rapid prototyping: principles and applications*. John Wiley & Sons Incorporated.

Petrovic, V., Vicente Haro Gonzalez, J., Jorda Ferrando, O., Delgado Gordillo, J., Ramon Blasco Puchades, J., & Portoles Grinan, L. (2011). Additive layered manufacturing: Sectors of industrial application shown through case studies. *International Journal of Production Research*, *49*(4), 1061–1079. doi:10.1080/00207540903479786

Polifke, M. (1998). *Lost Core Technology On The Way To New Application*. Spe-Antec.

Salmoria, G. V., Paggi, R. A., Lago, A., & Beal, V. E. (2011). Microstructural and mechanical characterization of PA12/MWCNTs nanocomposite manufactured by selective laser sintering. *Polymer Testing*, *30*(6), 611–615. doi:10.1016/j.polymertesting.2011.04.007

Semetay, C. (2007). *Laser Engineered Net Shaping (LENS) modeling using welding simulation concepts*. Lehigh University.

Singh, R. (2011). Process capability study of polyjet printing for plastic components. *Journal of Mechanical Science and Technology, 25*(4), 1011-1015.

Szilvśi-Nagy, M., & Matyasi, G. Y. (2003). Analysis of STL files. *Mathematical and Computer Modelling*, *38*(7-9), 945–960. doi:10.1016/S0895-7177(03)90079-3

Vaupotic, B., Brezocnik, M., & Balic, J. (2006). Use of PolyJet technology in manufacture of new product. *Journal of Achievements in Materials and Manufacturing Engineering*, *18*(1-2), 319–322.

Wohlers, T. (2010). *Wohlers Report 2010*. Wholers Associates.

Chapter 3
Recent Advancements in Customized Investment Castings Through Additive Manufacturing:
Implication of Additive Manufacturing in Investment Casting

Sunpreet Singh
Lovely Professional University, India

Chander Prakash
Lovely Professional University, India

M. Uthayakumar
Kalasalingam University, India

ABSTRACT

Conventional investment casting (IC) has suffered from numerous limitations such as rigidity of the process, longer production cycles, higher tooling cost, and waste during different manufacturing stages. With the invent of additive manufacturing (AM) technologies, it is now possible to overcome the aforesaid issues along with additional benefits in terms of comparatively better quality characteristics of the resulting castings. The collaboration of AM and IC provided numerous avenues, specifically in biomedical, aerospace, and automobile sectors. AM technologies supported the IC process both in direct and indirect ways where these systems can be used for both job and mass production applications, respectively. In the chapter, the author will try to discuss the assistance of AM process to IC in detail. Each and every step to be followed will be supported with the practical findings, either by the contributing author or published somewhere else. Moreover, some of the case studies will be discussed in detail to highlight the practical importance of the duo.

DOI: 10.4018/978-1-5225-5445-5.ch003

INTRODUCTION

The casting industry is a vital segment of the manufacturing sector which produces intricate parts, blends, internal features and varying thickness with excellent surface finish with negligible metallurgical limitations, (Pal et al., 2002). In this today's manufacturing era, we rely on conventional investment casting (IC) process for hundreds and thousands of objects that we need for various end user applications. This method of casting is one of the most ever-present manufacturing processes we are utilizing, since 300 BC, before our civilization. The ancient expertise in casting resided among several cultures such as Mesopotamians, Egyptians, Chinese, Thai and Indians (Greer, 2008). However, the basic principles of IC can be traced back to 5000 BC when the early-men employed this method to produce rudimentary tools (Taylor, 1983), followed by jewellery and artistic products (Träger & Bührig-Polaczek, 2002), and the development of aerospace and subsequently engineering components during the 2nd world war (Barnett, 1988). The IC process, often known as lost-wax or precision casting process, is one of the manufacturing processes that have capability to produce high accuracy, intricate shapes and fine impression with tight geometrical tolerance. This is an alternative approach to produce parts that are hard to machine (Singh & Singh, 2016). IC process has numerous applications that includes agricultural equipments, automobile components, aircraft engines, air frames, fuel systems, computer hardware, electrical equipment, electronic hardware and radar, making jewellery, statues and art castings, dentistry and dental tools, prosthetics, guns and armament, hand tools, machine tool components etc. (Pattnaik et al., 2014). The working procedure of conventional IC process is given in Figure 1. The first step of IC process: (a) the injection of molten wax (at about 150°C) inside the precisely machined metallic die (usually split one) with help of an injection machine; (b) patterns are then safely ejected out of the metal cavity; (c) patterns let for strengthening in air conditioned environment so that the geometrical dimensions get fixed, afterwards, the pattern is assembled with another wax assembly that consisted of pouring sprue, runner and gate as integral parts; (d) this tree like structure is dipped into clay and stucco coatings are performed repeatedly to get the required mould is attained (e); (f) de-waxing is carried out to create a hollow cavity; (g) baked to increase hot strength; (h) molten metal is poured into the cavity and part is separated from the tree once solidify and (i) finally the refractory layers are removed on pneumatic vibration system (Singh & Singh, 2013, 2016). The key requirements of an investment casting mould are (Jones & Yuan, 2013): sufficient green (unfired) strength, sufficient fired strength, sufficiently weak to prevent hot-tearing, high thermal shock resistance, high chemical stability, and low reactivity with the metals, sufficient mould permeability and thermal conductivity, low thermal expansion to limit dimensional changes.

Problems associated with ceramic shell materials have been exacerbated following the introduction of the Environmental Protection Act (Jones & Yuan, 2013), wherein fix degree of emissions allowed and this has boosted the use of water-based shells within industry. Moreover, improvements in the quality, reduction in manufacturing costs and explores new markets for the process are the hot topics amongst current research area. Along with this, numerous tools and software are currently being used for the optimization of the mechanical and physical properties at various stages of the casting. The troubles like: long lead time and high tooling costs for low-volume IC production runs is still present in the conventional process. Wax, as pattern material, is suspect of shrinkage that affects the dimensional and surface features of the casting, and various researches has been carried out to control the shrinkage rate (Taşcıoğlu & Akar, 2003; 2007). However, the critical barrier in prototype development and production time could be easily reduced with the introduction of additive manufacturing (AM) technologies. The AM technologies allowing the designer to work more comfortably with the geometrical complexities,

Figure 1. Pictorial description of IC process
(Singh & Singh, 2013).

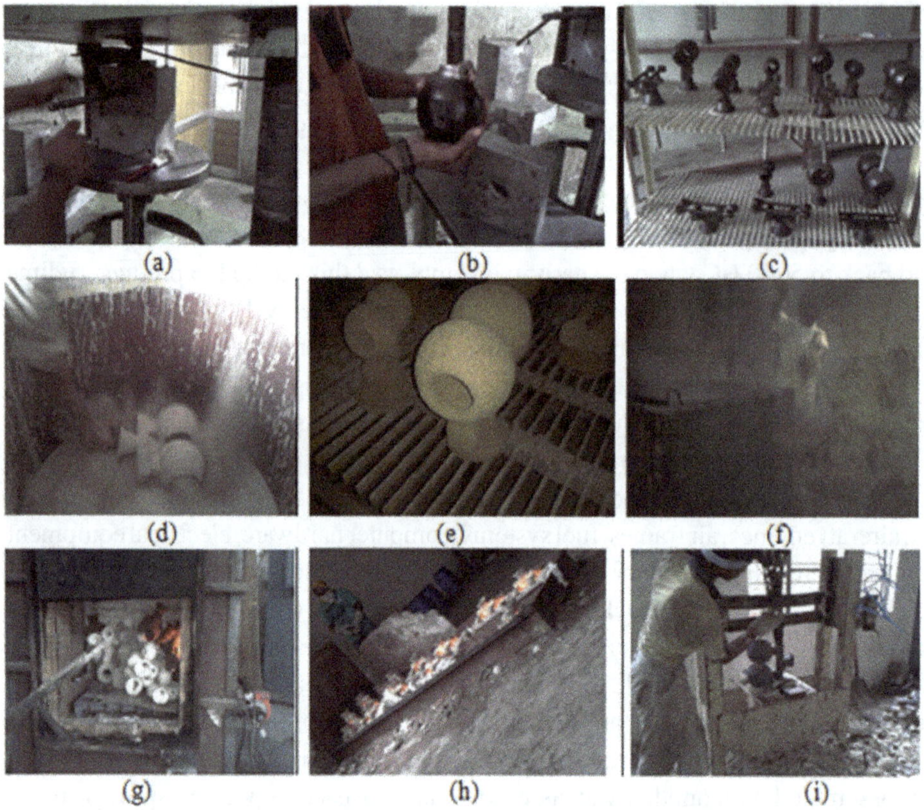

control over the dimensional tolerances and also to minimize the involved production times (Bidanda & Bártolo, 2007; Kumar & Kruth, 2010; Kruth et al., 2007).

Although traditional machining processes can be used for the development of desired pattern but in case of unusual shape or specific internal features, the machining becomes more difficult. AM allows the production of models directly from engineering designs. AM systems are limited only by the size of the model and not by its complexity (Gibson et al., 2010).

ADDITIVE MANUFACTURING TECHNOLOGIES

The AM technologies (refer Figure 2), often known as three dimensional printing (3DP) or rapid prototyping technology (RPT) or solid freeform fabrication (FFF), is a technology in which materials are correctly printed layer upon layer, converting digital 3D models into solid objects (Wendel et al., 2008; Chia & Wu, 2015; Tseng et al., 2014; Murphy & Atala, 2014). This is an emerging technology in manufacturing industry and this able to turn digital data into physical parts for different applications. In recent years, there has been a dramatic progress in 3D-printing technologies due to investments by governments and companies around the world. This emerging manufacturing technique comprises mainly of fused deposition modelling (FDM), selective laser sintering (SLS), laminated object manufacturing (LOM), and stereo-lithography (SLA). Depending on these manufacturing technologies, a 3D form of the material is created by precisely "printing" it layer by layer. The tangible parts could be created from the designed parts and it acts as a platform for an individual to get their requisite design. The three dimensional (3D) model can be drawn on different software systems, transmitted to AM building equipment and printed into a physical object (Figure 2). Using AM technologies, a rapid pattern substituted for conventional IC wax pattern can be easily prepared, with sustainable mechanical properties, as a wide range of feedstock materials are available that can be processed through different types of AM technologies to assist IC process. Acrylonitrile-butadiene-styrene (ABS), high density polythene (HDPE), low density polythene (LDPE), polyamide, poly-lactic-acid, rubber, resins, wax, photo curable polymers & resins, polymer composites, etc. (Cheah et al., 2015; Guo & Leu, 2013), are majorly used for rapid fabrication of sacrificial patterns. The ceramic moulds are generally baked in the furnace to evaporate these patterns for sufficient time duration, therefore chance of formation residuals are high and these can affect the chemical composition of the casting. The case may become worst when developing a biomedical device. According to Wang and Miranda 2010, carbon residues are mainly resulted due to incomplete combustion of the pattern material (Rosochowski & Matuszak, 2000). Sivadasan et al., 2012 found that in black tar and ash formed after flame burning of the moulds was negligible and can easily be removed by passing compressed air in the mould cavity (Chhabra & Singh, 2011). Further, it would be better to fabricate AM patterns at low density and appropriate design as thermal expansion during burn-out step can induce stresses in the mould and lead to breakage (Omar et al., 2012; Harun et al., 2013). Apart from all the limitations, AM systems are highly beneficial for low and medium volume production runs (Hague et al., 2004). Steps involved in creating a physical prototype with most of these technologies is same and includes: (1) create a CAD model of the design, (2) convert the CAD model to STL format, (3) slice the STL file into thin cross-sectional layers, (4) construct the model one layer atop another and (5) clean and finish the model.

AM technologies have been mainly used for (Madrazo et al., 2009; Dedoussis et al., 2008; Dedoussis et al., 2004; Lee et al., 2004):

- Physical verification of CAD model
- Creating models without regard to draft angles, parting lines, etc.
- Concept presentations and design reviews

Figure 2. Schematic for the AM process chain

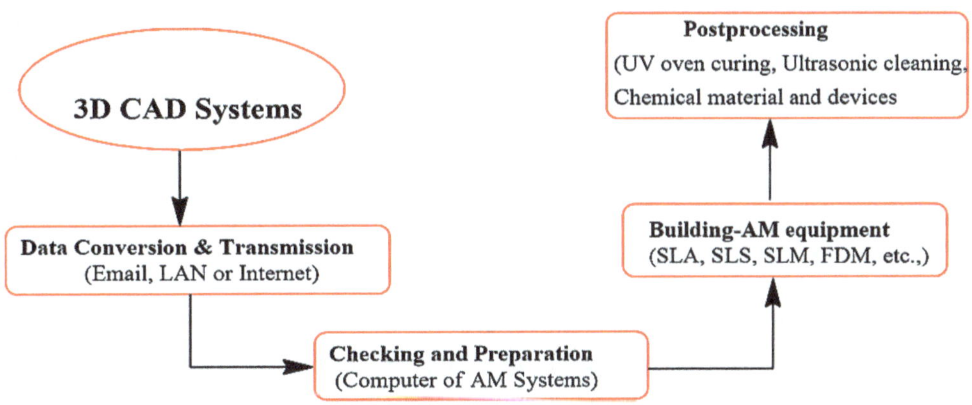

- Direct tooling
- Rapid tooling such as investment casting and silicone moulding
- Reducing time-to-market
- Creating anatomical models
- Producing relief models for geographical applications
- Creating 3D portraits

Additive manufacturing technologies can either assist IC process through a direct route or an indirect route (Bártolo & Gibson, 2011), as discussed below:

Direct Route of IC Pattern Making

In the direct route, the rapid patterns are directly used in IC process with minor or negligible intermediate processing. Many technologies can be employed here as given below:

Fused Deposition Modelling

The FDM, an extrusion based AM technology that, utilizes granular or pelletized feedstock with a screw-type extruder. The workhorse materials are usually the amorphous thermoplastic polymer filaments with a diameter of about 1.5-3mm (Turner et al., 2014). In the feed system, the filament is pushed through the system using a pinch roller mechanism like that illustrated in Figure 3(a). A stepper motor is connected to one of the rollers providing energy to move the filament through the system (Agarwala et al., 1996). One or both of the rollers may have a grooved surface like a gear to create sufficient friction. The pressure on the filament, between the rollers, is typically sufficient to slightly deform the filament, but these are designed to avoid crushing the filament. In FDM, the absorption of moisture by the thermoplastic filament raises significant problems when printing parts in an extrusion-based process. The material when extruded from the heated head the moisture vaporize and often lead to morphological changes in the material, blockages (Halidi & Abdullah, 2012). In FDM, layer thickness of 200-500pmm and resolution levels for the positioning of the nozzle of 25pm in x, y plane in line dimensions are achieved based on nozzle diameter (Turner et al., 2014). This method attempts to develop the structures with sharp overhangs or long, unsupported sections due to tough fabrication procedures. However, the extruded filaments have poor material strength during the extrusion process and thus result in drooping or total collapse of the unsupported segment. Filler materials used to support cavities, overhanging and lifted structures could be removed post printing have been developed to eliminate this issue (Chia & Wu, 2015). This extrusion type machine is a specific example to demonstrate the accelerated progress in the development toward a wide availability of 3D printers at an affordable cost; these include servo drives, precision rails, and various electronics items such as computers and software which deal with difficult geometric data.

Stereo-Lithography

The SLA system (refer Figure 3(b)) uses an ultraviolet laser beam to cure a photosensitive monomer resin layer upon layer in order to build prototypes having complex shapes. In the SLA process, a tessellated STL file is first sliced into very thin cross sections, as in case of other instruments. This technology uses

Figure 3. Schematic of FDM (a) (Turner et al., 2014), SLA (b) (Chockalingam et al., 2006), SLA (Kruth et al., 2005) and LOM (Kruth et al., 2003).

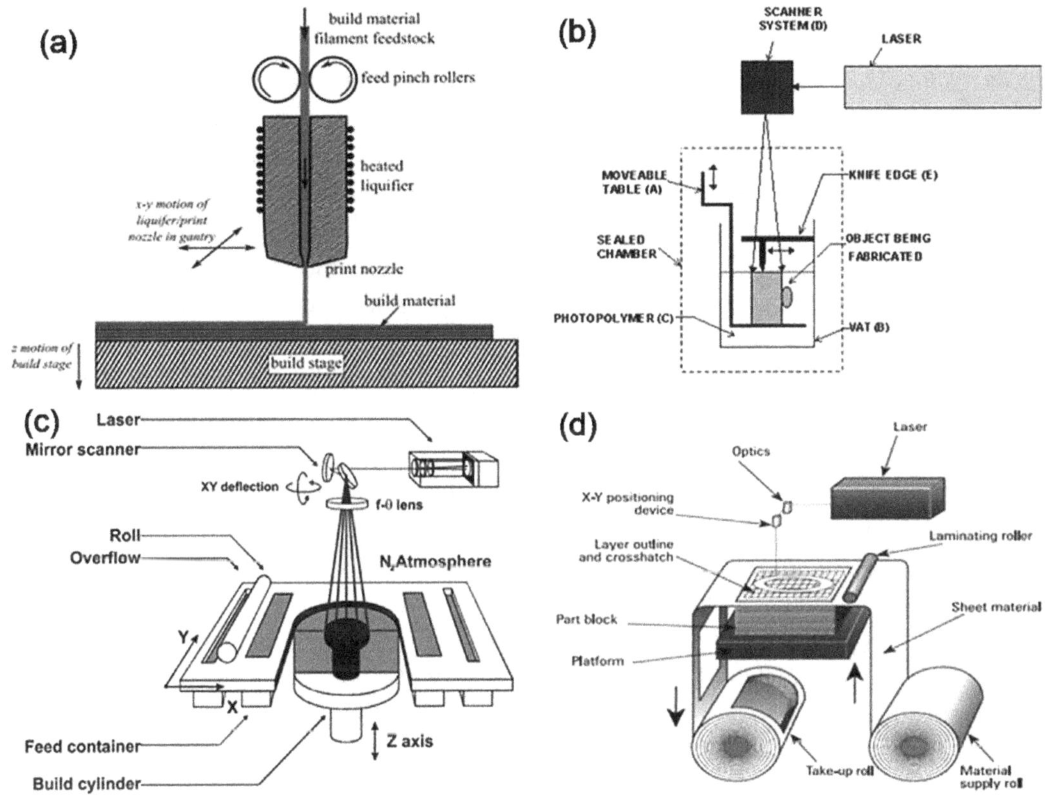

a computer-controlled laser to cure a photosensitive epoxy or acrylate resin layer by layer to create a 3D part (Chockalingam et al., 2006). The prototype is built on a platform positioned just under the exterior of fluid epoxy or acrylate resin enclosed in a vat. A low-power highly focused UV laser beam traces out the first layer, selectively solidifying the resin while leaving the excess area liquid. The elevator incrementally lowers the platform into the vat by a distance equal to the layer thickness value chosen. The photo-curable resin is relatively viscous and does not permit the liquid to instantly coat the whole of the upper surface of the cured part in a consistent way upon moving elevator downside. A recoating mechanism is therefore required to facilitate this process, and a wiper arm traverses over its surface to speedily level the surplus viscous material. The process is continued till the final layer is built. The solid part is removed from the vat and rinsed to remove excess liquid. In the post-curing process, the part is placed in a UV oven to complete the curing after nearly 97% of liquid polymer gets solidified during the build process.

Selective Laser Sintering

The SLS is another type of AM process that allows the manufacturing of both polymer and metallic based workhorse materials (both in powder form) to generate complex 3D parts (Kruth et al., 2005). In this system, the consolidation is obtained by processing the selected areas lying within the geometry of the parts by using highly localized laser beam. A beam deflection system (generally galvano mirrors) scans individual layers as per the cross section, calculated from the CAD model. Figure 3(c) shows a schematic description of SLS setup. There is a wide range of commercially available SLS machines that basically differ in the way the powder is deposited (i.e. roller or scraper), the atmosphere (argon or nitrogen) and in the laser type (CO_2 laser, lamp or diode pumped Nd:YAG laser, disk or fibre laser) (Kruth et al., 2003). The printing is based on hardening or sintering of powders on the platform in terms of successive layers. SLS is one of the common examples of powder bed fusion techniques. These techniques are used to develop industrial prototyping suitable to print polymers, ceramics, metals and combination of these materials into multifaceted and distinctive geometric shapes (Turner & Gold, 2015). The different forms of the powder such as metal or ceramic and thermoplastic polymer are sintered or hardened using a laser and each of the constructed platforms is lowered (up to a pre-defined level) and an additional powder layer is applied and sintered. There is no waste in this manufacturing process, as the un-sintered powder used for building support structures can be re-used for next printing. Appropriate ink materials for SLS have to be a powder ranging from 10 to 100pm, with a reasonable melting point. An additional property is that the particles must have excellent particle flow rate in the bed surface, which is essential to reduce electrostatic forces and surface fictionalisation (Vermeulen et al., 2013). Usually, SLS devices are bulky, sluggish, costly, and necessitate an enormous quantity of material. But the capability to deal with multiple materials in the single bed has been confirmed as most effective in various manufacturing industries. This technique follows similar principles as SLS except for the use of the liquid binding solution for binding of particles within each layer than the use of melting particles along with laser beam in SLS. This technique is widely utilized to design devices with applications in different areas of rapid tooling (Yang et al., 2009).

Laminated Object Manufacturing

The LOM (refer Figure 3(d)) was ever defined in a narrow sense in AM classification. A generalized LOM may be defined here in a wider sense that includes classical LOM, functionally graded materials and coatings on substrate material. LOM has some common features in structures and thermo-elastic responses. It needs high manufacturing temperature and/or uses more than one material, which result in thermal residual phenomenon and reduce workpiece quality (Mekonnen et al., 2016). Effective control of thermal residual phenomenon is a main goal to improve LOM techniques. It allows obtaining desired shapes by sticking successive sheets of a certain rolled material. Each layer is cut with a tool, allows removing the undesired material (Shu & Wang, 2017). The layer is cut with CO2 laser beam to a cross-section determined by CAD file and it cuts the excess material (Das et al., 2003). The excess material provides support for subsequent layers. Originally, LOM was used to make laminated paper models (Park et al., 2000); however now functional parts out of metals, ceramics and polymers are also build using it (Marchelli et al., 2011; Chiu et al., 2006). Literature demonstrated that LOM can successfully produce functional parts and patterns, for its dimensional accuracy (Yi et al., 2004), and low surface average roughness (Ahn et al., 2012). The device, its positioning control systems and dispensing glue system are similar to a traditional printer. When building is completed, a compact layered block is formed.

Indirect Route of IC Pattern Making

In case of indirect route of AM assisted IC process, the master pattern prepared from the printer is used in replication processes, for example silicon moulding, from where number of wax replicas can be generated for IC process. Figure 4 shows the classification of indirect route of pattern replicating process. This collaboration of AM and silicon moulding further speed up the process and cuts the cost, significantly.

Silicon Rubber Tooling

In this process, a liquid silicone and hardener mixture is cast/mould around AM pattern contained in a box, as shown in Figure 5. Initially master patter in hanged inside a plastic, wooden or metallic split box. The runner and gating channels are incorporated by embedding ABS or Perspex rods into the liquid silicone or by cutting the channels in the cured silicone block. Upon curing, the rods are removed to form through channels and the pattern is subsequently removed to form the mould cavity by cutting along the parting line (Cheah et al., 2005). Silicone rubber tooling allows quick production of inexpensive multiple moulds for small and large parts with good part cosmetics (Smith et al., 1999). One mould box can be used for the production of patterns 100-300; however the maximum reported are 1000 wax patterns (Vickers, 2001). Zhang et al. (1999) produced ice patterns at Tsinghua University for silicone moulding of wax pattern production.

Epoxy Tooling

In this process, the AM pattern is surrounded in clay up to the pre-determined parting line and integral runner and gating channels are created by attaching ABS rods on the pattern.

Then the assembly is spray coated with release agent before liquid resin or powder mixture is cast around the pattern. The system is left for curing and then the pattern is separated from the hardened mould half, cleaned and re-coated with release agent in preparation for casting the second mould half (Cheah et al., 2005). Holes form the alignment and locking features for the mould whereas matching pegs form on the second mould half occupying the empty volume of these holes during casting. Epoxy resin mould has been successfully utilized for both plastic and wax injection moulding (Warner 1993). The metal filled epoxy tooling increases the durability and heat transfer characteristics of the mould, for example, PolySteel moulds are reportedly much stronger than aluminium resin tooling.

Figure 4. Classification of indirect pattern making

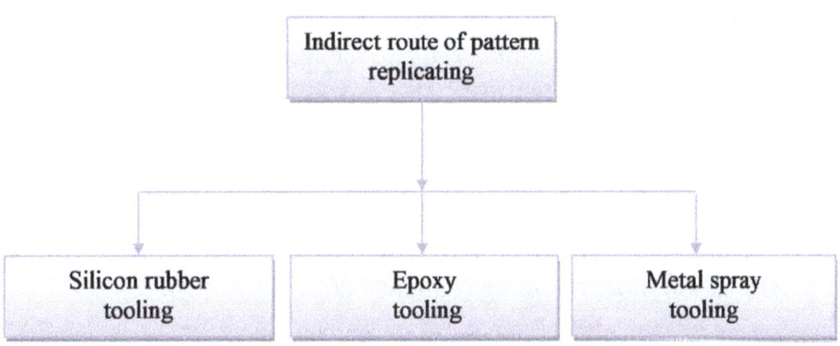

Figure 5. Description of silicon moulding process
(image courtesy: Fracktal Works).

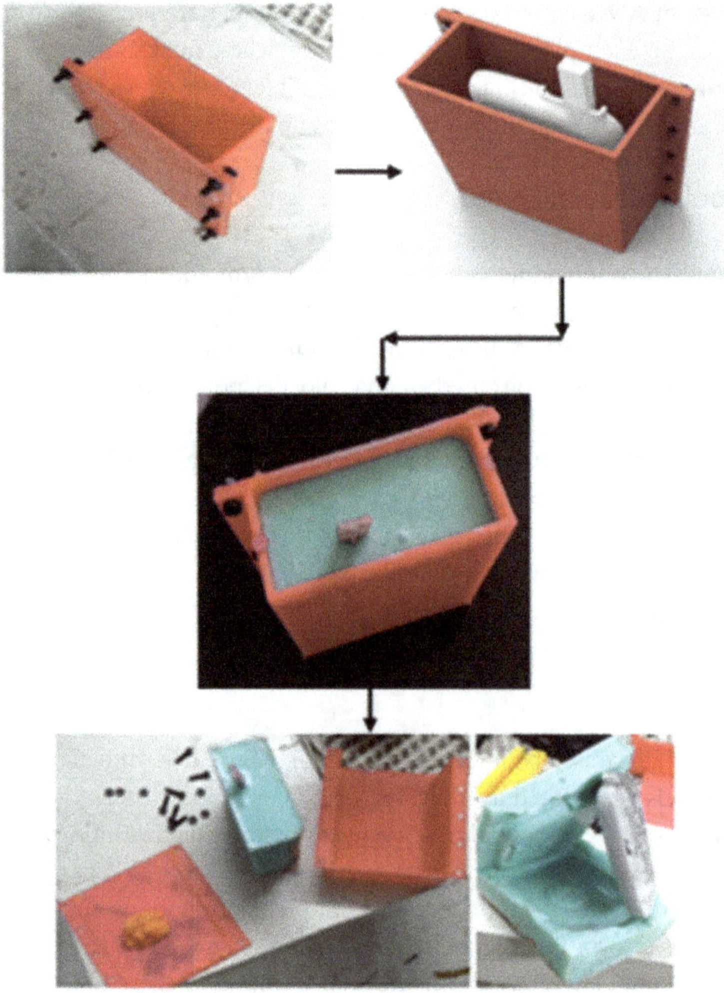

Metal Spray Tooling

To create a spray metal mould, the tiny droplets of tin-zinc or steel melt are sprayed onto AM pattern using an arc spray process (Chua et al., 1999). Further to strengthen the shell, a solid backing is cast around the shell using pure epoxy, metal-filled epoxy or low-melt alloy backfill materials (Cheah et al., 2005). With a proper choice of backfill material and the incorporation of cooling channels, a spray metal mould exhibits good injection cycle times. This process offers numerous advantageous such as low costs, high tolerances and good tooling life of approximately 10,000-100,000 injections (Heine et al., 1955). Spray metal tools have been used in many applications and various plastics have been molded including polypropylene, ABS, polystyrene and difficult process materials such as reinforced nylon and polycarbonate. Spray metal tooling processes were developed by Ford's Spray-Form (Chalmers, 2001) and INEEL rapid solidification process tooling (Knights, 2001).

TREATMENTS OF AM PATTERN SURFACES

Poor surface quality of finally produced AM based IC patterns is one of the most critical barrier restricting the commercialization of AM systems within foundries. All the different technologies are working on a same principle of layer manufacturing; hence it is difficult to avoid the stair-case stepping (Novakova-Marcincinova, 2012). The final casted product will occupy the same surface and geometrical features as of sacrificial pattern, thereby previous study highlighted the use of post processing of the AM patterns prior to casting. There are many methods that can be employed for improving the surface topography as given in Figure 7. Table 1 show the various research attempts in order to improve the surface finish of AM parts.

It has been found from the literature review that the required level of surface finish is achievable through several different attempts including: optimizing process parameters (such as part orientation, built orientation and layer thickness), machining operation (like: CNC machining), chemical machining and barrel finishing laser surface finishing operations.

CASE STUDIES

In this section, a brief discussion of few interesting studies of the various AM technologies assisted IC process has been made in order to outline the feasibility of the combined route for attaining better quality characteristics in-terms of the finally obtained component.

FDM Assisted IC

FDM-ABS Patterns for Investment Casting Process (Harun et al., 2009)

The authors established the physical and collapsibility characteristics of H shaped FDM pattern produced as hollow and solid as shown in Figure 7. They have evaluated the various properties such as roughness, geometrical and distortion occurred and established these characteristics. The surface roughness was

Figure 6. Methods of improving surface finish of AM parts.

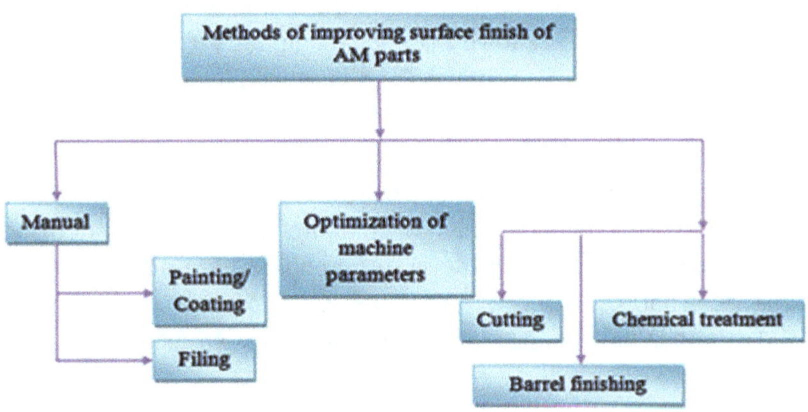

Table 1. Various attempts to improve surface finish of AM parts.

S. No.	AM System	Material	Method Adopted	Ref.
1	FDM	ABS	Chemical vapor treatment	Singh et al., 2017; Singh & Singh, 2017
2			Barrel finishing	Boschetto & Bottini, 2015; Boschetto et al., 2016
3			Parametric optimization	Vijay et al., 2011; Lan et al., 1997; Alexander et al., 1998
4			Hot cutter machining	Pham et al., 1999
5	SLA	Polyethylene wax emulsion	Micro-ultrahigh pressure atomizing coating	Yang et al., 2017
6		E-shell and SI500	Process optimization-meniscus approach	Kunjan et al., 2016
7		CIBA TOOL 5530	Process optimization	Wood, 2016
8		Acrylats, epoxies and vinylethers	New layerless fabrication strategy	Onuh & Hon, 1998
10	SLA	Polyamide 2200	Parametric optimization	Bacchewar et al., 2007
11		Ciba Geigy XB5143	Mass burnishing media finishing	Delfs et al., 2015
12		Silicone, polyurethane, Vinyl-acryl polymers and divers	Vibratory grinder	Schmid et al., 2009
13		Polyamide 6 and polyamide 12	Process parameter optimization	Arjun et al., 2012
14	LOM	Poly(vinyl chloride)	Machine parameter regulation	Pilipović et al., 2011

recorded highest at the interface of support and model material as compared to other surfaces, and highest roughness value of 32.13µm. Other surfaces of both solid and hollow patterns were having a constant trend between 17.00µm to 19.00µm. The lowest surface roughness was recorded as 6.17µm in case of hollow patterns. In case of dimensional accuracy, the highest and least percentage of deviation found at location 22 and 2, respectively, as 1.23% and 1.14%. Distortion results refer Figure 8, showed that hollow construction distorted 33.11% higher solid patterns. Hollow specimens recorded lower deviation value than solid. It is suggested that the higher percentage of deviations occurred on solid patterns as compared to hollow pattern. From the graph plotted in Figure 9, 90% of the patterns were observed to be burned-off between 300°C and 400°C. However, the complete burn-off occurred at 500°C and ash free burn was visible above 550°C. At 300°C, 400°C, 450°C and 500°C, some cracks were observed on the shell of the solid pattern at the burning temperatures of due to expansion.

Figure 7. H shaped FDM pattern; (a) solid and (b) hollow

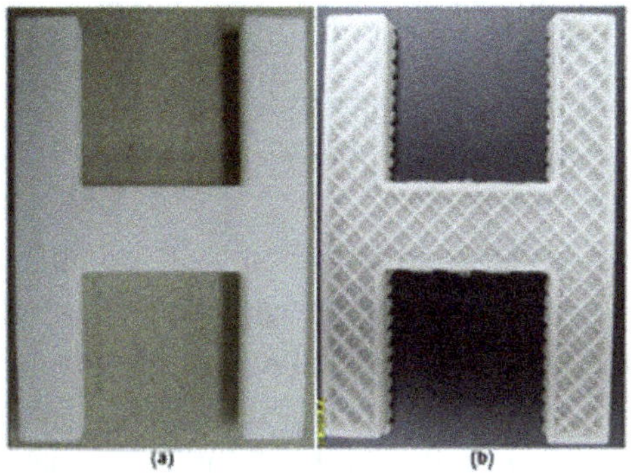

Figure 8. Distortion plot

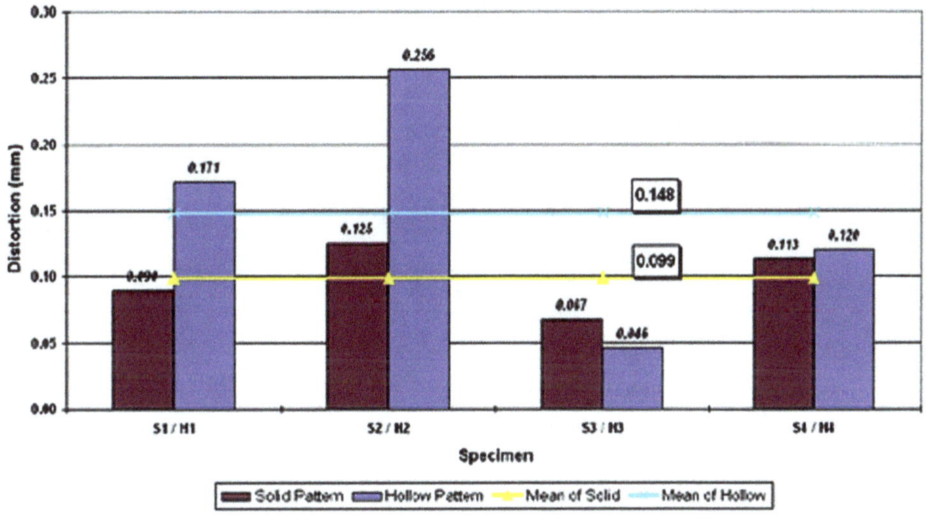

Figure 9. Collapsibility analysis of pattern

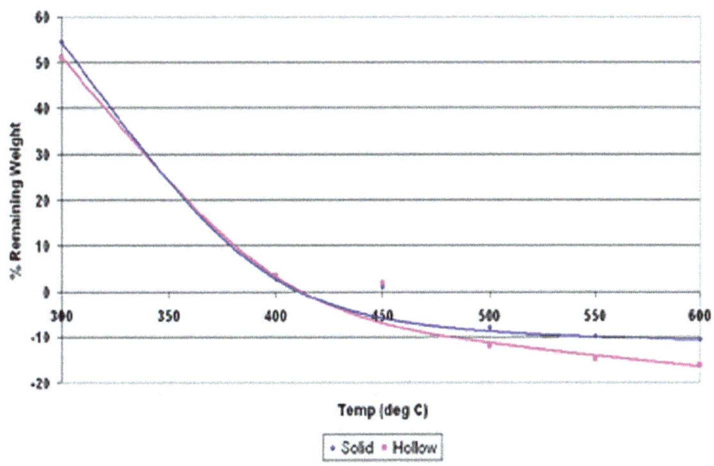

Overall, it has been found that hollow patterns were proved as viable in-terms of dimensional accuracy, mould cleanliness, pattern collapsibility and no mould cracking at all temperatures.

Surface Finish of Biomedical Implant Fabricated by Rapid Investment Casting (Singh et al., 2017)

In this study, the authors have verified the feasibility of four stage combined processes that includes FDM, chemical smoothing, vacuum casting and IC for the development of customized implant with appropriate surface and dimensional accuracy. They have first conducted the pilot study first and proceeded towards IC with the best settings as given in Table 2. Table 3 shows the third stage, i.e. IC stage, as per Taguchi L9 array and Figure 10 shows the S/N response. Table 2 Input parameters and their level.

From Figure 10, it has been found that accuracy decreased within crease in mould wall thickness and was least at 9 layers. In case of 7 layers, accuracy was highest because of less thickness of mould wall that resulted into higher rate of heat transfer and thereby shrinkage occurs. The tolerance grades for the dimensions obtained were found to be acceptable as per the ISO standard UNI EN 20286-I (1995).

Table 2. Input parameters and their level

Stage	Process	Parameter	Level 1	Level 2	Level 3	Optimum Setting
1	FDM	Orientation (degree)	0	90	-	90
		Part density	Low	Medium	High	High
	Chemical smoothing	Number of cycle	3	4	5	5
		Cycle time (sec)	4	6	8	4
2	VC	PU material	6130	PX223	-	PX223
		% Hardener	90	95	100	100
		Curing temperature (degree)	70	80	90	70
		Curing time (sec)	45	55	65	45

Table 3. Control log of experiment

S. No.	Dry Time (min)	Number of Layers
1	330	7
2	330	8
3	330	9
4	360	7
5	360	8
6	360	9
7	390	7
8	390	8
9	390	9

Figure 10. S/N response to dimensional accuracy (a) and surface roughness (b)

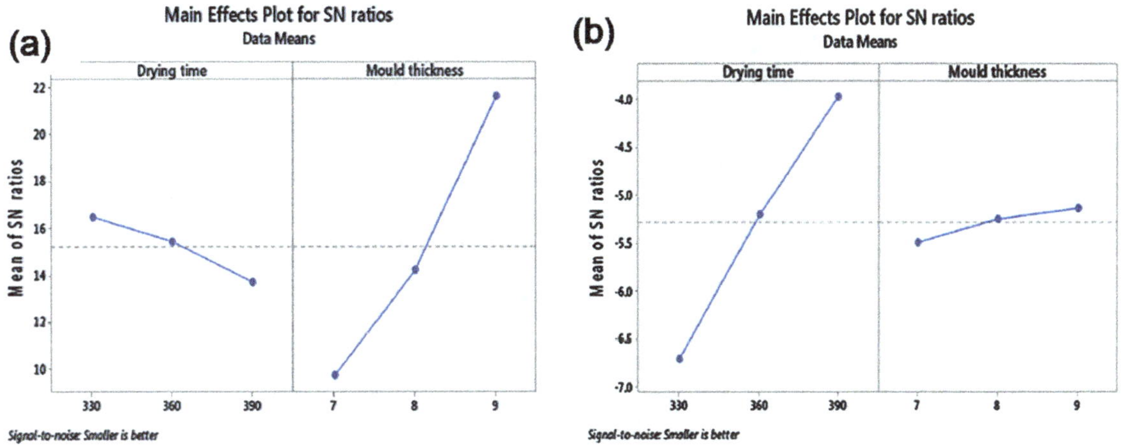

Moreover, this combined route is statistically controlled as the resulting Cpk value is 1.44, means non-confirming parts per million are just 20.27.

Hardness of Al–Al_2O_3 Composite Prepared Using Alternative Reinforced FDM Pattern (Singh & Singh, 2016)

Here in this work, FDM assisted IC route was explored for the development of Al-Al2O3 based functionally graded material. Initially, alternative reinforced FDM filaments were fabricated on single screw extruder (refer Figure 11a) by using commercial available Al2O3 powder, Al and nylon-6 granules as filament ingredients. Then cubical patterns (shown in Figure 11b and 11c) of volume 17576mm3, 27000mm3 and 39304mm3 were prepared with FDM at low density, high density and solid density. Barrel finishing was also performed on the resulting patterns in order to improve their surface finish and used for IC process.

Taguchi method used to optimize input parameters without increasing the manufacturing. It has been found that increasing Al2O3 content in FDM filament have increased the hardness value of the casted specimens.

SLA Assisted IC

Investment Casting of Hollow Turbine Blade Based on Stereo-Lithography (Wu et al., 2009)

The major purpose of this research study was to develop a new process of preparing integral ceramic molds for IC process by using SLA. Aqueous gel casting process was utilized to fill the resin pattern with low viscosity ceramic slurry. At last, IC of hollow turbine blade was obtained by vacuum drying, pyrolyzing and sintering. The process was outlined as stable due to guaranteed accuracy but difficult of measuring the inner dimensions and the relative position accuracy of cores and shell are reportedly required to test further. Figure 12 shows the finally produced hollow turbine blade. This novel approach enhanced the adaptability of SLA prototype in the manufacture of integral ceramic mold for IC process.

Figure 11. Extrusion of filament (a), prototyping of pattern (b) and reinforced prototypes (c)

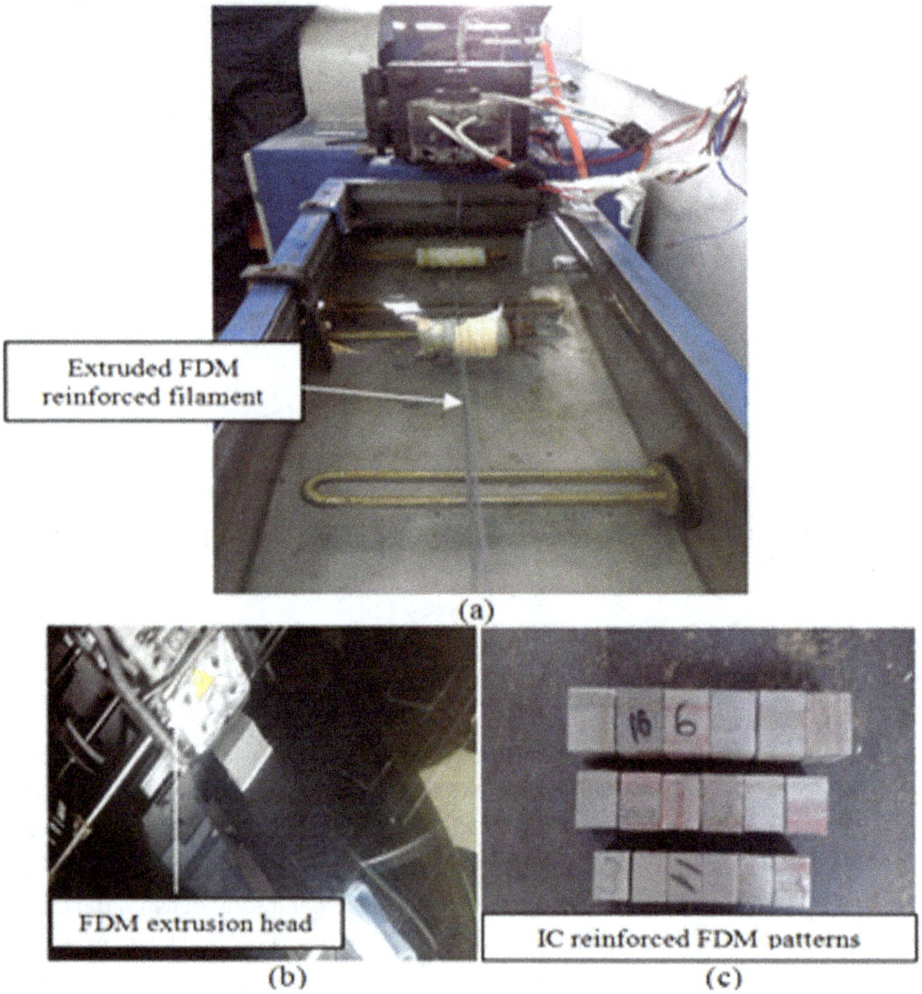

Ceramic Shell Cracking in Stereo-Lithography-Based Rapid Casting Parts (Chen et al., 2011)

Similar to above, Chen et al. (2011) simplified a thermo-mechanical model for reducing the chances of cracks in ceramic mould prepared with using SLA patterns. In their work, the distribution rule of circumferential stress on the contact boundary between SL patterns and ceramic shell was theoretically derived using displacement method to predict dangerous area of freeform-surface parts. Then the variation rule of circumferential stress at the predicted dangerous area with changing temperature was revealed by applying transient thermo-mechanical finite element analysis on a turbine blade cross section. Circumferential stress prior to 70°C was occurred majorly due to thermal expansion-induced interaction, however relatively low stress and high strength in early stage did not ensure the preventing of shell cracking in the entire pyrolyzing process. By adding high-temperature polymers, strength at higher temperature can be increased to prevent shell cracking.

Figure 12. Developed hollow turbine blade

Lattice Structure for SL Investment Casting Patterns (Norouzi et al., 2009)

This study presented importance of octagonal lattice structure (refer Figure 13a) of SLA patterns for IC process in comparison of former hexagonal structure. The application of 3D-coupled thermal-mechanical FE analysis, magnitudes of hoop stresses on ceramic shell during burnout can be computed. Additionally, practical results obtained by the real-time strain gauging (see Figure 13b) of the cylindrical models during the heating process up to 1008°C.

Numerical results for new octagonal structure indicate that, there was 62% reduction in maximum hoop stresses exerted on the ceramic shell as well as reduction of both axial and hoop strain (near to

Figure 13. New proposed octagonal structure (a), strain measuring systems (b).

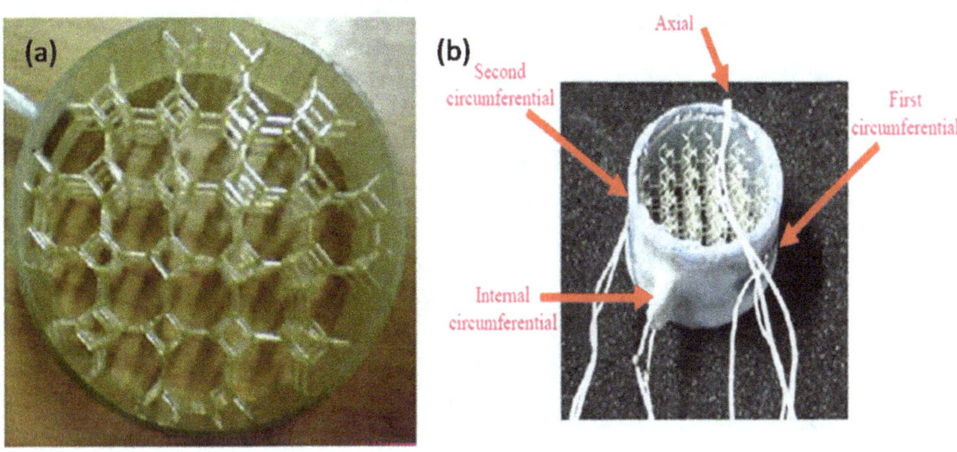

50%). Furthermore, 49% of internal mass expressed the probability of dimensional accuracy improvement for the octagonal structure.

LOM Assisted IC

Rapid Tooling and Patternmaking in Foundry Industry (Mueller & Kochan, 1999)

The LOM process in rapid tooling and pattern making is especially advantageous due to the robustness, wood-like properties and their low material costs. Explicit application examples in sand casting, investment casting and ceramics processing show how a reduction of necessary process steps and cycle times can be achieved by the application of LOM models. The axle bracket (refer Figure 14) was a typical pressure die casting part used in the automotive industry. To reduce the effort for the wax injection tool a LOM mold was used instead of a machined metal tool.

SLS Assisted IC

Selective Laser-Sintered of Zirconia Molds (Harlan et al., 2001)

The authors presented an approach of making zirconia and low temperature based copolymer moulds, with SLS, to create titanium casting molds. In this process, SLS shapes were infiltrated with colloidal zirconium and subsequently fired to pyrolyze the copolymer and crystallize the colloidal particles. The resulting mold, consisted of cubic and monoclinic zirconia phases, increased infiltrated weight, improved flexural strength and density, and reduced roughness of the mold material. The process resulted into

Figure 14. Axle bracket of automobile

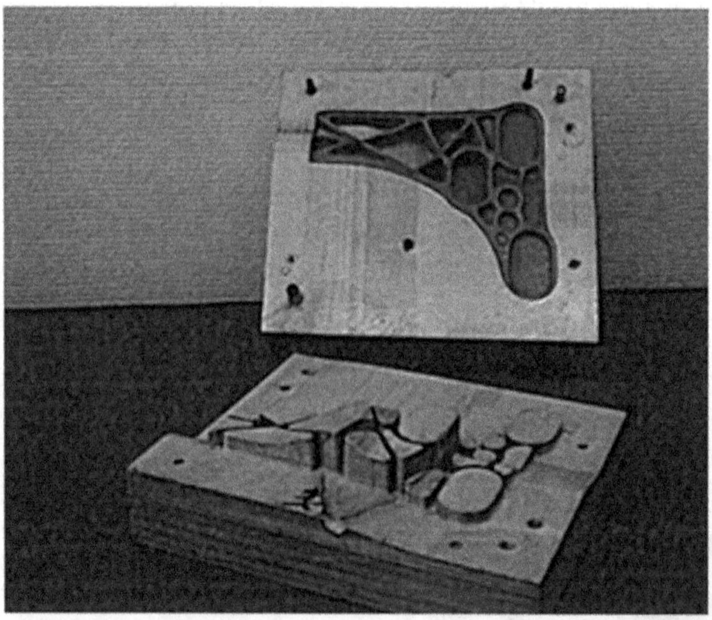

flexural strength of 5MPa, average density close to 60% shrinkage occurred (during the first firing cycle) to 13%. Metallographic examination of titanium casting revealed a narrow alpha case and an interior consisting of acicular. This work demonstrated the feasibility of producing tailor titanium parts with complex geometries directly from the scanned data without using any wax tooling or pattern.

Quality and Success Rate of the Patterns (Dotchev & Soe, 2006)

The main objective of their work was to analyze all stages of the CastForme polystyrene (CF) pattern fabrication through laser sintering, to identify their inferior quality and to find out the scope for quality improvement and failures reduction. This paper described manufacturing of patterns for shell or flask investment casting. The process involved data preparation, laser sintering fabrication of "green" part, cleaning and wax infiltration. All process stages were equally important for successful project completion in terms of pattern quality and delivery time. The paper proposed a new approach for wax infiltration. For this the green parts were heater at about 70-80°C and the plate and part was immersed into a vat of molten wax. The infiltration process was observed through a window in the oven door and as soon as all part surfaces was covered with wax; the plate was lifted in order to drain any extra wax from the part. The preheating and infiltration time varied from 30 min to 1 hr as per the geometrical and volumetric requirements.

CONCLUSION

In the light of literature reviewed in this chapter with respect to various innovations and research efforts made with different types of AM technologies, following outcomes can be drawn:

The combination of AM and IC is a mature process that is continuously advancing to achieve the goal of digitization of the foundry workshops in order to eliminate the clutters of injection tools, metallic dies as well as to strengthen their economy by producing superior quality castings in a robust and flexible manner. The outcomes of the literature are encouraging commercialization of the AM solutions in investment and sand casting technologies. It has been found that the utmost contributions of AM in foundry industry is the facilitation of concurrent engineering progresses in simulating, designing, developing, analysis and then production of any geometrical shape. The AM application, discussed here, would serve the purpose of researchers, industrialist, and academicians to explore the viability of the using established practices in their regular productions as well as to look into the areas where further development is required. The current approaches such as FDM, SLA and SLS have been more on applications in novel product development and specialty construction.

AM technologies are doing well enough in production of rapid patterns as per requirements. However, research efforts are still required to make these technologies suitable for mass production applications, aside of indirect routes. Also it is the demand of the time to integrate these methods into manufacturing oriented approaches that could include core, supports and mold production via sand printing and dies and tools.

In literature, numerous metals and alloys are successfully casted through AM assisted IC technology and there is no restriction over the use of materials if is suitable for casting. Further, this hybrid manufacturing system is highly popular in medical sector hence it is utmost importance to study the counter effects of foreign elements on the biological life and performance of the devices. Along with this, four key tasks should also be addressed in the near future: (i) integrate AM into current metal casting, (ii) develop higher speed milling, turning and CNC approaches for pre or post operations, (iii) print pattern like disposable tooling by using AM and (iv) sustain and change traditional manufacturing with AM.

REFERENCES

Agarwala, M. K., Jamalabad, V. R., Langrana, N. A., Safari, A., Whalen, P. J., & Danforth, S. C. (1996). Structural quality of parts processed by fused deposition. *Rapid Prototyping Journal*, *2*(4), 4–19. doi:10.1108/13552549610732034

Ahn, D., Kweon, J. H., Choi, J., & Lee, S. (2012). Quantification of surface roughness of parts processed by laminated object manufacturing. *Journal of Materials Processing Technology*, *212*(2), 339–346. doi:10.1016/j.jmatprotec.2011.08.013

Alexander, P., Allen, S., & Dutta, D. (1998). Part orientation and build cost determination in layered manufacturing. *Computer Aided Design*, *30*(5), 343–356. doi:10.1016/S0010-4485(97)00083-3

Arjun, N. R., Ganiger, M., & Idris, M. (2012). *Studies on optimization of Selective Laser Sintering process to manufacture Fuel tanks*. Available at http://amsi.org.in

Bacchewar, P. B., Singhal, S. K., & Pandey, P. M. (2007). Statistical modelling and optimization of surface roughness in the selective laser sintering process. *Proceedings of the Institution of Mechanical Engineers. Part B, Journal of Engineering Manufacture*, *221*(1), 35–52. doi:10.1243/09544054JEM670

Barnett, S.O. (n.d.). Investment casting - the multi-process technology. *Foundry Trade Journal International*, 1-11.

Bártolo, P. J., & Gibson, I. (2011). History of stereolithographic processes. In Stereolithography (pp. 37-56). Springer US.

Bidanda, B., & Bártolo, P. J. (Eds.). (2007). *Virtual prototyping & bio manufacturing in medical applications*. Springer Science & Business Media.

Boschetto, A., & Bottini, L. (2015). Roughness prediction in coupled operations of fused deposition modeling and barrel finishing. *Journal of Materials Processing Technology*, *219*, 181–192. doi:10.1016/j.jmatprotec.2014.12.021

Boschetto, A., Bottini, L., & Veniali, F. (2016). Finishing of fused deposition modeling parts by CNC machining. *Robotics and Computer-integrated Manufacturing*, *41*, 92–101. doi:10.1016/j.rcim.2016.03.004

Chalmers, R. E. (2001). Rapid tooling technology from Ford country. *Manufacturing Engineering, 127*(5), 36.

Cheah, C. M., Chua, C. K., Lee, C. W., Feng, C., & Totong, K. (2015). Rapid prototyping and tooling techniques: A review of applications for rapid investment casting. *International Journal of Advanced Manufacturing Technology, 25*(3-4), 308–320. doi:10.100700170-003-1840-6

Chen, X., Li, D., Wu, H., Tang, Y., & Zhao, L. (2011). Analysis of ceramic shell cracking in stereolithography-based rapid casting of turbine blade. *International Journal of Advanced Manufacturing Technology, 55*(5), 447–455. doi:10.100700170-010-3064-x

Chia, H. N., & Wu, B. M. (2015). Recent advances in 3D printing of biomaterials. *Journal of Biological Engineering, 9*(1), 4. doi:10.118613036-015-0001-4 PMID:25866560

Chiu, Y. Y., Liao, Y. S., & Hou, C. C. (2003). Automatic fabrication for bridged laminated object manufacturing (LOM) process. *Journal of Materials Processing Technology, 140*(1), 179–184. doi:10.1016/S0924-0136(03)00710-6

Chockalingam, K., Jawahar, N., Ramanathan, K. N., & Banerjee, P. S. (2006). Optimization of stereolithography process parameters for part strength using design of experiments. *International Journal of Advanced Manufacturing Technology, 29*(1), 79–88. doi:10.100700170-004-2307-0

Chua, C. K., Hong, K. H., & Ho, S. L. (1999). Rapid tooling technology. Part 2. A case study using arc spray metal tooling. *International Journal of Advanced Manufacturing Technology, 15*(8), 609–614. doi:10.1007001700050109

Das, A., Madras, G., Dasgupta, N., & Umarji, A. M. (2003). Binder removal studies in ceramic thick shapes made by laminated object manufacturing. *Journal of the European Ceramic Society, 23*(7), 1013–1017. doi:10.1016/S0955-2219(02)00266-2

Dedoussis, V., Canellidis, V., & Mathioudakis, K. (2008). Aerodynamic experimental investigation using stereolithography fabricated test models: The case of a linear compressor blading cascade. *Virtual and Physical Prototyping, 3*(3), 151–157. doi:10.1080/17452750802120201

Dedoussis, V., & Giannatsis, J. (2004). Stereolithography assisted redesign and optimisation of a dishwasher spraying arm. *Rapid Prototyping Journal, 10*(4), 255–260. doi:10.1108/13552540410551388

Delfs, P., Li, Z., & Schmid, H. J. (2015). Mass finishing of laser sintered parts. *Proceedings of the 26th Annual International Solid Freeform Fabrication Symposium*, 514-526.

Dotchev, K., & Soe, S. (2006). Rapid manufacturing of patterns for investment casting: Improvement of quality and success rate. *Rapid Prototyping Journal, 12*(3), 156–164. doi:10.1108/13552540610670735

Gibson, I., Rosen, D. W., & Stucker, B. (2010). *Additive Manufacturing Technologies Rapid Prototyping to Direct Digital Manufacturing*. Springer-Verlag New York.

Greer, S. E. (2009). A comparison of the ancient metal casting materials and processes to modern metal casting materials and processes. Hartford, CT: Rensselaer Polytechnic Institute.

Guo, N., & Leu, M. C. (2013, September 1). Additive manufacturing: Technology, applications and research needs. *Frontiers of Mechanical Engineering*, *8*(3), 215–243. doi:10.100711465-013-0248-8

Hague, R., Mansour, S., & Saleh, N. (2004). Material and design considerations for rapid manufacturing. *International Journal of Production Research*, *42*(22), 4691–4708. doi:10.1080/00207840410001733940

Halidi, S. N., & Abdullah, J. (2012). Moisture effects on the ABS used for fused deposition modeling rapid prototyping machine. In *Humanities, Science and Engineering Research (SHUSER), IEEE Symposium* (pp. 839-843). IEEE. 10.1109/SHUSER.2012.6268999

Harlan, N. R., Bourell, D. L., Beaman, J. J., & Reyes, R. (2001). Titanium castings using laser-scanned data and selective laser-sintered zirconia molds. *Journal of Materials Engineering and Performance*, *10*(4), 410–413. doi:10.1361/105994901770344818

Harun, W.S., Safian, S., & Idris, M.H. (2009). Evaluation of ABS patterns produced from FDM for investment casting process. *Computational Method and Experiments in Materials Characterisation IV*, *1*, 319-28.

Heine, R. W., Loper, C. R., & Rosenthal, P. C. (1955). *Principles of metal casting*. Tata McGraw-Hill Education.

Jones, S., & Yuan, C. (2003). Advances in shell moulding for investment casting. *Journal of Materials Processing Technology*, *135*(2), 258–265. doi:10.1016/S0924-0136(02)00907-X

Knights, M. (2001). Rapid tooling is ready for prime time. *Plastics Technology*, *47*(1), 46–53.

Kruth, J. P., Levy, G., Klocke, F., & Childs, T. H. (2007). Consolidation phenomena in laser and powder-bed based layered manufacturing. *CIRP Annals-Manufacturing Technology*, *56*(2), 730–759. doi:10.1016/j.cirp.2007.10.004

Kruth, J. P., Mercelis, P., Van Vaerenbergh, J., Froyen, L., & Rombouts, M. (2005). Binding mechanisms in selective laser sintering and selective laser melting. *Rapid Prototyping Journal*, *11*(1), 26–36. doi:10.1108/13552540510573365

Kruth, J. P., Wang, X., Laoui, T., & Froyen, L. (2003). Lasers and materials in selective laser sintering. *Assembly Automation*, *23*(4), 357–371. doi:10.1108/01445150310698652

Kumar, S., & Kruth, J. P. (2010). Composites by rapid prototyping technology. *Materials & Design*, *31*(2), 850–856. doi:10.1016/j.matdes.2009.07.045

Kunjan, C., Jawahar, N., Chandrasekhar, U., Praveen, J., & Karthic, M. (2016). Development of Process Model for Optimal Selection of Process Parameters for Geometric Tolerances and Surface Roughness in Stereolithography. *International Journal of Advanced Design and Manufacturing Technology*, *9*(3), 103–107.

Lan, P. T., Chou, S. Y., Chen, L. L., & Gemmill, D. (1997). Determining fabrication orientations for rapid prototyping with stereolithography apparatus. *Computer Aided Design*, *29*(1), 53–62. doi:10.1016/S0010-4485(96)00049-8

Lee, C. W., Chua, C. K., Cheah, C. M., Tan, L. H., & Feng, C. (2004). Rapid investment casting: Direct and indirect approaches via fused deposition modelling. *International Journal of Advanced Manufacturing Technology*, *23*(1-2), 93–101. doi:10.100700170-003-1694-y

Madrazo, I., Zamorano, C., Magallón, E., Valenzuela, T., Ibarra, A., Salgado-Ceballos, H., ... Guízar-Sahagún, G. (2009). Stereolithography in spine pathology: A 2-case report. *Surgical Neurology*, *72*(3), 272–275. doi:10.1016/j.surneu.2008.04.034 PMID:18614210

Mahesh, M., Wong, Y. S., Fuh, J. Y., & Loh, H. T. (2004). Benchmarking for comparative evaluation of RP systems and processes. *Rapid Prototyping Journal*, *10*(2), 123–135. doi:10.1108/13552540410526999

Marchelli, G., Prabhakar, R., Storti, D., & Ganter, M. (2011). The guide to glass 3D printing: Developments, methods, diagnostics and results. *Rapid Prototyping Journal*, *17*(3), 187–194. doi:10.1108/13552541111124761

Mekonnen, B. G., Bright, G., & Walker, A. (2016). *A Study on State of the Art Technology of Laminated Object Manufacturing (LOM). In CAD/CAM, Robotics and Factories of the Future* (pp. 207–216). New Delhi: Springer.

Mueller, B., & Kochan, D. (1999). Laminated object manufacturing for rapid tooling and patternmaking in foundry industry. *Computers in Industry*, *39*(1), 47–53. doi:10.1016/S0166-3615(98)00127-4

Murphy, S. V., & Atala, A. (2014). 3D bioprinting of tissues and organs. *Nature Biotechnology*, *32*(8), 773–785. doi:10.1038/nbt.2958 PMID:25093879

Norouzi, Y., Rahmati, S., & Hojjat, Y. (2009). A novel lattice structure for SL investment casting patterns. *Rapid Prototyping Journal*, *15*(4), 255–263. doi:10.1108/13552540910979776

Novakova-Marcincinova, L. (2012). Application of fused deposition modeling technology in 3D printing rapid prototyping area. *Manuf. and Ind. Engg.*, *11*(4), 35–37.

Omar, M. F., Sharif, S., Ibrahim, M., Hehsan, H., Busari, M. N., & Hafsa, M. N. (2012). Evaluation of direct rapid prototyping pattern for investment casting. *Advanced Materials Research*, *463*, 226–233. doi:10.4028/www.scientific.net/AMR.463-464.226

Onuh, S. O., & Hon, K. K. (1998). Optimising build parameters for improved surface finish in stereolithography. *International Journal of Machine Tools & Manufacture*, *38*(4), 329–342. doi:10.1016/S0890-6955(97)00068-0

Pal, D. K., Ravi, B., & Bhargava, L. S. (2002). E-manufacturing one-off intricate castings using rapid prototyping technology. In *International Conference on e-Manufacturing, Bhopal* (pp. 259-63). American Foundry Society.

Pandey, P. M., Thrimurthulu, K., & Reddy, N. V. (2004). Optimal part deposition orientation in FDM by using a multicriteria genetic algorithm. *International Journal of Production Research*, *42*(19), 4069–4089. doi:10.1080/00207540410001708470

Park, J., Tari, M. J., & Hahn, H. T. (2000). Characterization of the laminated object manufacturing (LOM) process. *Rapid Prototyping Journal*, *6*(1), 36–50. doi:10.1108/13552540010309868

Pattnaik, S., Jha, P. K., & Karunakar, D. B. (2014). A review of rapid prototyping integrated investment casting processes. *Proceedings of the Institution of Mechanical Engineers, Part L: Journal of Materials: Design and Applications, 228*(4), 249-77.

Pham, D. T., Dimov, S. S., & Gault, R. S. (1999). Part orientation in stereolithography. *International Journal of Advanced Manufacturing Technology, 15*(9), 674–682. doi:10.1007001700050118

Pilipović, A., Raos, P., & Šercer, M. (2011). Experimental testing of quality of polymer parts produced by laminated object manufacturing–LOM. *Tehnicki Vjesnik (Strojarski Fakultet), 18*, 253–260.

Rosochowski, A., & Matuszak, A. (2000). Rapid tooling: The state of the art. *Journal of Materials Processing Technology, 106*(1-3), 191–198. doi:10.1016/S0924-0136(00)00613-0

Schmid, M., Simon, C., & Levy, G. N. (2009). Finishing of SLS-parts for rapid manufacturing (RM)–a comprehensive approach. *Proceedings SFF*, 1-10.

Shu, X., & Wang, R. (2017). Thermal residual solutions of beams, plates and shells due to laminated object manufacturing with gradient cooling. *Composite Structures, 15*(174), 366–374. doi:10.1016/j.compstruct.2017.04.060

Singh, J., Singh, R., & Singh, H. (2017). Dimensional accuracy and surface finish of biomedical implant fabricated as rapid investment casting for small to medium quantity production. *Journal of Manufacturing Processes, 25*, 201–211. doi:10.1016/j.jmapro.2016.11.012

Singh, R., & Singh, M. (2017). Surface roughness improvement of cast components in vacuum moulding by intermediate barrel finishing of fused deposition modelling patterns. *Proceedings of the Institution of Mechanical Engineers. Part E, Journal of Process Mechanical Engineering, 231*(2), 309–316. doi:10.1177/0954408915595576

Singh, R., & Singh, S. (2013). Effect of process parameters on surface hardness, dimensional accuracy and surface roughness of investment cast components. *Journal of Mechanical Science and Technology, 27*(1), 91–97. doi:10.100712206-012-1218-5

Singh, R., Singh, S., Singh, I. P., Fabbrocino, F., & Fraternali, F. (2017). Investigation for surface finish improvement of FDM parts by vapor smoothing process. *Composites. Part B, Engineering, 111*, 228–234. doi:10.1016/j.compositesb.2016.11.062

Singh, S., & Singh, R. (2016). Effect of process parameters on micro hardness of Al–Al2O3 composite prepared using an alternative reinforced pattern in fused deposition modelling assisted investment casting. *Robotics and Computer-integrated Manufacturing, 37*, 162–169. doi:10.1016/j.rcim.2015.09.009

Singh, S., & Singh, R. (2016). Fused deposition modelling based rapid patterns for investment casting applications: A review. *Rapid Prototyping Journal, 22*(1), 123–143. doi:10.1108/RPJ-02-2014-0017

Singh, S., & Singh, R. (2016). Precision investment casting: A state of art review and future trends. *Proceedings of the Institution of Mechanical Engineers. Part B, Journal of Engineering Manufacture, 230*(12), 2143–2164. doi:10.1177/0954405415597844

Sivadasan, M., Singh, N.K. & Sood, A.K. (2012). Use of fused deposition modeling process in investment precision casting and risk of using selective laser sintering process. *International Journal of Applied Research in Mechanical Engineering.*

Smith, B. J., St Jean, P., & Duquette, M. L. (1996). A comparison of rapid prototype techniques for investment casting Be-Al. In *Proceedings of Rapid Prototyping and Manufacturing'96* (Vol. 2, pp. 1–11). Dearborn, MI: Conference.

Taşcıoğlu, S., & Akar, N. (2003). A novel alternative to the additives in investment casting pattern wax compositions. *Materials & Design*, *24*(8), 693–698. doi:10.1016/S0261-3069(03)00097-9

Taşcıoğlu, S., & Akar, N. (2007). Conversion of an investment casting sprue wax to a pattern wax by chemical agents. *Materials and Manufacturing Processes*, *18*(5), 753–768. doi:10.1081/AMP-120024973

Taylor, P. R. (1983). An illustrated history of lost wax casting. *Proceedings of the 17th Annual BICTA Conference.*

Tiwary, V., Arunkumar, P., Deshpande, A. S., & Khorate, V. (2015). Studying the effect of chemical treatment and fused deposition modelling process parameters on surface roughness to make acrylonitrile butadiene styrene patterns for investment casting process. *International Journal of Rapid Manufacturing*, *5*(3-4), 276–288. doi:10.1504/IJRAPIDM.2015.074807

Träger, H., & Bührig-Polaczek, A. (2000). *Foundry technology*. Ullmann's Encyclopedia of Industrial Chemistry. doi:10.1002/14356007.a12_035

Tseng, P., Murray, C., Kim, D., & Di Carlo, D. (2004). Research highlights: Printing the future of microfabrication. *Lab on a Chip*, *14*(9), 1491–1495. doi:10.1039/c4lc90023e PMID:24671475

Turner, B., Strong, R., & Gold, S. (2014). A review of melt extrusion additive manufacturing processes: I. Process design and modeling. *Rapid Prototyping Journal*, *20*(3), 192–204. doi:10.1108/RPJ-01-2013-0012

Turner, B. N., & Gold, S. A. (2015). A review of melt extrusion additive manufacturing processes: II. Materials, dimensional accuracy, and surface roughness. *Rapid Prototyping Journal*, *21*(3), 250–261. doi:10.1108/RPJ-02-2013-0017

Vermeulen, M., Claessens, T., Van Der Smissen, B., Van Holsbeke, C. S., De Backer, J. W., Van Ransbeeck, P., & Verdonck, P. (2013). Manufacturing of patient-specific optically accessible airway models by fused deposition modeling. *Rapid Prototyping Journal*, *19*(5), 312–318. doi:10.1108/RPJ-11-2011-0118

Vickers C. (n.d.). *An alternative route to metal components for prototype and low-volume production. Rapid prototyping casebook.* Academic Press.

Vijay, P., Danaiah, P., & Rajesh, K. V. (2011). Critical parameters effecting the rapid prototyping surface finish. *Journal of Mechanical Engineering and Automation*, *1*, 17–20. doi:10.5923/j.jmea.20110101.03

Wang, S., Miranda, A. G., & Shih, C. (2010). A study of investment casting with plastic patterns. *Materials and Manufacturing Processes*, *25*(12), 1482–1488. doi:10.1080/10426914.2010.529585

Warner, M. C. (1993). Rapid prototyping methods to manufacture functional metal and plastic parts. Rapid Prototyping System: Fast Track to Product Realisation, 137-44.

Wendel, B., Rietzel, D., Kühnlein, F., Feulner, R., Hülder, G., & Schmachtenberg, E. (2008). Additive processing of polymers. *Macromolecular Materials and Engineering*, *293*(10), 799–809. doi:10.1002/mame.200800121

Wood, I. (2016). *Development of a layerless additive manufacturing stereolithography machine to improve surface quality and dimensional accuracy* (Doctoral dissertation). University of Ontario Institute of Technology.

Wu, H., Li, D., & Guo, N. (2009). Fabrication of integral ceramic mold for investment casting of hollow turbine blade based on stereolithography. *Rapid Prototyping Journal*, *15*(4), 232–237. doi:10.1108/13552540910979749

Yang, J., Shi, Y., Shen, Q., & Yan, C. (2009). Selective laser sintering of HIPS and investment casting technology. *Journal of Materials Processing Technology*, *209*(4), 1901–1908. doi:10.1016/j.jmatprotec.2008.04.056

Yang, Q., Lu, Z., Zhou, J., Miao, K., & Li, D. (2017). A novel method for improving surface finish of stereolithography apparatus. *International Journal of Advanced Manufacturing Technology*, *393*(5-8), 1537–1544. doi:10.100700170-017-0529-1

Yi, S., Liu, F., Zhang, J., & Xiong, S. (2004). Study of the key technologies of LOM for functional metal parts. *Journal of Materials Processing Technology*, *150*(1), 175–181. doi:10.1016/j.jmatprotec.2004.01.035

Zhang, W., Leu, M. C., Ji, Z., & Yan, Y. (1999). Rapid freezing prototyping with water. *Materials & Design*, *20*(2), 139–145. doi:10.1016/S0261-3069(99)00020-5

Chapter 4
Miniaturization of Test Specimen for Composites

Saood Ali
Yeungnam University, South Korea

V. Murari
Motilal Nehru National Institute of Technology Allahabad, India

ABSTRACT

The objective behind the development of miniaturization or small specimen test technology is to reduce the cost and quantity of material involved during the characterization of the material. The idea of the development of miniaturization took attention when the nuclear industry starts developing as these materials are very costly and it is not economically feasible to waste large amount of these materials for the sole purpose of testing. The second factor which promotes the miniaturization is that the working of machine is not affected while at the same time its material is being tested. At present, the idea of miniaturization is being applied to other materials also. The miniaturization of standards for metals has been done successfully in the past. For composites, not much work has been done. In the chapter, the specimen size effects on tensile properties of glass fiber composite have been identified by varying the length and width simultaneously and have established a relationship between the ASTM standard specimen and the small size specimen.

INTRODUCTION

Background and Motivation

Fiber reinforced composites have an extensive array of applications. They range from structural to recreational use. The aerospace and automotive industries look to composites to improve fuel economy due to their high strength to weight ratio. The sports industry looks to composites to improve sports equipment technologies. The fact that composites offer increased strength without sacrificing additional weight is what gives composites the advantage from most structural and recreational materials. There have been continuous efforts, to miniaturize test specimens for various reasons like scope for deriving

DOI: 10.4018/978-1-5225-5445-5.ch004

more number of specimens from the sample removed, reduction in size of test equipment, saving in waste handling requirements etc. The present work highlights the development of glass fiber composite, layout of miniature test specimens, preparation of test specimens, miniature specimen testing procedures and co-relationship with conventional procedures.

Problem Definition

The more recent introduction of FRP composite technology, together with the large range of materials used and being introduced, brings, a broad design base, which was available for many metals, yet not been compiled for FRP materials, which means, a much testing of composite specimens has to be carried out either on full-scale prototypes, or, in order to save both time and expense, on small-scale models by use of the principles of dimensional analysis. And if any discrepancies encountered whilst scaling from model to full size (i.e. any size effects) should be both identified and understood.

Materials are subjected to various types of tests like tensile, impact and fatigue-fracture characterization. Sub-sized conventional tests, which are essentially a scaled down version of conventional testing, utilize specimens of similar geometry loaded in a similar manner, to produce results equivalent to that obtained from larger specimens. Miniature specimen tests are employed for determination of residual service life of the operating component, by extrapolating the results of evaluation of small specimen.

In a review of literature, it has been shown that the majority of the existing work in the field of small specimen test technology used high quality, pre-preg carbon/epoxy laminates which have been used in the aero- space industries. It is shown that a number of authors have come to the conclusion that the scale/size phenomenon exists. But very few statistical and experimental analyses of the results and trends are reported.

The scaling problem is very complex for composites due to intricate nature of their micro-structure. In addition the many possible material properties that may be considered, such as manufacturing technique and conditions, and fiber and matrix materials, further complicate the problem. Further, the mechanical testing of composite materials is a very wide subject, with many variables that would affect the observed material properties.

The objective of the present study is to identify specimen size effects on tensile properties of glass fiber composite by varying the in-plane dimensions (length and width) simultaneously and establish a relationship between the ASTM standard specimen and the small size specimen. Development of miniature specimen testing technique involves two aspects: namely, development of methods for preparation of miniature specimens and development of techniques to extract useful mechanical properties from such specimens. Hand lay-up method have been used to prepare glass fiber composite. ASTM D3039 standard have been used to prepare standardized specimen and the number of specimens prepared for each size is ten. ASTM standard specimen have been tested on nano plug machine and the sub-specimen have been tested tinius olsen machine. Regression analysis have been done with the help of minitab software.

ORGANIZATION OF CHAPTER

The current report is divided into six parts.

Part 1 describes the objective and problem of the current work. Literature reviewed during the current work has been described in part 2. Literature gap and the justification of the objective has been given at the end of part 2. Testing conditions, standard used, testing equipment, material properties, fabrication method and problems encountered during experiments have been described in part 3. Results obtained by the testing of specimens for each sample is been given in part 4. Young modulus for all the samples, along with the correlation and regression analysis between ASTM standard specimen and the sub size specimens have also been described in part 4. In the end conclusion and future work have been given in part 5.

LITERATURE REVIEW

The question of size effect in composites has been around since 1960's. The fact that much work is still underway shows that conclusive evidence has not yet arise to answer the question. The work done in the field of miniaturization which has been reviewed is described below under the separate heading of metals and composites.

Work Done in the Field of Metals

Reed et al. (1883) compare the results of two methods (miniature specimens machined from the plate and short transverse specimens with welded prolongations) to calculate the short transverse tensile properties of light gage (<50mm) steel plate. He found that the advantage of miniature specimen is that, they are ideally suited for light gage plates and have the ability to test specific regions of a plate such as surface, quarterline or centerline through positioning of the specimen. In the end he concluded that the miniature specimen provides valuable test data unobtainable from the welded specimens.

Manhan et al. (1995) presented the result of a study which is focused on designing an optimized miniature notch test (MNT) specimen and machine. Experiments are performed on one sided grooved miniaturized notch test specimens which shows that 1/16 scale miniature specimens can be designed to yield transitional fracture behavior and the fracture appearance and energy temperature curves can be quantitatively related to the conventional ASTM E23 specimen data.

Fleur et al. (1998) use small punch test techniques to evaluate mechanical properties of components using miniaturized sized specimens and correleations between mechanical characterstics determined from small punch test tests and uniaxial tensile tests for low alloy steels have been obtained. Analytical formulations have also been used to estimate the uniaxial tensile stress-strain behavior for low alloy steels from experimental small punch load deflection curves.

Rosinski and Corwin et al. (1998) study the result of an international testing exercise which was conducted under ASTM committee E10.02 to obtain a cross comparison of material properties which are obtained from various sub-size specimen testing techniques. The testing material used in the exercise is ASTM A533 grade B class 1 plate designated as HSST plate 03 privided by Oakridge national laboratory.

Yuanchao et al. (2000) uses the modified miniature specimen test (MMST) technique to determine the yield strength, tensile strength and uniform elongation of unirradiated and irradiated reactor pressure vessel steels and also describe the conditions for irradiating the miniature specimensand compare the results with conventional tensile test data. He also describe the use of MMST technique to radiation surveillance for pressurized water reactor components.

Kim et al. (2011) studies the constitutive data and widely obtained fracture toughness data sets at transition temperature of F82H to establish master curve method for reduced-activation ferritic (RAF) steels from view point of fracture mechanics and for effective production of irradiation database. Small specimen test are used with different levels of phosphorus addition to F82H steels. Effect of specimen size and phosphorus on fracture toughness is also evaluated using different size specimen.

Volak et al. (2012) design and manufacture the special miniature test specimens for the purpose of fatigue testing of seven different steels and an Al alloy. Also describe the results of fatigue tests performed on miniature test specimens in comparison with traditional fatigue tests.

Songa et al. (2012) performed conventional tensile test and small punch test on a failed ASTM 350 forging flange to study specimen size effect on mechanical behavior and also study the effect of anisotropy on mechanical properties by machining the specimens from failed flange along circumferential and longitudinal directions.

Shin et al. (2012) evaluate fatigue crack propagation properties using miniature specimens and find out that fatigue crack propagation rate in sub-size specimens is slightly but consistently slower than that in the standard specimens.

Ermi et al. (1986) developed miniature specimens to characterize the effect of neutron irradiation on the fatigue crack growth behavior of candidate fusionreactor first wall materials. The miniature centre-cracked-tension specimens show the same crack growth rate results as the larger specimens and occupying a fraction of the volume of the larger specimens.

Lucon et al. (1993) evaluate service exposed components mechanical properties by means of sub-size and miniature specimens. Different sub-size or miniature specimens have been used for various mechanical testing, as, cylindrical specimens with circumferential crack for fracture toughness testing & miniature disk shaped specimens for upper shelf ductile regime, small size specimens for low cycle fatigue testing with enhanced cycle frequency.

Manhan et al. (1999) study the results of the two standards which are developed by ASTM subcommittee E 28.07.07. The first standard focused on test procedures for instrumented impact testing and ASTM E23 will reference the instrumented test standard for tests conducting conventional Charpy V-notch (CVN) specimens. The second standard covers miniaturized charpy V-notch (MCVN) testing. Six specimens each are tested in upper shelf region and in transition region.

Kundan et al. (2006) studies the boat sampling developed at BHABA ATOMIC RESEARCH CENTER (BARC) and explain the method to prepare miniature specimens and develop co-releationship with conventional tets.

Ksada et al. (2006) find out the fracture toughness of blanket structural materials using master curve (MC) methodology and evaluate change in fracture toughness in the cleavage regime of a reduced activation ferrite (RAF) before and after thermal embrittlement treatment has been applied by using sub sized compact tension test specimens.

Wanga et al. (2008) performed small punch tests on round specimens at room temperature with six kind of thicknesses and experiments results show that the small punch energy, fracture strain and fracture toughness increase with the thickness of the small punch samples.

Kurtz et al. (2002) developed the miniature specimen shear punch test to evaluate the shear behavior of ultra high molecular weight polyethylene used in total joint replacement components. Also investigated the shear punch behavior of virgin and crosslinked materials as well as of ultra high molecular weight polyethylene from tibial implants which were gamma irradiated in air and shelf aged for upto 8.5 years.

Work Done in the Field of Composite

Mcloughlin et al. (1972) described a method for preparing small carbon fiber reinforced composite test specimens which results in high quality specimens of fiber volume fraction. The small size test specimens will serve as standard against which large specimens can be compared and they are also useful for evaluating new fibers, resins and molding conditions. Also the effect of specimen geometry, such as span to thickness ratio and width to thickness ratio, has been studied for carbon fiber/epoxy composite.

Funk et al. (1986) developed a miniature double cantilever beam (DCB) and edge delamination test (EDT) specimens to determine the Mode I critical strain energy release rate G_{Ic} and the mixed mode I and II critical strain energy release rate G_c and the results from miniature specimens are compared with that of standard specimens used for DCB and EDT test.

Crasto et al. (1994) uses a miniature sandwich specimen to study the effect of matrix modulus and fiber matrix interfacial bond strength on unidirectional carbon/epoxy composite. The miniature specimens have higher compression strengths as compared to conventional specimens for unidirectional carbon/epoxy composites.

Kim et al. (1994) used finite element method to study the effect of stress distribution at the ends of gage section of a miniature sandwich specimens and a conventional all composite specimen. The influence of parameters such as tab material, thickness and the taper angle is also investigated to optimize the geometry of miniature sandwich specimen with respect to stress concentrations.

Nozawa et al. (2002) investigated miniaturization of tensile specimen methodology to evaluate tensile properties of ceramic matrix composites such as SiC/SiC composites and for the establishment of small specimen test technique and specimen size effects on the tensile properties of plain weave, satin weave and 3-D SiC/SiC composites. There are dissimilarities between the dendencies of tensile properties for 2-D SiC/SiC composites to that for 3-D ones.

Nozawa et al. (2005b) identify specimen size effects on tensile properties of SiC/SiC composites and find out that both 2D and 3D SiC/SiC composites with the same axial fiber volume content hardly exhibit meaningful size effect on the tensile properties in fiber longitudinal direction.

Nozawa et al. (2005a) uses the small specimen test technique to determine the tensile, flexural and shear properties of unidirectional SiC/SiC composites fabricated with highly crystalline and stochiometric fibers and matrix but with three different fiber matrix interfaces and then irradiated to compare the role of different interfaces on neutron irradiation behavior.

Lee et al. (2008) study the scaling effects in determining the effect of notch size, ply and thickness on the strength of composites with open holes which are prepared from carbon/epoxy pre-impregnated tapes by hand lay-up method in the form of unidirectional and multidirectional laminates.

Bing et al. (2008) studies the specimen size effect in off-axis compression test by conducting experiments on small blocks off-axis specimens of low modulus S2 glass fiber reinforced composites and high modulus AS4 carbon fiber reinforced composites and found that the off-axis compressive strength of glass/epoxy composite decreased by small amount while there is a considerable reduction in off-axis compressive strength of high modulus carbon/epoxy composites when specimen width or thickness has increased.

Literature Gap

Small size specimens have been prepared for carbon/epoxy composites and SiC/SiC composites and have been tested to determine various properties. But the correlation between the small size specimen properties and the ASTM standard specimen have not been identified.

While preparing the small size specimens, the dimensions of the ASTM standard specimen have been varied one at a time while keeping the other two dimensions kept constant. In the current problem, two dimensions (length and width) have been varied simultaneously while the third dimension (i.e thickness) will remain constant.

EXPERIMENTAL PROCEDURE

ASTM Standard

ASTM standard used during the experimental procedure is ASTM D3039.

ASTM D3039 determines the in-plane tensile properties of polymer matrix composite materials reinforced by high modulus fibers. The composite material forms are limited to continuous fiber or discontinuous fiber reinforced composites in which the laminate is balanced and symmetric with respect to the test direction.

Test Method

Thin flat strip of material having a constant rectangular cross section is mounted in the grips of a mechanical testing machine and monotonically loaded in tension while recording load. The ultimate tensile strength of the material can be determined from the maximum load carried prior to failure.

Properties Determined

The following properties can be determined by using the standard ASTM D3039:

- Ultimate tensile strength
- Ultimate tensile strain
- Tensile chord modulus of elasticity
- Poisson ratio
- Transition strain
- Test specimen geometry

Design recommendations for tensile test coupons for ASTM D3039 are given in table 1 and corresponding dimensions will be given in table 2.

Tensile test specimen drawing is been shown in figure 1.

Table 1. Tensile Specimen Geometry Requirements

Parameter	Requirement
Coupon requirements:	
Shape	rectangular cross section
Minimum length	Gripping + 2 times width+ gage length
Specimen width	As needed
Specimen width tolerance	±1% of width
Specimen thickness	As needed
Specimen thickness tolerance	±4% of thickness
Specimen flatness	Flat with light finger pressure
Tab requirements (if used)	
Tab material	As needed
Fiber orientation(composite tabs)	As needed
Tab thickness	As needed
Tab thickness variation between tabs	±1% tab thickness
Tab bevel angle	5 to 90°, inclusive

Table 2. Tensile specimen geometry recommendations

Fiber Orientation	Width mm(in.)	Overall Length mm(in.)	Thickness mm(in.)	Tab Length mm(in.)	Tab Thickness mm(in.)	Tab Bevel Angle, °
0° unidirectional	15(0.5)	250(10.0)	1.0(0.040)	56(2.25)	1.5(0.062)	7 or 90
90° unidirectional	25(1.0)	175(7.0)	2.0(0.080)	25(1.0)	1.5(0.062)	90
Balanced and symmetric	25(1.0)	250(10.0)	2.5(0.100)	Emery cloth	-	-
Random-discontinuous	25(1.0)	250(10.0)	2.5(0.100)	Emery cloth	-	-

Figure 1. Tensile Test Specimen

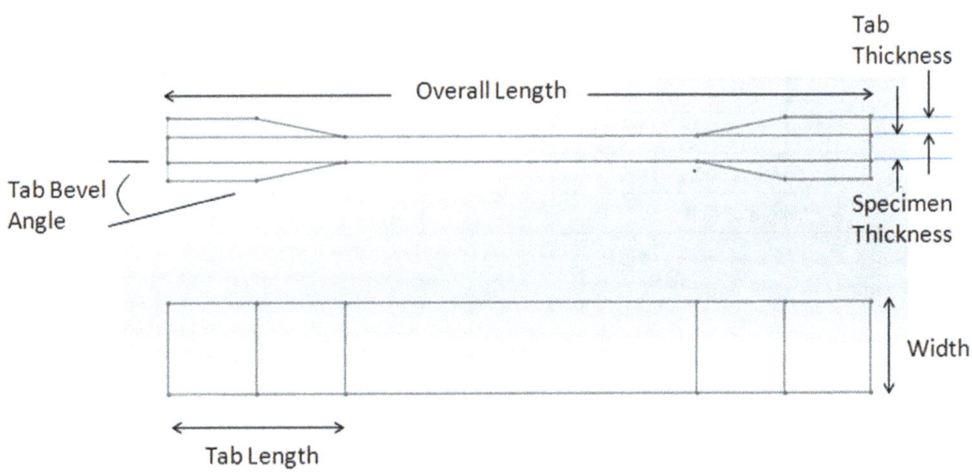

MATERIAL USED

Glass Fiber

Glass fiber, in fabric form, is been supplied by KE- Technical textiles pvt. Limited. The relevant properties of glass fiber fabric is given in table 3(refer figure 2).

Resin Material

Epoxy LY556 has been used as resin matrix material. Hardner (Arador HY 951) has been mixed in the epoxy in the ratio of 1:10.

Figure 2. Glass Fiber Fabric

Table 3. Glass Fiber Fabric Properties

Manufacturer		KE-Technical Textiles pvt. ltd.
Texture		Plain
No. of Threads per Inch	Warp	15
	Weft	12
Heat Loss at 600°c for ½ hr.(%)		0.82
Weight per Sq. Mtr. Gms.		362
Thickness(mm)		0.38±0.40
Tenslie Strength (kgf/cm.)	Warp	118-124
	Weft	NA
Width of Roll		40 inch
Length of Roll		20 meter

TEST SPECIMEN PREPARATION

Test Specimen Dimension

Dimension of the test specimen and number of specimens prepared for each dimension with a particular code is given in table 4.

Fabrication Method of Test Specimen

Hand lay-up method has been used to prepare the test specimen (Refer Figure 3)

Method for composite fabrication have been described in the following steps:

- First a smooth surface (steel plate or tile) has been placed horizontally and cleaned thoroughly with the acetone so that there will be no tiny particles present on the surface which otherwise would produce unwanted impressions on the composite surface.
- Then, a polythene sheet was placed on smooth surface and wax polish was applied on it which works as a releasing agent.
- Then, first fabric layer was put on the polythene sheet and resin has been applied on it all over the surface.
- After applying the resin on the top side of first layer, it was inverted and again the resin was applied all over it, so that both surfaces will become smooth.

Table 4. Test specimen dimensions

Test Specimen	Length(mm)	Width(mm)	Thickness(mm)	No. of Specimens	Code
Test specimen 1	250	15	1	10	UNID/G/A/1
Test specimen 2	250	12	1	10	UNID/G/A/2
Test specimen 3	250	9	1	10	UNID/G/A/3
Test specimen 4	250	6	1	10	UNID/G/A/4
Test specimen 5	200	15	1	10	UNID/G/B/1
Test specimen 6	200	12	1	10	UNID/G/B/2
Test specimen 7	200	9	1	10	UNID/G/B/3
Test specimen 8	200	6	1	10	UNID/G/B/4
Test specimen 9	150	15	1	10	UNID/G/C/1
Test specimen 10	150	12	1	10	UNID/G/C/2
Test specimen 11	150	9	1	10	UNID/G/C/3
Test specimen 12	150	6	1	10	UNID/G/C/4
Test specimen 13	100	15	1	10	UNID/G/D/1
Test specimen 14	100	12	1	10	UNID/G/D/2
Test specimen 15	100	9	1	10	UNID/G/D/3
Test specimen 16	100	6	1	10	UNID/G/D/4

- Subsequently layers were added till the required thickness was achieved with resin being applied after addition of each layer.
- Care have been taken to avoid any air getting trapped between the layers while adding the resin.
- After the addition of required number of fabric layers, the excess resin between the fiber layers were squeezed out with the help of a steel roller.
- After that, a second polythene sheet, with wax polish applied on it, is placed on top of the fabric layers.
- After that, second tile were placed on the polythene sheet.
- At last, a distributed weight is placed on the tile and the setup was allowed to cure under room temperature conditions for 48 to 72 hours.

Preparation of Samples From Fabricated Laminate

- The edges of the laminate should be cut out as they are of uneven thickness.
- Specimen dimensions were marked on the laminate with the help of soft point marker. Hard point marker should not be used as it would destroy the surface of laminate (refer figure 4).
- Margin of 3.0 mm has been given between markings of two consecutive samples to ease the cutting of samples from the laminate.
- Cut out samples were ground on the surface grinder to smooth the edges of the sample.
- At the end, the edges of the sample were rubbed with the emery paper to smooth out the edges.
- One of the main problems encountered during experiment was the slipping of specimens from the grip of machine. To avoid the slipping of specimen emery paper have been wrapped around the tab length of the specimen and then gripped in the machine (refer figure 5).

TESTING EQUIPMENT

Tinius Olsen machine with a load cell of range 10KN has been used to test sub size specimens. Nano Plug'n'Play machine manufactured by BISS with a load cell of range 0 to 25KN and maximum stroke of 60mm has been used to test ASTM standard specimen. The grip provided in Tinius Olsen machine is of mechanical type while grip in Nano Plug'n'Play machine is of hydraulic type (refer figure 6).

Figure 3. Hand Lay-up Method

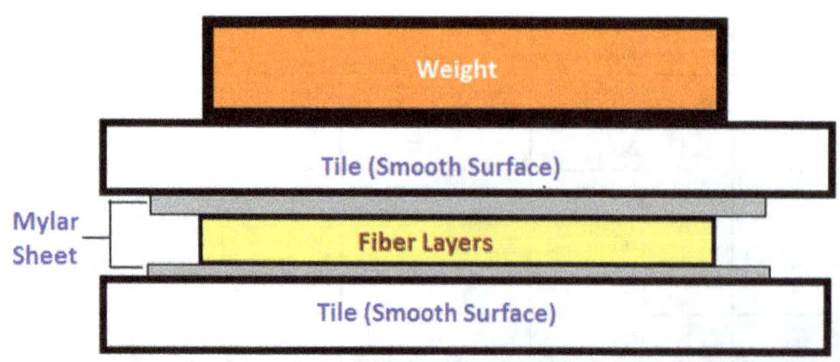

Figure 4. Composite Laminate with marks for cutting the test specimen

Figure 5. Gripping of specimen

Figure 6. Tinius Olsen and Nano Plug'n'Play Machine

RESULTS AND DISCUSSION

Tensile Test Result

Tensile test result for all the specimens with stress-strain curves, normal distribution of young modulus and tensile modulus for each sample has been shown from figure 7 to figure 22.

Mean value of tensile modulus for each sample and the standard deviation for all the specimen has been given in table 5.

CORRELATIONS BETWEEN ASTM STANDARD SPECIMEN AND SUB SIZE SPECIMEN

Pearson correlation coefficient is used to measure the degree of linear relationship between two variables. The correlation coefficient assumes a value between -1 and +1. If one variable tends to increase as the other decreases, the correlation coefficient is negative. Conversely, if the two variables tend to increase together the correlation coefficient is positive. Correlation between young modulus of ASTM standard specimen and the sub size specimens is given in table 5 and the nature of relationship based on the value of correlation coefficient is been given in table 6.

Table 5. Mean value of tensile modulus and standard deviation

Specimen	Tensile Modulus (GPa)	Standard Deviation
UD/G/A/1	26.38	1.265
UD/G/A/2	27.57	1.507
UD/G/A/3	6.216	1.513
UD/G/A/4	7.413	2.301
UD/G/B/1	4.007	1.113
UD/G/B/2	6.723	2.612
UD/G/B/3	4.943	1.366
UD/G/B/4	7.834	2.873
UD/G/C/1	4.259	1.200
UD/G/C/2	5.217	0.8126
UD/G/C/3	6.297	1.007
UD/G/C/4	6.839	1.789
UD/G/D/1	1.044	0.1904
UD/G/D/2	1.773	0.6014
UD/G/D/3	2.716	0.4662
UD/G/D/4	3.106	0.4982

Figure 7. Tensile Test result for Specimen UD/G/A/1

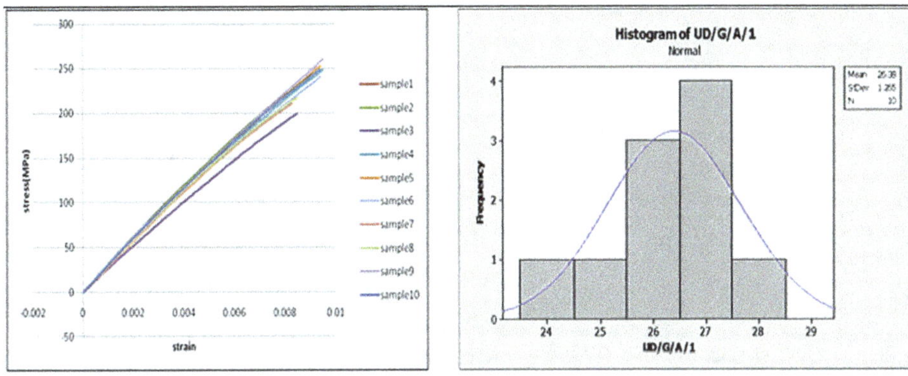

(a) Stress-strain curve

(b) Normal distribution curve for tensile modulus

sample	1	2	3	4	5	6	7	8	9	10
young modulus(GPa)	27.339	28.348	23.728	26.2222	27.009	25.49	25.826	25.882	27.225	26.705

(c) Tensile modulus for each sample

Figure 8. Tensile test result for specimen UD/G/A/2

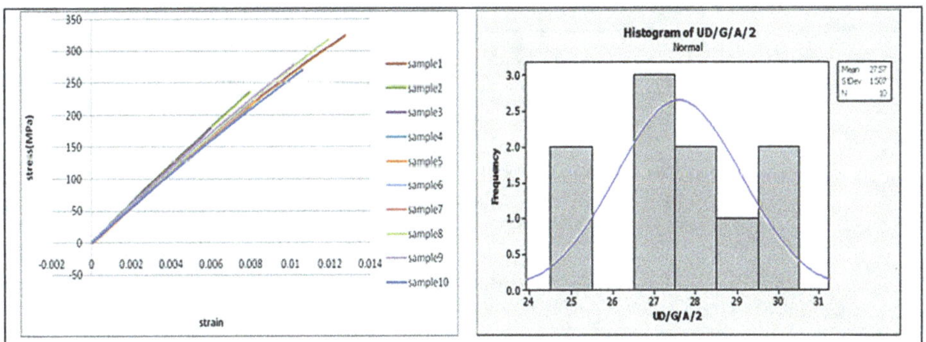

(a) Stress-strain curve

(c)Normal distribution curve for tensile modulus

sample	1	2	3	4	5	6	7	8	9	10
young modulus(GPa)	25.493	29.565	29.82	28.908	27.016	27.704	27.36	26.802	27.59	25.487

(c) Tensile modulus for each sample

Correlation Coefficient Value Interpretation

From table 6 and 7, it is clear that there is a substantial relationship between specimens UD/G/B/4 & UD/G/A/1 and specimens UD/G/C/3 & UD/G/A/1.

Figure 9. Tensile Test Result for specimen UD/G/A/3

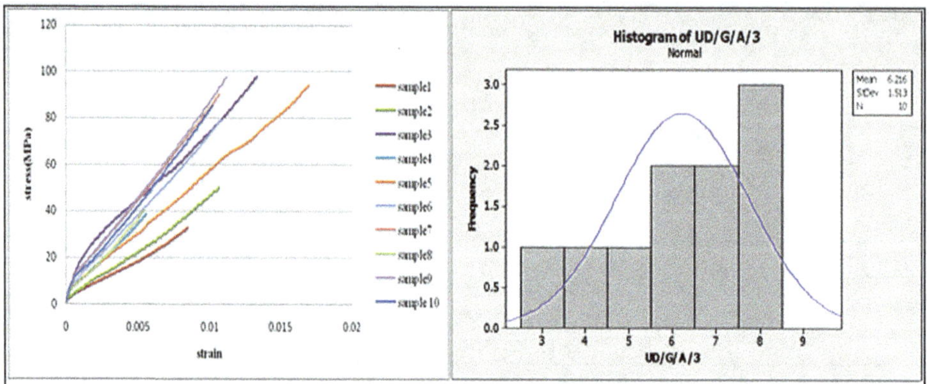

(a) Stress-strain curve

(b) Normal distribution curve for tensile modulus

sample	1	2	3	4	5	6	7	8	9	10
young modulus(GPa)	3.463	4.359	6.233	6.069	5.143	6.766	7.763	6.621	8.018	7.727

(c) Tensile modulus for each sample

Figure 10. Tensile Test result for Specimen UD/G/A/4

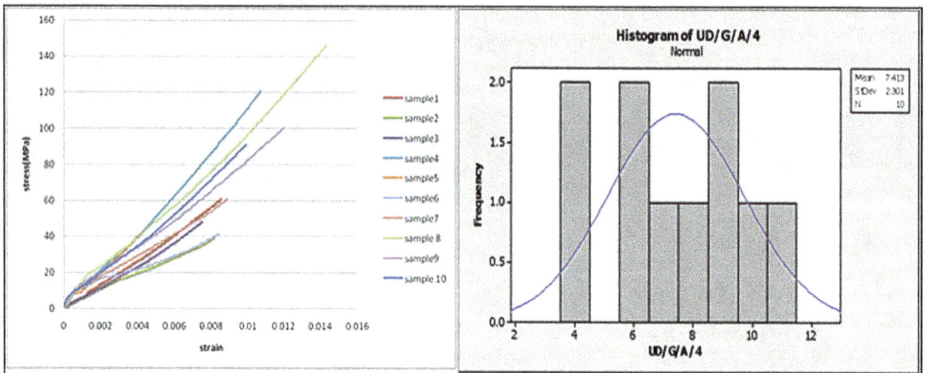

(a) Stress-strain curve

(b) Normal distribution curve for tensile modulus

sample	1	2	3	4	5	6	7	8	9	10
young modulus(GPa)	3.463	4.359	9.561	10.982	5.143	6.766	7.763	6.621	8.018	7.727

(c) Tensile modulus for each sample

REGRESSION ANALYSIS

Regression analysis has been done with the help of MINITAB software. In the regression analysis the predictor are length and width of specimens and the response variable is young modulus of the specimens. In the regression equation, the specimen codes represent the young modulus of that particular specimen.

Figure 11. Tensile Test result for Specimen UD/G/B/1

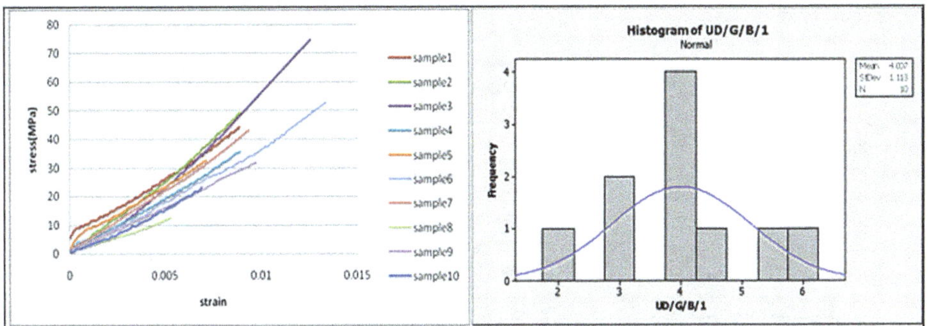

(a) Stress-strain curve

(b) Normal distribution curve for tensile modulus

sample	1	2	3	4	5	6	7	8	9	10
young modulus(GPa)	4.159	5.33	6.093	3.777	3.931	3.824	4.396	2.129	3.197	3.233

(c) Tensile modulus for each sample

Figure 12. Tensile Test result for Specimen UD/G/B/2

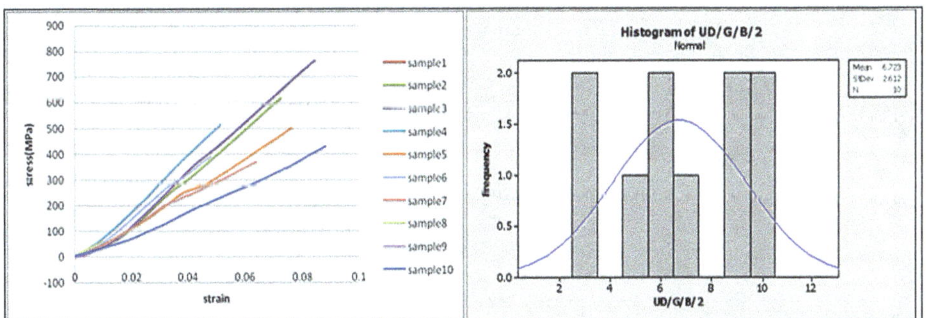

(a) Stress-strain curve

(b) Normal distribution curve for tensile modulus

sample	1	2	3	4	5	6	7	8	9	10
young modulus(GPa)	3.208	9.016	9.657	10.442	6.644	8.588	5.934	5.706	3.132	4.908

(c) Tensile modulus for each sample

Simple linear regression examines the linear relationship between two continuous variables: one dependent (y) and one independent (x). When the two variables are related, it is possible to predict a dependent value from an independent value with better than chance accuracy. Regression provides the line that "best" fits the data. This line can then be used to: how the dependent variable changes as the independent variable changes predict the value of a dependent variable (y) for any independent variable (x).The method used to draw this 'best line" is called the least-squares criterion. The least- squares criterion requires that the best-fitting regression line is the one with the smallest sum of the squared error terms (the distance of the points from the line).

Figure 13. Tensile Test result for Specimen UD/G/B/3

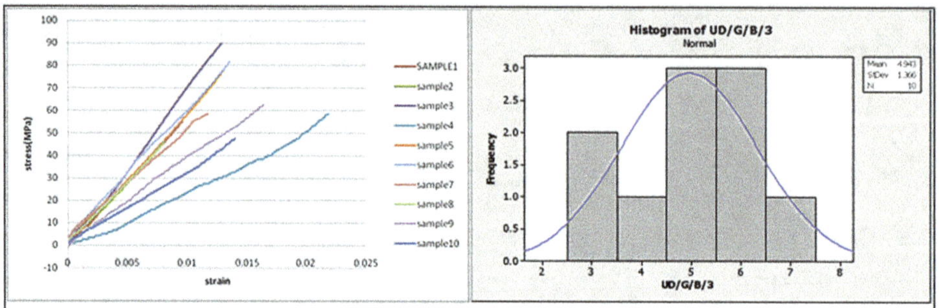

(a) stress-strain curve (b) Normal distribution curve for tensile modulus

sample	1	2	3	4	5	6	7	8	9	10
young modulus(GPa)	5.696	5.356	7.134	2.608	5.713	5.713	4.757	5.456	3.776	3.221

(c) Tensile modulus for each sample

Figure 14. Tensile Test result for Specimen UD/G/B/4

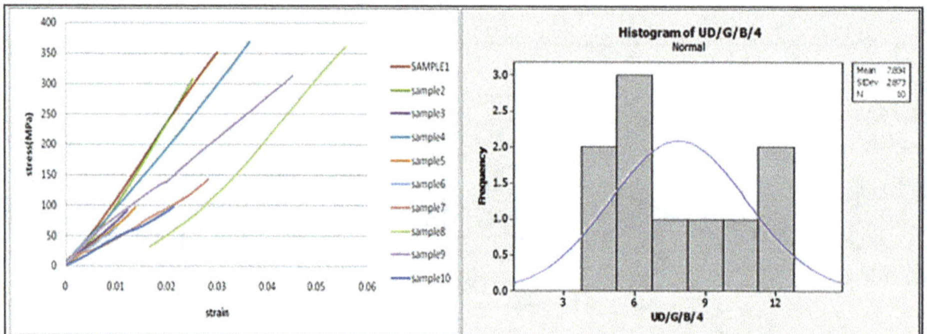

(a) Stress-strain curve (b) Normal distribution curve for tensile modulus

sample	1	2	3	4	5	6	7	8	9	10
young modulus(GPa)	11.97	12.572	6.794	9.978	6.734	5.796	4.679	8.727	6.652	4.436

(c) Tensile modulus for each sample

Regression Analysis for Change in Tensile Modulus (ΔE) vs. Change in Length (Δl) and Width (Δw)

The regression equation for change in tensile modulus (ΔE) is given by

$$\Delta E = 0.17544 \times \Delta l + 2.34285 \times \Delta w - 0.018711 \times \Delta l \times \Delta w$$

Table 9 gives the values of fits and residuals for regression analysis of change in tensile modulus.

Figure 15. Tensile Test result for Specimen UD/G/C/1

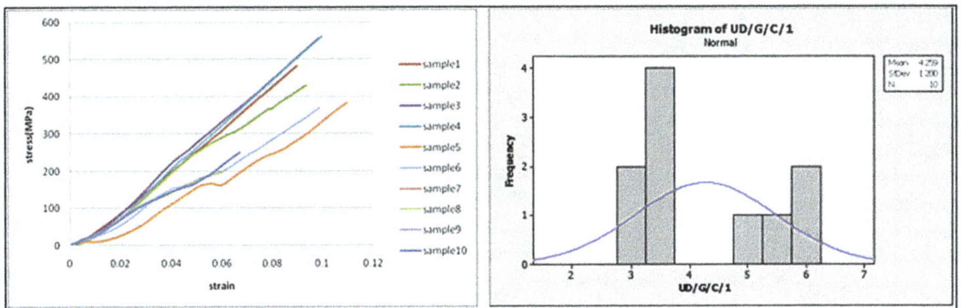

(a) stress-strain curve

(b) Normal distribution curve for tensile

sample	1	2	3	4	5	6	7	8	9	10
young modulus(GPa)	5.558	4.904	5.889	5.960	3.57	3.73	2.875	3.487	2.957	3.663

(c) Tensile modulus for each sample

Figure 16. Tensile Test result for Specimen UD/G/C/2

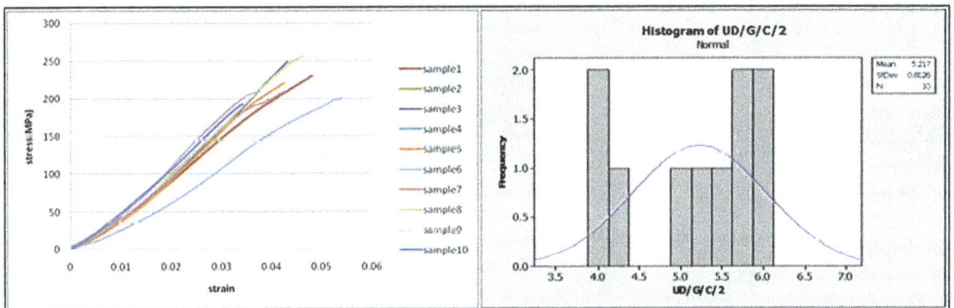

(a) stress-strain curve

(b) Normal distribution curve for tensile modulus

sample	1	2	3	4	5	6	7	8	9	10
young modulus(GPa)	5.115	4.049	6.018	4.232	5.619	4.071	5.368	5.774	6.097	5.115

(c) Tensile modulus for each sample

Figure 23 shows the residual plots for change in tensile modulus.

Regression Analysis for Tensile Modulus (E) vs. Length (l) and Width (w)

The regression equation for tensile modulus (E) is given by

$$E = -12.5 + 0.0889l + 0.438w$$

Figure 17. Tensile Test result for Specimen UD/G/C/3

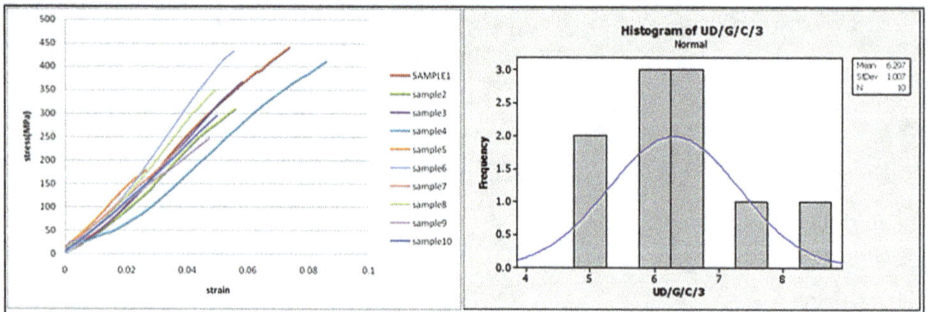

(a) stress-strain curve (b) Normal distribution curve for tensile modulus

sample	1	2	3	4	5	6	7	8	9	10
young modulus(GPa)	6.383	5.856	6.646	5.108	6.56	8.374	5.768	7.31	5.051	5.916

(c) Tensile modulus for each sample

Figure 18. Tensile Test result for Specimen UD/G/C/4

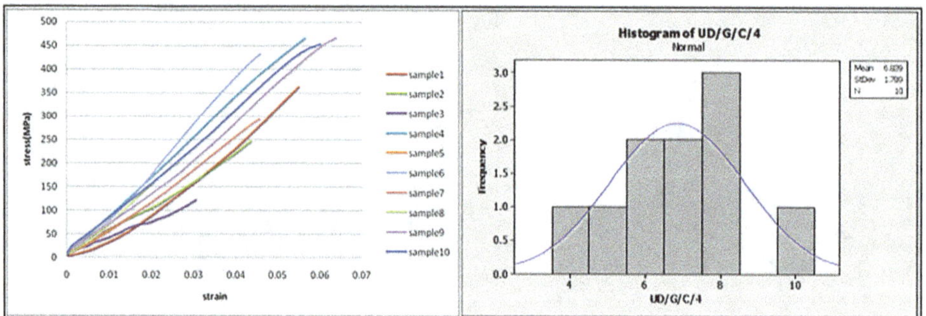

(a) stress-strain curve (b) Normal distribution curve for tensile modulus

sample	1	2	3	4	5	6	7	8	9	10
young modulus(GPa)	6.756	5.173	3.525	8.3	5.527	9.844	6.393	7.844	7.405	7.619

(c) Tensile modulus for each sample

COMPARISON OF RESULTS

Variation in tensile modulus for various combinations of length and width has been shown in figure 25.

From figure 25, it is clear that there is very little variation in the value of tensile modulus when width is change from 15mm to 12 mm while length is kept constant but after that, tensile modulus drops considerably when width is further reduced to 9mm. Also there is very little variation in the tensile modulus with further change in the value of length and width.

Figure 19. Tensile Test result for Specimen UD/G/D/1

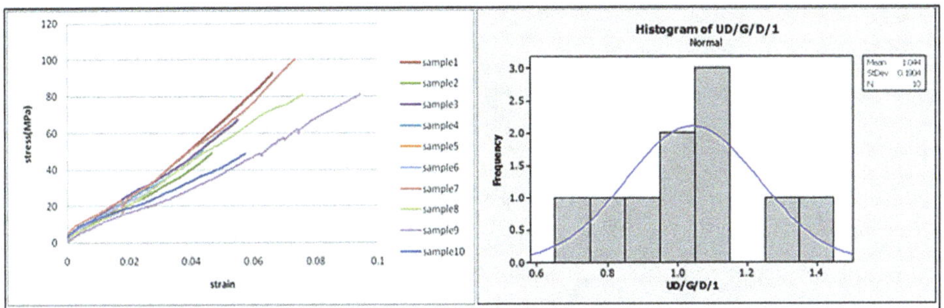

(a) stress-strain curve (b) Normal distribution curve for tensile modulus

sample	1	2	3	4	5	6	7	8	9	10
young modulus(GPa)	1.376	0.9777	1.13	1.055	1.021	0.9465	1.272	1.087	0.841	0.7291

(c) Tensile modulus for each sample

Figure 20. Tensile Test result for Specimen UD/G/D/2

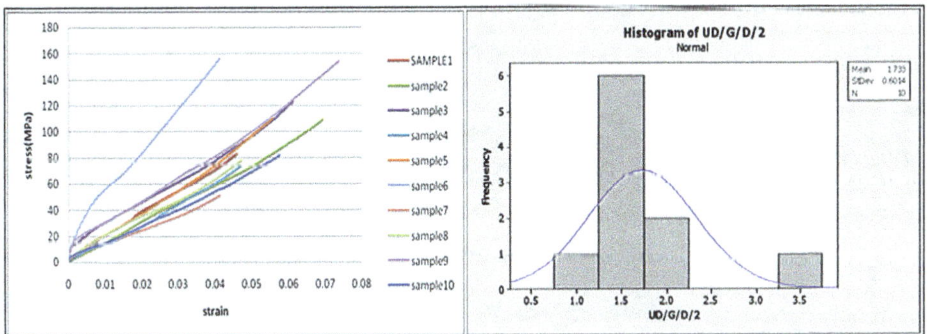

(a) Stress-strain curve (b) Normal distribution curve for tensile modulus

sample	1	2	3	4	5	6	7	8	9	10
young modulus(GPa)	1.707	1.503	1.721	1.518	1.864	3.296	1.068	1.465	1.859	1.327

(c) Tensile modulus for each sample

In figure 26 and 27, curve 'experimental' shows the values obtained of tensile modulus from experimental results while the curve 'fit1' & curve 'fit2' shows the values of tensile modulus obtained from regression equation respectively.

Figure 21. Tensile Test result for Specimen UD/G/D/3

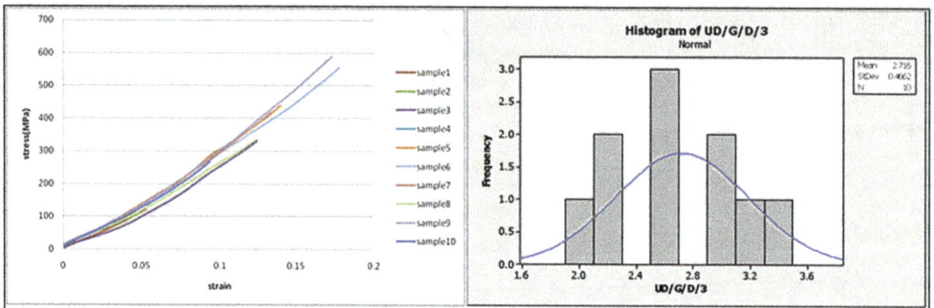

(a) stress-strain curve (b) Normal distribution curve for tensile modulus

sample	1	2	3	4	5	6	7	8	9	10
young modulus(GPa)	2.247	1.96	1.671	2.206	3.163	3.086	3.061	2.665	3.383	2.690

(c) Tensile modulus for each sample

Figure 22. Tensile Test result for Specimen UD/G/D/4

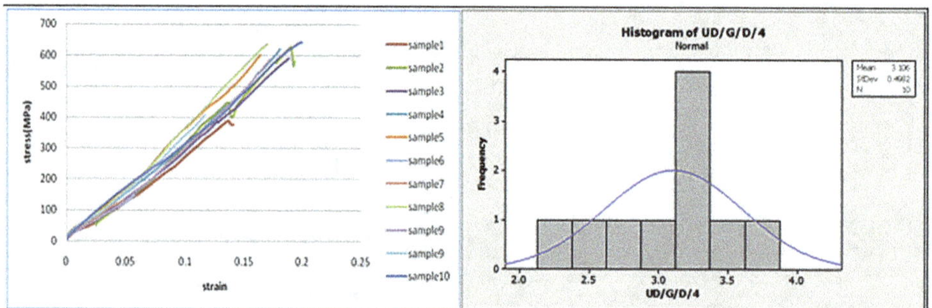

(a) stress-strain curve (b) Normal distribution curve for tensile modulus

sample	1	2	3	4	5	6	7	8	9	10
young modulus(GPa)	2.837	3.323	3.089	3.286	3.588	2.158	2.473	3.802	3.36	3.144

(c) Tensile modulus for each sample

CONCLUSION

In the present study, tensile test specimen, of glass fiber composite, of ASTM standard and various sub sizes have been tested and the data obtained from the experiments have been studied. The observation made from the experimental data and the regression analysis has been given below:

- There is very minor difference between tensile modulus of specimen UD/G/A/1 & UD/G/A/2 i.e when the width is reduced from 15mm to 12mm while length is kept constant.
- The value of tensile modulus decrease considerably when the width is further reduced from 12mm to 9mm but after that there is very small variation in the tensile modulus.

Table 6. Correlation between young modulus of ASTM standard specimen and the sub size specimen

Specimen	UD/G/A/1	
	Pearson Correlation	**P-Value**
UD/G/A/2	-0.284	0.427
UD/G/A/3	-0.379	0.280
UD/G/A/4	0.043	0.907
UD/G/B/1	-0.225	0.533
UD/G/B/2	-0.391	0.265
UD/G/B/3	-0.373	0.288
UD/G/B/4	0.490	0.151
UD/G/C/1	-0.160	0.660
UD/G/C/2	-0.271	0.448
UD/G/C/3	-0.401	0.251
UD/G/C/4	0.107	0.769
UD/G/D/1	-0.180	0.618
UD/G/D/2	-0.182	0.616
UD/G/D/3	-0.275	0.442
UD/G/D/4	0.276	0.441

Table 7. Correlation Coefficient Value Interpretation

Correlation Coefficient	Nature of Relationship
0.80 – 1.00	Very strong relationship
0.60 – 0.79	Strong relationship
0.40 – 0.59	Substantial/marked relationship
0.20 – 0.39	Low relationship
0.00 – 0.19	Negligible relationship

Box 1.

Predictor	Coef	SE Coef[1]	T[2]	P[3]
Δ l	0.17544	0.03584	4.46	0.001
Δw	2.34285	0.5973	2.78	0.017
Δ l Δw	-0.01871	0.006385	-2.59	0.024

S[4] = 4.7881 R-Sq[5] = 66.6% R-Sq(adj) = 58.8%

Table 8. Analysis of Variance for regression equation

Source	DF	SS[6]	MS[7]	F[8]	P
Regression	3	549.35	183.12	7.98	0.003
Residual Error	12	275.19	22.93		
Total	15	824.54			

Table 9. Fits and Residuals for Change in Modulus

Obs.	Δ l	Δ w	ΔE	fit[9]	Residual [10]
1	0	0	0.00	0	0
2	0	3	-1.20	7.02855	-5.82855
3	0	6	20.16	14.0571	6.1029
4	0	9	18.96	21.08565	-2.12565
5	50	0	22.37	8.77	13.6
6	50	3	19.65	12.99205	6.65795
7	50	6	21.43	17.2141	4.2159
8	50	9	18.54	21.43615	-2.89615
9	100	0	22.12	17.54	4.58
10	100	3	21.16	18.95555	2.20445
11	100	6	20.08	20.3711	-0.2911
12	100	9	19.54	21.78665	-2.24665
13	150	0	25.33	26.31	-0.98
14	150	3	24.64	24.91905	-0.27905
15	150	6	23.66	23.5281	0.1319
16	150	9	23.27	22.13715	1.13285

- Relationship between the change in tensile modulus & length and width and the relationship between the tensile modulus & length and width is given below:

$$\Delta E = 0.17544 \times \Delta l + 2.34285 \times \Delta w - 0.018711 \times \Delta l \times \Delta w$$
$$E = -12.5 + 0.0889 \times l + 0.438 \times w$$

Figure 23. Residual plots for change in tensile modulus

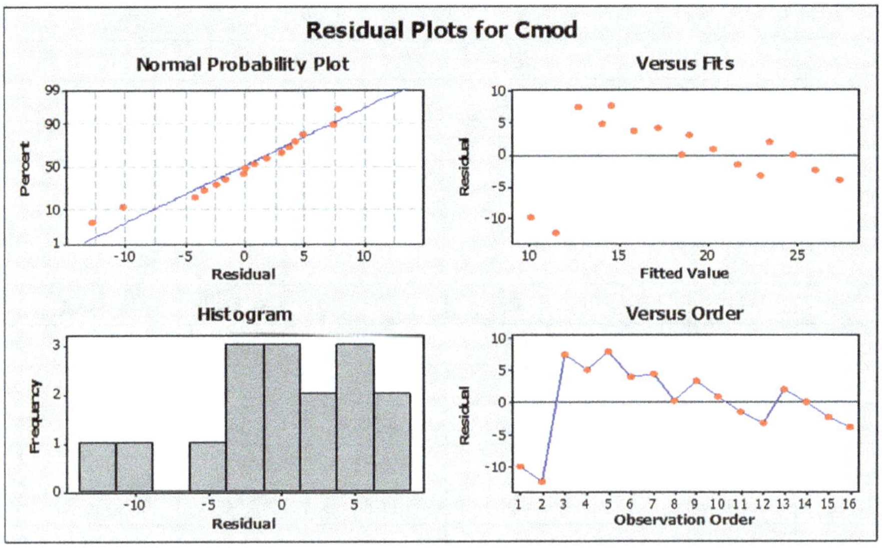

Figure 24.

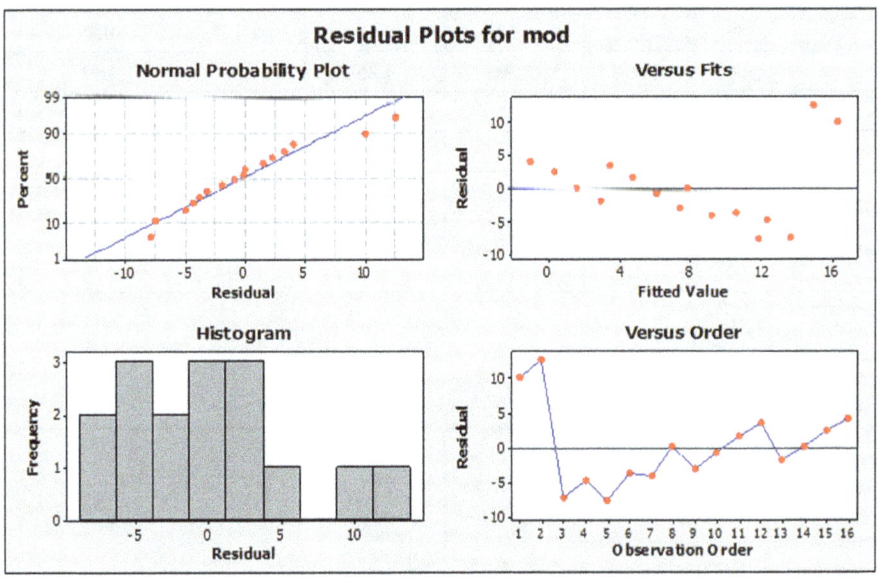

Box 2.

Predictor	Coef	SE Coef[11]	T[12]	P[13]
Constant	-12.517	6.930	-1.81	0.094
L	0.08892	0.02731	3.26	0.006
W	0.4380	0.4552	0.96	0.354

S[14] = 6.10780 R-Sq[15] = 47.0% R-Sq(adj) = 38.8%

Table 10. Analysis of Variance for regression equation

Source	DF	SS[16]	MS[17]	F[18]	P
Regression	2	429.90	214.95	5.76	0.016
Residual Error	13	484.97	37.31		
Total	15	914.86			

Table 11. Fits and Residuals for Tensile Modulus

Obs.	l	w	E	fit	SE Fit	Residual	St Residual
1	250	15	26.38	16.28	3.27	10.09	1.96
2	250	12	27.57	14.97	2.64	12.60	2.29R
3	250	9	6.22	13.66	2.64	-7.44	-1.35
4	250	6	7.41	12.34	3.27	-4.93	-0.96
5	200	15	4.01	11.84	2.64	-7.83	-1.42
6	200	12	6.72	10.52	1.81	-3.80	-0.65
7	200	9	4.94	9.21	1.81	-4.27	-0.73
8	200	6	7.83	7.90	2.64	-0.06	-0.01
9	150	15	4.26	7.39	2.64	-3.13	-0.57
10	150	12	5.22	6.08	1.81	-0.86	-0.15
11	150	9	6.30	4.76	1.81	1.53	0.26
12	150	6	6.84	3.45	2.64	3.39	0.62
13	100	15	1.04	2.95	3.27	-1.90	-0.37
14	100	12	1.73	1.63	2.64	0.10	0.02
15	100	9	2.72	0.32	2.64	2.40	0.44
16	100	6	3.11	-1.00	3.27	4.10	0.80

Figure 25.

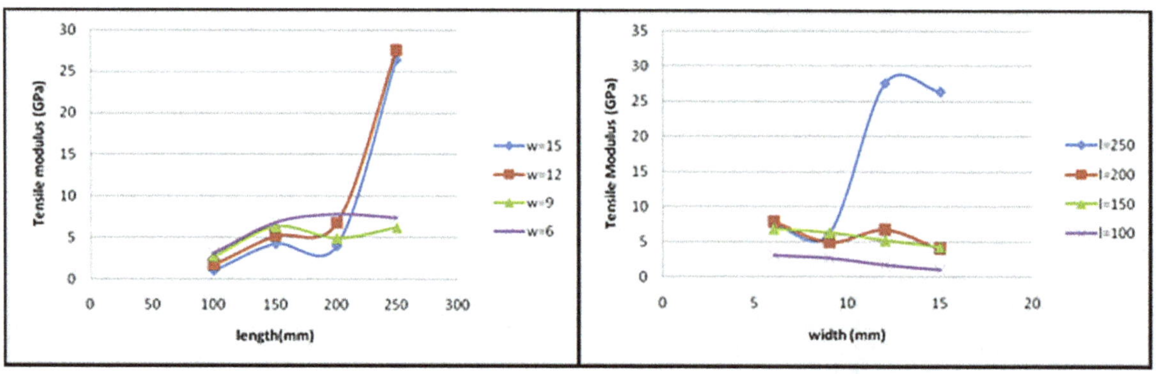

Figure 26.

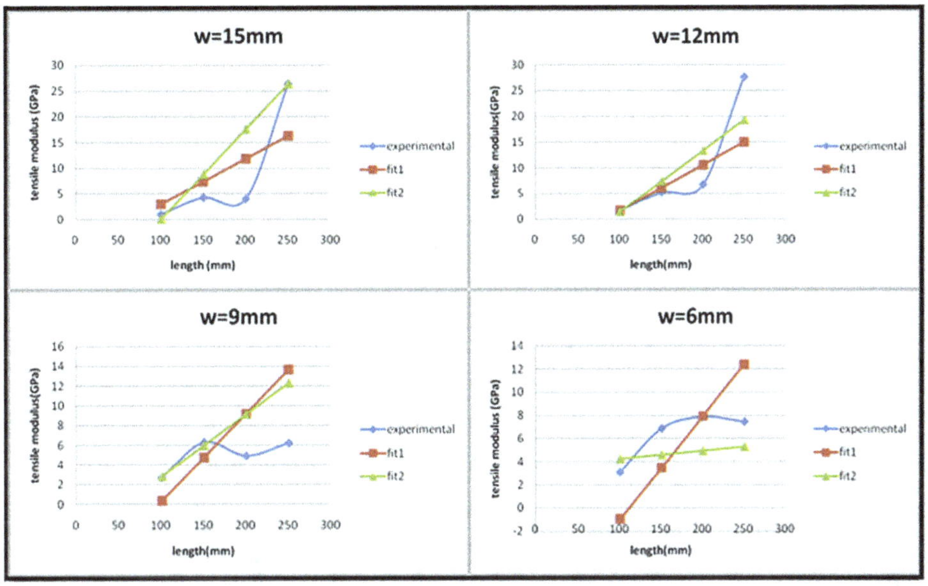

Figure 27.

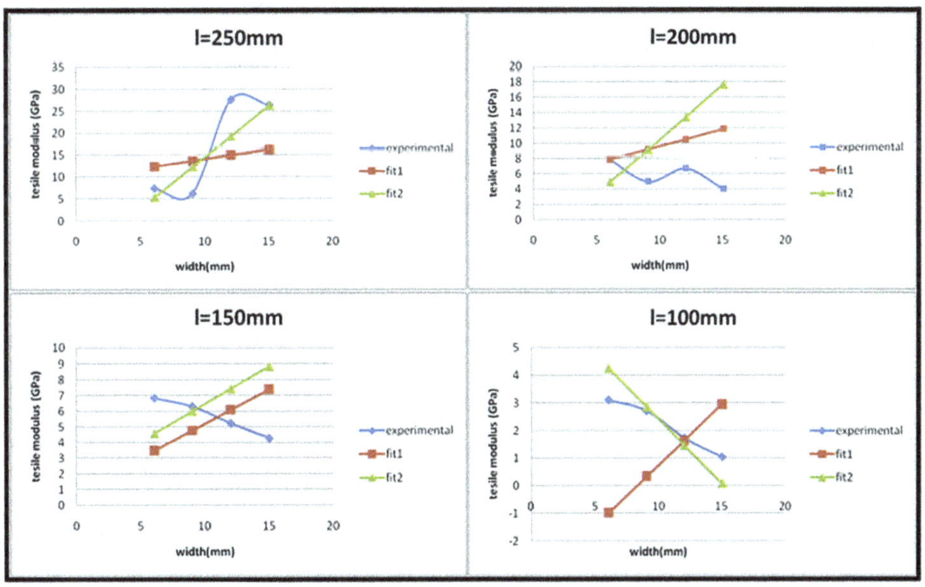

Future Work

1. Further reduction in the size of ASTM standard specimen should be done but up to the dimensions where tensile effect dominates.
2. Present work should be repeated for cross-ply and angle-ply laminates and the regression equations should be verified for cross-ply and angle-ply laminates.
3. Experimental data can be verified with the help of inverse finite element method.
4. Effect of miniaturization should also be inspected for three point bending test also.

REFERENCES

Bing, Q., & Sun, C. T. (2008). Specimen size effect in off-axis compression tests of fiber composites. *Composites. Part B, Engineering*, *39*(1), 20–26. doi:10.1016/j.compositesb.2007.02.010

Crasto, A. S., & Kim, R. Y. (1994). The Effects of Constituent Properties on the Compression Strength of Advanced Composites. In S. E. Groves & A. L. Highsmith (Eds.), *Compression Response of Composite Structures, ASTM STP 1185* (pp. 177–192). Philadelphia: American Society for Testing and Materials. doi:10.1520/STP24338S

Ermi, A. M., & James, L. A. (1986). Miniature Center-Cracked-Tension Specimen for Fatigue Crack Growth Testing. In W. R. Corwin & G. E. Lucas (Eds.), *The Use of Small-Scale Specimens for Testing Irradiated Material, ASTM STP 888* (pp. 261–275). Philadelphia: American Society for Testing and Materials. doi:10.1520/STP33009S

Fleury, E., & Ha, J. S. (1998). Small punch tests to estimate the mechanical properties of steels for steam power plant: I. Mechanical strength. *International Journal of Pressure Vessels and Piping*, *75*(9), 699–706. doi:10.1016/S0308-0161(98)00074-X

Funk, J. G., & Sykes, G. F. (1986, Fall). The Effects of Radiation on the lnterlaminar Fracture Toughness of a Graphite/Epoxy Composite. *Journal of Composites Technology & Research*, *8*(3), 92–97. doi:10.1520/CTR10328J

Kasada, R., Ono, H., & Kimura, A. (2006). Small specimen test technique for evaluating fracture toughness of blanket structural materials. *Fusion Engineering and Design*, *81*(8-14), 981–986. doi:10.1016/j.fusengdes.2005.08.088

Kim, R. Y., Crasto, A. S., & Yum, Y. J. (1994). Analysis of a Miniature Sandwich Compression Specimen. In S. E. Groves & A. L. Highsmith (Eds.), *Compression Response of Composite Structures, ASTM STP 1185* (pp. 338–350). Philadelphia: American Society for Testing and Materials. doi:10.1520/STP24347S

Kima, B. J., Kasadaa, R., Kimuraa, A., & Tanigawab, H. (2011). Effects of specimen size on fracture toughness of phosphorous added F82H steels. *Fusion Engineering and Design*, *86*(9-11), 2403–2408. doi:10.1016/j.fusengdes.2011.04.001

Kumar, Madhusoodanan, & Rupani. (2006). Miniature specimen technique as an NDT tool for estimation of service life of operating pressure equipment. *International Conference & Exhibition on Pressure Vessel and Piping*.

Kurtza, S. M., Jewetta, C. W., & Bergstr, J. S. (2002). Miniature specimen shear punch test for UHMWPE used in total joint replacements. *Biomaterials*, *23*(9), 1907–1919. doi:10.1016/S0142-9612(01)00316-7 PMID:11996031

Lee, J., & Soutis, C. (2008). Measuring the notched compressive strength of composite laminates: Specimen size effects. *Composites Science and Technology*, *68*(12), 2359–2366. doi:10.1016/j.compscitech.2007.09.003

Lucon, E., Bicego, V., D'Angelo, D., & Fossati, C. (1993). Evaluating a Service- Exposed Component's Mechanical Properties by Means of Subsized and Miniature Specimens. In Small Specimen Test Techniques Applied to Nuclear Reactor Vessel Thermal Annealing and Plant Life Extension. American Society for Testing and Materials. doi:10.1520/STP12738S

Manahan, M. P., Sr., Martin, F. J., & Stonesifer, R. B. (1999). Results of the ASTM Instrumented/Miniaturized Round Robin Test Program. In Pendulum Impact Testing: A Century of Progress. American Society for Testing and Materials.

Manahan, M. P., Stonesifer, R. B., Soong, Y., & Burger, J. M. (1995). Miniaturized Notch Test Specimen and Test Machine Design. In Pendulum Impact Machines: Procedures and Specimens for Verification. American Society for Testing and Materials. doi:10.1520/STP14656S

McLoughlin, J. R. (1972). The Preparation and Testing of Miniature Carbon Fiber Reinforced Composites. In *Composite Materials: Testing and Design*. American Society for Testing and Materials. 10.1520/STP27752S

Nozawa, Katoh, & Kohyama. (2005). Evaluation of Tensile Properties of SiC/SiC Composites with Miniaturized Specimens. *Materials Transactions, 46*(3), 543-551.

Nozawa, Ozawa, Kondo, Hinoki, Katoh, Snead, & Kohyama. (2005). Tensile, Flexural, and Shear Properties of Neutron Irradiated SiC/SiC Composites with Different Fiber-Matrix Interfaces. *Journal of ASTM International, 2*(3).

Nozawa, T., Hinoki, T., Katoh, Y., Kohyama, A., & Lara-Curzio, E. (2002). Specimen Size Effects on Tensile Properties of 2D/3D SiC/SiC Composites. In M. A. Sokolov, J. D. Landes, & G. E. Lucas (Eds.), *Small Specimens Test Techniques: Fourth Volume, ASTM STP 1418*. West Conshohocken, PA: ASTM International. doi:10.1520/STP10828S

Reed, D. N., Smith, R. P., Strattan, J. K., & Swift, R. A. (1983). A Comparison of Short Transverse Tension Test Methods. In R. J. Glodowski (Ed.), *Through-Thickness Tension Testing of Steel ASTM STP 794* (pp. 25–39). American Society for Testing and Materials. doi:10.1520/STP10004S

Rosinski, S. T., & Corwin, W. R. (1998). ASTM Cross-Comparison Exercise on Determination of Material Properties Through Miniature Sample Testing. In Small Specimen Test Techniques. American Society for Testing and Materials. doi:10.1520/STP37979S

Shin, C.-S., & Lin, S.-W. (2012). Evaluating fatigue crack propagation properties using miniature specimens. *International Journal of Fatigue*, *43*, 105–110. doi:10.1016/j.ijfatigue.2012.02.018

Songa, M., Guana, K., Qinb, W., & Szpunar, J. A. (2012). Comparison of mechanical properties in conventional and small punch tests of fractured anisotropic A350 alloy forging flange. *Nuclear Engineering and Design*, *247*, 58–65. doi:10.1016/j.nucengdes.2012.03.023

Sutherland, L. S., Shenoi, R. A., & Lewis, S. M. (1999). Size and scale effects in composites: II. Unidirectional laminates. *Composites Science and Technology*, *59*(2), 221–233. doi:10.1016/S0266-3538(98)00083-9

Volak, J., Novak, M., Kaiser, J., & Mentl, V. (2012, December). Fatigue testing by means of miniature test specimens. *Journal of Achievement in Materials and Manufacturing Engineering*, *55*(2), 386–389.

Wang, Z.-X., Shi, H.-J., Lu, J., Shi, P., & Ma, X.-F. (2008). Small punch testing for assessing the fracture properties of the reactor vessel steel with different thicknesses. *Nuclear Engineering and Design*, *238*(12), 3186–3193. doi:10.1016/j.nucengdes.2008.07.013

Xu, Y., Ning, G., Zhang, C., Yu, Q., & Xu, Y. (2000). Application of the miniature specimen technique to material irradiation tests and surveillance for reactor components. *International Journal of Pressure Vessels and Piping*, *77*(12), 715–721. doi:10.1016/S0308-0161(00)00066-1

ENDNOTES

1 SE coef = gives estimated standard deviations (standard errors).
2 T= gives T-statistics which is used in significance test for null hypothesis.
3 P= gives Pearson correlation value.
4 S= standard deviation of error terms.
5 R-sq=coefficient of determination (gives amount of variation in the response which is explained by the model).
6 SS= sum of squares of error terms.
7 MS= mean of square of error terms.
8 F-value= test the hypothesis that all the coefficient in a regression model are zero.
9 Fit= value of change in modulus obtained from regression equation.
10 Residual= difference between experimental value and fitted value.
11 SE coef = gives estimated standard deviations (standard errors).
12 T= gives T-statistics which is used in significance test for null hypothesis.
13 P= gives Pearson correlation value.
14 S= standard deviation of error terms.
15 R-sq=coefficient of determination (gives amount of variation in the response which is explained by the model).
16 SS= sum of squares of error terms.
17 MS= mean of square of error terms.
18 F-value= test the hypothesis that all the coefficient in a regression model are zero.

Chapter 5
Kinematic Modelling and Simulation of 8 Degrees of Freedom SCARA Robot

Saravana Mohan M.
Kumaraguru College of Technology, India

Anbumalar V.
Velammal College of Engineering and Technology, India

ABSTRACT

Robots are electromechanical systems that need mechatronic approach before manufacturing to reduce the development cost. In this chapter, the modelling of the 8 degrees of freedom (DOF) SCARA robot with a multiple gripper using SolidWorks CAD software and the dynamic study with the aid of MATLAB/SimMechanics is presented. The SCARA with multiple gripper is used for pick and place operation in manufacturing industries. The SolidWorks CAD model of SCARA with multiple grippers is converted into SimMechanics block diagram by exporting the 3D CAD model to the MATLAB/SimMechanics second generation technology environment. The motion sensing capability of the SimMechanics is used for determining the dynamic parameters of the manipulators. The SimMechanics block diagrams and the results of the dynamic study presented in this chapter infer that the structure of the robot can be changed to get the required dynamic parameters.

INTRODUCTION

The SCARA (Selective Compliance Articulated Robot Arm) is a extensively applicable robot manipulator in this industrial developed age. It is a popular configuration with RRP (Revolute Revolute Prismatic) structure with four degrees of freedom. It has two revolute and one prismatic joint. The gripper is attached to end of the prismatic arm.The prototype of SCARA robot is introduced in the year 1978 in Japan (Siciliano & Khatib, 2008). SCARA is compact and the working envelopes are relatively limited. Today SCARA robots are very widely used in manufacturing industries for its high speed, short cycle time, advanced control for path precision and controlled compliance to perform the necessary light duty

DOI: 10.4018/978-1-5225-5445-5.ch005

tasks to achieve high flexibility, dexterity and productivity. Few light duty applications of SCARA are: product inspection, touch panel evaluation, conveying masks for wafers, Screw tightening, stacking electronics components, and inserting components in printed circuit boards, tapping, and chamfering, deburring, drilling, welding, soldering, gluing, packing, loading and unloading parts of an automated line.

BACKGROUND

Nowadays automotive industries are utilizing SCARA robots for handling the body works, engines, chassis, and other components (Jazar, 2009). The flexibility in workspace and the usage of multiple tool is very essential for the above task. This can be achieved by the redundancy in the design of the manipulator. The SCARA with redundant characteristics can be achieved by kinematic modelling of the configuration followed by dynamic study with the help of simulation environment in aid with CAD modelling tools. Rehiara (2011) worked and authored an article explaining the forward kinematics and inverse kinematic approach to find the position of the SCARA robot end effector position using D-H convention and corresponding transformation matrices. Spong, Hutchinson and Vidyasagar (2005) explained the forward and inverse kinematics of various robot configurations, including SCARA comprehensively in his publication. Hernandez, Bravo, de Jesus Rubio and Pacheco (2011) studied forward and inverse kinematics for SCARA, Cylindrical robot with four degrees of freedom to find the end-effector position and orientation which is applicable for TIG or MIG welding. The researchers like Wijesekara Arachchige and Salem Abderrahmane (2013) worked on reconfigurable end effectors. The SCARA robot was reconfigured from 4 DOF to 6 DOF. The state of the joint was selected by the motion of the end effector, and the constraints. This methodology is applied to the SCARA robot manipulator to improve its last joint capability. The researchers replaced the last joint with new reconfigurable joint and robot kinematic theory is applied for model evaluation. Patel and Sobh (2014) made a comprehensive study of manipulator performance measures that are very essential to design and study the applications of robotic manipulators. The kinematic indices, manipulability indices and important performance parameters are referred in his chapter to develop a robot with improvised configuration. These researches facilitates the development of new kinematic model for the SCARA robot with multiple tool which is mentioned in this chapter.

Fang and Li (2013) observed and verified the correctness of the SCARA robot model problem in terms of motion of each joint. The researchers used the kinematic modelling and simulation techniques. Alshamasin, Ionesco and Taha Al-Kasasbeh (2009) developed a complete mathematical model of the SCARA robot which includes servo actuator dynamics and presented together the dynamic simulation in the research. The forward and inverse kinematics equations are derived by using D-H convention. The researchers Ionescu, Chojnowski and Constantin (2002, 2007) proposed that simulation is important for robot programmers in allowing them to evaluate and predict the behaviour of a robot, and in addition to verify and optimize the path planning of the process. Michel (2004) emphasized in his chapter the need and the application of modelling and simulation software's to predict the accuracy and computational efficiency of the manipulator dynamics. Zlajpah (2008) found that the simulation facilitates designing visualization, testing robots and solve many problems before making it a reality. Al Mashagbeh and Khamesee (2015) developed a multi-body model of four degrees of freedom SCARA for pick and place application using MapleSim software and evaluated the robot performance. Wood and Kennedy (2015, April 24) presented the mathematical and software developments needed for efficient simulation of mechanical systems in the Simulink simulation environment. Talib, Swadi, Abed, Abed and Karim (2013) studied and investigated the kinematics of SCARA robots using kinematic and vibration methods in dynamic conditions of the manipulator. The position, velocity and acceleration were analysed using the simulation study. Das and Dulger (2005) presented the simulation technique in his work for a vertical revolute configuration 2-R robot with two degrees of freedom is for small parts insertion tasks in electronic component assembly lines. Ma, Zheng Lin Yu Guo, Hua Cao, Yan Bin Zheng & Liu (2014) developed the Kinematic equations for a high speed robot for material handling to analyse forward and inverse kinematics problems based on modified D-H coordinate system theory for their proposed SCARA. MATLAB Simulation was used to validate that the robot parameter design is reasonable and the trajectory planning by interpolation calculation in joint space is feasible. Kumar Jha, Dutta and Saha (2014) analysed the dynamics of the SCORA -ER14 robot with a single gripper end effector using MATLAB simulation.

Gouasmi, Ouali, Fernini and Meghatria (2012) compared the two robot postures for the same length of time with the same trajectory to obtain the kinematic and dynamic parameters using SolidWorks and MATLAB/Simulink. Urrea and Kern (2012) had modelled a redundant SCARA robot for pick and place application with five degrees of freedom. The authors developed the dynamic model of their proposed robot by means of MATLAB/Simulink programming and performed several tests like actuator dynamics with different controllers under path tracking requirements. Schlotter (2003) described about simulating the dynamics of multibody systems with SimMechanics, a gripperbox for the MATLAB/Simulink environment. The use of 3D CAD model and conversion of mechanical systems into a block diagram. The dynamic property of the robot mechanical system was well explained. He also mentioned the flexibility to change the structure, optimize system parameters and to analyse the results within the Simulink environment in much lesser time. Udai, Rajeevlochana and Saha (2011) developed a 3D CAD model of KUKA KR5 robot applicable for peg-in-hole insertion using Autodesk Inventor. Further, he performed the dynamic simulation using MATLAB/SimMechanics and verified the inverse dynamics. Umar and Bakarb (2014), performed the identical simulation, which results between derived direct dynamic mechanical system of robot manipulator and SimMechanics first generation model. It was found that both methods satisfied the principle of two link manipulator model. Zi (2011) investigated dynamic simulation and trajectory tracking control of hybrid driven planar five bar parallel mechanism using a simulation model of dynamics based on MATLAB/SimMechanics. The simulation based on SimMechanics was carried out and angular velocity, angular acceleration of two driving links of kinematic pairs at any time was acquired. Kütük, Halicioglu and Dulger (2015) studied the hybrid driven mechanical system mechanism characteristics using MATLAB/SimMechanics platform. The authors of this chapter, Saravanamohan and Anbumalar (2016) in one of their publication presented the modelling and simulation of the redundant SCARA robot with multiple drilling tool using MATLAB/SimMechanics. Fedak, Durovsky, and Uveges (2014) presented an application of robot mechanics modelling and its dynamic simulation using add-on SimMechanics modules of MATLAB and Graphical User Interface (GUI) in their work. The above discussed research work by various researchers, indicate that modern simulation approach is very important and essential to develop SCARA Robot in reduced time and cost. To give impetus to the application of the SCARA, a newness in the design and development which is very cost effective is to be incorporated. Hence, many researchers adopted modelling and simulation methodology to develop and study the dynamic performance of conventional non redundant SCARA with various types of conventional end effectors to meet the industrial requirements. If the redundancy in robot structure and the multiple tool end effector is used simultaneously, it will be expected to give dexterity to the manipulator. It is essential to meet with manufacturing competence today's industries.

This chapter discusses the Kinematic Modelling and simulation of 8 degrees of freedom SCARA Robot with multiple gripper shown in the Figure 1. The method of developing a new generation and configuration of SCARA robots using kinematic modelling using Denavit Hartenberg method and MATLAB/SimMechanics simulation incorporated with SolidWorks CAD modelling to achieve the objective of this chapter is explained in the following subsections.

KINEMATIC MODELLING OF SCARA WITH MULTIPLE GRIPPERS

Robot kinematics is a geometric study of motion of a robotic manipulator with respect to the datum coordinates system. The dynamics is described in terms of the time rate of change of the robot configuration in relation to the joint torques. In this chapter kinematic model is developed using the Denavit-Hartenberg (D-H) forward kinematic approach. Conventionally dynamic parameters are computed using the laborious equations, but in this present work the simulation is used to determine the dynamics of our system. The forward kinematics deals with computing the position and orientation of the end effector for the given joint variables. The kinematic model to find the position of the Multiple gripper end effector attached to the SCARA robotics derived using the Denavit-Hartenberg (D-H) forward kinematic approach. The coordinate frames are assigned based on D-H convention to each joint as shown in Figure 2 and its parameters are given in the Table 1.

The homogeneous transformation Matrix A_i (Tsai, 1999) is shown in the Equation (1) and it is represented as a product of four basic transformations given in the Equations (1) to (6). It expresses the position of the gripper with respect to the reference frame.

Figure 1. SCARA Robot with multiple gripper

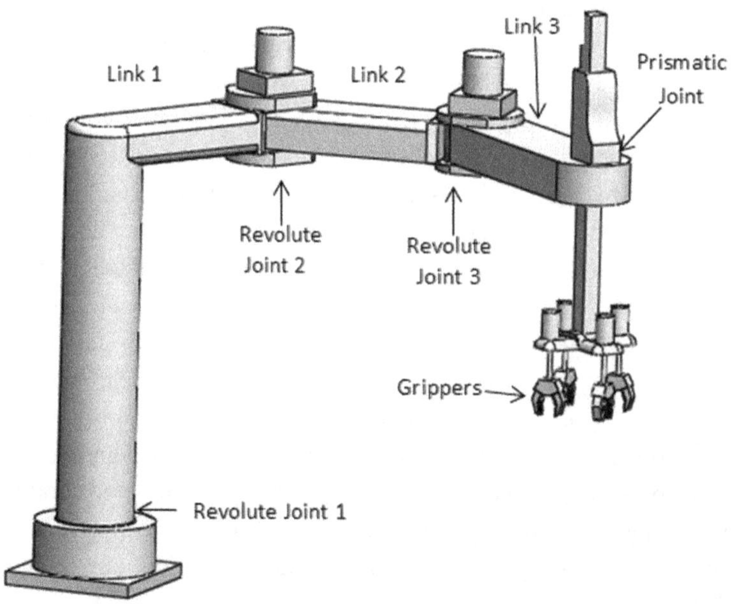

Figure 2. D-H parameters of SCARA with multiple grippers

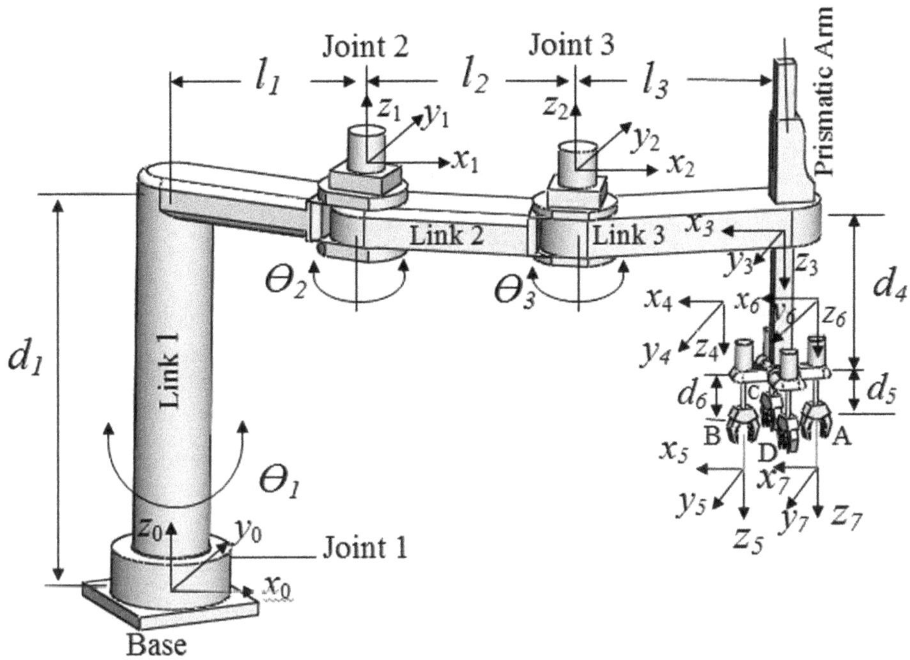

$$A_i = T(z, d)\, T(z, \Theta)\, T(x, a)\, T(x, \alpha) \tag{1}$$

$$T(z,d) = \begin{bmatrix} 1 & 0 & 0 & 0 \\ 0 & 1 & 0 & 0 \\ 0 & 0 & 1 & d_i \\ 0 & 0 & 0 & 1 \end{bmatrix} \tag{2}$$

$$T(z,\theta) = \begin{bmatrix} \mathrm{Cos}\,\theta_i & -\mathrm{Sin}\,\theta_i & 0 & 0 \\ \mathrm{Sin}\,\theta_i & \mathrm{Cos}\,\theta_i & 0 & 0 \\ 0 & 0 & 1 & 0 \\ 0 & 0 & 0 & 1 \end{bmatrix} \tag{3}$$

$$T(x,a)=\begin{bmatrix}1 & 0 & 0 & a_i\\ 0 & 1 & 0 & 0\\ 0 & 0 & 1 & 0\\ 0 & 0 & 0 & 1\end{bmatrix} \tag{4}$$

$$T(x,\alpha)=\begin{bmatrix}0 & 0 & 0 & 0\\ 0 & \operatorname{Cos}\alpha_i & -\sin\alpha_i & 0\\ 0 & \sin\alpha_i & \operatorname{Cos}\alpha_i & 0\\ 0 & 0 & 0 & 0\end{bmatrix} \tag{5}$$

The rotation and the translation motion by robotic the manipulator is mentioned in the equations (7) and (8).

$$A_i=\begin{bmatrix}\operatorname{Cos}\theta_i & -\operatorname{Cos}\alpha_i\operatorname{Sin}\theta_i & \sin\alpha_i\sin\theta_i & a_i\operatorname{Cos}\theta_i\\ \operatorname{Sin}\theta_i & \operatorname{Cos}\alpha_i\operatorname{Cos}\theta_i & -\sin\alpha_i\operatorname{Cos}\theta_i & a_i\sin\theta_i\\ 0 & \sin\alpha_i & \operatorname{Cos}\alpha_i & d_i\\ 0 & 0 & 0 & 1\end{bmatrix} \tag{6}$$

$$Ai=\begin{bmatrix}R_i^{i-1} & O_i^{i-1}\\ 0 & 1\end{bmatrix} \tag{7}$$

The 4 x 4 matrix given in the equations (6) and (7) is partitioned into two sub matrices as shown in the equations (8) and (9), which represent rotation combined with translation to produce the motion required by the manipulator. Where

$$R_i^{i-1}=\begin{bmatrix}\operatorname{Cos}\theta i & -\sin\theta i\operatorname{Cos}\alpha i & \operatorname{Sin}\alpha i\operatorname{Sin}\theta i\\ \operatorname{Sin}\theta i & \operatorname{Cos}\theta i\operatorname{Cos}\alpha i & -\operatorname{Sin}\alpha i\operatorname{Cos}\theta i\\ 0 & \operatorname{Sin}\alpha i & \operatorname{Cos}\alpha i\end{bmatrix} \tag{8}$$

$$O_i^{i-1}=\begin{bmatrix}a_i\operatorname{Cos}\theta_i\\ a_i\sin\theta_i\\ d_i\end{bmatrix} \tag{9}$$

The (4x4) rigid homogeneous transformation matrices A_1, A_2, A_3, A_4, A_5, A_6, A_7 shown in the equations (10) to (16) are computed by applying the D-H parameters listed in Table 1 in the equation (6)

Where l_1, l_2, l_3 are link lengths in mm, l_a is length of the Gripper A from the prismatic arm axis (Z_3), l_b is length of the Gripper B from the prismatic arm axis (Z_3), it is in the x axis but in negative direction. So it is assumed as - l_b, d_1, d_2, d_3, d_4, d_5, d_5 are link offset length between the successive links, C_2, C_3, C_6, C_7 are Cosine function of joint angles, S_2, S_3, S_6, S_7 are Sine function of joint angles.

$$A_1 = \begin{bmatrix} C_1 & -S_1 & 0 & l_1C_1 \\ S_1 & C_1 & 0 & l_1S_1 \\ 0 & 0 & 1 & 0 \\ 0 & 0 & 0 & 1 \end{bmatrix} \quad (10)$$

$$A_2 = \begin{bmatrix} C_2 & -S_2 & 0 & l_2C_2 \\ S_2 & C_2 & 0 & l_2S_2 \\ 0 & 0 & 1 & 0 \\ 0 & 0 & 0 & 1 \end{bmatrix} \quad (11)$$

$$A_3 = \begin{bmatrix} C_3 & S_3 & 0 & l_3C_3 \\ S_3 & -C_3 & 0 & l_3S_3 \\ 0 & 0 & 1 & 0 \\ 0 & 0 & 0 & 1 \end{bmatrix} \quad (12)$$

Table 1. D-H parameters of the proposed SCARA robot with multiple grippers

Axis Number	Joint Angle (θ_i)	Link Offset (d_i)	Link Length (a_i)	Twist Angle (α_i)
1	θ_1	d_1	l_1	0
2	θ_2	0	l_2	0
3	θ_3	0	l_3	180^0
4	0	d_4	l_a	0
5	0	d_4	l_b	0
6	θ_6	d_5	0	0
7	θ_7	d_6	0	0

$$A_4 = \begin{bmatrix} 1 & 0 & 0 & l_a \\ 0 & 1 & 0 & 0 \\ 0 & 0 & 1 & d_4 \\ 0 & 0 & 0 & 1 \end{bmatrix} \tag{13}$$

$$A_5 = \begin{bmatrix} 1 & 0 & 0 & -l_b \\ 0 & 1 & 0 & 0 \\ 0 & 0 & 1 & d_5 \\ 0 & 0 & 0 & 1 \end{bmatrix} \tag{14}$$

$$A_6 = \begin{bmatrix} C_6 & -S_6 & 0 & 0 \\ S_6 & C_6 & 0 & 0 \\ 0 & 0 & 1 & d_5 \\ 0 & 0 & 0 & 1 \end{bmatrix} \tag{15}$$

$$A_7 = \begin{bmatrix} C_7 & -S_7 & 0 & 0 \\ S_7 & C_7 & 0 & 0 \\ 0 & 0 & 1 & d_6 \\ 0 & 0 & 0 & 1 \end{bmatrix} \tag{16}$$

Kinematic Model for Positioning Gripper A and Gripper B

The direct kinematic model to find the position of the Gripper A is obtained by multiplying the successive transformation matrices as shown in the equation (17) to get the coordinate equations in the transformation matrix given in the equation (18).

$$T_A = A_{1.}A_{2}.A_{3}.A_{4.}A_{6.} \tag{17}$$

$$T_A = \begin{bmatrix} C_6C_{123} + S_6S_{123} & -S_6C_{123} + C_6S_{123} & 0 & (l_a + l_3)C_{123} + l_1C_2 + l_2C_{12} \\ C_6S_{123} - S_6C_{123} & -S_6S_{123} - C_6C_{123} & 0 & (l_a + l_3)S_{123} + l_1S_2 + l_2S_{12} \\ 0 & 0 & -1 & d_1 - d_5 - d_4 \\ 0 & 0 & 0 & 1 \end{bmatrix} \tag{18}$$

The Equations (19) to (21) represent the Gripper A position with reference to the base frame

$$x_A = (l_a + l_3)C_{123} + l_1C_2 + l_2C_{12} \tag{19}$$

$$y_A = (l_a + l_3)S_{123} + l_1S_2 + l_2S_{12} \tag{20}$$

$$z_A = d_1 - d_5 - d_4 \tag{21}$$

The kinematic model in the homogeneous transformation matrix form for Gripper B is T_B given in the Equations (22) and (23).

$$T_B = A_1.A_2.A_3.A_5.A_7 \tag{22}$$

$$T_A = \begin{bmatrix} C_6C_{123} + S_6S_{123} & -S_6C_{123} + C_6S_{123} & 0 & (l_b - l_3)C_{123} + l_1C_2 + l_2C_{12} \\ C_6S_{123} - S_6C_{123} & -S_6S_{123} - C_6C_{123} & 0 & (l_b - l_3)S_{123} + l_1S_2 + l_2S_{12} \\ 0 & 0 & -1 & d_1 - d_5 - d_4 \\ 0 & 0 & 0 & 1 \end{bmatrix} \tag{23}$$

Gripper B position is indicated by the Eqs.(17÷19)

$$x_B = (l_b - l_3)C_{123} + l_1C_2 + l_2C_{12} \tag{24}$$

$$y_B = (l_b - l_3)S_{123} + l_1S_2 + l_2S_{12} \tag{25}$$

$$Z_B = d_1 - d_5 - d_4 \tag{26}$$

where C_{23} denotes Cos $(\theta_2 + \theta_3)$, S_{23} denotes Sin $(\theta_2 + \theta_3)$

Determination of Gripper C and D Position

The position of Gripper C and Gripper D are found out geometrically. By using the principle of a right angle triangle between the grippers and the links are shown in the Figure 3.

The Gripper C and gripper D position can be computed using the coordinate position equations of the midpoint 'M' of the gripper head given in the equations (27) to (29)

Figure 3. Geometrical representation of the SCARA with multiple grippers

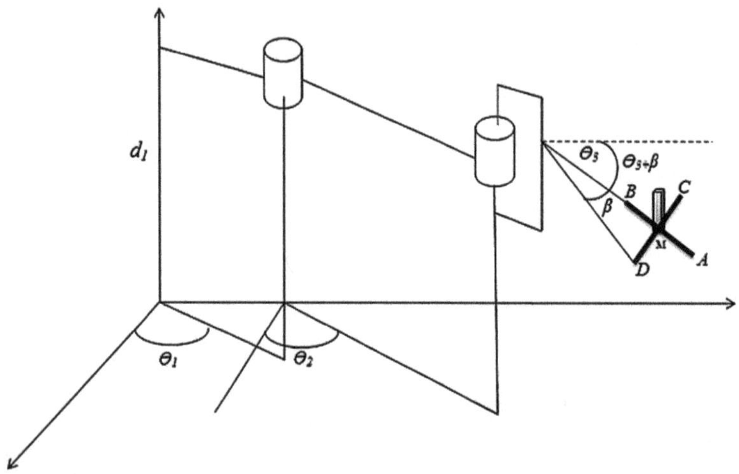

$$x_M = l_3 C_{123} + l_1 C_2 + l_2 C_{12} \tag{27}$$

$$y_M = l_3 S_{23} + l_1 S_2 + l_2 S_{12} \tag{28}$$

$$z_M = d_1 - d_4 - d_M \tag{29}$$

From the geometrical representation in Figure 3, the Gripper C position coordinates x_c and y_c can be predicted by substituting the value $\theta_3 = \theta_3 - \beta$ in the equations (30) and (31). Z_c coordinate can be found out by using the equation (32).

$$x_C = l_3 C_{123} + l_1 C_2 + l_2 C_{12} \tag{30}$$

$$y_C = l_3 S_{23} + l_1 S_2 + l_2 S_{12} \tag{31}$$

$$z_C = d_1 - d_4 - d_M \tag{32}$$

Here $d_M = d_5 = d_6$

Similarly substituting $\theta_3 = \theta_3 + \beta$ in the equations (33) and (34). The position coordinates x_d and y_d of gripper D can be determined. The z_d coordinate can be determined by using the Equation (35).

$$x_D = l_3 C_{123} + l_1 C_2 + l_2 C_{12} \tag{33}$$

$$y_D = l_3 S_{23} + l_1 S_2 + l_2 S_{12} \tag{34}$$

$$z_D = d_1 - d_4 - d_M \tag{35}$$

The values assigned for the D-H parameters $l_1 = l_2 = l_3 = 250mm$, $l_a = l_b = 50mm$, $d_1 = 320mm$, $d_2 = d_3 = 0$, $d_4 = 5mm$ *to* $300mm$, $d_5 = d_6 = 80mm$ are substituted in the coordinate equations to obtain the position coordinates of each gripper of the end effector .

INVERSE KINEMATICS

In the inverse kinematics the joint variables are determined for the desired position of the end effector of the robot. The algebraic methods of inverse kinematics are used to verify the joint angles (Asada, 2016).

After simplifying the terms of x_A and y_A coordinates given in the equation (19) and (20), θ_1, θ_2 and θ_3 can be obtained for Gripper A position as shown in the equations (36) to (40). Similarly θ_2 and θ_3 are found out for gripper B, gripper C and gripper D positions.

$$x_{J2} = x_p - l_3 \cos\theta_p \tag{36}$$

$$y_{J2} = y_p - l_3 \sin\theta_p \tag{37}$$

$$\theta_2 = \pi - \cos^{-1}\frac{l_1^2 + l_2^2 - x_{J2}^2 - y_{J2}^2}{2l_1 l_2} \tag{38}$$

$$\theta_2 = \tan^{-1}\frac{y_p}{x_p} - \cos^{-1}\frac{l_1^2 - l_2^2 + x_{J2}^2 + y_{J2}^2}{2l_1\sqrt{x_{J2}^2 + y_{J2}^2}} \tag{39}$$

$$\theta_3 = \theta_p - \theta_1 - \theta_2 \tag{40}$$

where the symbol θ_1,θ_2 and θ_3 denotes the angular displacement of revolute joint 1, 2 and 3.

DYNAMIC STUDY BY SIMMECHANICS SIMULATION

The dynamic behavior is in terms of the time rate of change of the robot configuration in relation to the joint torques exerted by the actuators is studied in this section instead of tedious equations, the simulation of the CAD models gives the results. The CAD model of the SCARA was exported from Solidworks environment to MATLAB/SimMechanics environment in the form of XML and STL file through SimMechanics second generation link .The XML file of the model was executed using the MATLAB command window.The CAD model of the robot was converted into a block diagram with the connecting blocks representing the revolute and prismatic joints. The inputs assigned to joint primitives to get the desired output through scope and Simout mentioned in the Figures 4 and 8.

Forward Dynamics and Motion Computation

In Forward Dynamic study using SimMechanics, the torque is specified by input and the motion is automatically computed using SimMechanics block diagram shown in Figure 4. The automatically derived motion of the joint 1 and the joint 2 is simulated and viewed in MATLAB/Mechanics Explorer as shown in the Figure 5. The plots shown in Figure 6 and Figure 7 shows the simulation results of the forward dynamics in obtaining the motion of the joints automatically by the given input.

The simulated elbow up and elbow down path by the manipulator links for the assigned joint primitives can be visualized in the MATLAB Explorer window as shown in Figure 5.

Figure 4. SimMechanics block diagram for forward dynamics of SCARA with Multiple Gripper

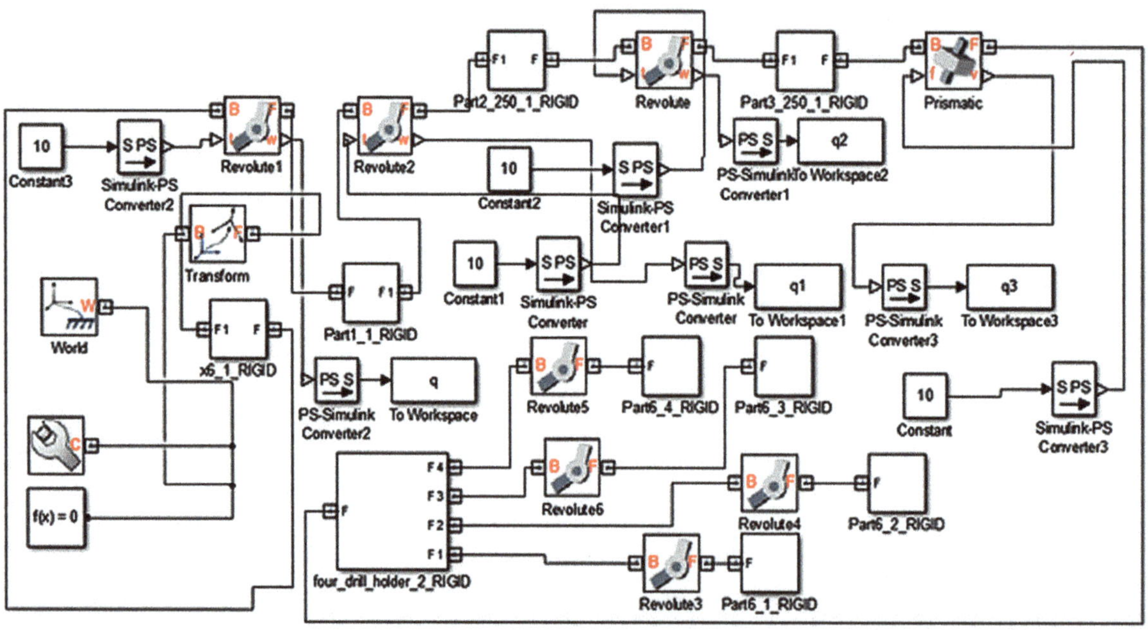

Figure 5. Simulated view of the proposed SCARA in MATLAB/SimMechanics Explorer

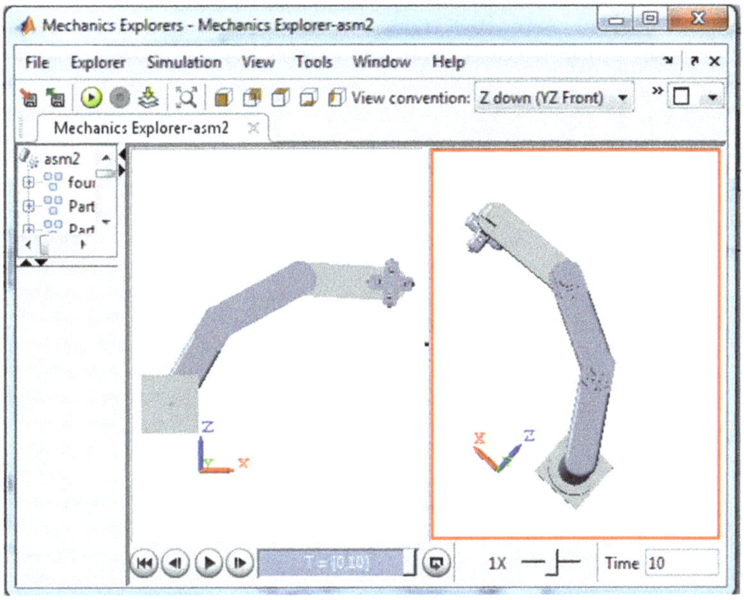

Determination of Velocity by SimMechanics

The forward dynamics of SimMechanics pave the way to find the dynamic parameter, velocity. In the Figure 4, adding PS-Simulink converter blocks and configuring the joint primitives. It will provide the velocity of the joints output through the workspace blocks. The angular position of joint 1, joint 2 and joint 3 with change in time t = 10s is shown in the Figure 12 and Figure 13 of the SCARA model with multiple gripper. By enabling the joint velocity sensing primitive, the angular velocity in the joints are determined by the plots shown in the Figure 6, 7 and 8 respectively. The Figure 6, 7 and 8 indicates the graphical simulation results of angular velocity at the revolute joints if $l_1 = l_2 = l_3 = 200$mm and mass of the links $l_1 = l_2 = l_3 = 2.3$Kg. For the revolute joint 1 joint 2 and joint 3, the angular displacement of 90^0 was set in the joint actuation menu. The maximum angular velocity observed at the joint 1, joint 2 and joint 3 are 14.37deg/s, 11.37deg/s and 12.14deg/s. The Figure 9 shows the linear velocity vs. time plot generated for prismatic joint 2. The linear velocity of the prismatic arm is observed as 9.91m/s.

Inverse Dynamics and Torque Computation

In inverse dynamics study the motion input is given in the joint primitive and thereby the torque is automatically Computed. The automatically computed torque in the joint 1, joint 2 and the joint 3 is simulated using the block diagram generated from the CAD model of the SCARA in MATLAB/SimMechanics as shown in the Figure 10. The sine wave block and Simulink - PS converter block shown in the Figure 11 provide the input motion to the revolute joint 1, joint 2 and joint 2.

The Figure 11, 12 and 13 shows the simulation results of the inverse dynamics. The simulation results in the Figure 11,12 and 13 indicates the automatically computed torque for the input motion for the joint 1, joint 2 and joint 3 in the predetermined trajectory with respect to time t = 10s. The observed torque values are$\tau_1 = 619.1$Nm, $\tau_2=360.1$Nm and $\tau_3=1135.6$ Nm respectively if $l_1 = l_2 = l_3 = 200mm$.

Figure 6. Angular velocity at Joint 1

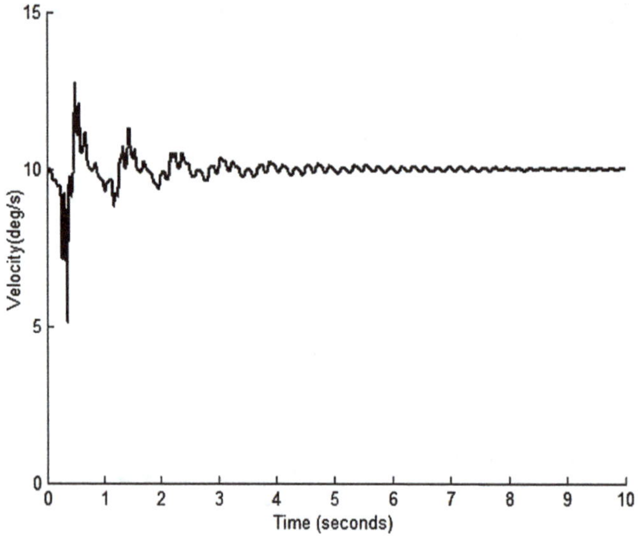

Figure 7. Angular velocity at Joint 2

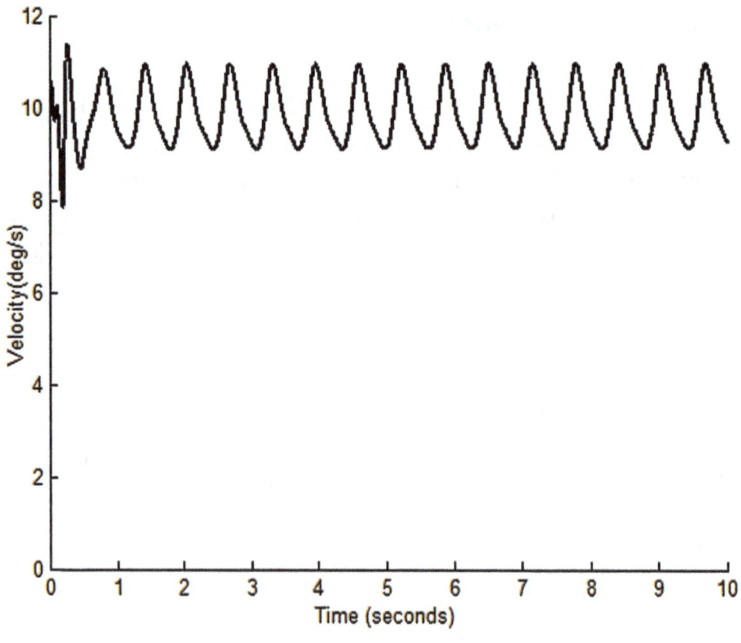

Figure 8. Angular velocity at joint 3

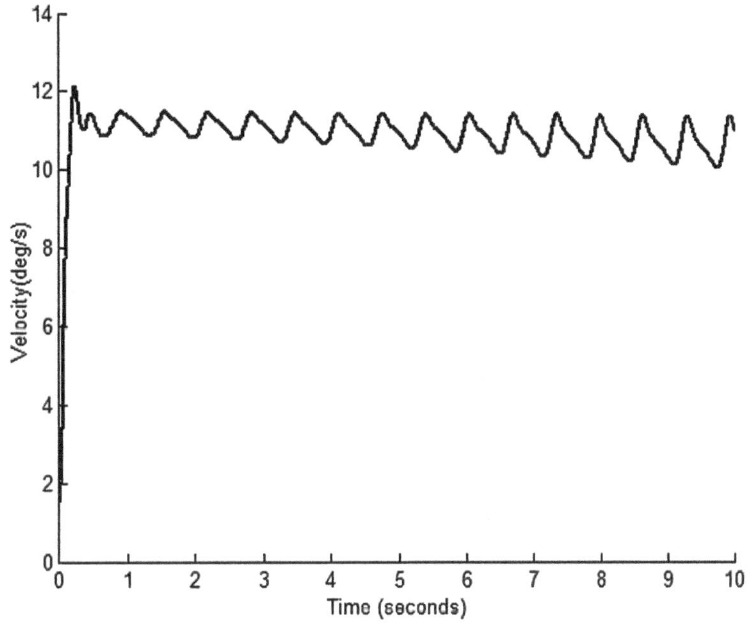

Figure 9. Linear velocity at prismatic joint

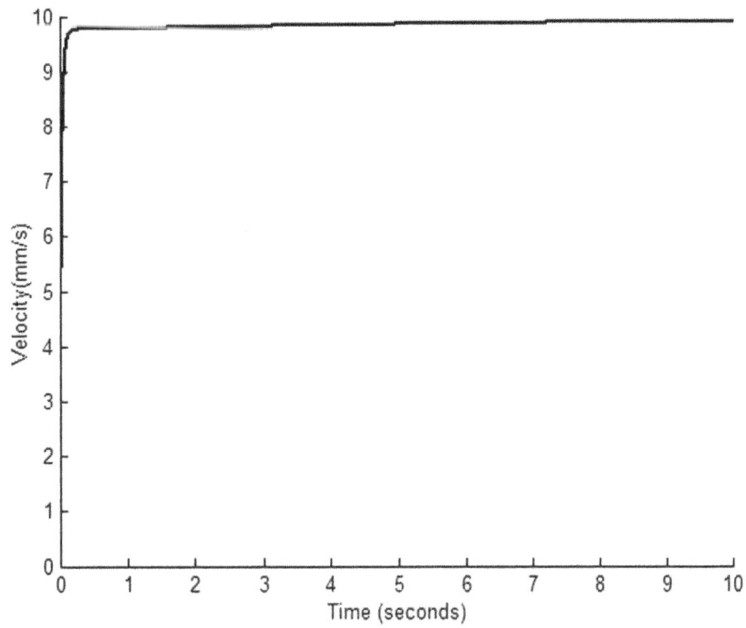

Figure 10. SimMechanics model for inverse dynamics of the SCARA

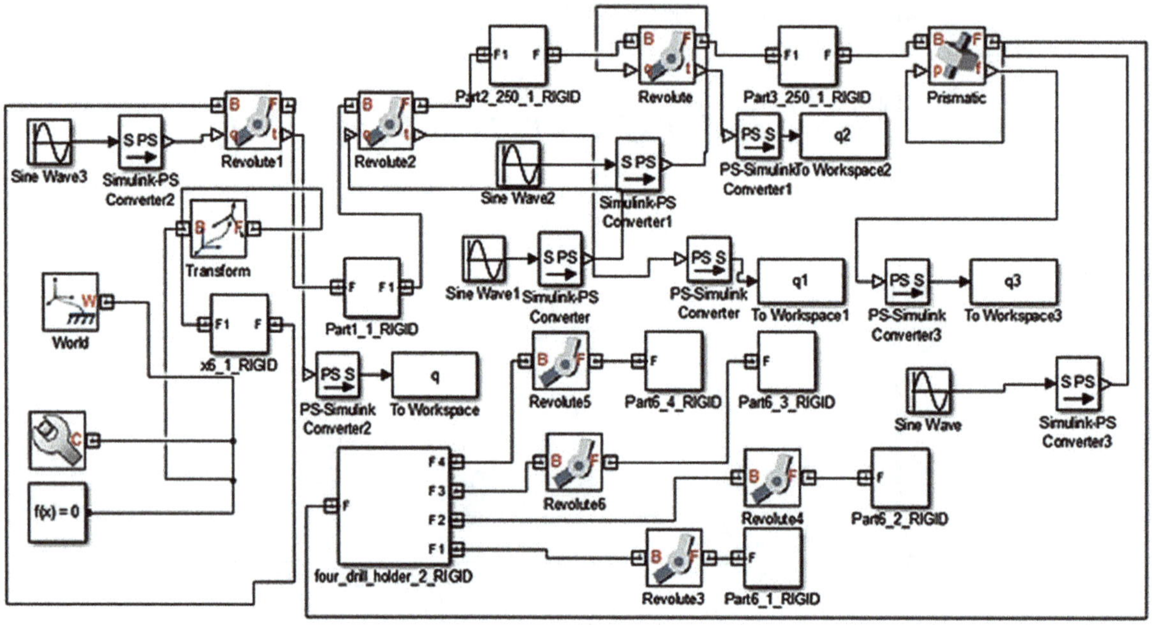

Figure 11. Torque at joint 1

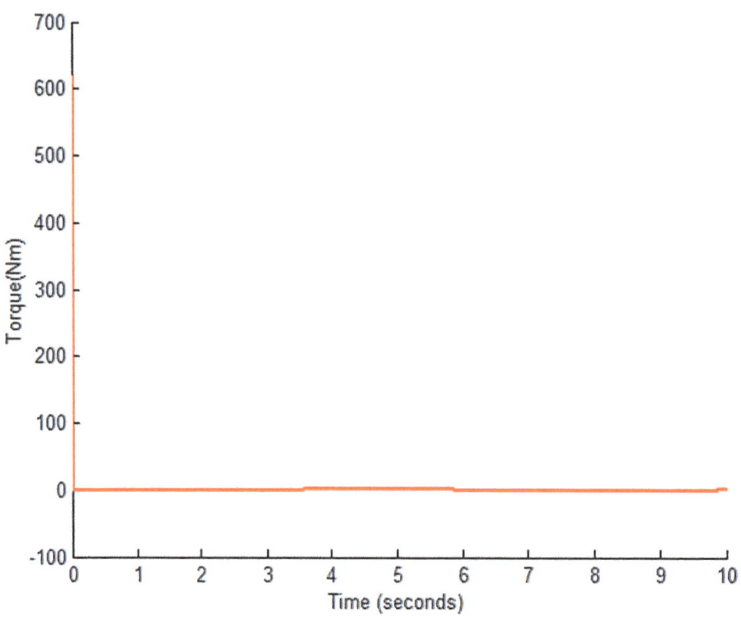

Figure 12. Torque at joint 2

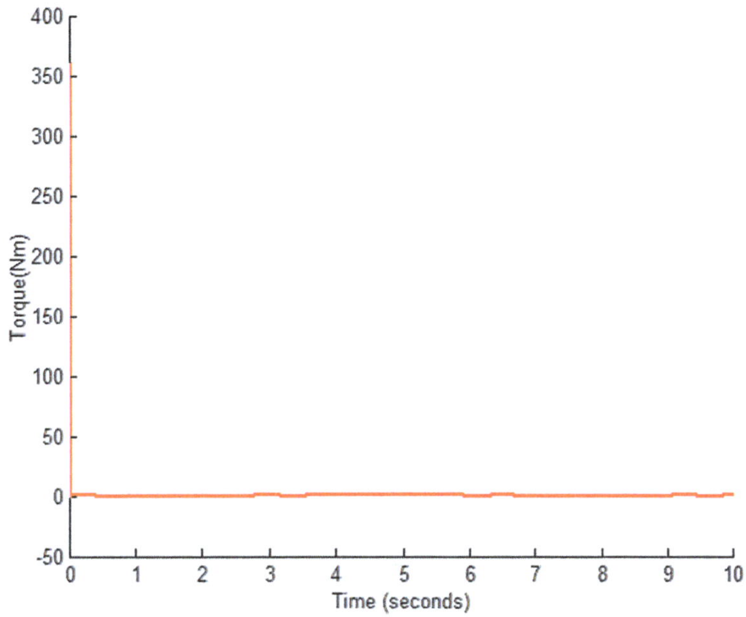

Figure 13. Torque at revolute joint 3

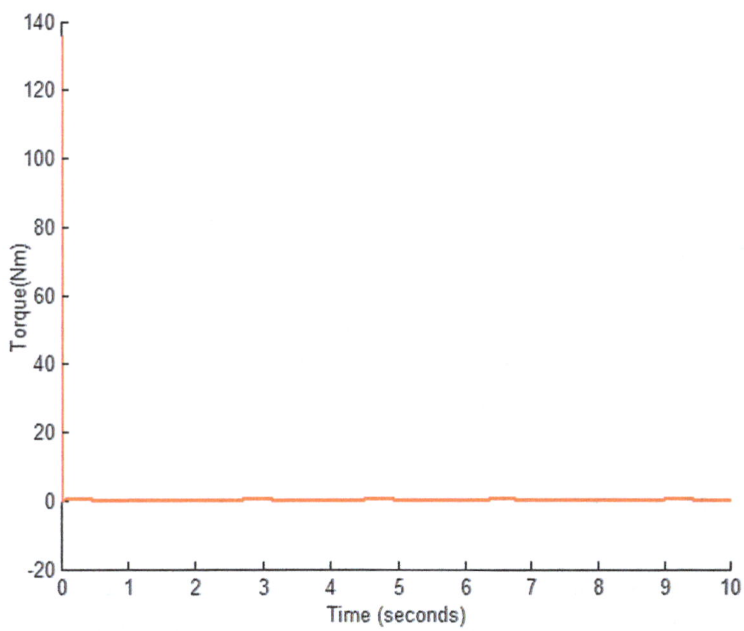

Figure 14. Force at prismatic joint

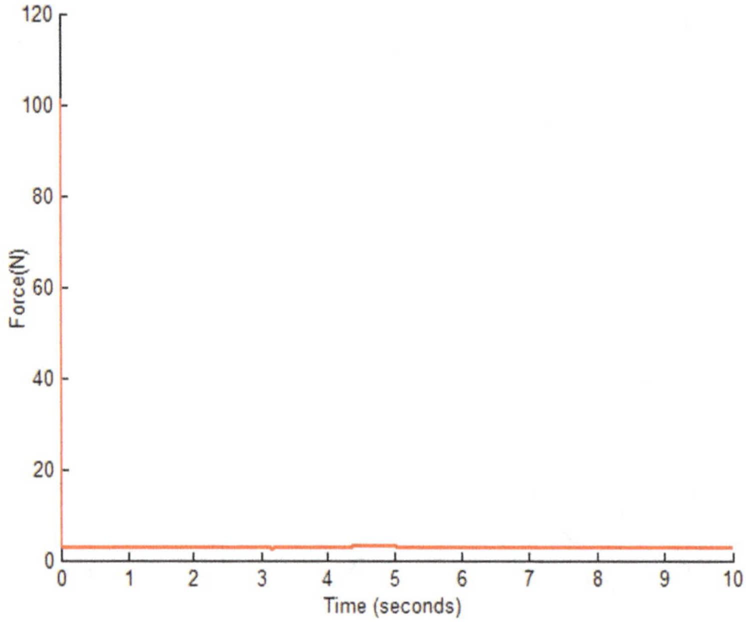

For the prismatic joint the force is computed as $F_1 = 101.1N$.

CONCLUSION

In this chapter the modeling of the SCARA robot with the Multiple gripper is carried out by SolidWorks CAD modeling software and simulated in the MATLAB/SimMechanics environment. Without having complicated mathematical calculations the joint torques are computed automatically for a predetermined trajectory of the manipulator for the given input motion using SimMechanics software with second generation technology. The modeling and simulation of robot using this methodology reveal that the design and structural changes can be done with great ease based on the dynamic study. Thus developing a robot with desired configuration will be economical and easier.

FUTURE RESEARCH DIRECTIONS

This work can be extended in future to predict the contact force acting on each gripper based on the job performed. Also the influence of the contact force on the joint dynamic parameters can also be investigated. Various types of controllers can be implemented and position control can be studied. The number of links and joints of the manipulator can be increased and the dynamic characteristics can be studied. The work space study can also be performed for the improved redundant characteristics of a SCARA robot. The modelling and simulation methodology suggested in this research can be implemented for any kind of robot development.

REFERENCES

Al Mashagbeh, M., & Khamesee, M. B. (2015). Virtual performance evaluation of an industrial SCARA robot prior to real-world task. *Microsystem Technologies*, *21*(12), 2605–2609. doi:10.100700542-015-2502-y

Alshamasin, M. S., Ionesco, F., & Al-Kasasbeh, R. T. (2012). Kinematic modelling and Simulation of a SCARA Robot by Using Solid Dynamics and Verification by MATLAB/Simulink.International Journal of Modelling. *Identification and Control*, *15*(1), 28–38. doi:10.1504/IJMIC.2012.043938

Arachchige & Abderrahmane. (2013). Design of reconfigurable joints for the advanced robotic systems. In *Proceedings of the National Conference on Undergraduate Research (NCUR)*. La Crosse, WI: University of Wisconsin.

Das, M. T., & Dulger, L. C. (2004). Mathematical modelling, simulation and experimental verification of a scara robot. *Simulation Modelling Practice and Theory*, *13*(3), 257–271. doi:10.1016/j.simpat.2004.11.004

Fang & Li. (2013). Four degrees of freedom SCARA robot kinematics modelling and simulation analysis. *International Journal of Computer, Consumer and Control*, *2*(4), 20-27.

Fedak, Durovsky, & Uveges. (2014). Analysis of Robotic System Motion in SimMechanics and MATLAB GUI Environment. *MATLAB Applications for the Practical Engineer*, 565-581. doi:10.5772/58371

Gouasmi, M., Ouali, M., Fernini, B., & Meghatria, M. (2012). Kinematic Modelling and Simulation of a 2-R Robot Using SolidWorks and Verification by MATLAB/Simulink. *International Journal of Advanced Robotic Systems*, *9*(6), 245. doi:10.5772/50203

Harry Asada, H. (2016, March 12). *Introduction to Robotics*. Retrieved from https://ocw.mit.edu/courses/mechanical-engineering/2-12-introduction-to-robotics-fall-2005/lecture-notes/chapter4.pdf

Hernandez, V., Bravo, G., Jose de Jesus, R., & Pacheco, J. (2011). Kinematics for the SCARA and the Cylindrical Manipulators. *ICIC Express Letters. Part B, Applications*, *2*(2), 421–425.

Ionescu, F. (2007). Modelling and Simulation in Mechatronics. In IFAS inter.confer.MCPL2007 (pp. 26-29). Sibiu, Romania: Academic Press.

Ionescu, F., Chojnowski, F., & Constantin, G. (2002). Virtual Reality in Mechanical Engineering: Modelling and Simulation with Solid Dynamics. *ARA-Journal*, *1*, 27.

Jazar, R. N. (2009). *Theory of applied robotics*. New York: Springer.

Jha, Dutta, & Saha. (2014). Analysis of Dynamics of SCORA-ER14 Robot in MATLAB. *International Journal of Innovative Research in Advanced Engineering*, *1*(4), 145–150.

Kütük, M. E., Halicioglu, R., & Dulger, L. C. (2015). Kinematics and Simulation of a Hybrid Mechanism: MATLAB/ SimMechanics. *Journal of Physics: Conference Series*, *574*, 451–458. doi:10.1088/1742-6596/574/1/012016

Ma, Yu, Cao, Zheng, & Liu. (2014). The Kinematic Analysis and Trajectory Planning Study of High-Speed SCARA Robot Handling Operation. *Applied Mechanics and Materials*, *687-691*, 294-299. Retrieved from www.scientific.net/AMM.687-691.294

Michel, O. (2004). Professional Mobile Robot Simulation. *International Journal of Advanced Robotic Systems*, *1*(1), 39–42. doi:10.5772/5618

Patel, S., & Sobh, T. (2014). Manipulator Performance Measures - A Comprehensive Literature Survey. *Journal of Intelligent & Robotic Systems*, *77*(3-4), 547–570. doi:10.100710846-014-0024-y

Paul. (1981). *Robot Manipulators: Mathematics, Programming, and Control*. MIT Press.

Rehiara. (2011). Kinematics of Adept Three Robot Arm. In Robot Arms (pp. 21-38). Rijeka, Croatia: InTech.

Saravanamohan, M., & Anbumalar, V. (2016). Modelling and simulation of multi spindle drilling redundant SCARA robot using SolidWorks and MATLAB/SimMechanics. *Revista de la Facultad de Ingeniería*, *81*, 63–72.

Schlotter, M. (2003). *Multibody System Simulation with SimMechanics*. Retrieved from http://www.imac.unavarra.es/DSM/download/trabajos/

Siciliano, B., & Khatib, O. (2008). *Handbook of Robotics*. Heidelberg, Germany: Springer. doi:10.1007/978-3-540-30301-5

SimMechanics Getting Started Guide. (2015, March 16). Retrieved from https://fenix.tecnico.ulisboa.pt/downloadFile/845043405443232/sl_using_r2015a.pdf

Spong, Hutchinson, & Vidyasagar. (2005). Robot Modelling and Control. New York: John Wiley & Sons.

Tsai. (1999). *Robot Analysis: The Mechanics of Serial and Parallel Manipulators*. New York: John Wiley & Sons.

Udai, Rajeevlochana, & Saha (2011). Dynamic Simulation of a KUKA KR5 Industrial Robot using MATLAB SimMechanics. In *15th National Conference on Machines and Mechanisms* (pp. 1-8). Chennai, India: IIT Chennai.

Umar & Bakarb. (2014). Study on Trajectory Motion and Computational Analysis of Robot Manipulator. *Jurnal Teknologi (Sciences & Engineering)*, *67*(1), 53-59. doi:10.11113/jt.v67.2206

Urrea, C., & Kern, J. (2012). Modelling, Simulation and Control of a Redundant SCARA Type Manipulator Robot. *International Journal of Advanced Robotic Systems*, *9*(58), 58–72. doi:10.5772/51701

Talib, Swadi, Abed, Abed, & Karim. (2013). Vibration and Kinematic Analysis of SCARA Robot Structure. *Diyala Journal of Engineering Sciences*, *6*(3), 127–143.

Wood & Kennedy. (2015, April 24). *Simulating Mechanical Systems in Simulink with SimMechanics The MathWorks*. Retrieved from http://cn.mathworks.comtagteam/12634_SimMechanics.pdf

Zi, B., Cao, J., & Zhu, Z. (2011). Dynamic Simulation of Hybrid-driven Planar Five-bar Parallel Mechanism Based on SimMechanics and Tracking Control. *International Journal of Advanced Robotic Systems*, *8*(4), 28–33. doi:10.5772/45683

Zlajpah, L. (2008). Simulation in robotics. *Mathematics and Computers in Simulation*, *79*(4), 879–897. doi:10.1016/j.matcom.2008.02.017

KEY TERMS AND DEFINITIONS

Denavit-Hartenberg Method: In this convention, coordinate frames are attached to the joints between two links. One transformation is associated with the joint and the another one is associated with the link.

Forward Kinematics: The forward kinematics is when the kinematical data are known for the joint coordinates. The variables of the end-effector in a given Cartesian space are to be computed.

Inverse Kinematics: The inverse kinematics is when the kinematics data are known for the end-effecter in Cartesian space. The joint variables are to be computed.

Kinematic Model: It is a mathematical representation to define the orientation and translation of the manipulator is in the form of homogeneous transformation matrix.

Kinematics: Is the science of geometry in motion. It is restricted to a pure geometrical description of motion by means of position, orientation, and their time derivatives.

SCARA: SCARA means selective compliance assembly robot arm. The robotic arm is rigid in the Z-axis and flexible in the XY-axes.

SimMechanics: It is a set of add on block libraries and special simulation features for modeling physical systems in the Simulink environment.

Section 2

Optimization Techniques and Material

Chapter 6
PageRank Algorithm-Based Recommender System Using Uniformly Average Rating Matrix

Bathrinath S.
Kalasalingam University, India

Saranyadevi S.
Kalasalingam University, India

Thirumalai Kumaran S.
Kalasalingam University, India

Saravanasankar S.
Kalasalingam University, India

ABSTRACT

Applications of web data mining is the prediction of user behavior with respect to items. Recommender systems are being applied in knowledge discovery techniques to the problem of making decisions on personalized recommendation of information. Traditional CF approaches involve the amount of effort increases with number of users. Hence, new recommender systems need to be developed to process high quality recommendations for large-scale networks. In this chapter, a model for UAR matrix construction method for item rank calculations, a Page Rank-based item ranking approach are proposed. The analysis of various techniques for computing item-item similarities to identify relationship between the selected items and to produce a qualified recommendation for users to acquire the items as their wish. As a result, the new item rank-based approaches improve the quality of recommendation outcome. Results show that the proposed UAR method outperforms than the existing method. The same method is applied for the large real-time rating dataset like Movie Lens.

DOI: 10.4018/978-1-5225-5445-5.ch006

1. INTRODUCTION

Most of the real world applications use the web. Web mining is one of the techniques of Data mining – a process of extracting knowledge or information from large volume of data. Web mining concept is broadly divided into three groups namely Web Content Mining, Web Usage Mining and Web Structure Mining.

Web Content Mining is the process of discovering interesting knowledge from the contents of the web documents. The content may have text, images, audio, video, or structured records.

Web Usage Mining is the process of extracting interesting usage pattern from the web data.

Web structure mining is the process of extracting structure information from the Web. This type of mining uses some of the concepts of graph theory.The structure of a typical web graph consists of web pages as nodes, and hyperlinks as edges connecting related pages. Based on the type of structure information used, it can be further divided into hyperlinks and document structures. Here our focus is only on Web Structure Mining (Srivastava et al., 2005).

Web Structure analysis was first done by Larry Page and Sergey Brin who were the doctoral students of Stanford University, introduced the famous PageRank algorithm (Brin & Page, 2012) for performing link analysis of web pages for web search. It is the heart of Google search engine. Initially, the algorithm was designed to find out the significance of web pages over the web by allocating the rank score. Later, the applications of page rank expanded to some other areas which include rating as an important factor. The notable applications related to ratings are: movie ratings, product ratings for online shopping, article ratings, health recommendations, social networking and many more business applications. Those systems support user to attain a satisfactory thing or item which is fit for their requirements. More over recommender systems provide suggestions in some other applications like share market, medical suggestions, hotel/restaurant bookings, music, web pages, search queries in the web and etc,. There are various recommendation systems available, namely MovieLens, Netflix, Jester, eBay and Amazon (Gao, Wu, & Jiang, 2010) The purpose of page rank is to rank any products or web pages which finally provide valid recommendations to the user.

Personalization is known to be an important factor in recommendation system. It is the facility to make content and services always available, which can be customized to each individual users according as their personal liking behaviors (Gao, Liu, & Wu, 2010; Gao & Wu, 2010; Gao, Wu, & Jiang, 2011). Personalized recommendation system is applied in several academic services including digital libraries, e-learning, news filtering, e-commerce, and search engine domains and applications (Gao, Liu, & Wu, 2010). Recommendation system uses Collaborative Filtering (CF), an important concept, to collect and evaluate large amount of data based on the user selection (Zhao et al., 2016; Barbieri et al, 2011). An extension of CF model, a new recommendation engine (Lee, Kuo, & Lin, 2016)was developed to work with course recommendation. The proposal merged a two stage CF representation regularized by course dependency with a graph based recommender support. According to this paper the students are considered as users and the courses are considered as items.

Graph theory can be applied for web page ranking since the web can be related to a graph with nodes and edges. Here the nodes are considered as web pages and the link between web pages are considered as edges (Abedin & Sohrabi, 2009). There are several web link structure evaluation approaches were discussed like cognitive walk through for web, markov chains, survey methods, graph theory. Graph theory concept has been used in collaborative ranking to rank items founded on the consistent recommendation paths (Shams & Haratizadeh, 2017). An iterative network-oriented approach for collaborative ranking has been proposed [19] to consider the data as a bipartite graph where one part has number of users and the other part has pair wise likings. Our research work is mainly focused on item ranking based on user preferences using graph theory concepts.

This paper is organized as follows: In section2 the related work is presented. In section3 we discuss about the web graph, Rating Matrix (RM), Correlation Symmetric Matrix (CSM), Normalization of CSM, and Page Rank algorithm. In section4, we present an overview of Item-Based Collaborative Filtering (CF) and Uniformly Average Rating Matrix (UAR Matrix) and an algorithm for UAR Matrix Construction are given. Section 5 describes an experimental analysis part using the MovieLens Dataset and the performance evaluation. Concluding remarks and future research directions are given in the last section.

2. RELATED WORK

While searching over the web, it is becoming hard to identify the interesting knowledge of user. Recommender System (Barbieri et al., 2011) is a type of filtering system that suggests people in selecting the preferred product or item. In recent years recommender systems are utilized in variety of popular fields such as movies, music, news, books, research articles, search queries, social tags, and products in general. There are some recommendation algorithms, well-known for their usage on e-commerce web sites, where the input of a single customer is used to produce a list of recommended items or products (Linden et al., 2003) For instance, we can consider "Amazon.com" where the customer provides as input a particular product, and then the system displays a list of related products to the customer.

The widely used approach for designing recommendation systems is Collaborative Filtering. Collaborative Filtering (Zhao et al., 2016; Barbieri et al, 2011) is the process of gathering and evaluating a huge amount of data on user behaviors, preferences and predicting what the users will prefer to use based on the similarity to other users. There are two types of CF techniques; one is user based CF and the other one is item-based CF. User-based CF method calculates the user's similarities based on user ratings and recommend the items whereas the Item-based CF method considers similarities between the items and recommends the most similar items. In Sarwar et al. (2001), various Item-based recommendation algorithms for measuring item-to-item similarities and techniques for obtaining recommendations are analyzed.

A contextual item –based CF method is proposed in Tan & Pan (2012), which computes the similarity between the current context of an active user with other contexts. In a particular context, rating of previously given items by the active users is compared with user's current context to decide whether it is similar or not. Then it predicts the item that the active user will like to choose in the current context.

Personalization is one of the important concepts used in recommendation systems. Personalized recommendation is a widely accepted approach and has achieved much interest in business and research fields (Gao, Liu, & Wu, 2010; Gao & Wu, 2010; Gao, Wu, & Jiang, 2011). Personalization is the process of collecting and evaluating user information for producing the accurate information at right time. The purpose of incorporating personalized contextual information with item-based collaborative filtering recommendation is to propose a context-based item difference analysis, a personalized context analysis, and a prediction of rating estimation method based on Slope One algorithm. The method chooses only one contextual parameter for a single user, and the user's weight is not taken into consideration. This is the major limitation of this method (Jiang & Gao, 2009).

In the item-based CF method with user ranking approach was proposed to find the relative weight of the users. The quality of recommendations is improved by incorporating the weight of a user, user-rank into the calculation of item similarities and differentials. The calculations are done by using the Page-Rank based user ranking approach, user-rank based item similarities or differentials approach. One of the well-known ranking algorithms in web search is the PageRank algorithm. PageRank algorithm measures the significance of the web pages.

A random walk based scoring algorithm was proposed (Gori et al., 2007) to recommend high-ranking items to potentially attracted users by ranking the products based on the anticipation of user's liking. The algorithm works better in terms of memory usage and computational cost when compared to other ranking algorithms. Application of graph theory concepts with page rank algorithm was discussed in Abedin & Sohrabi, 2009). The study presented a new proposal by using page rank concept for the analysis and improvement of website link structure mining by means of web usage mining. The improvements were: discovery of target web pages by performing ranking and prioritizing the web pages and by applying graph theory definitions some direct links from other web pages were added to the target page. Also the study indicated that by applying graph theory concepts for evaluating the website usability is a better way than cognitive walk through, markov chains and survey-based methods. Two main drawbacks of the method were: the proposed ranking technique was unproductive with more than 40 pages and there was no upper and lower bound average connectivity level of edges.

Many research works dealt with the homogenous implicit feedback of buying preferences of various users in different transactions which did not include the implicit feedback like browsing activities of individuals (Pan et al., 2015). Hence the problem of such vagueness could be simply handled by means of heterogeneous implicit feedback method of adaptive Bayesian personalized ranking (ABPR). The proposed method avoided the uncertainty using pair wise preference analysis on implicit feedback. A hierarchical itemspace rank (HIR) method has been proposed (Nikolakopoulos, 2015) to reduce the problem of sparsity. The study proposed a disintegration method to divide the itemspace into item blocks by defining the indirect closeness among the elements. The proposed approach is scalable and efficacy because of not only the dimension is irrespective of the number of users but also the smart mathematical properties of the hierarchical closeness matrix which allow easy computational management.

A reliable graph-based collaborative ranking approach named ReGRank (Shams & Haratizadeh, 2017a) has been presented to represent users' priorities and evaluates it to produce a recommendation list directly. ReGRank ranked the items according as the reliable recommendation paths which are synchronized with the semantics following different techniques in neighborhood collaborative ranking. The proposal used the neighborhood collaborative ranking method which enhanced the traditional collaborative ranking method. A collaborative ranking technique using an iterative network-oriented approach has been proposed (Shams & Haratizadeh, 2017b) which used a bipartite graph where one part has number of users and the other part has pair wise likings. The proposed approach filtered the user similarities and likes by means of a random walk method on the graph arrangement.

3. PRELIMINARIES

A directed graph $D = \left(V, E\right)$ consists of a finite nonempty set V whose elements are called vertices or nodes and a set E of ordered pairs of distinct elements of V are called edges or links.

Web Graph: Web graph is a directed graph where V is a set of web pages and elements of E are hyperlinks.

Let $U = \{U_1,...,U_m\}$ be a set of users and $I = \{I_1,...,I_n\}$ be a set of items, and let r_{ij} denotes the rating value of item j by user i. The $m \times n$ matrix $R = (r_{ij})$ is called the Rating Matrix.

Suppose each pair of items i, j is rated by a single user u and $r(u,i)$ be the rating value of item i by user j.

Then the correlation similarity [4] of i and j is defined by,

$$Sim(i,j) = \frac{\sum_{u \in U}(R_{u,i} - R_i)(R_{u,j} - R_j)}{\sqrt{\sum_{u \in U}(R_{u,i} - R_i)^2}\sqrt{\sum_{u \in U}(R_{u,j} - R_j)^2}} \tag{1}$$

where,

R_{ui} and R_{uj} are the user's rating on items *i* and *j* respectively.
R_i and R_j are the rating values of particular items *i* and *j* respectively.

This gives an $n \times n$ symmetric matrix and is called as the Correlation Symmetric Matrix (CSM). The CSM can be normalized by dividing each value of $Sim(i,j)$ by the sum of the elements of jth column. Hence in the normalized CSM, the sum of the elements in each column is 1.

3.1. Concept of PageRank Algorithm

PageRank is an important concept of link analysis, for evaluating the significance of web pages. The main objective of PageRank algorithm is to calculate the importance of web pages depending on the interconnection of links over the web. The more the number of incoming links to a page, the more the importance is.

Let G= (V, E) be a web graph (Gao, Wu, & Jiang, 2011). For any $v \in V$, any hyperlink of the form (u,v) is called an in-link at v and any hyperlink from (v,u) is called an out-link at v. The in-degree $id(v)$ is the number of in-links at v and the out-degree $od(v)$ is the number of out-links at v.

Let us consider an example of 4 web pages which are interconnected by hyperlinks among them. The web pages are: P_a, P_b, P_c, P_d.

From the above example, the web pages are connected by using hyperlinks with each other. For any two web pages it is not necessary that a link must exist between them. The in-link and out-link may or may not exist for any specific web page with other web pages. It is understandable that there is no self loop exists for a particular web page in the web graph i.e. the hyperlink of any web page does not connect to the same web page.

The rank (Liu, 2007) of a specific page, i is denoted by P(i) and is defined by the formula,

Figure 1. A web graph with interconnected links

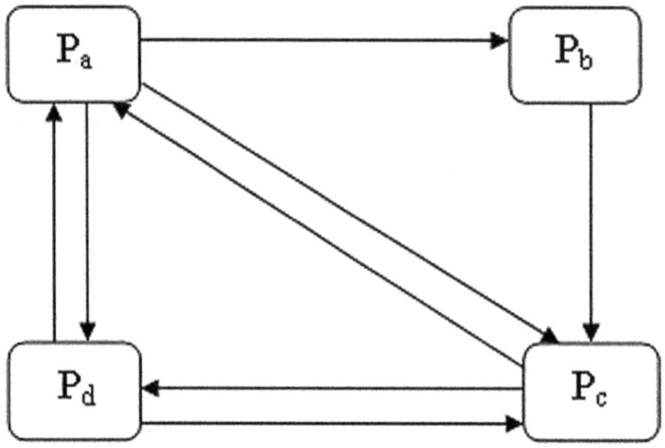

$$P(i) = \sum_{(j,i)\in E} \frac{P(j)}{od_j} \tag{2}$$

The above formula is the fraction of the summation of incoming links of all the connected pages with $P(i)$ to the out-degree of that particular page, $P(j)$.

Let $P = \left(\left(P(1), P(2), \ldots. P(n)\right)^T\right)$. Then for each page i, PageRank formula is also given by,

$$P(i) = (1-d) + d\sum_{(j,i)\in E} \frac{P(j)}{od(j)} \tag{3}$$

The parameter d is termed as the damping factor, the value of d is between 0 and 1. A Transition Probability Distribution (TPD) matrix, $M = M_{ij}$ is defined by,

$$M_{ij} = \begin{cases} \frac{1}{n}, & if\, od(i) = 0 \\ \frac{1}{od_i}, & if\,(i,j) \in E \\ 0, & otherwise \end{cases} \tag{4}$$

The sum of all elements in a column of the TPD matrix must be 1.
Clearly,

$$\sum_{i=1}^{n} Mij = 1 \tag{5}$$

For any fixed j.

A TPD matrix of the sample web graph (Figure 1) is formed as follows:

$$M_{ij} = \begin{pmatrix} 0 & 0 & 1/2 & 1/2 \\ 1/3 & 0 & 0 & 0 \\ 1/3 & 0 & 0 & 1/2 \\ 1/3 & 1 & 1/2 & 0 \end{pmatrix}$$

Matrix 1: TPD Matrix of the Web Graph

Note that in Matrix1, sum of all the elements of a single column is equal to 1.

An Initial Probability Distribution (IPD) matrix with n-dimensional column vector be,

$$P_0 = \left(\left(P_0(1), P_0(2), \ldots . P_0(n) \right)^T \right)$$

where $\sum_{i=1}^{n} P_0(i) = 1$, be an initial probability distribution. IPD matrix can be formed as,

$$P_0 = \begin{pmatrix} 1/4 \\ 1/4 \\ 1/4 \\ 1/4 \end{pmatrix}$$

Matrix 2: IPD Matrix

Let,

$$P(i) = \sum_{i=1}^{n} M_{ij} P_0(j) \tag{6}$$

For the first transition, Equation (6) takes the form,

$$P_1 = M_{ij} \times P_0 \tag{7}$$

Hence the probability distribution after n transitions is given by,

$$P_n = M_{ij} \times P_{n-1} \tag{8}$$

The number P_i is called the page rank value of the page i.

Talking about the applications of PageRank algorithm, it is widely used in many of the business applications such as e-commerce (marketing areas, online shopping/tradings), social networking. While considering the marketing area, the concept of item – user transaction needs to be focused. There the user rates an item or product based on his/her previous history of experience. Based on the importance of item, it is rated by the user and could be ranked. Since the web is grown explosively, it is becoming hard to discover the desired item by simply searching over the web. In that situation, recommender system plays a vital role. Recommender System is a type of filtering system that suggests people to make choice of the preferred product or item. The most commonly used approach for designing recommendation systems is the Collaborative Filtering (CF). Collaborative Filtering is a filtering system of gathering and evaluating a huge amount of data from user likings, preferences and predicting what the users will prefer to use in future based on the similarity to other users. There are two types of CF techniques; one is user-based CF and the other one is item-based CF.

4. ITEM-BASED COLLABORATIVE FILTERING

The main concept of traditional (user-based) collaborative filtering approach is to predict ratings of a user based on the view of similar users. Though this approach has been very successful, there have been some potential problems such as sparsity and scalability (Sarwar et al., 2001) Hence item-based CF got importance in rating products to overcome the above problems.

4.1. Illustration of Item-Based CF

Consider the rating matrix R where $U = \{U_1, \ldots, U_m$ is the set of m users and $I = \left\{I_1, \ldots, I_n\right\}$ is the set of n items. For instance, if there are 7 users and 6 items, a rating matrix is given below. The rating values are taken in the 5-point scale and the entry ɸ represents that the item is not rated by the particular user (Jiang & Gao, 2009).

Matrix 3: The Rating Matrix (RM)

$$\begin{pmatrix} 5 & & 3 & 4 & 2 & \\ 2 & & 4 & & 3 & 1 \\ & 4 & & 4 & & 4 \\ 3 & 2 & 3 & 2 & 1 & 3 \\ & 4 & 3 & & 2 & \\ 3 & & 1 & 1 & 1 & 5 \\ 2 & 4 & 2 & 2 & & 4 \end{pmatrix}$$

4.2. Uniformly Average Rating Matrix: (UAR Matrix)

UAR matrix of a rating matrix is obtained by the following steps:

- If α is the largest number in a column, which is repeating n times where $n > 1$ then replace all the other entries by zero except α.
- If the maximum number in a column is not repeated more than once, let x and y be the two largest numbers in that column, and let

$$\alpha = \frac{x + y}{2} \tag{9}$$

Replace x and y by α and replace all the other entries in the column by zero.

- If x is the first largest number which is not repeated more than once but the second largest number y is repeating more than once in that column then we have,

$$\alpha = \frac{x + n\left(y\right)}{\left(n + 1\right)} \tag{10}$$

Here, we replace x and all the respective places of y by α.

Matrix 4: UAR Matrix of RM

The UAR Matrix of RM is given below.

$$\begin{pmatrix} 3.67 & 0 & 3.25 & 3 & 2.33 & 0 \\ 0 & 0 & 3.25 & 0 & 2.33 & 0 \\ 0 & 4 & 0 & 3 & 0 & 4.33 \\ 3.67 & 0 & 3.25 & 3 & 0 & 0 \\ 0 & 4 & 3.25 & 0 & 2.33 & 0 \\ 3.67 & 0 & 0 & 0 & 0 & 4.33 \\ 0 & 4 & 0 & 3 & 0 & 4.33 \end{pmatrix}$$

Matrix 5: CSM From UAR Matrix

Next step is to find the correlation between the items of the UAR Matrix and CSM matrix of the above UAR matrix is given below.

$$\begin{pmatrix} 0 & 0 & 2 & 1 & 1 & 1 \\ 0 & 0 & 1 & 1 & 1 & 1 \\ 2 & 1 & 0 & 1 & 3 & 0 \\ 1 & 1 & 1 & 0 & 1 & 1 \\ 1 & 1 & 3 & 1 & 0 & 0 \\ 1 & 1 & 0 & 1 & 0 & 0 \end{pmatrix}$$

Matrix 6: Normalization of CSM

The normalized matrix of the CSM matrix is given below

$$\begin{pmatrix} 0 & 0 & 2/7 & 1/5 & 1/6 & 1/3 \\ 0 & 0 & 1/7 & 1/5 & 1/6 & 1/3 \\ 2/5 & 1/4 & 0 & 1/5 & 3/6 & 0 \\ 1/5 & 1/4 & 1/7 & 0 & 1/6 & 1/3 \\ 1/5 & 1/4 & 3/7 & 1/5 & 0 & 0 \\ 1/5 & 1/4 & 0 & 1/5 & 0 & 0 \end{pmatrix}$$

5. EXPERIMENTAL ANALYSIS

5.1. The MovieLens Dataset

The commonly used popular dataset for the recommender systems is the MovieLens dataset that comes under MovieLens project (Sarwar et al., 2001). MovieLens is a freely available dataset by GroupLens Research Project at the University of Minnesota. GroupLens Research currently works on a movie recommender system based on collaborative filtering (http://www.movielens.org). The dataset consists of 6040 users and 3952 movies. The ratings are made on the scale of 1 – 5. Each user has at least 20 ratings from 3952 movies. The dataset can be divided into two parts. i.e., training set and test set. 80% of the data was considered as training set and 20% of the data was considered as test set. The data set is represented as a user-item rating matrix, which has 6040 users in the row and 3952 movies in the column of the matrix.

5.2. Experimental Results

In this study, an algorithm is developed to remove the least significant items with lower ratings. For this, the top rating items can be kept as it is and the lower rating items must be ignored in the beginning itself. This process can be achieved by using the concept of UAR matrix construction method. As a result a symmetric matrix is obtained based on the column value of the initial matrix. Once the symmetric matrix is formed then the item ranking process can be carried out.

The whole process of ranking is as follows: First we consider a web graph and the adjacency matrix of the same graph can be obtained. The Initial Probability matrix is formed first and also then Transition Probability Matrix is formed based on the existing link analysis algorithms. Next the rank of each item is calculated using Equations 4–8. Item ranking algorithm is executed for 10 iterations and the damping factor is set as 0.5 for both without UAR and with UAR matrix construction methods.

Replacement of Uniform Probability Distribution Value (1/n)

Uniform probability distribution value is applicable if any column has all the entry as 0. The same can be replaced by the uniform probability distribution values 1/n. And then the ranking calculations take place for further analysis.

Implementation of Large Real Time Dataset

The implementation of sample web graph has been explained with our proposed UAR matrix construction method of item ranking algorithm. Next we can do the same process for analyzing the large volume of real time data – MovieLens data. The values for number of iterations and the damping factor are set as 50 and 0.5 respectively. From the experimental results we come to know that the ranking values have significant changes than the existing ranking algorithm. UAR matrix construction method produces some significant changes in both sample and real time data. Because our process considers only the high rated items. Though the low rated items are ignored in the beginning of process, it is easy to find correlations depending upon the average values calculated by our UAR method.

5.3. Efficiency of Proposed UAR Matrix Calculation

The purpose of obtaining UAR matrix is to neglect the low rated items for further process. This step has the benefit that only the relevant- high rated products will be listed out for user selection. Then the correlation matrix for the high rated item's ratings is obtained. Finally Item ranking is calculated for those identified item's ratings based on the PageRank algorithm. In the proposed UAR matrix construction method, the consideration of only high rated items plays a vital role in performing the item ranking. This is an efficient way of reducing the number of correlations.

Figure 2 explains about the ranking score of both existing and proposed methods of item ranking. In the figure IR represents Item Ranking. Here the increase in existing IR also shows the increase in proposed IR. Therefore the proposed ranking based on Uniformly Average Rating method proves to be efficient than the existing one. Similarly for the real time dataset, the experimental results show the significant improvement of the proposed method of ranking.

5.4. Evaluation Metrics

The most extensively used metrics for measuring recommendation deviations from their true values are the Mean Absolute Error (MAE) (Gao, Wu, & Jiang, 2011) and Root Mean Square Error (RMSE) (Zhao, Wang, & Lai, 2016).

Figure 2. Experimental Results of Existing IR and Proposed UAR Based IR

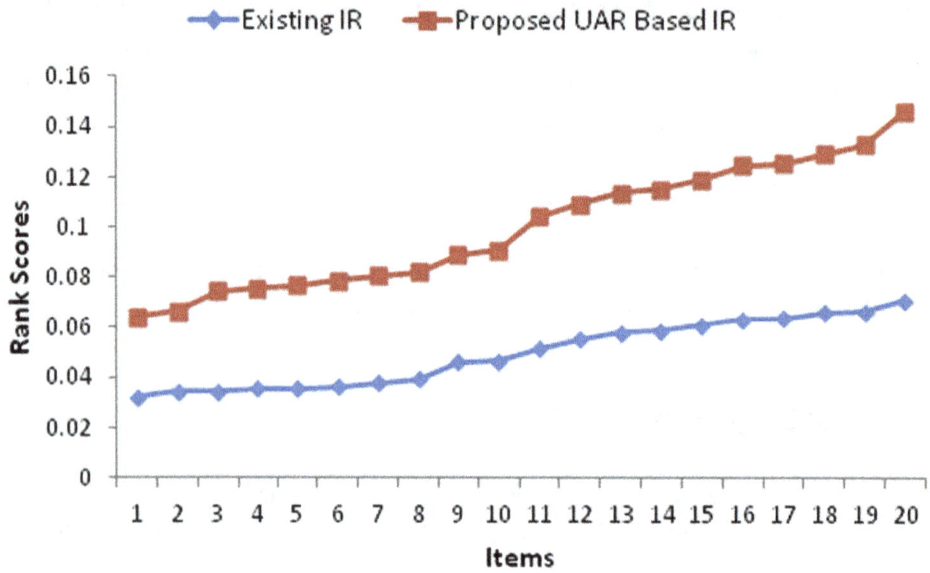

MAE can be calculated as follows: Let $\{p_1, p_2, \ldots p_n\}$ be the set of all predictions and $\{r_1, r_2, \ldots r_n\}$ be the set of real ratings of predictions. MAE is the average of absolute error for each prediction-rating pair, i.e., $\{p_i, r_i\}$. The cardinality of the prediction-rating pair can be denoted by $|p_i - r_i|$. Formula for measuring MAE is given by,

$$MAE = \frac{\sum_{i=1}^{n} |p_i - r_i|}{n} \tag{11}$$

RMSE can be calculated by using the below formula.

$$RMSE = \sqrt{\frac{\sum_{i=1}^{n} (p_i - r_i)^2}{n}} \tag{12}$$

If the resulting MAE and RMSE values are lower, then the predictions will be more accurate and the recommendation will be better [6]. From the statistical analysis, results of MAE and RMSE in the proposed UAR method are lower than the existing IR method which proves that the proposed method produces comparatively better results than the existing one.

5.5. Performance Evaluation Using the Tests of Significance

A very important aspect of the sampling theory is the study of tests of significance which enable us to decide on the basis of sample results. To make the efficiency analysis of the proposed method more precise, significant tests can be conducted. We have used Paired T-test for evaluating the efficiency of UAR based IR method.

- **Null Hypothesis H_0:** Proposed UAR method has no significant improvement than the existing IR method.
- **Alternative Hypothesis H_1:** Proposed UAR method has a significant improvement than the existing IR method.

The calculated value of paired T-test method is greater than the tabulated value, for 5% level of significance with 19 degree of freedom (since no of items taken for calculation here are 20).

Hence we accept H_1 and reject H_0. So the proposed UAR based IR method has significant improvement than the existing IR method.

6. CONCLUSION

Recommender System suggests users to find items based on their interest. In this paper we have presented Uniformly Average Rating method to strengthen the recommendation system by neglecting the low rated items. Low ratings are identified and ignored, and the uniformly average rating value of each item given by the individual user is calculated by the UAR method. Then the item ranking is performed on the PageRank-based item ranking approach. The experimental results prove that the proposed method gives a considerable enhancement of the existing method. More over the results of paired T-test provides us there is a significant improvement than the existing IR method. In this paper the low-rated items were ignored for the calculations, so in the future work we will make use of the respective items for preparing recommender system such that even the low-rated items are also sold out. In this way we will be able to reduce the loss risk of the organization by making the user to buy the low-rated items using our recommender system.

REFERENCES

Abedin, B., & Sohrabi, B. (2009). Graph theory application and web page ranking for website link structure improvement. *Behaviour & Information Technology*, *28*(1), 63–72. doi:10.1080/01449290701840948

Barbieri, N., Manco, G., & Ritacco, E. (2011, April). A probabilistic hierarchical approach for pattern discovery in collaborative filtering data. In *Proceedings of the 2011 SIAM International Conference on Data Mining* (pp. 630-641). Society for Industrial and Applied Mathematics. 10.1137/1.9781611972818.54

Brin, S., & Page, L. (2012). Reprint of: The anatomy of a large-scale hypertextual web search engine. *Computer Networks*, *56*(18), 3825–3833. doi:10.1016/j.comnet.2012.10.007

Gao, M., & Wu, Z. (2010). Incorporating Personalized Contextual Information in Item-based Collaborative Filtering Recommendation. *JSW*, *5*(7), 729–736. doi:10.4304/jsw.5.7.729-736

Gao, M., Wu, Z., & Jiang, F. (2011). Userrank for item-based collaborative filtering recommendation. *Information Processing Letters*, *111*(9), 440–446. doi:10.1016/j.ipl.2011.02.003

Gao, M., Liu, K., & Wu, Z. (2010). Personalisation in web computing and informatics: Theories, techniques, applications, and future research. *Information Systems Frontiers, 12*(5), 607-629.

Gori, M., Pucci, A., Roma, V., & Siena, I. (2007, January). ItemRank: A Random-Walk Based Scoring Algorithm for Recommender Engines. *IJCAI (United States)*, *7*, 2766–2771.

Jiang, F., & Gao, M. (2009). Collaborative filtering approach based on item and personalized contextual information. *Proceedings of the International Symposium on Intelligent Information Systems and Applications Qingdao*, 63-66.

Lee, E. L., Kuo, T. T., & Lin, S. D. (2016). *A Collaborative Filtering-Based Two Stage Model with Item Dependency for Course Recommendation*. Academic Press.

Linden, G., Smith, B., & York, J. (2003). Amazon. com recommendations: Item-to-item collaborative filtering. *IEEE Internet Computing*, *7*(1), 76–80. doi:10.1109/MIC.2003.1167344

Liu, B. (2007). *Web data mining: exploring hyperlinks, contents, and usage data*. Springer Science & Business Media.

Nikolakopoulos, A. N., Kouneli, M. A., & Garofalakis, J. D. (2015). Hierarchical itemspace rank: Exploiting hierarchy to alleviate sparsity in ranking-based recommendation. *Neurocomputing*, *163*, 126–136. doi:10.1016/j.neucom.2014.09.082

Pan, W., Zhong, H., Xu, C., & Ming, Z. (2015). Adaptive Bayesian personalized ranking for heterogeneous implicit feedbacks. *Knowledge-Based Systems*, *73*, 173–180. doi:10.1016/j.knosys.2014.09.013

Sarwar, B., Karypis, G., Konstan, J., & Riedl, J. (2001, April). Item-based collaborative filtering recommendation algorithms. In *Proceedings of the 10th international conference on World Wide Web* (pp. 285-295). ACM.

Shams, B., & Haratizadeh, S. (2017a). Reliable Graph-based Collaborative Ranking. *Information Sciences*.

Shams, B., & Haratizadeh, S. (2017b). IteRank: An iterative network-oriented approach to neighbor-based collaborative ranking. *Knowledge-Based Systems*, *128*, 102–114. doi:10.1016/j.knosys.2017.05.002

Srivastava, T., Desikan, P., & Kumar, V. (2005). Web mining–concepts, applications and research directions. *Foundations and Advances in Data Mining*, 275-307.

Tan, X., & Pan, P. (2012). A contextual item-based collaborative filtering technology. *Intelligent Information Management Journal, 4*, 85-88.

Zhao, Z. L., Wang, C. D., & Lai, J. H. (2016). AUI&GIV: Recommendation with asymmetric user influence and global importance value. *PLoS One*, *11*(2), e0147944. doi:10.1371/journal.pone.0147944 PMID:26828803

Chapter 7
Performance of PM Linear Generator Under Various Ferromagnetic Materials for Wave Energy Conversion

Izzeldin Idris Abdalla Yagoube
Universiti Teknologi PETRONAS, Malaysia

Taib Ibrahim
Universiti Teknologi PETRONAS, Malaysia

Nursyarizal Mohd Nor
Universiti Teknologi PETRONAS, Malaysia

Perumal Nallagownden
Universiti Teknologi PETRONAS, Malaysia

ABSTRACT

This chapter examines the influence of the various ferromagnetic materials on the performance of a single-phase tubular permanent-magnet linear generator (TPMLG) for wave energy conversion. Four ferromagnetic materials were considered in this study. They are non-oriented electrical steel, Permalloy (Ni-Fe-Mn), Accucore, and Somaloy 700. The generator equipped with a tubular stator carries a single coil and employs a quasi-Halbach magnetized moving-magnet translator. Therefore, in order to obtain an accurate performance analysis, the nonlinear time-stepping finite-element analysis (FEA) technique has been used. The electromagnetic characteristics, including the magnetic field distributions, flux-linkage, winding inductance, electromagnetic force, and electromotive force (EMF) have been investigated. It is shown that a generator whose stator is fabricated from soft magnetic composite (SMC) materials has potential advantages in terms of ease of manufacture, highest force capability, lower cost, and minimum eddy-current loss.

DOI: 10.4018/978-1-5225-5445-5.ch007

INTRODUCTION

Nowadays, the consensus on the electric linear machines (ELMs) to replace their rotary counterparts has been gradually fostered. By using ELMs numerous advantages are possible, such as high efficient energy generation, the direct-drive system as well as simple structure (Lee, Kim, Jun, & Lee, 2011).

ELMs produce linear motions directly without mechanical transmission means. Thus, it can significantly simplify the structure of the system and improve the working efficiency. These machines have been used in widespread applications, such as in wave energy generation, automotive applications, robotics, medical operation, and reciprocating compressors (Feng et al., 2015; Zheng, Huang, Gao, & Chang, 2015; Abdalla, Ibrahim, & Mohd Nor, 2014). Among the ELMs, the tubular permanent-magnet linear machine (TPMLM) has superior merits, such as high efficiency, high power density, and remarkable force capability due to the lack of end-windings. Moreover, the net attractive force between the stator and translator is zero (Niu, Ho, & Fu, 2011).

The TPMLM can possibly be broken down into three categories, such as moving-coil, moving-iron, and moving-magnet. However, moving-coil TPMLM suffers from numerous demerits, such as limited access to moving-coil, difficulty in dissipating the heat from the coil and the fragility of the connections and flying leads, as well as the limited stroke. The moving-iron type is rarely used due to the heavy moving mass, relatively poor energy conversion, and low force density. Therefore, because of the copper coil directly wound around the yoke of the moving-magnet TPMLM accordingly the high force density can be obtained, thus, this kind of TPMLM it seems to be more suitable for the linear energy conversion (Ibrahim, 2009; Si, Feng, Su, & Zhang, 2014; Wang, Howe, & Lin, 2008).

The appropriate selection of the materials for the ELMs represents a significant role in the development and performance of the machine. The hard-magnetic materials of both high coercivity and high remanence at a wide range of temperature and affordable cost have an important role in the performance of ELMs. The rare-earth elements, such as neodymium-iron-boron (NdFeB) and samarium-cobalt (SmCo), can be considered as the best choice for the permanent-magnet (PM). As a comparison, the SmCo have better chemical properties, whereas, NdFeB is superior in terms of physical properties. The fact that, the PMs of the two groups have complementary characteristics to each other as well as their energy product is high, thus; they are possible to be used in the new technical designs of TPMLMs. Moreover, rare-earth PM materials have superior merits as compared with the conventional ferrite PMs; they can reduce the volume of the required PMs and have significantly higher flux densities (Gieras, 2008; Shahgholian & Shafaghi, 2010).

The soft magnetic composite (SMC) materials are gaining much attention when the high efficiency and manufacturing of the complex design of electrical machine are required. When comparing SMC materials with non-oriented electrical steel, the SMC materials have lower total magnetic losses at high frequency due to the iron particle insulation, which lowers the eddy current loss component. The use of the SMC materials offers near-net-shape, low-cost manufacture and good utilization of the available space to achieve a compact design. Besides, decreasing the hysteresis loss is considered as an important objective in the SMC development (Castro & Landgraf; Ibrahim, 2009).

Therefore, in order to have an efficient energy conversion system, the simplicity, affordability and force capability, are highly desirable features. Basically, the performance of the TPMLMs can be well improved by using a right selected materials (Chen, Fan, & Lu, 2008; Tavana, Shoulaie, & Dinavahi, 2012). Besides, the usage of quasi-Halbach magnetization in the TPMLM offers numerous attractive features, such as a sinusoidal field distribution, which results in a minimum EMF ripple (Lee, Choi, Lee, Lim, & Gweon, 2006; Meessen, Gysen, Paulides, & Lomonova, 2008).

This research aims to study the effect of the ferromagnetic materials on the performance of a single-phase TPMLM that used for wave energy conversion. Four ferromagnetic materials have been considered in this study namely, non-oriented electrical steel, Permalloy (Ni-Fe-Mn) (Gieras, Piech, & Tomczuk, 2011), Accucore (Gieras, 2008; Gieras et al., 2011; Gieras, Wang, & Kamper, 2008) and Somaloy 700 (Gieras, 2008; Ibrahim, 2009). Therefore, Four analysis models for single-phase TPMLM were presented and compared. The conditions of the Four analysis models are the same, except for the ferromagnetic materials. Therefore, the electromagnetic quantities have been computed to investigate that the performance of the machine is depending on the characteristics of the core material. The non-linear time-stepping Finite-Element Analysis (FEA) has been successfully utilized in this study for the design and analysis.

BACKGROUND

The three most well-developed technologies for deriving electrical power from the ocean include tidal power, wave power and ocean thermal energy conversion. From these possibilities, the wave energy conversation seems to have the greatest general application (Szabo, Oprea, Viorel, & Biró, 2007).

Figure 1 depicts the simplest configuration of the wave energy conversation system that consists of the linear generator, the mechanical accessory, and the floating buoy lifted-dropped by the waves connected directly to the linear generator fixed on a concrete foundation, which stands on the bottom of the sea (Li, Chau, & Jiang, 2011)

MAIN FOCUS OF THE CHAPTER

The study is based on the TPMLM that its initial design and operational specifications are listed in Table 1, and its schematic representation for 2-D and 3-D is shown in Figure 2. The generator consists of a slotted stator and a 2-pole PMs translator with a trapezoidal PMs configuration.

Figure 1. Configuration of the floating buoy system with the linear generator

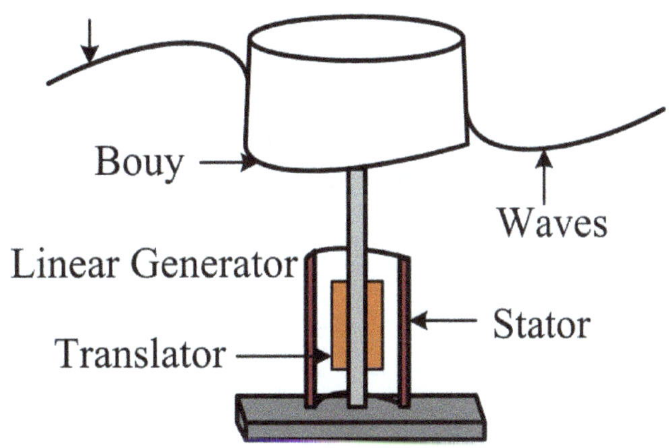

Table 1. Design parameters and operational specifications of a TPMLM

No	Parameter	Value
1	Stator undercut angle	30°
2	Tooth tip height	1.0 mm
3	Thickness of the yoke	3.30 mm
4	Air gap length	0.8 mm
5	Pole-pitch	25 mm
6	Inner radius of the supporting tube	11.1 mm
7	Length of permanent-magnet	50.0 mm
8	Ferromagnetic height	3.9 mm
9	Outer diameter of the stator core	100.0 mm
10	Opening of the slot	10.0 mm
11	Outer radius of the magnet	26.0 mm
12	Magnet height	5.0 mm

Figure 2. The structure of the proposed TPMLM (a) 2-D (b) 3-D

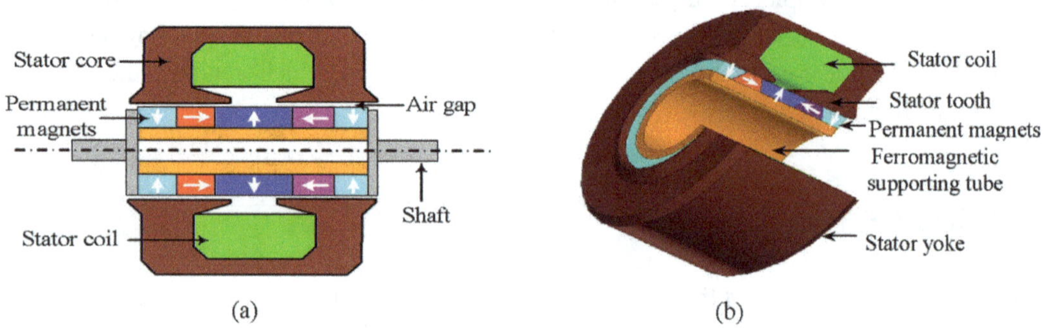

The fundamental principles of electromagnetic machines are Lorentz law, Faraday law, and the Biot-savart law. The force produced in the proposed TPMLM is based on Lorentz's law as expressed in (1)

$$F = B \times l \times i$$

where B is the air gap magnetic flux density, l is the length of the conductor and i is the current in the winding. Based on Faraday's law, the induced voltage when the flux passes through of an N-turn coil is expressed as in (2) and (3) (Polinder, Mueller, Scuotto, & Goden de Sousa Prado, 2007)

$$e = -\frac{d\phi N_c}{dt}$$

$$e = B \times l \times v_t$$

where e is the induced voltage in the winding and ϕ is the flux passing through the coil, N_c is the number of coil turns, t is the time and v_t is the speed of the translator. Generally, the flux density can be computed as

$$B = \frac{\mu_r \times \mu_0}{2 \times \pi \times r} \times i$$

where μ_0 the permeability of the free space and its value is, $\mu_0 = 4\pi \times 10^{(-7)}$ H/m and μ_r is the relative permeability, and for any material is given by the ratio of the permeability to the free space permeability, and expressed as the following:

$$\mu_r = \frac{\mu}{\mu_0}$$

The magnitude of the magnetic field intensity, H in the core due to the applied current is, $H = i / l$, and then the equation (4) can be re-written as:

$$B = \mu_r \times \mu_0 \times H$$

It can be noted that the strength of the magnetic field flux produced in the electromagnetic machines also depends on the properties of the materials (Feng et al., 2015).

The mechanical equation governing the motion of the proposed TPMLM can be expressed as in (7), while the resonant frequency, f_r is expressed as in (8).

$$M\ddot{z}_d(t) = K_T i_a - kz_d(t) - D\dot{z}_d$$

$$f_r = \frac{1}{2\pi}\sqrt{\frac{k}{M}}$$

where M, K_T, k and D, are the mass of the translator, the thrust force coefficient, the effective spring stiffness and the damping coefficient, respectively. $z_d(t)$, $\dot{z}_d(t)$ and $\ddot{z}_d(t)$, are the displacement, the velocity of the translator and the acceleration, respectively (Al-Otaibi & Jack, 2008; Ibrahim, 2009; Jang, Choi, Cho, & Lee, 2005).

Normally, the stator cores of the electrical machines are fabricated from electrical steel sheets. The ferromagnetic sheets their thicknesses are from 0.5 to 0.6 mm, at frequency 50 Hz, the skin effect practically does not exist. The laminated cores of small ELMs are more difficult to fabricate, whereas SMC materials simplify the manufacturing process for the complicated shapes (Gieras et al., 2011; Gieras et al., 2008).

Figure 3 compares the B-H properties of non-oriented electrical steel sheets, Permalloy, Accucore and Somaloy 700 (Gieras, 2008; Gieras et al., 2011; Ibrahim, 2009). Actually, non-oriented electrical steel sheets are used for the construction of large, medium and low-power electrical machines as well as small transformers and magnetic amplifiers. The silicon content increases the resistivity (Gieras et al., 2011). Permalloy (Ni-Fe-Mn), is used for the construction of small electrical machines and micro-machines working in humid or chemical active atmospheres, also it is a good ferromagnetic material for small transformers used in electronic devices and electromagnetic converters. The Permalloy has the best corrosion-resistant, but its saturation for magnetic flux density is lower than that of non-oriented electrical steel sheets (Gieras et al., 2011).

Accucore is a new SMC material that is competitive with traditional steel laminations. Accucore is used in the production of small electrical machines and electrical machines with complicated shapes. The components of SMC are iron powder, dielectric (epoxy resin) and filler (glass or carbon fibers) for mechanical strengthening, the specific density is 7550 to 7700 kg/m^3 (Gieras, 2008; Gieras et al., 2011; Gieras et al., 2008). Somaloy 700 is an SMC material produced by Hoganas, Sweden, that is a surface-coated metal powder with an excellent compressibility. Somaloy 700 can be used for developing 3-D electrical machines, transformers, and sensors (Ibrahim, 2009).

Therefore, in order to assess the performances of the proposed TPMLM, an axisymmetrical calculation model has been created by using the Maxwell 2-D FEA software, and the nonlinear characteristics of the materials were included. The machine was analyzed by using aforementioned different ferromagnetic materials. The NdFeB PM has been used and modeled with a remanent flux density of 1.14 T, a coercive force of -860 kA/m and a relative permeability, μ_r of 1.05 as well as the magnetization vector direction has been adopted.

Therefore, the output results can be given in various forms, including a plot of vector magnetic potential, flux density map, and flux path plot. Figure 4 compares the magnetic flux density distributions in the TPMLM with PMs excitation only and zero current in the winding. The results plotted at

Figure 3. B-H curves of the four ferromagnetic materials

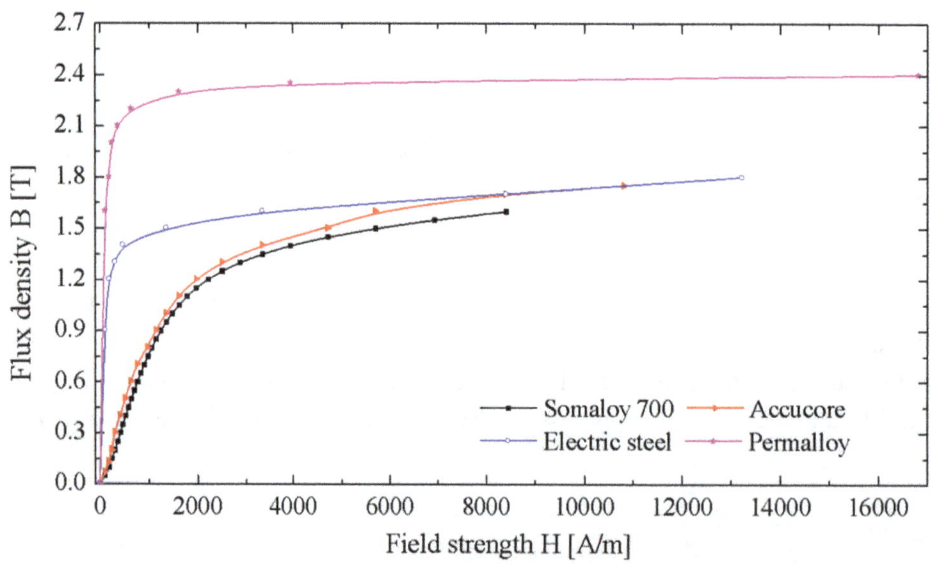

Figure 4. Magnetic flux density distribution on a TPMLM at zero current in the winding (a) Non-oriented electrical steel (b) Permalloy (c) Accucore (d) Somaloy 700.

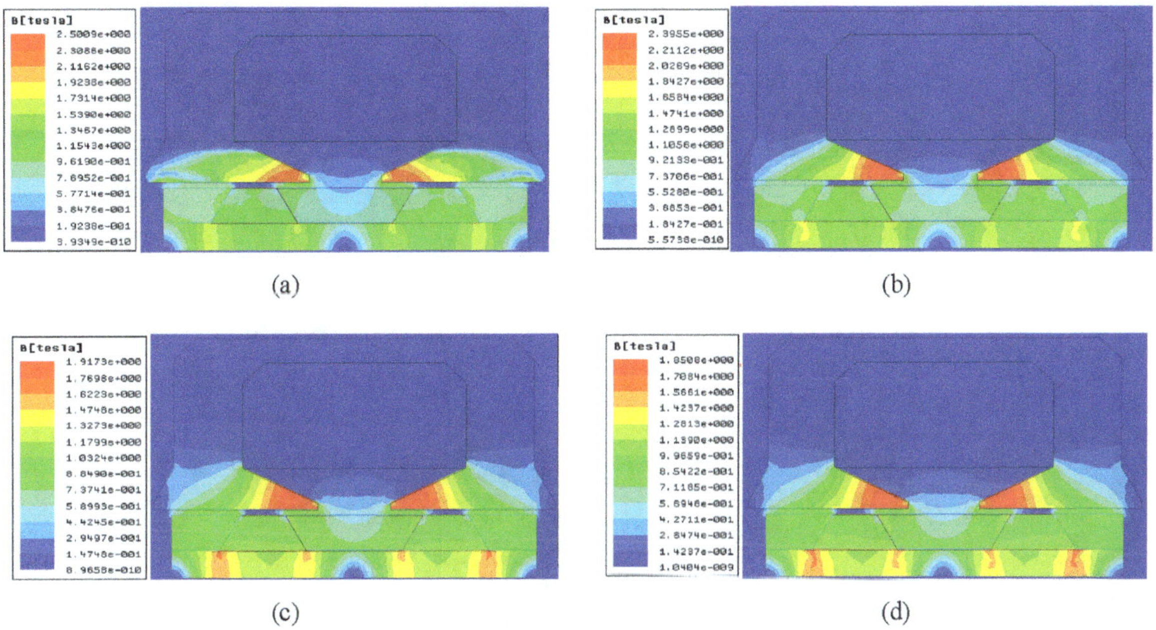

zero displacements between the stator and translator, this serves to illustrate that how the magnetic flux density flows in the proposed machine.

2-D FEA-calculated flux lines distributions in the TPMLM are presented in Figure 5. The results presented at the no-load and initial position of the translator ($z_d = 0.0$ mm). It can be observed that at zero displacement the magnetic flux distributions is uniformed and symmetrical with respect to the axial center. Hence, the flux-linkage of the winding is zero.

Figure 6 and Figure 7, respectively, show the FEA calculated open-circuit radial and the axial component flux density in the air gap of the proposed machine. The axial air gap flux density waveforms due only to the magnets of the TPMLM is shown in Figure 7. As can be been observed from Figure 7, the peak value of the flux density is due to the flux focus effect in the machine. Therefore, there are two kinds of peak values in the axial air gap flux density waveforms of the machine, where one is called the local peak value and the other is called the absolute peak value, that is due to the intersections between the translator and the stator teeth. Thus, the local peak value corresponds to the true peak value of the PM flux in the machine and it is more important. As it was proven from the plot representation, the non-oriented electrical steel and Permalloy are exhibiting a higher magnetic flux density distribution in the air gap of the proposed machine, consequently higher loss and resulted in low efficiency.

Figure 8 shows the moving speed versus time of the translator. Whilst the magnetic flux-linkage that is resulting in the winding is shown in Figure 9. From the result, it can be observed that maximum flux-linkage has been obtained for Somaloy 700.

Figure 10 compares the EMF waveforms of the four models at a constant armature velocity of 1 m/s. The PMs in the translator generates a magnetic field excitation. Whereas, the second magnetic field is generated from the excitation can be sinusoidal or non-sinusoidal including voltage and current applied to the winding or an external circuit attached to the winding of the machine.

Figure 5. Magnetic flux lines illustrated by the FEA for TPMLM at open-circuit (a) Non-oriented electrical steel (b) Permalloy (c) Accucore (d) Somaloy 700.

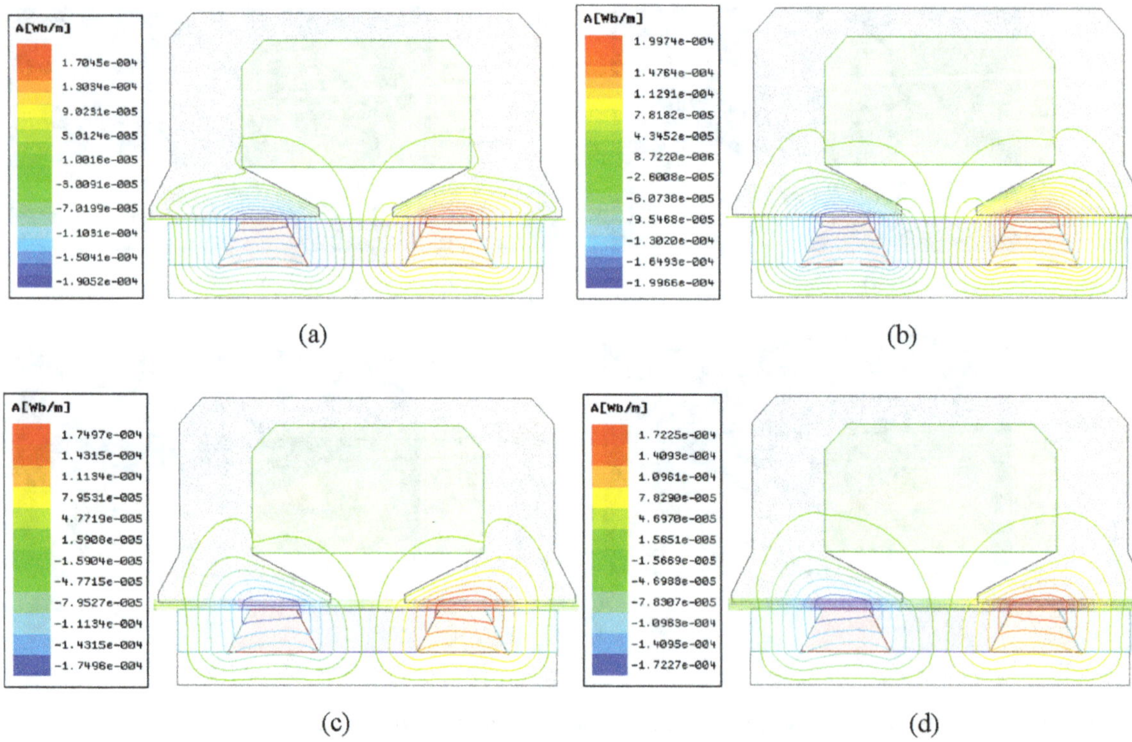

Figure 6. Effect of the ferromagnetic materials on the air gap radial flux density distribution

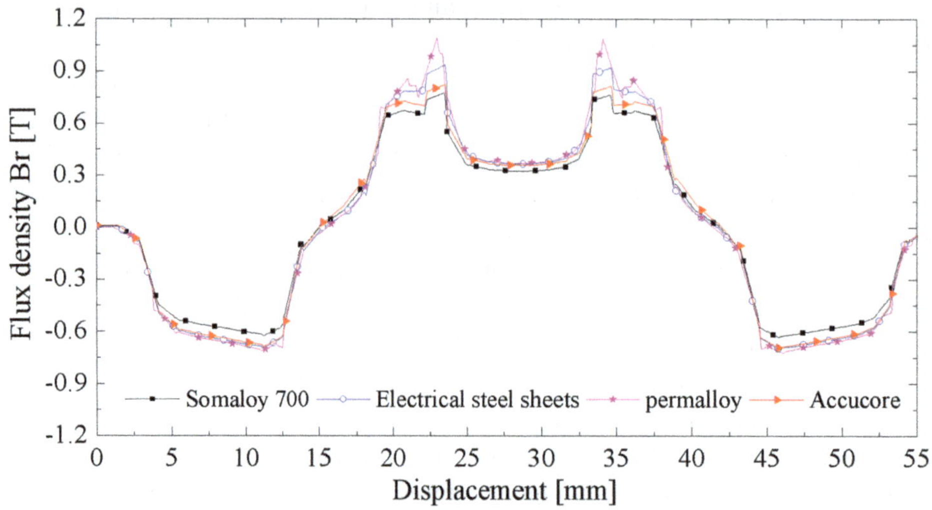

Figure 7. Effect of the ferromagnetic materials on the air gap axial flux density distribution

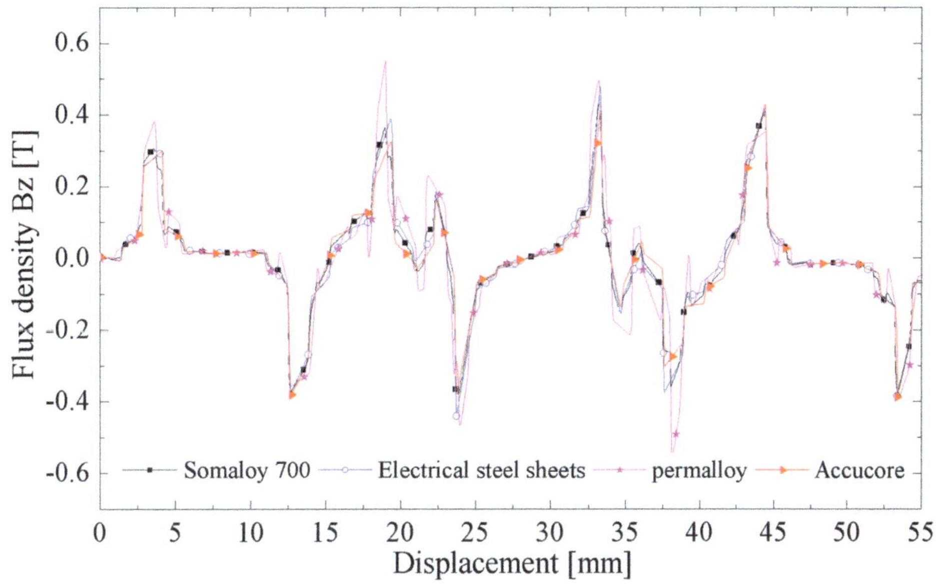

Figure 8. Translator moving speed versus time

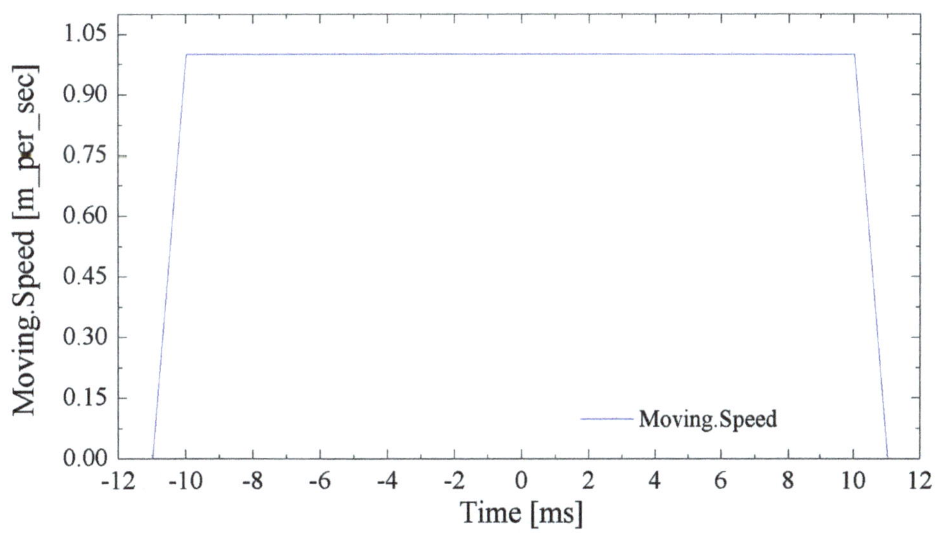

The force in the machine produces as a result of the interaction of the fields from the PMs and fields from the winding, or from the interaction of the PMs and the stator ferromagnetic material. The first force is called the electromagnetic or magnetic force, and the second one is called a cogging force (Jiabin Wang, Ibrahim, & Howe, 2010). The magnetic force value, and thereby the sense of the no-load movement, depending on the relative position of the stator and mover magnets as well as the current sense in the stator coils. The calculated field distribution is used to estimate the electromagnetic force, which acts on the translator (Tomczuk, Schroder, & Waindok, 2007).

Figure 9. Effect of the ferromagnetic materials on the winding flux-linkages

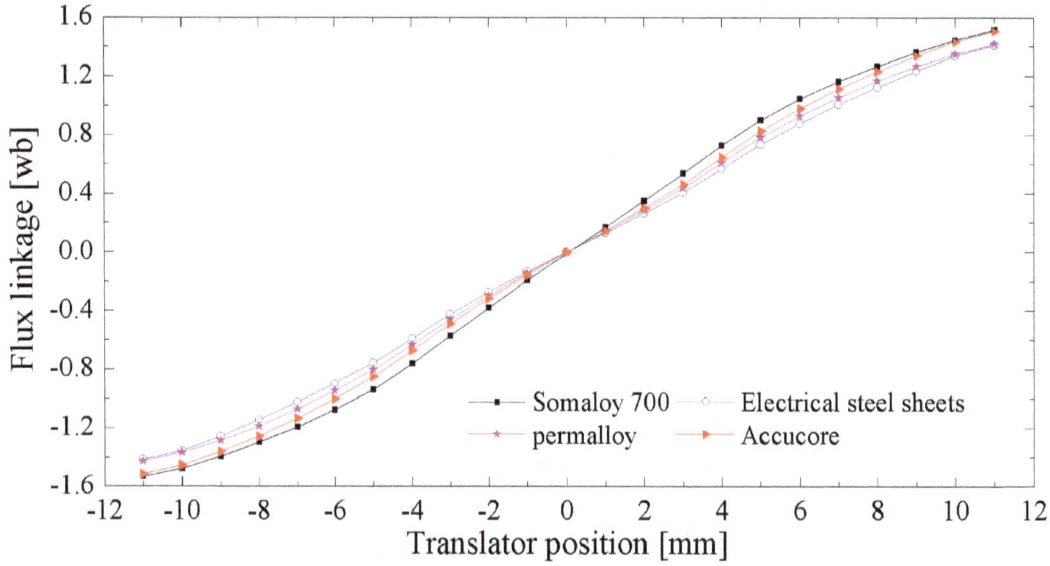

Figure 10. Comparison of EMFs using four different ferromagnetic materials

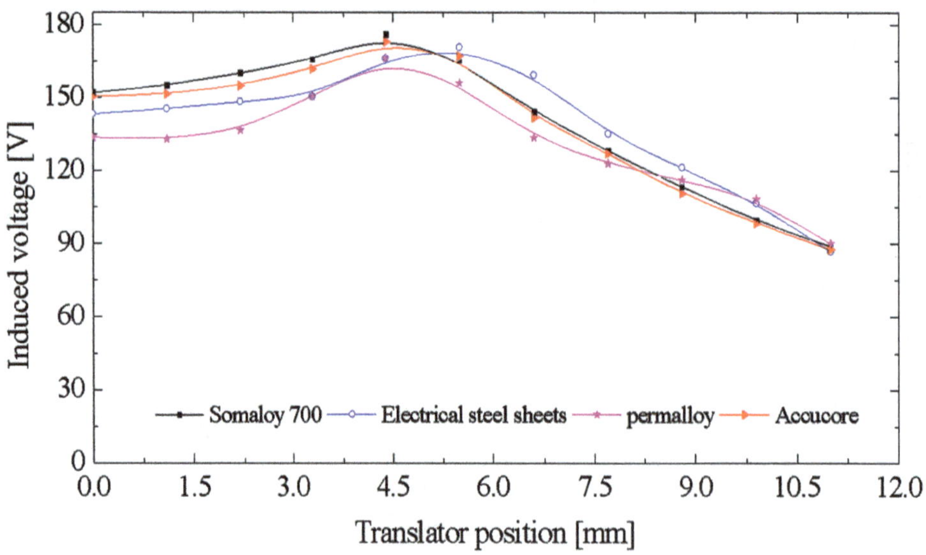

Figure 11 depicts the thrust force versus time characteristics of the four models which are dealt according to the stator core material at rated armature current. The stator winding inductance represents one of the most important key parameters for the motor modeling and performance. Figure 12 compares the FE calculated winding inductances of the four models.

Figure 11. The magnetic forces of TPMLM using four different stator core materials

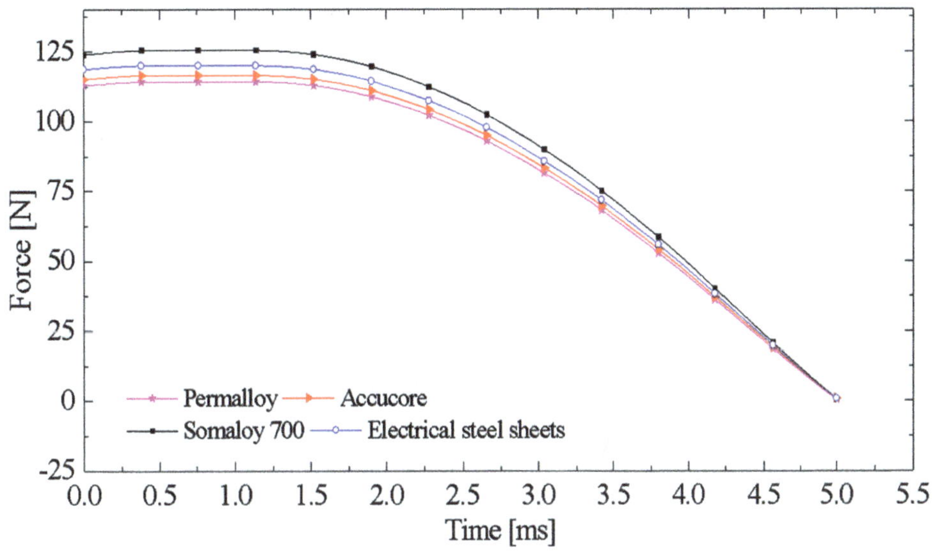

Figure 12. Comparison of winding inductances

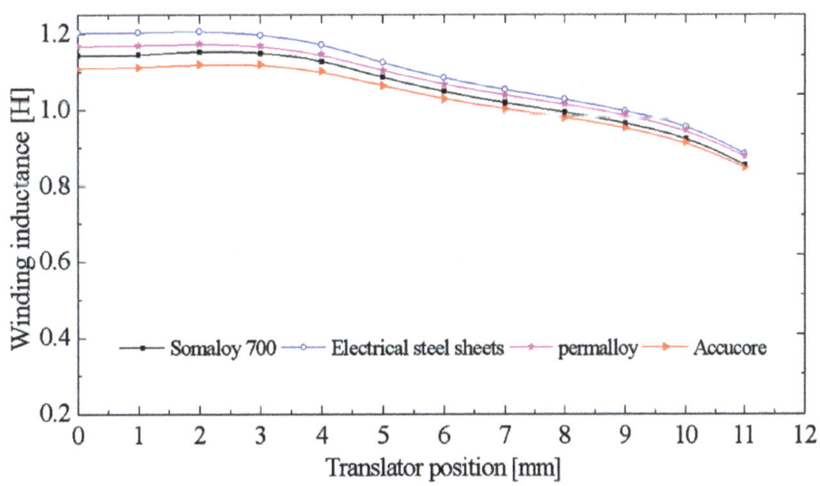

CONCLUSION

The influence of four ferromagnetic materials, such as non-oriented electrical steel sheets, Permalloy (Ni-Fe-Mn), Accucore and Somaloy 700, on the design and performance of single-phase tubular permanent-magnet linear generator (TPMLG) has been analyzed, and their relative merits have been evaluated. It has been shown that the designs which employ non-oriented electrical steel and Permalloy have the highest magnetic flux density at the air gap, despite its stator having poorer space utilization. The TPMLG with SMC materials offers potential features of lower cost manufacture, easy to be formed to any complicated shape and has a high force density and low eddy-current loss. The time-stepping

Finite-Element Analysis (FEA) has been successfully employed in this study. As the machine with SMC materials has low loss, the high efficiency is expected, which conducive to an energy-efficient generator. Computer simulations for the magnetic field distributions, thrust force, winding inductance and EMF of the four models have been established. The SMC materials have lower saturation induction than the electrical steel sheet. Besides, decreasing the hysteresis loss is considered as an important objective in the SMC materials development.

ACKNOWLEDGMENT

The authors would like to thank Universiti Teknologi PETRONAS for providing facilities for the work, and Ministry of Education, Malaysia, for the financial support.

REFERENCES

Abdalla, I. I., Ibrahim, T., & Mohd Nor, N. (2014). Analysis of a Tubular Linear Permanent Magnet Motor for Reciprocating Compressor Applications. *Applied Mechanics and Materials*, *448*, 2114–2119.

Al-Otaibi, Z., & Jack, A. (2008). *Utilising SMC in single phase permanent magnet linear motors for compressor applications*. Academic Press.

Chen, Y.-M., Fan, S.-Y., & Lu, W.-S. (2008). Performance analysis of linear permanent-magnet motors with finite-element analysis. *IEEE Transactions on Magnetics*, *44*(3), 377–385. doi:10.1109/TMAG.2008.915618

Feng, H., Song, Y., Zuo, Z., Shang, J., Wang, Y., & Roskilly, A. P. (2015). Stable Operation and Electricity Generating Characteristics of a Single-Cylinder Free Piston Engine Linear Generator: Simulation and Experiments. *Energies*, *8*(2), 765–785. doi:10.3390/en8020765

Gieras, J. F. (2008). *Advancements in electric machines*. Springer.

Gieras, J. F., Piech, Z. J., & Tomczuk, B. (2011). *Linear synchronous motors: transportation and automation systems*. CRC Press.

Gieras, J. F., Wang, R.-J., & Kamper, M. J. (2008). *Axial flux permanent magnet brushless machines* (Vol. 1). Springer. doi:10.1007/978-1-4020-8227-6

Ibrahim, T. (2009). *Short-stroke, single-phase tubular permanent magnet motors for refrigeration applications*. The University of Sheffield.

Jang, S.-M., Choi, J.-Y., Cho, H.-W., & Lee, S.-H. (2005). Dynamic characteristic analysis and experiments of moving-magnet linear actuator with cylindrical Halbach array. *IEEE Transactions on Magnetics*, *41*(10), 3814–3816. doi:10.1109/TMAG.2005.854931

Lee, J. H., Kim, H. Y., Jun, M. J., & Lee, S. C. (2011). *Optimum shape design of single-sided linear induction motors using response surface methodology and finite element method.* Paper presented at the Electrical Machines and Systems (ICEMS), 2011 International Conference on. 10.1109/ICEMS.2011.6073769

Lee, M. G., Choi, Y.-M., Lee, S. Q., Lim, D., & Gweon, D.-G. (2006). Design of high precision linear stage with double-sided multi-segmented trapezoidal magnet array and its compensations for force ripples. *Mechatronics*, *16*(6), 331–340. doi:10.1016/j.mechatronics.2006.01.006

Li, W., Chau, K., & Jiang, J. (2011). Application of linear magnetic gears for pseudo-direct-drive oceanic wave energy harvesting. *IEEE Transactions on Magnetics*, *47*(10), 2624–2627. doi:10.1109/TMAG.2011.2146233

Meessen, K. J., Gysen, B., Paulides, J., & Lomonova, E. A. (2008). Halbach permanent magnet shape selection for slotless tubular actuators. *IEEE Transactions on Magnetics*, *44*(11), 4305–4308. doi:10.1109/TMAG.2008.2001536

Niu, S., Ho, S., & Fu, W. (2011). Performance Analysis of a Novel Magnetic-Geared Tubular Linear Permanent Magnet Machine. *IEEE Transactions on Magnetics*, *47*(10), 3598–3601. doi:10.1109/TMAG.2011.2148167

Okonkwo, R. (2006). Design and performance of permanent-magnet DC linear motors. *IEEE Transactions on Magnetics*, *42*(9), 2179–2183. doi:10.1109/TMAG.2006.880397

Polinder, H., Mueller, M., Scuotto, M., & Goden de Sousa Prado, M. (2007). Linear generator systems for wave energy conversion. *Proceedings of the 7th European Wave and Tidal Energy Conference.*

Shahgholian, G., & Shafaghi, P. (2010). *State space modeling and eigenvalue analysis of the permanent magnet DC motor drive system.* Paper presented at the 2010 International Conference on Electronic Computer Technology (ICECT). 10.1109/ICECTECH.2010.5479987

Si, J., Feng, H., Su, P., & Zhang, L. (2014). Design and analysis of tubular permanent magnet linear wave generator. *The Scientific World Journal.* PMID:25050388

Szabo, L., Oprea, C., Viorel, I.-A., & Biró, K. Á. (2007). *Novel permanent magnet tubular linear generator for wave energy converters.* Paper presented at the Electric Machines & Drives Conference. 10.1109/IEMDC.2007.382809

Tavana, N. R., Shoulaie, A., & Dinavahi, V. (2012). Analytical Modeling and Design Optimization of Linear Synchronous Motor With Stair-Step-Shaped Magnetic Poles for Electromagnetic Launch Applications. *Plasma Science. IEEE Transactions on*, *40*(2), 519–527.

Tomczuk, B., Schroder, G., & Waindok, A. (2007). Finite-element analysis of the magnetic field and electromechanical parameters calculation for a slotted permanent-magnet tubular linear motor. *IEEE Transactions on Magnetics*, *43*(7), 3229–3236. doi:10.1109/TMAG.2007.894216

Wang, J., Howe, D., & Lin, Z. (2008). Analysis of a short-stroke, single-phase, quasi-Halbach magnetised tubular permanent magnet motor for linear compressor applications. *IET Electric Power Applications*, *2*(3), 193–200. doi:10.1049/iet-epa:20070281

Wang, J., Ibrahim, T., & Howe, D. (2010). Prediction and measurement of iron loss in a short-stroke, single-phase, tubular permanent magnet machine. *IEEE Transactions on Magnetics*, *46*(6), 1315–1318. doi:10.1109/TMAG.2010.2042685

Zheng, Z.-Q., Huang, P., Gao, D.-X., & Chang, Z.-Y. (2015). Analysis of Electromagnetic Force of the Linear Generator in Point Absorber Wave Energy Converters. *Journal of Marine Science and Technology*, *23*(4), 475–480.

Chapter 8
Optimization of Process Parameters for Electro-Chemical Machining of EN19:
Using Particle Swamp Optimization

Divya Zindani
National Institute of Technology Silchar, India

Nadeem Faisal
Birla Institute of Technology, India

Kaushik Kumar
Birla Institute of Technology, India

ABSTRACT

Electrochemical machining (ECM) is a non-conventional machining process that is used for machining of hard-to-machine materials. The ECM process is widely used for the machining of metal matrix composites. However, it is very essential to select optimum values of input process parameters to maximize the machining performance. However, the optimization of the output process parameters and hence the machining performance is a difficult task. In this chapter an attempt has been made to carry out single and multiple optimization of the material removal rate (MRR) and the surface roughness (SR) for the ECM process of EN19 using the particle swarm optimization (PSO) technique. The input parameter considered for the optimization are electrolyte concentration (%), voltage (V), feed rate (mm/min), and inter-electrode gap (mm). The optimum value of MRR and SR as found using the PSO algorithm are 0.1847 cm3/min and 25.0612, respectively.

DOI: 10.4018/978-1-5225-5445-5.ch008

INTRODUCTION

The traditional processes such as grinding, milling, turning, drilling etc., remove material by mechanical abrasion, micro chipping or chip formation. However due to the following reasons the traditional processes are not economical and even possible (Kalpakjain, 1984):

- Hardness is very high i.e., above 400 HB.
- Material is too brittle
- The machining forces are too high for the delicate and slender workpiece specimen.
- The complexity of the part to be machined.
- The residual stresses in the machined component which is not at all acceptable.

The above disadvantages have led to the development of other material removal mechanisms such as chemical, thermal, electrochemical and other hybrid mechanisms. These material removal mechanisms have therefore resulted in machining processes referred to as non-traditional machining processes. Owing to the advantages offered by the non-traditional machining processes, these are available for a wide range of industrial applications. The source of energy used differ from process to process and therefore can be categorised accordingly: thermal and electrothermal processes such as laser beam machining, ion beam machining, electric discharge machining etc., chemical and electrochemical processes such as electrochemical machining, electrochemical honing etc., mechanical processes for instance ultrasonic machining water jet machining etc. and hybrid processes as for instance abrasive electrical discharge machining etc.

Amongst the different processes, electrochemical machining (ECM) process is one of the most promising methods and is the prime focus in the present study. The electrolysis process is the working principle of the ECM process (Rajurkar et al., 1999)the laws of which were formulated by Faraday in 1983. In the ECM process tool and the workpiece are respectively the cathode and the anode. A high current density ranging 10-200 A/cm2 results on application of a constant potential difference across the two electrodes. A suitable electrolytic solution is used such that the shape of the cathodic tool remains unchanged. The electrolytic solution is pumped at rates ranging 3-60 m/s and serves to remove the unwanted machining waste and minimize the effects arising due to electrical heating and generation of gas at the electrodes. The gap width along the electrode length reaches a steady value as the cathodic tool approaches the workpiece. Under these conditions the shape of the cathodic tool is produced as a mirror image on the anodic workpiece.

The ECM process possess a number of advantages such as the capability to handle a large number of materials which are not limited by their mechanical properties but by their electrochemical properties. The difficult to machine materials and high strength alloys are machined by the ECM process with higher material removal rates. Further, the ECM process can easily machine and shape the fragile parts. The advantages such as the machining of the 3D curved surfaces free from striation marks, burrs and stress, no tool wear makes the ECM process a widely accepted machining processes for a number of applications. A wide range of sophisticated parts such as rifle bores, turbine blades, micro components etc., are being manufactured using the ECM process.

However the ECM process suffers from a number of limitations such as higher power consumption, high cost on initial investment and large floor space. The problem is further aggravated with the issues of toxicity and the corrosion from the etchants. Further the changes taking place at the inter electrode gap are difficult to predict because of the complexity of the process involved. The properties of the electrolytic solution varies considerably because of the continual heat and gas bubbles. The analysis becomes further complicated as a result of the variation of hydrodynamic parameters such as pressure during the machining process. Therefore a careful selection of the different parameters and careful planning is recommended before commencing the machining process. The selection of suitable machining parameters is critical to achieve optimum machining results. Further, the production cost as well as the greenhouse gases increases with the increasing power consumption. Therefore, the optimization of machining parameters can lead to efficient machining operation and hence saving of energy resulting in green manufacturing.

In order to take note of the above limitations, number of reports has been published on optimization of the ECM process. A two dimensional inter electrode gap model was proposed by Bhattacharyya et al. (1973) with the objective of maximizing the material removal rate. The design variables considered were velocity of the electrolytic flow and the tool feed rate. Choking, passivity and temperature were considered as the constraints while optimizing the material removal rate. The authors had considered only single objective optimization problem and the optimization problem was solved using the less accurate graphical technique. Simplified assumptions such as conductivity of the electrolyte as a function of void fraction and constant electrolytic pressure were considered.

A cost model considering the different costs involved was proposed for the ECM process by El-Dardery (1982). The optimum values of the different variables considered in the cost equation i.e., feed rate, voltage and electrolytic flow rate, were obtained by partially differentiating the cost equation with respect to the variables considered. However, the values of the variables were not of practical importance since no constraints were considered in solving the problem.

A multi-objective optimization technique was investigated for the ECM process by Acharya et al. (1986). The material removal rate and the tool life were maximized while the dimensional inaccuracies were minimized.The decision variables considered were velocity of the electrolytic flow rate, applied voltage and tool feed rate. The choking constraints, passivity constraints and the temperature constraints were considered as the constraints for the multi-objective optimization problem. Goal programming was used for solving the problem. However, the bounds for the feed rate and differences in the inter electrode gap wasn't considered for the model.

Choobineh and Jain (1993) overcame the limitations of the model proposed by Acharya et al. (1986). The authors of the work considered maximization of dimensional accuracy and material removal rate and adopted the vertex method to solve the multi-objective problem.

The optimization model proposed by Acharya et al. (1986) was modified by Jain and Jain (2007) by expanding the bound ranges for tool feed rate and velocity of electrolytic flow. The objective was to minimize the dimensional inaccuracy. The constraints and the objective function wasn't linearized. Genetic algorithm was used to address the optimization model.However, the authors of the work neglected the passivity constraint. Further, the single objective function was only considered by the authors. There are number of disadvantages associated with the genetic algorithm such as low speed of convergence, difficulty in selection of parameters such as crossover rate, rate of mutation and the population size.

The grey relational analysis (GRA) was used by Asokan et al (2008)to simultaneously optimize material removal rate and surface response. Optimization of material removal rate, cylindricity error, overcut and surface response was considered by Chakradhar and Gopal (2011). The authors of the work adopted the GRA method to solve the optimization problem with feed rate, concentration of electrolyte and applied voltage as input parameters.

Grey-Taguchi approach was adopted by Das et al. (2014) for optimization of material removal rate and surface roughness of EN31machined using the ECM process. The process parameters considered for the optimization were voltage, electrolytic concentration, voltage and interelectrode gap. The optimization of material removal rate for various process parameters for the ECM machining of EN41 and EN19 was carried out by Vikas et al. (2014) using the Taguchi method.

RESEARCH GAP

It can be observed from the above literature survey that traditional methods such as graphical solution technique, GRA, linear programming technique etc. have mostly been used for the parametric optimization of the ECM process. However the traditional methods are restricted over a broad range of spectrum of problem domains. The complexity of the optimization problem further limits the application of the traditional optimization methods. The use of non-traditional methods such as GRA provides a near optimal solution. Therefore to overcome the drawbacks of the traditional optimization techniques and non-traditional methods such as GRA, evolutionary optimization techniques are being used by the scientific community. The evolutionary optimization algorithm is based on biological genetics. These optimization techniques are more robust in comparison to the traditional techniques because instead of making use of functional derivatives, fitness information is used by the evolutionary optimization techniques. Therefore efforts are being made to use such optimization techniques to achieve a more accurate solution.

Therefore in this chapter, particle swarm optimization (PSO) was adopted for the optimization of surface roughness and material removal rate for the electro-Chemical Machining of EN19. Electrolyte concentration (%), Voltage (V), Feed rate (mm/min) and Inter-electrode gap (mm) were considered as the process parameters. Further, the multi-objective optimization was also performed for an objective function comprising of both the surface roughness and the material removal rate. The optimal process parameters would be identified for best output conditions.

EXPERIMENTATION FOR ECM PROCESS

Machine Set Up for ECM Experimentation

METATECH electrochemical set up was used for conducting the various experiments. The experimental set up is shown in the Figure 1. There are three important machining components for ECM machining set up: the control panel, machining chamber and the circulation system for the electrolytic flow. The anodic workpiece is fixed on a fixture inside the machining chamber whereas the cathodic tool is attached to the servo controlled main screw. A current sensing circuit is provided to take care of the shorty circuit. The current sensing circuit sends a signal to the stepper motor in case the machining current exceeds the acceptable limit. The controller circuit of the stepper motor immediately reverses the downward motion of the tool and thereby shutting the ECM machine. The electrolytic tank contains the electrolyte (KCl) which is pumped to the machining chamber.

Figure 1. ECM process setup
(Tiwari et al., 2016)

Material for ECM Experimentation

The anodic workpiece used for carrying out the ECM experimentation was EN 19 steel block with cylindrical cross-section. The EN19 steel has good ductility and shock absorbing capacity. Further the wear resistance of the tool is high and is known for its high strength property. The tool had the diameter of 25 mm and height of 15 mm. Figure 2 shows the EN19 steel tool used for carrying out the ECM experimentation. The EN19 steel tool lands itself for a wide range of applications such as drive shaft, axels, crankshaft, stud, high tensile bolt, rifle barrels and propeller shaft joints.

A copper tool with diameter of 17 mm was used for the ECM experimentation. The tool used is shown in the Figure 2. The electrolytic solution of KCL was fed to the machining zone through the centrally drilled hole in the tool. KCl was used because it has no passivation effect on the workpiece surface.

Figure 2. EN19 tool for ECM experimentation
(Tiwari et al., 2016)

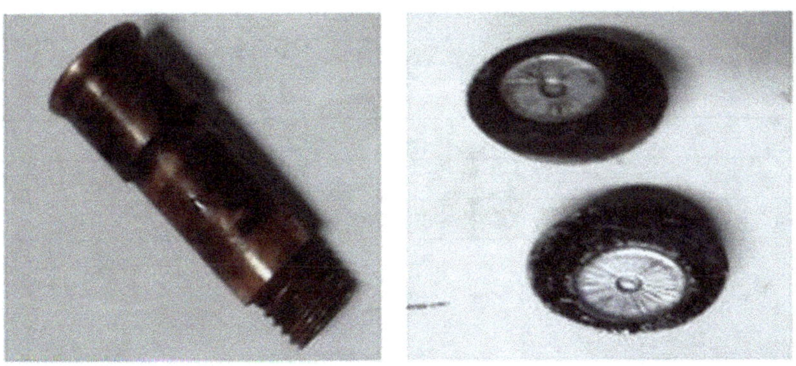

Inputs Parameters

The four input parameters considered for the ECM experimentation were: Voltage (V), Electrolytic concentration (%), interelectrode gap (mm) and feed rate (mm/min).The above parameters were selected on the basis of extensive literature surveys that have reported these parameters to be most influential for the material removal rate (MRR) and the surface roughness (SR). The input parameters with their levels have been tabulated in the table 1.

Output Parameters

The two output parameters considered were: MRR and SR.

Material Removal Rate (MRR)

Equation 1 can be used for the determination of the MRR (cubic centimetre/min) in the ECM process:

$$MRR = C \times I \times H \tag{1}$$

where, C is the specific removal rate (cm3/amp-min); I is the current (amp) and h is the current efficiency.

Therefore, MRR depends upon the amount of current passed and the elapsed machining time. Besides these critical factors the MRR is also dependent on the type of electrolytic solution used, electrolytic flow rate etc.

Surface Roughness (SRR)

The deviation of a surface from its ideal level is defined in terms of surface roughness. The surface roughness is defined according to ISO 4287:1997 international standard. The term surface roughness is often referred to as roughness and determines the surface texture. The surface roughness is calculated by the deviations I i.e., deviation of surface from a theoretical centre line. If the deviations are large, the surface roughness is high. Whereas the surface is considered to be smooth for small deviations. This is known as arithmetic mean surface roughness and is depicted in Figure 3.

Table 1. Input parameters with their levels

Process Parameters (Unit)	Symbol	Level 1	Level 2	Level 3
Electrolytic Concentration (%)	A	10	15	20
Voltage (V)	B	8	10	12
Feed Rate (mm/min)	C	0.1	0.21	0.32
Inter-electrode gap (mm)	D	0.2	0.25	0.3

Figure 3. Arithmetic mean surface roughness

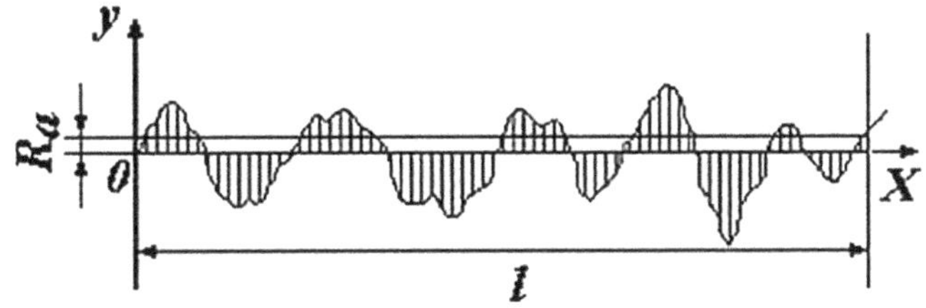

Design of Experiments

One of the systematic processes to relate the input process parameters with that of the output process parameters is the design of experiments (DOE). The DOE aids in optimizing the output by processing the input process parameters. There are certain input factors that can be altered with respect to the time and are termed as the controllable factors, whereas there are uncontrollable factors that cannot be modified throughout an experiment. The DOE speeds up the designing process and minimizes the manufacturing cost. The DOE can be divided into five main phases: Conceptual, Discovery, Breakthrough, Optimization and Validation.

Terms commonly used in DOE are:

- **Response:** A dependent variable which is the outcome in an experiment. There may be one or more than one response variables under investigation.
- **Factors:** Variable contributing and affecting the response variable.
- **Levels:** The conditions of the factors under study.
- **Orthogonality:** It is a desirable property that can be thought of as factors independence. In an orthogonal experiment the factors are independently varied. The summary of the results obtained by variation of the factors can be depicted graphically. However, orthogonality is no longer a requirement because of the availability of powerful software and computers.
- **Blocks/Blocking:** there may be a situation in experimentation where there may be large number of runs that cannot be completed in homogeneous conditions. In such situations the inclusion of nuisance factors into investigation is inevitable. The nuisance factors are of not primary interest. Blocking can be used to account for the nuisance factor by separating the experimental runs into separate blocks carrying out experimental runs in homogeneous conditions.
- **Randomization:** it is the process of recording the observations in a random order i.e., assignment of the different level of the input process parameters to the experimental output in random order. The process of randomization involves the start and stop technique for each observation performed.
- **Replica/Repetition:** the results obtained with the repetition of the experimental runs incepting from the initial step. However, the signal-to-noise ratio increases arising from the nuisance variables inevitable to be avoided. Replication is therefore the method for increasing the level of precision of the experiment.

- **Treatments:** the condition that is associated with a specific level in a specific experimental run.
- **Factorial Experimentation:** the effect on the output process parameters of the individual factor or combination of factors is estimated using the method of factorial experimentation. The data is collected at the vertices of a cube in k-dimensions and if the data is collected from each of the vertices of a cube then it is known as full factorial.

Some of the techniques for design of experiments are: Taguchi method, Sieve DOE, Split-Plot DOE, and group method of data handling (GMDH). In the present work Taguchi method has been used for the formulation of DOE. According to Taguchi, the total degrees of freedom of the selected orthogonal array (OA) must be greater than the DOF for the standard OA. In the present the number of factors are four i.e., Voltage (V), Electrolytic concentration (%), interelectrode gap (mm) and feed rate (mm/min). The total number of interactions and the DOF of four factors is 20 (4x 2 + 3 x 4). Since the standard OA for four factors with three levels are L9 and L27 having DOF of 8 and 26 respectively, L27 orthogonal array was selected for the problem under consideration. The DOE with L27 orthogonal array has been tabulated in the Table 2. The results obtained for MRR and SR with the design of experiments have been tabulated in the Table 3.

Regression Model

The relationship between the input parameters and the output parameters can be obtained using the statistical tool of regression analysis. Minitab 16 was used for obtaining the relationship between MRR and input variables and also for obtaining the relationship between SR and the input variables. Equations 1 and 2 are the relations for MRR and SR respectively.

Table 2. DOE using L27 Orthogonal array

Exp. No.	A	B	C	D	Exp. No.	A	B	C	D
1.	1	1	1	1	15.	2	2	3	2
2.	1	1	2	2	16.	2	3	1	1
3.	1	1	3	3	17.	2	3	2	2
4.	1	2	1	2	18.	2	3	3	3
5.	1	2	2	3	19.	3	1	1	3
6.	1	2	3	1	20.	3	1	2	1
7.	1	3	1	3	21.	3	1	3	2
8.	1	3	2	1	22.	3	2	1	1
9.	1	3	3	2	23.	3	2	2	2
10.	2	1	1	2	24.	3	2	3	3
11.	2	1	2	3	25.	3	3	1	2
12.	2	1	3	1	26.	3	3	2	3
13.	2	2	1	3	27.	3	3	3	1
14	2	2	2	1					

Table 3. DOE results

Exp No.	MRR	SR	Exp No.	MRR	SR
1.	0.1033	13.5576	15.	0.1169	16.5729
2.	0.0917	13.6241	16.	0.1506	19.2462
3.	0.0801	13.6906	17.	0.1390	19.3127
4.	0.1138	16.3639	18.	0.1274	19.3792
5.	0.1022	16.4304	19.	0.0809	13.8626
6.	0.1281	16.4204	20.	0.1067	13.8526
7.	0.1243	19.1702	21.	0.0951	13.9191
8.	0.1502	19.1602	22.	0.1289	16.9524
9.	0.1386	19.2267	23.	0.1172	16.6589
10.	0.0921	13.7101	24.	0.1056	16.7254
11.	0.0805	13.7766	25.	0.1394	19.3987
12.	0.1064	13.7666	26.	0.1277	19.4652
13.	0.1026	16.5164	27.	0.1536	19.4552
14	0.1285	16.5064			

$$MRR = 0.000253\text{A} + 0.011500\text{B} + 0.008014\text{C} - 0.250000\text{D} + 0.058007 \quad (1)$$

$$SR = 0.0254\text{A} + 1.3904\text{B} + 0.3726\text{C} + 0.5101\text{D} + 2.0411 \quad (2)$$

OPTIMIZATION OF MATERIAL REMOVAL RATE USING PSO

Particle Swarm Optimization Technique

Kennedy and Eberhart (1995) developed the particle swarm optimization (PSO) technique which is one of the evolutionary optimization techniques. The common evolutionary computational attributes exhibited by the PSO technique simulates graphically the choreography of the bird flock. The PSO technique incepts with the initialization with population of random solution and then updating the generations to achieve the optimal solution. The potential solutions which are referred to as particles are then flown through the space of the problem which follows the current optimum solution. The particles in the problem space keeps track of its position. The coordinate of each particle is associated with the best solution achieved so far. The best position achieved by the particle is referred to as 'pBest'. The best position achieved by any particle in the problem space is known as the 'gBest'. Thus the PSO technique is based on accelerating the particles towards their 'pBest' and 'gBest' locations. The random terms weights the acceleration and therefore separate set of random numbers are generated for acceleration towards 'gBest' and 'pBest'. The updated velocity i.e., the new velocity of the particle is obtained using the Equation 3. Equation 4 on the other hand gives the updated position of the particle in the problem space.

$$V_{i+1} = wV_i + c_1 r_1 (pBest_i - X_i) + c_2 r_2 (gBest_i - X_i) \tag{3}$$

$$X_{i+1} = X_i + V_{i+1} \tag{4}$$

The random numbers r1 and r2 are generated randomly and lies in the range [0, 1]. c1and c2 are the acceleration constants that weights the acceleration terms. The confidence of the particle in itself is represented by c1 whereas c2 represents the confidence a particle has in a swarm. c1 and c2 are referred to as cognitive and social parameters respectively. The values of these two parameters determines the change in amount of tension in the system. The low values of the parameters results in the particles roaming far away from the target regions whereas a higher value results in an abrupt movement towards the target solution (Dong et al., 2005). The exploration abilities of the swarm particles are controlled by the inertia weight w and is therefore very critical in determining the convergence behaviour of PSO. The lower values of w restricts the velocity updates to nearby region in the search space, whereas higher values results in velocity updates for a wider space in the problem space. Berg and Engelbrecht (2006) have investigated the effect on convergence if benchmark functions of w, c1 and c2.

Heuristics have been developed to obtain the values of c1 c2 and w but these are applicable for single-objective optimization problems only. Determination of the parameters and the inertia weights for multi-objective optimization problems is relatively difficult and therefore a time variant of PSO has been described by Tripathi et al. (2007). In this variant of PSO, the parameters and the inertia weights are adaptive and changes with the iterations. The search space can be explored more efficiently with the adaptive nature of the time variant PSO technique. The premature convergence was taken care off by the mutation operator.

The PSO technique doesn't require complex encoding and decoding process as required by genetic algorithm. The real number represents a particle which update their internal velocity in search of the best solution. In PSO technique the particles tend to converge towards the best solution by iteratively looking for the best position which comprises of the evolution phase. It is very important to handle the non-linear equation constraints and evaluate the infeasible particles. This is mainly due to the fact that the particles generated during the process may violate the constraints of the system and thereby resulting in infeasible particles. Various approaches such as repair of infeasible particles, rejection of the infeasible particles, penalty function methods etc. can be adopted to cope with the evolutionary problems with constraints. The recent developments however suggest the use of penalty method for addressing the same (Dong et al., 2005).

Optimization of Material Removal Rate and Surface Roughness Using PSO

The PSO technique incepts with the setting of PSO parameters which determines the performance of the PSO algorithm. The values of parameters were taken from the work of Hu and Eberhart (2002) Therefore, w = (0.5+ (rand/2), c1 =c2= 1.49445, number of iterations =100 and population size = 27 particles.

As a second step to PSO technique, initialization of the random position and velocity vectors is followed. The fitness values are derived from the following function given by Equation 5.

$$MinF(x) = -w_1 f_1 + w_2 f_2 \tag{5}$$

f_1 and f_2 are normalized values of MRR and SR. The values of inertia weights considered are 0.7 for MRR and 0.3 for SR. a higher value of inertia weight for the MRR signifies that a MRR is given relatively higher significance than SR. The initial values have been tabulated in the Table 4.

Finding pBest and gBest is a third step to the PSO technique. In this chapter 3 particles have been considered for each parameter. The lowest fitness value is selected as the gBest and for the first iteration the pBest will be same as gBest. The change in position is then obtained using Equations 3 and 4 for each of the parameters. Table 5 shows the new position of particles after first iteration and the change in position of each particle has been tabulated in the Table 6.

As can be observed from Table 5, the particle 1 has the lowest fitness value of 7.7726. Therefore the pBest for each parameter will be the values corresponding to 7.7726. However, it is higher than the initial gBest value which was 5.7618. Hence the gBest achieve is not updated. The process is continued for 100 iterations and the final optimised value of gBest gives the final solution.

Similar process was adopted for the single objective optimization of MRR and SR.

Table 4. Initial population and fitness value

Particle	A	B	C	D	MRR	SR	Fitness Value
1	10	8	0.1	0.2	0.1033	13.5576	3.9949
2	15	10	0.21	0.25	0.1160	16.5319	4.8784
3	20	12	0.32	0.3	0.1286	19.5062	5.7618

Table 5. New position of particles after the first iteration

Particle	A	B	C	D	MRR	SR	Fitness Value
1	12.3728	16.9602	0.3374	0.4118	0.1559	26.2726	7.7726
2	21.8046	18.6454	0.1986	0.2630	0.2138	28.7277	8.4686
3	24.5809	17.3213	0.6141	0.6934	0.095	27.2726	8.1330

Table 6. Changed position of particles after the first iteration

Particle	A	B	C	D
1	2.3728	8.9602	0.2374	0.2118
2	6.8046	8.6454	-0.0114	0.0130
3	4.5809	5.3213	0.2941	0.3934

RESULTS AND DISCUSSION

The results obtained from the PSO algorithm have been tabulated in the Table 7. As can be observed from Table 7 that the maximum value of MRR is corresponding to experimental number 13 and is 0.1847 cubic centimetre/min. The values corresponding to the maximum MRR for the various input parameters are: A=15; B=10; C=0.1; D=0.3. Figure 4 shows the variation of optimized values of MRR for the different experimental runs.

On the other hand the minimum value of SR is 25.0612 corresponding to the experiment number 3. The values corresponding to the minimum SR for the various input parameters are at A=10; B=8; C=0.32; D=0.3. Figure 5 shows the variation of optimized values of SR for the different experimental runs.

The fitness calculation for the multi-objective optimization shows that experiment number 2 is the best experimental run with the minimum fitness value of 7.4251. The values of input parameters for the minimum value of fitness function are: corresponding to the input parameters are: A=10; B=8; C=0.32; D=0.3. The variation of fitness values for different experimental runs is depicted in Figure 6.

CONCLUSION

In the present chapter the single and multi-objective optimization of the MRR and SR for the ECM process of EN19 was carried out using the PSO algorithm. The optimum value of MRR and SR were established. The single objective optimization can be used to fulfil the specific requirement of the manufacturing unit. For instance if MRR plays a critical role in production of goods, then the manufacturer can go for maximizing of the MRR value.

Table 7. PSO results

Exp No.	A	B	C	D	MRR	SR	Fitness Value	Exp No.	A	B	C	D	MRR	SR	Fitness Value
1.	1	1	1	1	0.1311	25.1447	7.4517	15.	2	2	3	2	0.1791	29.2040	8.6358
2.	1	1	2	2	0.1451	25.0888	7.4251	16.	2	3	1	1	0.1470	27.0418	8.0097
3.	1	1	3	3	0.1493	25.0612	7.4138	17.	2	3	2	2	0.1610	26.9859	7.9831
4.	1	2	1	2	0.1786	29.1024	8.6057	18.	2	3	3	3	0.1653	26.9583	7.9718
5.	1	2	2	3	0.1827	29.0667	8.5921	19.	3	1	1	3	0.1512	25.1968	7.4532
6.	1	2	3	1	0.1631	29.0879	8.6122	20.	3	1	2	1	0.1314	25.2099	7.4710
7.	1	3	1	3	0.1648	26.8567	7.9417	21.	3	1	3	2	0.1456	25.1621	7.4467
8.	1	3	2	1	0.1450	26.8698	7.9594	22.	3	2	1	1	0.1650	29.2235	8.6516
9.	1	3	3	2	0.1592	26.8220	7.9352	23.	3	2	2	2	0.1790	29.1676	8.6250
10.	2	1	1	2	0.1471	25.2608	7.4753	24.	3	2	3	3	0.1832	29.1400	8.6137
11.	2	1	2	3	0.1512	25.2251	7.4617	25.	3	3	1	2	0.1610	26.9576	7.9746
12.	2	1	3	1	0.1316	25.2463	7.4818	26.	3	3	2	3	0.1651	26.9219	7.9610
13.	2	2	1	3	0.1847	29.2388	8.6423	27.	3	3	3	1	0.1455	26.9431	7.9811
14	2	2	2	1	0.1649	29.2518	8.6601								

Figure 4. Variation of optimized value of MRR.

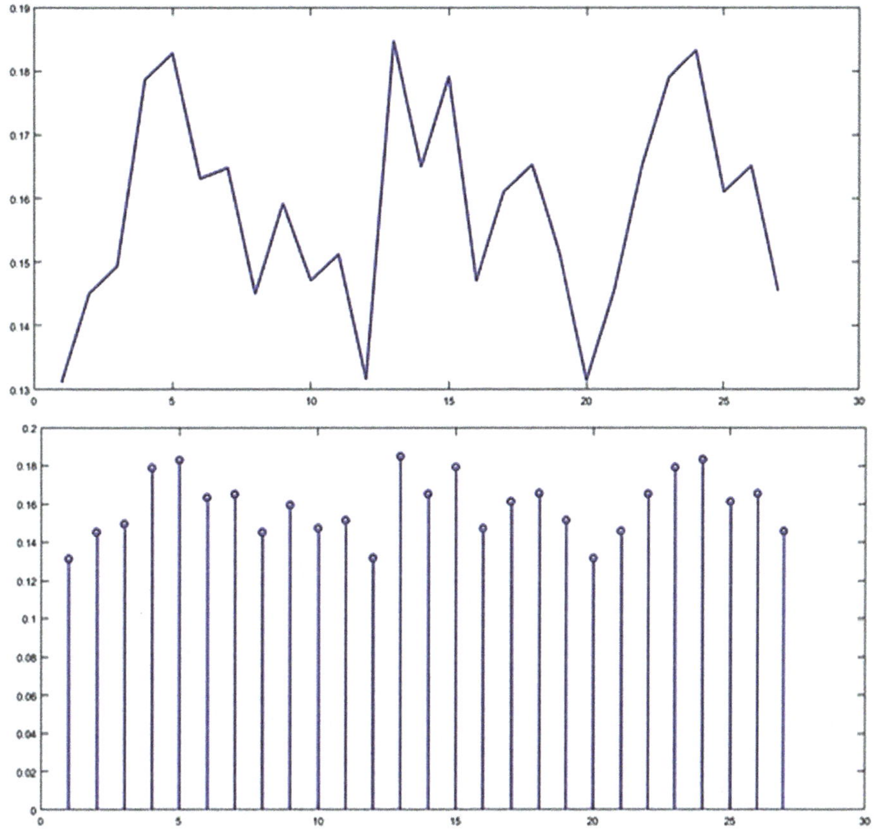

The multi-objective optimization was based on the concept of fitness function and the experimental run with the minimum value of fitness function was found to be best amongst the different runs considered. Therefore the PSO algorithm proved to be a cost effective solution for establishing the optimum values of input process parameters for enhancing the machining performance of the ECM process. The multi-objective optimization thus aids in selecting the best experimental run for the production of goods and services under the conflicting output process parameters such as MRR and SR.

SCOPES OF FUTURE RESERACH

The PSO algorithm can be used for the optimization of other output process parameters such as accuracy. Further, the optimization can be performed by taking into account more input parameters. The machining optimization of ECM for composite materials can be investigated with the PSO. The PSO algorithm can be used for the optimization of process parameters for other non-conventional machining processes such as ultrasonic machining, electro discharge machining etc.

Figure 5. Variation of optimized value of SR.

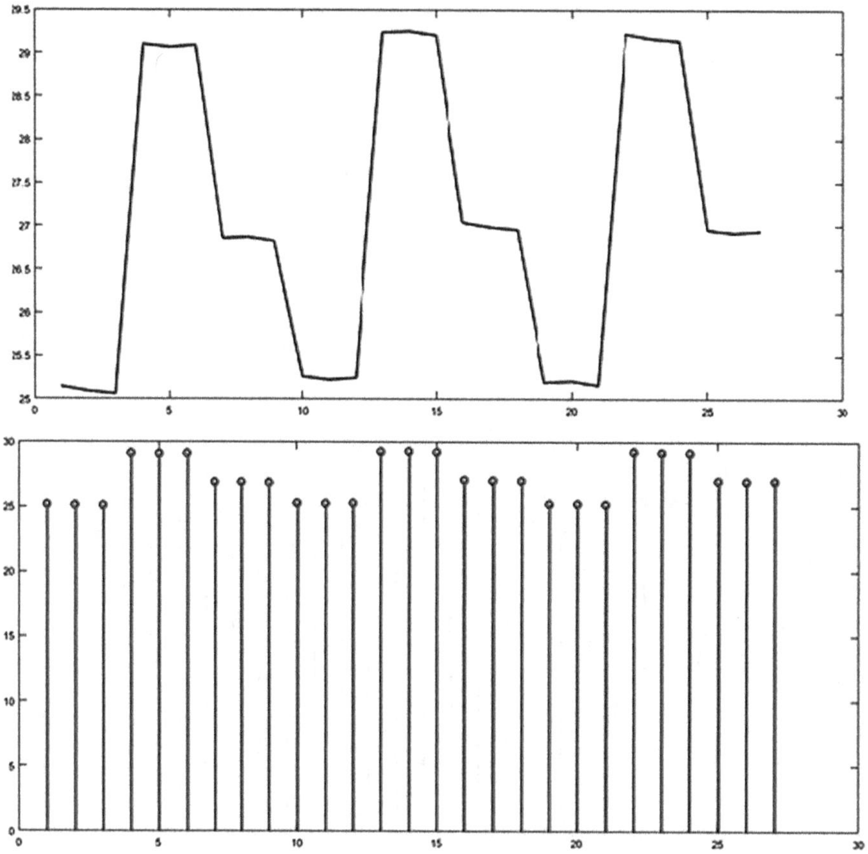

Figure 6. Variation of optimized value of SR.

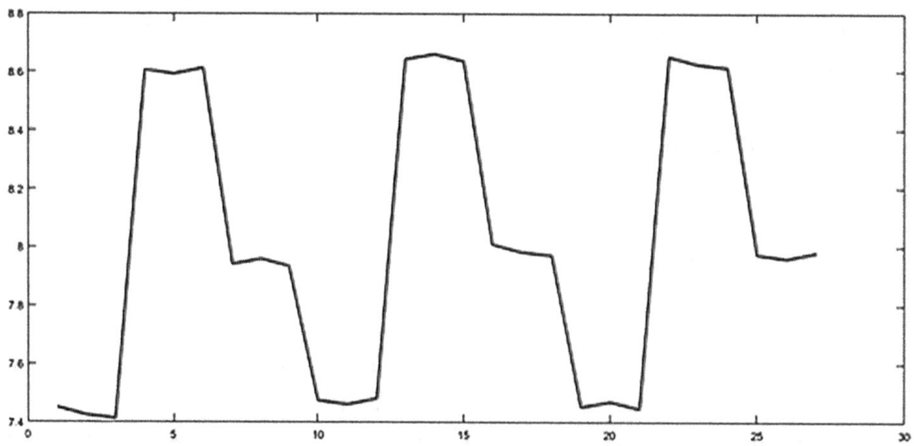

ACKNOWLEDGMENT

The authors sincerely acknowledge the comments and suggestions of the reviewers that have been instrumental for improving and upgrading the paper in its final form.

REFERENCES

Acharya, B. G., Jain, V. K., & Batra, J. L. (1986). Multi-objective optimization of the ECM process. *Precision Engineering*, *8*(2), 88–96. doi:10.1016/0141-6359(86)90091-7

Asokan, P., Kumar, R. R., Jeyapaul, R., & Santhi, M. (2008). Development of multi-objective optimization models for electrochemical machining process. *International Journal of Advanced Manufacturing Technology*, *39*(1-2), 55–63. doi:10.100700170-007-1204-8

Bhattacharyya, A., Sur, B., & Sorkhel, S. K. (1973). Analysis of optimum parametric combination in electro-chemical machining. *Ann. CIRP*, *22*(1), 59–60.

Chakradhar, D., & Gopal, A. V. (2011). Multi-objective optimization of electrochemical machining of EN31 steel by grey relational analysis. *International Journal of Modeling and Optimization, 1*(2), 113.

Choobineh, F., & Jain, V. K. (1993). A fuzzy sets approach for selecting optimum parameters of an ECM process. *Processing of Advanced Materials*, *3*, 225–232.

Das, M. K., Kumar, K., Barman, T. K., & Sahoo, P. (2014). Optimization of surface roughness and MRR in electrochemical machining of EN31 tool steel using grey-Taguchi approach. *Procedia Materials Science*, *6*, 729–740. doi:10.1016/j.mspro.2014.07.089

Dong, Y., Tang, J., Xu, B., & Wang, D. (2005). An application of swarm optimization to nonlinear programming. *Computers & Mathematics with Applications (Oxford, England)*, *49*(11-12), 1655–1668. doi:10.1016/j.camwa.2005.02.006

El-Dardery, M. A. (1982). Economic study of electrochemical machining. *International Journal of Machine Tool Design and Research*, *22*(3), 147–158. doi:10.1016/0020-7357(82)90023-3

Hu, X., & Eberhart, R. (2002). Multiobjective optimization using dynamic neighborhood particle swarm optimization. In *Evolutionary Computation, 2002. CEC'02. Proceedings of the 2002 Congress on* (*Vol. 2*, pp. 1677-1681). IEEE.

Jain, N. K., & Jain, V. K. (2007). Optimization of electro-chemical machining process parameters using genetic algorithms. *Machining Science and Technology*, *11*(2), 235–258. doi:10.1080/10910340701350108

Kalpakjian, S. (1984). *Manufacturing processes for engineering materials*. Pearson Education India.

Kennedy, R. (1995, November). J. and Eberhart, Particle swarm optimization. *Proceedings of IEEE International Conference on Neural Networks IV,* 1000. 10.1109/ICNN.1995.488968

Rajurkar, K. P., Zhu, D., McGeough, J. A., Kozak, J., & De Silva, A. (1999). New developments in electro-chemical machining. *CIRP Annals-Manufacturing Technology*, *48*(2), 567–579. doi:10.1016/S0007-8506(07)63235-1

Sekar, T., & Marappan, R. (2008). Experimental investigations into the influencing parameters of electrochemical machining of AISI 202. *Journal of Advanced Manufacturing Systems*, *7*(02), 337–343. doi:10.1142/S0219686708001486

Tripathi, P. K., Bandyopadhyay, S., & Pal, S. K. (2007). Multi-objective particle swarm optimization with time variant inertia and acceleration coefficients. *Information Sciences*, *177*(22), 5033–5049. doi:10.1016/j.ins.2007.06.018

Van den Bergh, F., & Engelbrecht, A. P. (2006). A study of particle swarm optimization particle trajectories. *Information Sciences*, *176*(8), 937–971. doi:10.1016/j.ins.2005.02.003

Vikas, S., Shashikant, Roy, A. K., & Kumar, K. (2014). Effect and Optimization of Machine Process Parameters on MRR for EN19 and EN41 materials using Taguchi. *Procedia Technology*, *14*, 204–210. doi:10.1016/j.protcy.2014.08.027

Chapter 9
Performance Study of $LaPO_4$–Y_2O_3 Composite Fabricated by Sol–Gel Process Using Abrasive Waterjet Machining

M. Uthayakumar
Kalasalingam University, India

Balamurugan Karnan
Vignan's University, India

Adam Slota
Cracow University of Technology, Poland

Jerzy Zajac
Cracow University of Technology, Poland

J. Paulo Davim
University of Aveiro, Portugal

ABSTRACT

This chapter presents an effective approach to assess the abrasive water jet machining of lanthanum phosphate reinforced with yttrium composite. A novel composite is prepared with the mixture of lanthanum phosphate sol and yttrium nitrate hexalate with a ratio of 80/20 by aqueous sol-gel process. Silicon carbide of 80 mesh size is used as abrasive. The effects of each input parameter of abrasive water jet machining are studied with an objective to improve the material removal rate with reduced kerf angle and surface roughness. The observations show that the jet pressure contributes by 77.6% and 45.15% in determining material removal rate and kerf angle, respectively. Through analysis of variance, an equal contribution of jet pressure (38.18%) and traverse speed (40.97%) on surface roughness is recorded. Microscopic examination shows the internal stress developed by silicon carbide which tends to get plastic deformation over the cut surface.

DOI: 10.4018/978-1-5225-5445-5.ch009

INTRODUCTION

Ceramic materials are well known for its high temperature stability, high strength, relatively low density, high hardness, excellent wear and corrosion resistance. In present days the efforts have been taken to improve their toughness so that better machineability can be achieved. However, the growing demand of new materials for use in a variety of environments has become necessary to investigate and identify new materials which will suit to specific requirements. The rare earth phosphate material because of its poor toughness and high brittleness, it is limited in its wide practical applications.

Development of rare earth composites is of great interest to the researchers owing to their advantages over conventional materials. Among them, rare earth lanthanum phosphates had attracted many researchers as Sujith (2014), Chenghao (2015), Min (2001) whose study explores the unique characteristics of the composite like: high temperature withstanding capacity, non-reactant, chemically inert and no phase transformation at elevated sintering temperature and it also proved that the Lanthanum Phosphate ($LaPO_4$) acts as an interfacial material in machining composite. Gong (2006), had proved that the $LaPO_4$ in the composite mixture will enhance to form an interfacial layer while machining. Hence it induces the toughness property to the hard materials and leads to increase the machiniability of the materials. Abdul Majeed (2008), prepared Al_2O_3-$LaPO_4$ composite, with an addition of 30% of $LaPO_4$. This composite was identified to have least hardness value and improves machinability.

Alangi (2011), had developed a Y_2O_3 thin film deposition on tantalum substrate. The developed thin film admirably act as a corrosion resistance layer against the liquid uranium attack, even at elevated working conditions of 1300^0 C. Study made on different materials likely CaO, MgO and Y_2O_3 by Alberdi (2013) upon developing of a stabilized compound with Zirconia, Yttrium Oxide (Y_2O_3) has identified to have a least hardness value. Composite with the addition of Y_2O_3 particles is found that an increase in densification of the composite. Hence it is stated that the yttria particles have a tendency to reduce the porosity of the particle. To validate the Alberdi etal work, Gunduz (2008), observed a similar report while doping Yttria with Bovine Hydroxyapatite. Further it is confirmed that the composite has found to have least porosity even at the elevated sintering temperature of about 1200^0 C. Maiti (2006), report on microstructural property of Al_2O_3 based Y_2O_3 composite delivers that an equiaxed grain structure with fine grain size and grain boundary is obtained with addition of yttria elements. According to Sun (2007), the additions of yttria in the silicate ceramic enhance the machiniability of the composite. The formation of weak interface bonds of the atomic planes (Y-O) with silicate improves the toughness of the composite material. Mogilevsky (2007), had identified that the solubility of the yttrium in the $LaPO_4$ was superior with increase in the working temperatures. It is further reported that the solubility of yttrium is found to be maximum when the atomic composition of yttrium is below 42%. Kim (2006), states that at elevated sintering temperature reduction in porosity of W-Y_2O_3 composite is achieved with increase of Y_2O_3 particles. Sankar (2011, 2012) in his studies on $LaPO_4/Y_2O_3$ composite states that the properties Y_2O_3 and $LaPO_4$ are found to be similar in nature moreover, it is further stated that when these particles are formed as composite, they are proven to be in a stabilized state at ambient condition. Hence this enhances the researchers to develop a composite material that possess an excellent property. Also, this composite has been found to withstand high temperature and the sankar et al., proposed that the $LaPO_4/Y_2O_3$ composite could be a suitable material for high thermal barrier, melting and also for functional material applications.

Use of conventional methods to machine these ceramic composites are often difficult is due to their superior physical and mechanical properties. To perform the machining operation, it requires a very high cutting energy and temperature. The removal of material in Abrasive Water Jet Machining (AWJM) is purely by erosion caused by the high accelerated abrasive particles from the jet stream. Xu (1995) reported that the erosion wear on hard surface causes an increase in temperature on sample surface. Further the simultaneous cooling by water in AWJM will induces residual stresses on the cut surface. This would results in the formation of micro-crack on the machined surface of the composites. Zhang (2011) had conducted a comparative study on various unconventional machining practices like laser cutting machine and the wire electrical discharge machine, it is concluded that AWJM is a cold working process and the surfaces are free from heat affected zones. AWJM studies by Wang (2004) limits by kerf width. Kerf taper is characterized by a wider entry at the top than at the bottom. It is one of the major barriers that limit AWJM applications. Early research studies by Shanmugam (2009, 2008) had described various methods to reduce the Kerf Angle (KA) on machining ceramic composites in AWJM. The effect of out-put performance characteristics on different operating conditions in AWJM with various abrasives on different materials shows diverse results which has been evaluated and reported by Gudimetla (2002), Ghosh (2015), Chen (1996), Ahmet Hascalik. (2007). According to Kopac (2007), exclusive of input parameters material properties predominantly determine the machinability of the material.

According to Pirso (2010), the ratio of hardness of an abrasive and the hardness of the structured material (Hardness Ratio: H_a/H_m) creates an impact in machining. Selection of abrasive particle will enhance the quality of the machining. Further, it is also stated that SiC abrasive is broadly used for the manufacturing of highly engineered parts because of its high hardness, low coefficient of friction, wear resistance and high decomposition temperature. Inspite of the report, Srinivas (2011), had suggested that apart from the hardness of abrasives, the hardness of the particles present in the matrix composition will also significantly affect the machining performance. Investigation on abrasive erosion in AWJM by Fang (1992, 1993), proves that based on the movement of abrasive particles in the jet, the nature of wear will be in the form of micro-cutting, grain fracture or plastic deformation on the cut surface.

Khan (2007) had performed the AWJM on glass substrate material with different abrasives. Through the observation, it is suggested that SiC abrasives will exhibits a superior performance characteristic in AWJM machining compared to alumina and garnet. The abrasive wear mechanism mainly depends on the relative hardness between the machinable material and abrasives. When Xiaoyong (2013), made a review on use of different abrasive in AWJM and it is identified that on use of varied abrasives creates different sign on the kerf region. Al_2O_3 abrasive enhance to get kerf surface such as ploughing, when SiC is used as abrasive, it creates micro-cutting, grain fracture and plastic deformation, on use of SiO_2 as abrasives tend to produces extrusion and removal of binder phase and also, a slight plastic deformation surface with grooves are identified on the cut surface. The hard abrasive particles intensively determine MRR and Surface Roughness (Ra) but according to Fowler (2009), they do not have any influence on the surface waviness. Zhao (2014), performed an experimental analysis in AWJM with four different materials. The authors concluded that for hard materials the cut surface was found to be smooth whereas rough wear tracks and wear scare are found in ductile type of materials. The AWJM will change depending upon the material properties and cutting conditions. Selvan (2012), on cutting the cast iron using AWJM, on considering the surface roughness alone, it is found that JP significantly determines the output response than SOD and TS. Whereas in Derzija (2015), study in AWJM on aluminium material yields a diverse result showing that the TS has significant effect than the other independent parameters. Hence the machining effects will change based on the working condition and the material property. Sharma (2011), suggest that to evaluate the machining performance of a newly fabricated composite material and also, to predict the various independent parameters limitations on AWJM while machining the composite, a necessity of consideration of an integrated taguchy approach will reduce the complicity. This could be done by optimizing the series of experiments.

Hloch (2009) has conducted a series of experiments with varied sample thickness. The surface topography and the mechanism of material removal in AWJM are studied. The study reveals that with increase in depth of cut the surface tend to have more wear track, also the kerf taper increases with increase in the thickness of the sample. Hutyrova (2015), had extended the AWJM applications as beside cut through operations, it is also now used in turning operations. This kind of operations will completely eliminate the tool wear, thermal induced surface and is also not limited with the material properties. It could perform operations on wood, plastics and alloys etc. Carach (2016), had performed the turning operation on hard materials and reported that there is no evidence of melt surface over the kerf regions.

Every material varies in their mechanical behavior because of their physical and chemical nature. So it became important to investigate the mechanical properties before advocating the use of a ceramic matrix composite for a particular application. In general large numbers of combination of matrix and reinforcements are available to develop ceramic matrix composite. In present work, a series of experiments have been conducted on $LaPO_4/Y_2O_3$ composite which is prepared by the Sol-Gel process. The composite is cut for taguchy's L27 orthogonal array using AWJM. The input parameters such as Jet Pressure (JP), Stand-Off Distance (SOD) and Traverse Speed (TS) are considered to be the affecting machining parameters and observations are conducted with an objective to minimize the KA, Ra and to improve the MRR. Further, ANOVA study is performed on each output parameters to measure the performance deviation of each input parameter over the individual output parameters.

MATERIALS AND METHODS

Preparation of Composite

Synthesis of these nano-structured gels requires significantly milder conditions and temperatures than we encountered in the "heat and beat" method. Wet gels in which the pores are filled with a fluid can be synthesized by a number of different so-called "sol-gel" routes involving the controlled hydrolysis and condensation of molecular precursors.

Synthesis of Lanthanum Phosphate

A Wide Variety of synthesis of lanthanum phosphate powders are performed by the researchers. An attempt has been made to synthesis the lanthanum powder by Aqueous Sol-Gel process. The commercial available materials are used as the starting component for the process. The obtained powders are sintered to the determined temperature to get the nanosize particles. The prepared powders are characterized and reported in detail. From the observations lanthanum and the lanthanum phosphate powders obtained through the sol-gel process are in the range of 25-80 nm.

Commercially available lanthanum chloride is mixed with deionized water to form the solution and further addition of orthophosphoric acid in a defined stoichiometric estimation in a ultrasonic bath and as a supplementary product 25% of ammonia is flocculated to $LaPO_4$ particle. With a pH range of 6.8 $\pm$ 0.2, a complete flocculated precipitate of $LaPO_4$ is obtained. The precipitate is rinsed several times in hot water followed by centrifuged force to remove the chlorine content. To convert the precipitate in to a colloidal sol, 20% (Vol) of nitric acid is added with continuous stirring condition at pH in a range of 1.75 to 1.85 and it is followed by ultrasonication for 15 min. $LaPO_4$ gel is obtained in the ammonia atmosphere and is dried followed by ball milling. The gained powder is calcinated to1400°C for 2 hours. A systematic flowchart had shown in Figure 1 will delivers the preparatory steps of lanthanum phosphate powder.

For a typical case phenomenon, to fabricate 20 gms of lanthanum phosphate powder, commercially available lanthanum chloride hexa hydrate ($LaCl_3.6H_2O$) is taken as the starting material. The deionized water is added gently to the $LaCl_3.6H_2O$, as it converts to solution state. Nearly 1700 to 1800 ml of deionized water is required to acquire the 0.5 mole of $LaCl_3.6H_2O$ solution. To reduce the concentration of the prepared solution (pH), 88% concentrated orthophosphoric acid (H_3PO_4) is added as a supplementary product. To obtain the value of pH.2, the H_3PO_4 acid is added drop wise with continuous stirring condition.

Once the pH value is reached addition of the supplementary product is stopped. Further the stirring operation is performed continuously for about 2 hours for the complete mixing of the solution. To attain the lanthanum phosphate sol from the obtained solution, ammonia solution is also added like the H_3PO_4 to flocculate the sol. The addition of ammonia solution is performed until all the phosphate molecules in the solution get deposited. The pH value will change on the quantity of the ammonia solution. Once the clear solution is found in the top portion of the beaker make sure that all the phosphate molecules are deposited and then the addition of ammonia solution can be dropped. The precipitates formed are rinsed several times with warm water by centrifugal force to remove the chlorine content in the precipitate. By means of washing the precipitate through the centrifugal force will results in the removal of the excess amount of phosphate elements present in the medium.

A conventional silver test is carried out to confirm the complete removal of chlorine atoms in the medium. The rinsed precipitate is then mixed with the deionized water to form the sol. Nitric acid with 20% in volume percentage of sol is added with the pH in the range of 1.75 to 1.85. Under vigorous stirring condition of about 5 hours, the lanthanum phosphate sol is peptized. Using Ultra sonic processor P2, Vibronics, India, Ultrasonification is done for 15 minutes. By placing the ammonia solution closer to the dessicator, the ammonia atmosphere is created. In this atmosphere, the sol is allowed to settle down. After 24 hours of settlement, a thick lanthanum phosphate gel is formed. The obtained gel is calcinated to 70°C for about 24 hours and then it is grinded. To remove the water content, the grinded powders are further heated to about 1400°C.

Synthesis of Lanthanum Phosphate: Yttria Composite

To synthesis 80% lanthanum phosphate with 20% yttria composite, the stoichiometrically prepared lanthanum phosphate sol is taken as the primary material. Lanthanum phosphate sol is obtained when preparation of $LaPO_4$ without yttria which is explained in the above topic. The commercially available Yttrium nitrate hexahydrate (99.9%) (Y (NO_3). $6H_2O$) is used as the starting material for the yttrium oxide. The systemic procedure for the preparation of the lanthanum phosphate is revealed on preparation of the lanthanum phosphate powder.

The calculated quantity of the starting element of the yttrium oxide is mixed with the distilled water to acquire a sol form. The acquire sol is mixed with the primarily obtained lanthanum phosphate sol with the continuous string condition. Mechanical stirrer is used for the stirring operating. Ammonia solution is taken as a supplementary product and is added to the mixer to maintain the pH value. The addition of ammonia solution will rise the pH value also, it flocculated the $LaPO_4$ and yttrium oxide from the sol. The addition of ammonia is carefully monitored. Stirring is done continuously for 3 hours for the homogeneous mixture of the solution. The precipitate obtained is washed with the warm water and it is dried to 70°C to remove the excess water content. The obtained powder is furthered calcinated to 700°C for two hours.

To synthesis a 5gm of $LaPO_4$+20 wt%Y_2O_3 nano composite, 17gm of yttrium nitrate is dissolved in a minimum quantity of distilled water. The mixture is then added with 0.05 M lanthanum phosphate sol (preparation is stated in earlier stage) under constant stirring using a mechanical stirrer. The homogeneous mixture of lanthanum phosphate sol and the yttrium nitrate solution are obtained at the constant string condition of about 2 hours. The precipitation of Y(OH)3 as well as flocculation of $LaPO_4$ is achieved slowly by adjusting the pH to 8 using 25% NH3 solution and the composite precursor is kept stirring for 3 hours. The precipitate obtained was later dried over water bath and further calcined to different temperatures 1400°C and let for further characterization.

The obtained powder is consolidated to disk shape (36mm diameter and 7mm thickness) in titanium coated Oil Hardening Non Shrinking Die Steel (OHNS) with a uniaxial compression force of about 480 MPa at room temperature. The obtained disk is sintered to 1400°C for two hours.

EXPERIMENTAL STUDIES AND PROCEDURES

AWJM of model DIP 6D-2230 manufactured by Dardi International Corporation has a tungsten carbide nozzle of 0.67mm diameter with an inbuilt orifice of 0.25mm diameter is used for the observation. In the machining of these type of composite materials, selection of machining parameters is found to be a challenge for the researchers. To induce the usage of this technology over cutting these composites, its various input parameters are to be optimized. Table 2 shows the list of selected factors and levels used in the observation. The significance of each input parameter over the output parameter and its interaction are studied using Analysis of Variance.

Parameters of AWJM Process

There are many parameters involved in an AWJ machining process. The most significant factors of AWJM are considered in the study. The considered independent parameters are:

Table 1. Properties of $LaPO_4$-Y_2O_3 composite

S.No	Young's Modulus (N/m^2)	Flexural Strength (GPa)	Micro Vickers Hardness (GPa)	Theoretical Density (g/cm^3)	Experimental Density (g/cm^3)	Porosity (%)
1	4.96	96±4	5.2	4.95	4.87	1.1616

Table 2. Selected Factor and Levels

S.No	Factors	Levels			Units	Symbols
		1	2	3		
1	Jet Pressure	220	240	260	bar	A
2	Stand-off-distance	1	2	3	mm	B
3	Traverse Speed	20	30	40	mm/sec	C

- Jet Pressure (JP)
- Stand Off-Distance (SOD)
- Traverse Speed (TS)

The output responses are

- Material Removal rate (MRR)
- Kerf Angle (KA)
- Surface Roughness (Ra)

Material Removal Rate

The high precision weighing balance (AUX 220 make of shimadzu) is used to check the loss of material during machining and it has a least count of 10 mg. The quantity of material removed per minute is calculated using the following equation (1).

$$MRR = \frac{\left(Wf - Wi\right) * 1000}{Dw * t} \quad (1)$$

where, W_i - Initial Weight of work piece in grams before cutting, W_f - Final Weight of work piece in grams after cutting, D_w - Density of the Work piece (gm/cm^3) and t - Period of trial (min).

Kerf Angle

The generated cut surface in AWJM will always exhibit some poor surface characteristics. Among them kerf angle or kerf taper is a predominant problem exists on the finished end product. This alternatively moves to take the secondary operations to remove this kerf taper. The taper is formed by the impingement of the abrasives on the samples as the depth of cut on the sample increase will enhance to get larger taper. Burr formation on the cut section is due to the directional impingement of the water beam which leads to plastic deformation surface. This kerf taper defect will cause problems in assembling of components, formation of sharp edge which are dangerous for handling and this may affect the component's performance. The KA is measured using the profile projector. KA is measured by using equation (2).

$$\text{Kerf angle} = \tan\text{-}1 \frac{[X1 - X2]}{2t} \tag{2}$$

where x_1 = Top kerf width, x_2 = Bottom kerf width and t = Thickness of the material.

Surface Roughness

The cut surface can be divided into three regions. The initial deformation region (IDR) is where the jet initially enters the cut part the work piece. Continuous impingement of abrasives on the cut surface causes stray particles to create a rounding of the upper edges. The second region is the smooth cutting region (SCR), also called as cutting wear dominates. During cutting operation, the shear force of the sharp edges abrasive particles will chip away material from the work piece. A small amount of materials are removed by each abrasives at the low angles of incidence hence the surface finish in this regions are considerably higher than rough cut region. This may lead to enhance the good surface finish in this region. The third region is the rough cut region (RCR). On further down the cut region, the jet curves create a higher angle face area. The collision of particles at higher angles will primarily change the mode of material removal. The wear is by deformation mechanism which will leads to striation and rough cutting marks which are generated along the diverged water beam. The increase in kerf taper is observed with increase in thickness of the cut samples. The surface roughness is measured by using surface roughness tester SJ-411which has a range of 350 μm with a probe speed of 0.25 mm/s over a span of 5mm. Three observations are made at top, middle and bottom of cut section and the averages of these are tabulated in Ra.

RESULTS AND DISCUSSION

Material Characterization

X-Ray diffraction of $LaPO_4$ with Yttria composite is shown in Figure 1. The pattern identified through XRD-X'Pert high score software indicates the crystal structure of the elements. The dominating elements were found to be in the combination of YPO_4. This has been drawn from the $LaPO_4$ and Y_2O_3 based on heat sintering process done at the elevated temperature of 1400°C ± 10°C. The mechanism behind this shows the agglomeration of yttria and phosphate reveals at elevated temperature and lanthanum can sustain temperatures above 2000°C. Thus these two materials possess high strength with good corrosion and oxidation resistance.

The objective is to identify the significance of each input parameters over various levels of individual output parameter. To get good surface finish with acceptable MRR and KA, a suitable orthogonal array is to be selected. The number of Degrees Of Freedom (DOF) for the selected three factors is determined by the total number of levels (n) in each parameter minus one (n-1). The L27 Orthogonal Array may well be computed to get the significant output. The experimental observations are presented in Table 3.

Figure 1. XRD of the $LaPO_4$ with 20% Y_2O_3

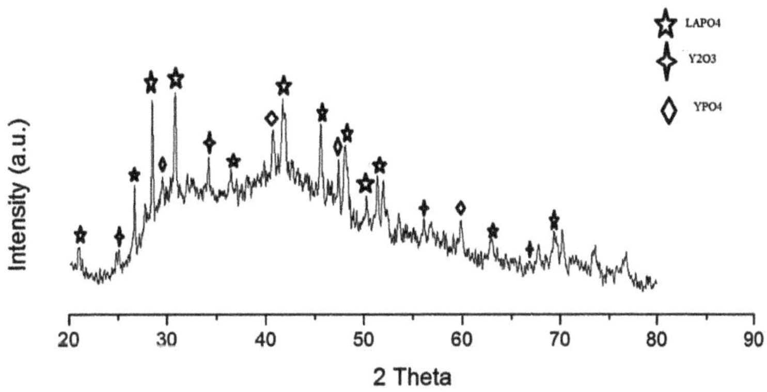

Table 3. Experimental observation of MRR, KA and Ra

Ex.No	JP (bar)	SOD (mm)	TS (mm/min)	MRR(g/s)	KA(Deg)	Ra (μm)
1	220	1	20	0.03189	0.215	1.191
2	220	1	30	0.02589	0.315	1.328
3	220	1	40	0.02152	0.364	1.486
4	220	2	20	0.03258	0.249	1.139
5	220	2	30	0.02998	0.332	1.289
6	220	2	40	0.02681	0.414	1.542
7	220	3	20	0.03471	0.298	1.344
8	220	3	30	0.03186	0.342	1.508
9	220	3	40	0.02895	0.398	1.675
10	240	1	20	0.04112	0.292	1.231
11	240	1	30	0.03412	0.364	1.494
12	240	1	40	0.03097	0.411	1.651
13	240	2	20	0.04526	0.293	1.331
14	240	2	30	0.04291	0.362	1.531
15	240	2	40	0.04017	0.454	1.682
16	240	3	20	0.05621	0.425	1.515
17	240	3	30	0.05014	0.468	1.656
18	240	3	40	0.04519	0.517	1.797
19	260	1	20	0.05495	0.348	1.423
20	260	1	30	0.04951	0.396	1.572
21	260	1	40	0.04652	0.492	1.692
22	260	2	20	0.06488	0.393	1.598
23	260	2	30	0.05715	0.462	1.731
24	260	2	40	0.05359	0.554	1.881
25	260	3	20	0.06887	0.516	1.645
26	260	3	30	0.06574	0.598	1.847
27	260	3	40	0.06252	0.681	1.966

Effect of Parameters on MRR

MRR is found to be a dominating factor in the determination of manufacturing cost and machining time. Figure 2 is plotted on the working condition of JP of 220 bar pressure for the three levels of TS and SOD, respectively. From Figure 2(a), MRR trends to increase with the increase in JP. An increase of JP accelerates the abrasive particle results and increases the width of jet beam before it hits the top surface of the specimen. It is noted that increase in TS gradually decreases MRR irrespective of JP. Due to the lack of machining time, the erosion of the abrasive particles at the rough cut section of the samples is reduced significantly. From Figure 2(b) it is clear that the increase in SOD increases MRR with respect to change in TS. The abrasive particles in a scattered beam losses its energy before it touches the rough cut region. The random movement of abrasive material is caused all over the bottom section of the cut surface due to high TS, which creates a large shift in MRR. At TS of 20 mm/min in varying levels of SOD, a least significance in MRR is obtained.

ANOVA for MRR is shown in Table 4 and it reveals that JP plays a most significant role of 77.6% in the determination of MRR, whereas a minor contribution of 13.36% for SOD and 6.39% for TS is observed. However, among the interfaces, JP x SOD contributes significantly higher than the other interactions (JP x TS and SOD x TS) of 2.02%. From Fisher test at 95% confidential level ($F_{0.05, 2, 8}= 4.46$), investigations done on these parameters reveals that JP, SOD and TS play a significant role in MRR.

Effect of Parameters on Kerf Angle

Figure 3(a) is plotted between JP and KA shows that a significant change in JP and TS which affects and lead to increase KA. The rapid movement of nozzle over the cut surface irrespective of JP produces the larger top kerf width and a narrow bottom kerf width. With increase in JP and increase in TS, a significant rise of KA is observed and this is due to increase in machining time which leads the water beam to propel through the entire thickness of the composite. The increase in JP improves the kinetic energy of the abrasive particle, which results in the bounce of particles within the jet and the back scat-

Figure 2. Effect of input parameters in MRR

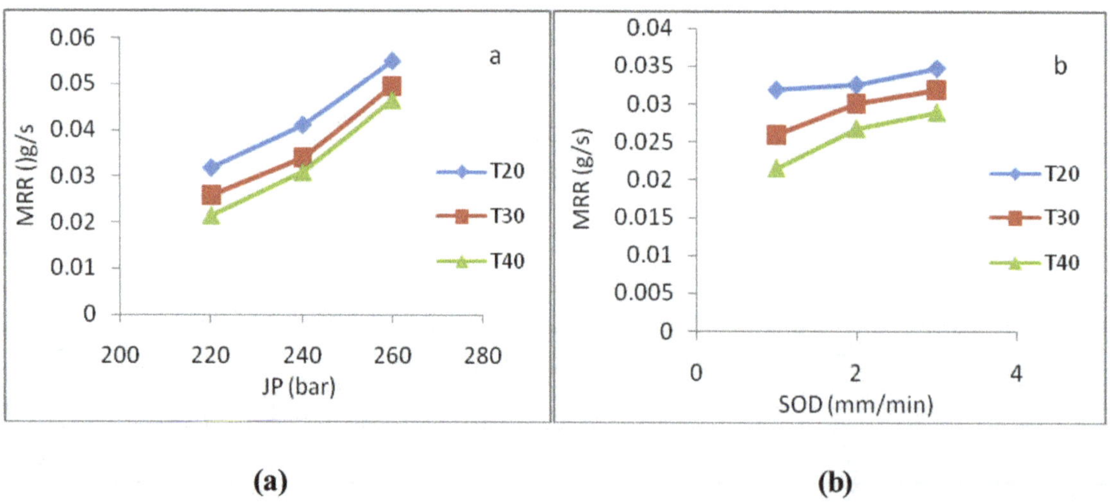

Table 4. ANOVA for MRR

Parameter	DOF	Sum of Squares	Mean Square	F-Value	Contribution (%)
JP	2	0.0037	0.0018	710.97	77.60
SOD	2	0.00065	0.0003	122.40	13.36
TS	2	0.00031	0.00015	58.60	6.39
JP x SOD	4	9.8E-05	2.44E-05	9.27	2.023
JP x TS	4	2.8E-06	7.24E-07	0.27	0.061
SOD x TS	4	5.9E-06	1.47E-06	0.56	0.122
Error	8	2.1E-05	2.63E-06		0.436
Total	26	0.0048	0.0023		100

tered abrasives affect the prior cut region of the composite which leads to jet divergence. A simultaneous increase of SOD wider the water beam and affects the newly cut region. This may create a wider kerf entry. In Figure 3(b), an unexpected increase in KA is observed, with the increase in SOD with respect to TS. The divergence of water beam, the collision of abrasive particle inside the jet, partial losses in sharp edges of the abrasive before impact and a significant reduction of machining time lead to the reduction of erosion effect.

From ANOVA of KA shown in Table 5, it is noted that JP influences greatly in affecting KA with the active contribution of nearly 45.15%. The other parameters namely; SOD and TS contribute nearly 22.10% and 25.98%, respectively. Among the interactions shown in Table 5, JP x SOD interaction alone significantly affects KA at a rate of 5.49%. F-test at 95% confidential level shows that the input parameters of AWJM (JP, SOD and TS) are greater than $F_{0.05, 2, 8} = 4.46$ and have greater significance in affecting the output KA. Excluding JP x SOD, the other two interactions have no significance in KA.

Figure 3. Effect of input parameters in KA

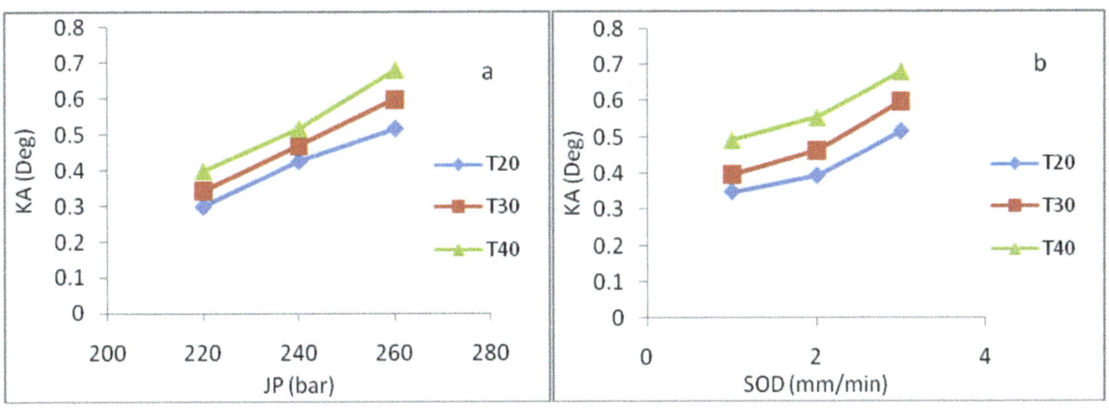

(a) (b)

Table 5. ANOVA for KA

Parameter	DOF	Sum of Squares	Mean Square	F-value	Contribution %
JP	2	0.135	0.0677	347.05	45.15
SOD	2	0.066	0.033	169.79	22.10
TS	2	0.078	0.039	199.69	25.98
JP x SOD	4	0.0165	0.004	21.09	5.49
JP x TS	4	0.0018	0.0005	2.42	0.63
SOD x TS	4	0.0004	0.0001	0.55	0.14
Error	8	0.002	0.0002		0.52
Total	26	0.299	0.144		100

Effect of Parameters on Surface Roughness

Surface finish is found to be a compromising factor in determining the quality and the end use of the product. Figure 4(a) is plotted with JP of 240 bar and SOD of 2mm for different levels of TS. The surface finish is found to be decreasing with the increase of JP and TS. The mechanism behind this is at high JP

Figure 4. Effect of input parameters in Ra

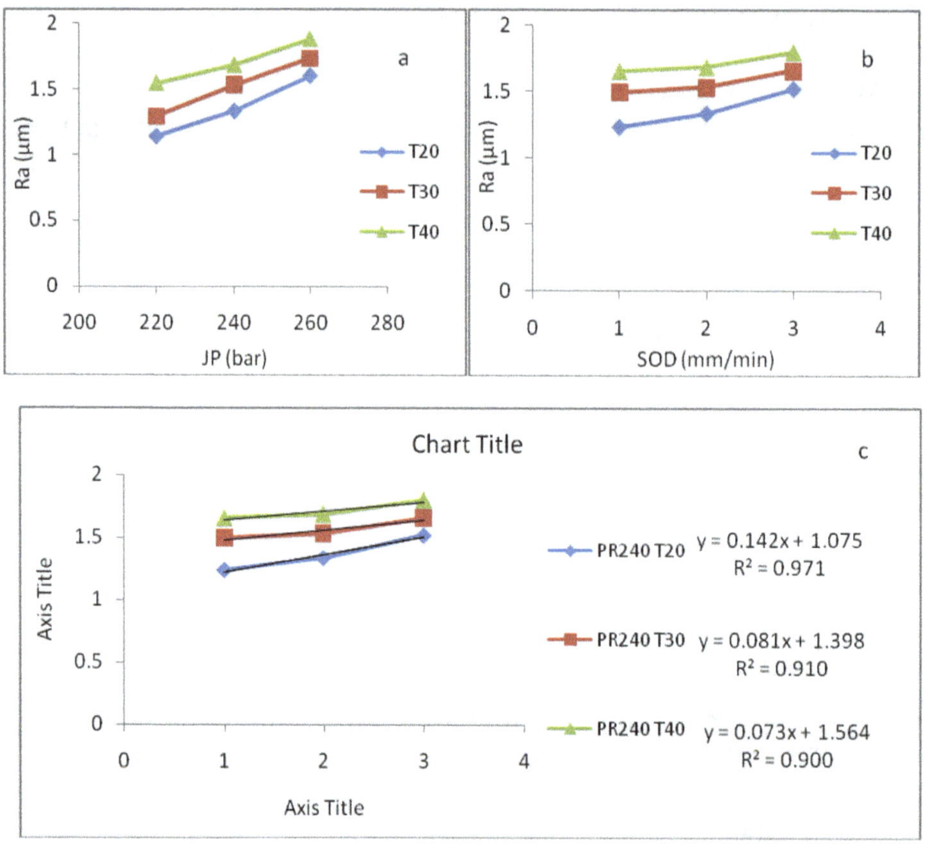

and TS, a superior removal of material is gained by accelerated abrasive particle. Owing to less machining time irrespective of high water pressure, the abrasives particles produce poor erosion over the cut surface of the composite. The dispersed water jet produces lower etching effect on the cut surface of the composite. In Figure 4(b) at the same machining condition, the rate of change in the Ra value is found to be negligible. It is believed that the influence of SOD and TS on surface finish of this composite has least significance. A linear trend line is drawn and it is shown in Figure 4 (c). From the obtained plots, a slight increase in Ra is noted but however, the range of increase is in an acceptable level. From the obtained linear equation at a varying SOD and TS machining condition, the residual square observation is found to be in acceptable range from 90% to 97%.

ANOVA of Ra is shown in Table 6. It is evident that there is an equal contribution of nearly 40% in JP and TS, where as in SOD, a contribution of 17.17% is obtained. The contribution level of JP x SOD is found to have a least value of 2.26% but its adverse effect on surface finish of the composites is verified by the F-test. Whereas the other two interactions (JP x TS and SOD x TS) give a partial influence less than 0.5% contribution and do have no effect in Ra. At 95% confident level in F-Test ($F_{0.05, 2, 8} = 4.46$), all the individual input parameters interpret to have a significant effect in Ra.

Analysis on Optimum Level

The manipulated level for each input parameters to enhance the optimistic output responses are shown in Table 7. JP alone shows a superior influence in MRR. High accelerated abrasives with less loss in kinetic energy of the particle, will erode a considerable amount of composite material at low SOD. The loss in kinetic energy of the particle at high SOD and increased width of water beam, results in larger top kerf width that cause an increase in MRR.

Each independent parameter in AWJM plays a significant role in the determination of the output responses. To reveal information about the machining effects on LaPO-Y_2O_3 composite in AWJM, the independent parameters are to be optimized to get the acceptable level of output responses. The data will provides the researcher in exploring the further addition properties of this composite material to a greater extent. From the observation, a minimum level of input parameters yields an affordable value to the composite.

Table 6. ANOVA for Ra

Parameter	DOF	Sum of Squares	Mean Square	F-Value	Contribution %
JP	2	0.452	0.226	195.23	38.18
SOD	2	0.203	0.102	87.83	17.17
TS	2	0.485	0.243	209.51	40.97
JP x SOD	4	0.0267	0.007	5.77	2.26
JP x TS	4	0.0056	0.001	1.22	0.48
SOD x TS	4	0.0019	0.0005	0.42	0.17
Error	8	0.0092	0.001		0.78
Total	26	1.184	0.579		100

Table 7. Parametric Optimistic Levels of each Input Parameters

Output Responses	Significant Factors	Proposed Levels to Optimistic Output
MRR	JP	JP=260bar, SOD=3mm and TS=20mm/min.
KA	JP	JP=220bar, SOD=1mm and TS=20mm/min.
Ra	JP, TS	JP=220bar, SOD=1mm and TS=20mm/min.

CUT SURFACE MICROSTRUCTURE CHARACTERIZATION STUDY

The microscopic images shown in Figure 5 under different operating conditions expose the presence of wear track, overlapping effect, weak grain boundary and erosion of composites by random movement of abrasives on cut surface. Figure 5 shows the microscopy image of the cut region of composite. The hard SiC particles with high acceleration energy impinge the top surface to perform the machining operations. With increase in machining time the rate of impingement of backscatter particles over the cut surface gets increased and it leads to get the plastic deformation surface. Figure 5(a) shows the surface roughness (Ra) on the top kerf surface. Fine cracks, direction of the motion of the abrasives and the crater formed during the machining are clearly visible. In Figure 5(b), it is noticed that the distribution of the spectrum on the Peak Roughness (Rp) and Valley Roughness (Rv) is uniform. The less energy backscatter abrasives erode the newly formed cut surface by increasing the kerf angle with superior surface finish.

Figure 5(c) shows the middle region of the kerf surface. A piece of composite about to tear apart on the kerf surface is noticed. The lack of machining time on these region reduces the abrasion flow rate may failed to remove this portion. Impinging abrasives after machining the composite with its remaining energy punches the nearby kerf surface. The continuous bombardment in the same portion results in the formation of large crater wear. Figure 5(d) shows the spectrum of the middle kerf surface. The irregular distribution of the peaks shows the random movement of the abrasives over the cut surface. The value of Rp is found to be less than Rv.

Figure 5(e) shows the rough kerf surface of the composite. The large wear and scar tracks are visible over the entire surface and the linear movement of the abrasives creates the curvy motion of the abrasives. The full acceleration gained abrasives require an additional time to reach the rough kerf surface than the top kerf surface at constant TS. This condition leads to get striations over these cut region. The deviation between the Rp and Rv in Figure 5(f), shows the irregular movement of hard abrasives while removing the composite. The amount of backscatter abrasives that hits the kerf surface is found to be negligible and this leads to have less wear track over the kerf surfaces.

Microscopy examinations of the kerf surface of the composite reveal that on SiC machined surface, the failure occurs by grain boundary deformation. The developed internal stress and repetitive cyclic impact load of hard SiC abrasives over the composite lead to get plastic deformation.

Figure 5. The three cut region in the composite with its profile spectrum on using SiC. a: SEM image of upper kerf surface, b: Kinematic profile of upper kerf surface, c: SEM image of middle kerf surface, d: kinematic profile of middle kerf surface, e: SEM image of low kerf surface, f: kinematic profile of lower kerf surface

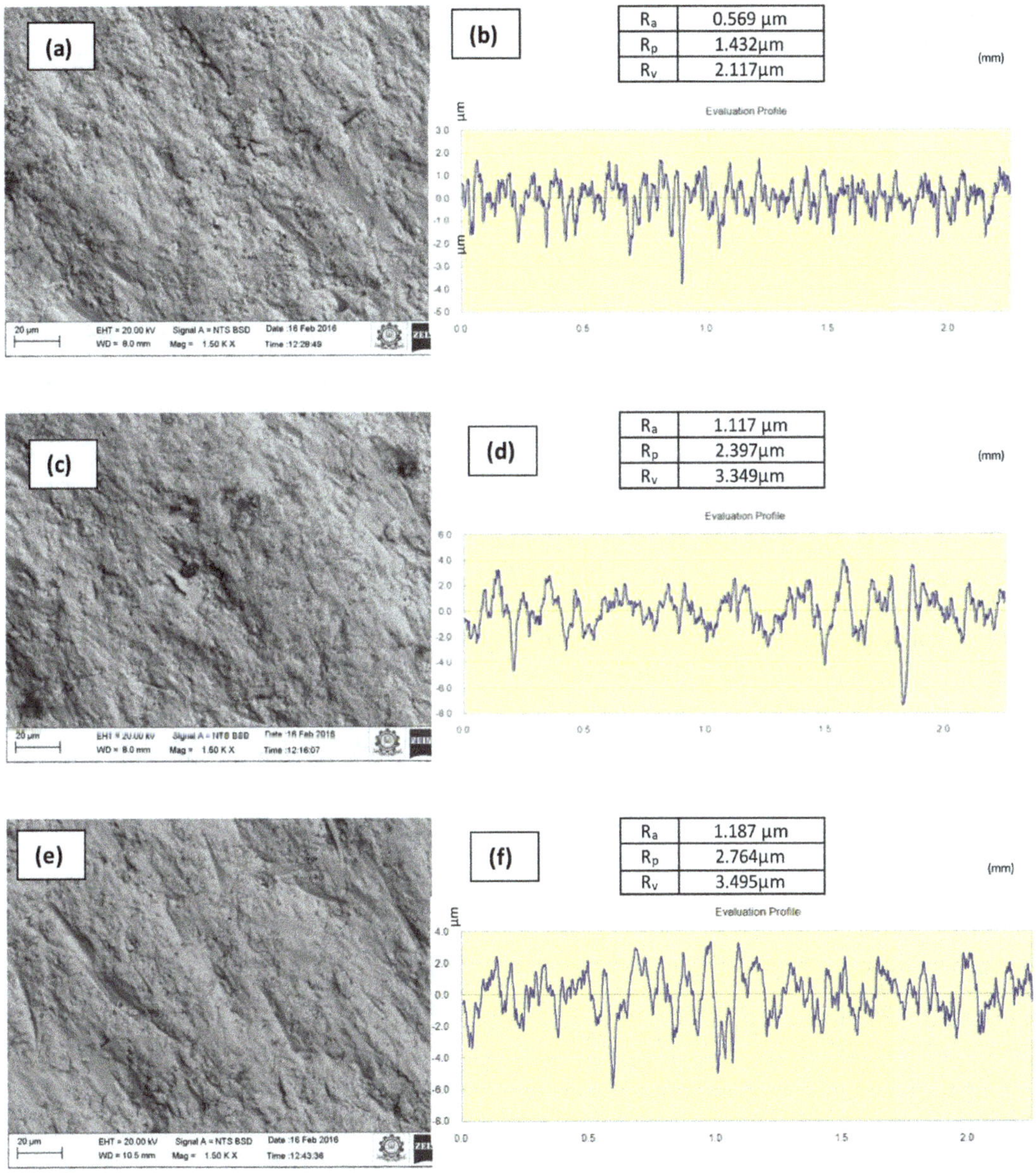

CONCLUSION

The $LaPO_4/Y_2O_3$ ceramic matrix composite with 80-20 ratio (Vol %) has been prepared by Aqueous Sol-Gel Process. The prepared powder is compacted to get the required shape. The composite is subjected to AWJM for interference study of each input process parameters over the individual output responses and the microstructural characterization studies on the cut region of composite are summarized as follows:

- Substantial increase in MRR is obtained on different input parameters. Increasing the SOD and TS, irrespective of JP produces an acceptable linear range in MRR with increase in KA.
- ANOVA of MRR shows that JP with a major contribution of 77.6% interprets MRR in the machinable composite.
- Increasing in JP and TS, irrespective of SOD produces an acceptable range of Ra with increase in KA.
- ANOVA of KA shows that JP of 45.14% followed by TS of 25.09% and SOD of 22.08% affects KA. In-spite of least contribution in interaction by F-test, a significant influence of JP x SOD with 5.5% on KA is noted.
- ANOVA of Ra shows that the JP and TS significantly affect with a contribution of nearly 40% each. From the F-test, it is identified that there is a significant effect in the interaction of JP x SOD in determining the surface finish of the composite.
- Among the input parameters, JP plays a vital role in the determination of each individual output responses.
- Microscopic examinations on the cut surface show that the tear of particles in the composite is by both transgranular and intergranular failure.
- The initiation and propagated cracks which are visualized along the grain boundary reveal that the impingement of hard abrasives on composite produces a break in bond and simultaneous hammering effect leads to the machining of the composite material.

ACKNOWLEDGMENT

The authors are thankful to the Centre for Advance Machining (DST-FIST Sponsored) Kalasalingam University, Krishnankoil for providing amenities and valuable support.

REFERENCES

Abdul Majeed, M., Vijayaraghavan, L., Malhotra, S. K., & Krishnamurthy, R. (2008). Ultrasonic machining of $Al_2O_3/LaPO_4$ composites. *Journal of Machine Tools & Manufacture*, *48*(1), 40–46. doi:10.1016/j.ijmachtools.2007.07.012

Alangi, N., Mukherjee, J., Anupama, P., Verma, M. K., Chakravarthy, Y., Padmanabhan, P. V. A., ... Gantayet, L. M. (2011). Liquid uranium corrosion studies of protective yttria coatings on tantalum substrate. *Journal of Nuclear Materials*, *10*(1-3), 39–45. doi:10.1016/j.jnucmat.2010.12.307

Alberdi, A., Suarez, A., Artaza, T., Escobar-Palafox, G. A., & Ridgway, K. (2013). Composite Cutting with Abrasive Water Jet. *Procedia Engineering*, *63*, 421–429. doi:10.1016/j.proeng.2013.08.217

Carach, J., Hloch, S., Hlavacek, P., Gombar, M., Klichova, D., Botko, F., ... Lehocka, D. (2016). Hydro-abrasive disintegration of alloy Monel K-500- the influence of technological and abrasive factors on the surface quality. *Procedia Engineering*, *149*, 17–23. doi:10.1016/j.proeng.2016.06.633

Chen, L., Siorest, T. E., & Wong, W. C. K. (1996). Kerf Characteristics in Abrasive Waterjet Cutting of Ceramic Materials. *Journal of Machine Tools & Manufacture*, *36*(11), 1201–1206. doi:10.1016/0890-6955(95)00108-5

Chenghao, L., Shusen, W., Naibao, H., Zhihong, Z., Shuchun, Z., & Jing, R. (2015). Effects of Lanthanum and Cerium Mixed Rare Earth Metal on Abrasion and Corrosion Resistance of AM60 Magnesium Alloy. *Rare Metal Materials and Engineering*, *44*(3), 521–526. doi:10.1016/S1875-5372(15)30031-X

Derzija, B. H., Ahmet, C., Muhamed, M., & Almina, D. (2015). Experimental study on surface roughness in abrasive water jet cutting. *Procedia Engineering*, *100*, 394–399. doi:10.1016/j.proeng.2015.01.383

Fang, L., Kong, X. L., Su, J. Y., & Zhou, Q. D. (1993). Movement patterns of abrasive particles in three-body abrasion. *Wear*, *162/164*, 782–789. doi:10.1016/0043-1648(93)90079-2

Fang, L., Kong, X. L., & Zhou, Q. D. (1992). A wear tester capable of monitoring and evaluating the movement pattern of abrasive particles in three-body abrasion. *Wear*, *159*(1), 115–120. doi:10.1016/0043-1648(92)90292-G

Fowler, G., Pashby, I. R., & Shipway, P. H. (2009). The effect of particle hardness and shape when abrasive water jet milling titanium alloy Ti6Al4V. *Wear*, *266*(7-8), 613–620. doi:10.1016/j.wear.2008.06.013

Ghosh, D., Doloi, B., & Das, P. K. (2015). Parametric analysis and optimization on abrasive water jet cutting of silicon nitride ceramics. *Journal of Precision Technology*, *5*(3/4), 294–311. doi:10.1504/IJPTECH.2015.073833

Gong, G., Zhang, B., Zhang, H., & Li, W. (2006). Pressure less sintering of machinable $Al_2O_3/LaPO_4$ composites in N_2 atmosphere. *International Journal of Ceramics International*, *32*(3), 349–352. doi:10.1016/j.ceramint.2005.03.002

Gudimetla, P., Wang, J., & Wong, W. (2002). Kerf formation analysis in the abrasive waterjet cutting of industrial ceramics. *Journal of Materials Processing Technology*, *128*(1-3), 123–129. doi:10.1016/S0924-0136(02)00437-5

Gunduz, O., Daglilar, S., Salman, S., Ekren, N., Agathopoulos, S., & Oktar, F. N. (2008). Effect of Yttria-doping on Mechanical Properties of Bovine Hydroxyapatite (BHA). *Journal of Composite Materials*, *42*(13), 1281–1287. doi:10.1177/0021998308092196

Hascalik, A., Çaydaş, U., & Gürün, H. (2007). Effect of traverse speed on abrasive waterjet machining of Ti–6Al–4V alloy. *Materials & Design*, *28*(6), 1953–1957. doi:10.1016/j.matdes.2006.04.020

Hloch, S., Valicek, J., & Simkulet, V. (2009). Estimation of the smooth zone maximal depth at surfaces created by Abrasive Waterjet. *Journal of Surface Science and Engineering*, *3*, 347–359.

Hutyrova, Z., Scuska, J., Hloch, S., Hlava, P., & Zele, M. (2016). Turning of wood plastic composites by water jet and abrasive water jet. *Journal of Advance Manufacturing Technology*, *84*, 1615–1624.

Khan, A. A., & Haque, M. M. (2007). Performance of different abrasive materials during abrasive water jet machining of glass. *Journal of Materials Processing Technology*, *191*(1-3), 404–407. doi:10.1016/j.jmatprotec.2007.03.071

Kim, Y., Hong, M.-H., Lee, S. H., Kim, E.-P., Lee, S., & Noh, J.-W. (2006). The Effect of Yttrium Oxide on the Sintering Behavior and Hardness of Tungsten. *Metals and Materials International*, *12*(3), 245–248. doi:10.1007/BF03027538

Kopac, J., & Krajnik, P. (2007). Robust design of flank milling parameters based on grey-taguchi method. *Journal of Materials Processing Technology*, *191*(1-3), 400–403. doi:10.1016/j.jmatprotec.2007.03.051

Maiti, K., & Sil, A. (2006). Preparation of Rare earth oxide doped alumina ceramics, their hardness and fracture toughness determinations. *Indian Journal of Engineering and Materials Sciences*, *13*, 443–450.

Min, W., Miyahara, D., Yokoi, K., Yamaguchi, T., Daimon, K., Hikichi, Y., ... Ota, T. (2001). Thermal and Mechanical Properties of Sintered $LaPO_4$–Al_2O_3 Composites. *Materials Research Bulletin*, *36*(5-6), 939–945. doi:10.1016/S0025-5408(01)00555-4

Mogilevsky, P., Boakye, E. E., & Hay, R. S. (2007). Solid solubility and thermal expansion in a $LaPO_4$–YPO_4 System. *Journal of the American Ceramic Society*, *90*(6), 1899–1907. doi:10.1111/j.1551-2916.2007.01653.x

Pirso, J., Viljus, M., Juhani, K., & Kuningas, M. (2010). Three-body abrasive wear of TiC–NiMo cermets. *Tribology International*, *43*(1-2), 340–346. doi:10.1016/j.triboint.2009.06.014

Sankar, S., & Warrier, K. G. K. (2011). Aqueous sol-gel synthesis of LaPO4 nano rods starting from lanthanum chloride precursor. *Journal of Sol-Gel Technology*, *58*(1), 195–200. doi:10.100710971-010-2377-4

Sankar, S., Athira, N., Raj, C.K., Jyothi, Warrier, K.G.K., & Padmanabhan, P.V.A. (2012). Room temperature synthesis of high temperature stable lanthanum phosphate–yttria nano composite. *Materials Research Bulletin*, 47.

Selvan, M. C. P., & Raju, N. M. S. (2012). Analysis of surface roughness in abrasive waterjet cutting of cast iron. *International Journal of Science Environment and Technology*, *1*, 174–182.

Shanmugam, D. K., & Masood, S. H. (2009). An investigation on kerf characteristics in abrasive waterjet cutting of layered composites. *Journal of Materials Processing Technology*, *209*(8), 3887–3893. doi:10.1016/j.jmatprotec.2008.09.001

Shanmugam, D. K., Wang, J., & Liu, H. (2008). Minimisation of kerf tapers in abrasive waterjet machining of alumina ceramics using a compensation technique. *Journal of Machine Tools & Manufacture*, *48*(14), 1527–1534. doi:10.1016/j.ijmachtools.2008.07.001

Sharma, V., Chattopadhyaya, S., & Hloch, S. (2011). Multi response optimization of process parameters based on Taguchi-Fuzzy model for coal cutting by water jet technology. *Journal of Advance Manufacturing Technology*, *56*(9-12), 1019–1025. doi:10.100700170-011-3258-x

Srinivas, S., & Babu, N. R. (2011). Role of garnet and silicon carbide abrasives in abrasive waterjet cutting of aluminum-silicon carbide particulate metal matrix composites. *Journal of Applied Research in Mechanical Engineering*, *1*, 109–122.

Sujith, S. S., Arunkumar, S. L., Mangalaraja, R. V., Mohamed, A. P., & Ananthakumar, S. (2014). Porous to dense $LaPO_4$ sintered ceramics for advanced refractories. *Ceramics International*, *40*(9), 15121–15129. doi:10.1016/j.ceramint.2014.06.125

Sun, Z., Zhou, Y., Wang, J., & Li, M. (2007). γ-Y2Si2O7, a Machinable Silicate Ceramic: Mechanical Properties and Machinability. *Journal of the American Ceramic Society*, *90*(8), 2535–2541. doi:10.1111/j.1551-2916.2007.01803.x

Wang, J. (2004). Techniques for enhancing the cutting performance of abrasive waterjets. *Key Engineering Materials*, *257/258*, 521–526. doi:10.4028/www.scientific.net/KEM.257-258.521

Xiaoyong, R., Peng, Z., Hu, Y., Wang, C., Fu, Z., Yue, W., ... Hezhuo, M. (2013). Abrasive wear behavior of TiCN cermets under water-based slurries with different abrasives. *Tribology International*, *66*, 35–43. doi:10.1016/j.triboint.2013.04.002

Xu, H. H. K., Wei, L., & Jahanmir, S. (1995). Grinding force and micro crack density in abrasive machining of silicon nitride. *Journal of Materials Research*, *10*(12), 3204–3209. doi:10.1557/JMR.1995.3204

Zhang, S., Wu, Y., & Wang, Y. A. (2011). A review on abrasive waterjet and wire electrical discharge machining – high speeds. *The Open Mechanical Engineering Journal*, *5*(1), 178–185. doi:10.2174/1874155X01105010178

Zhao, W., & Guo, C. (2014). Topography and microstructure of the cutting surface machined with abrasive waterjet. *Journal of Advance Manufacturing Technology*, *73*(5-8), 941–947. doi:10.100700170-014-5869-5

Section 3
Industrial Engineering and Management

Chapter 10
Methodology of Operationalization of KPIs for Shop-Floor

Mariana Raposo Oliveira
Universidade de Lisboa, Portugal

Diogo Jorge
Erising, Portugal

Paulo Peças
Universidade de Lisboa, Portugal

ABSTRACT

Key performance indicators (KPIs) are a critical tool to support activities and results' monitoring in any industrial organization. The published literature and the available approaches on KPIs focus on the business and administrative level, being computed with information retrieved at the shop-floor level. Despite that, there is a scarcity of structured and comprehensive approaches to support the generation of KPIs to be used at the shop-floor level (the few existent approaches are empiric-based). In this chapter, a methodology to support the selection and organization of KPIs at the shop-floor level is proposed. Departing from the Hoshin Kanri strategy deployment, it identifies the levels of decision and control in the company regarding the production activities and derives the most adequate KPIs for each level based on universal questions about "what performance to assess." The build-up of visual management boards for each level is also proposed.

INTRODUCTION

The Industry 4.0 and the KPIs

Nowadays Big Data is a widely used term to refer the generation and communication of data associated with new technologies (Fanning, 2016). This communication flow is based on the data collected

DOI: 10.4018/978-1-5225-5445-5.ch010

by the company itself and it is done with the final aim of allowing quick and informed decisions, with increased productivity in mind. This has become increasingly popular as a result of the Industry 4.0 concept, also known as fourth industrial revolution (Hermann, Pentek, & Otto, 2016). Industry 4.0 is associated with information technology, automation and control applied to production processes, i.e. through the "internet of things", the cyber-physical systems communicates and collaborates with each other and with humans, throughout the entire value chain and in real time (Hermann et al., 2016). This concept was born in Germany in 2011, under a German Government strategic project which aimed to promote the digitalization of production systems and to allow its in-line monitoring and from afar (Henning, Wolfgang, & Johannes, 2013; Hermann, Pentek, & Otto, 2015).

As this type of systems is implemented, the amount of data generated is expected to increase. If this data is to be used in a useful and efficient way, it requires an appropriate treatment. As a result, there is the need to transform "big data" into "smart data", nurturing to speed up the decision-making process, supplying vital and assertive information for management activities (Fanning, 2016; Hermann et al., 2016).

Taking into account such difficulties and the need to find how to filter the data (to make it useful) generated in companies, the concept of Key Performance Indicator (KPI) has become increasingly mentioned in the literature, as an essential instrument in the interface between data generated and decision makers (humans or decision algorithms) (Fanning, 2016; Hermann et al., 2016; Marr, 2015). A definition of KPIs can be given: a restricted group of indicators intended to gauge the performance of a system in a systematic and comprehensive manner (Kahn, 2013; Parmenter, 2010).

The (Mis)Use of KPIs

In order to derive the greatest benefits from KPIs, they should be linked to the company strategy, reflecting completely its key points (Kahn, 2013; Parmenter, 2010). Therefore, the KPIs can be used as a mean of communicating the company's strategy (inside the company) and monitor the effectiveness of its application. Usually, the adoption of Hoshin Kanri and Balanced Scorecard (BSC) approaches are recommended together with a well-established link with suitable KPIs for gauging the strategic objectives (Kaplan & Norton, 2005; Thürer, 2013; Witcher & Chau, 1993). Within the Hoshin Kanri approach, the strategy should result from a long process of negotiation creating an organic flow of information circulating throughout the entire company (Tennant & Roberts, 2001; Witcher & Butterworth, 1999). One of the purposes of using the BSC is to make it easier to define and communicate the strategy, in such a way as to help to define the KPIs (Kaplan & Norton, 2005). In order to be able to monitor all the work carried out, it is therefore essential to combine the generation of KPIs with these two approaches.

The effectiveness of the use of KPIs depends on alignment of the selected ones with the company strategy and on the ability of controlling the KPI. For this latter task, it is fundamental to define the target (the value that it intends to reach) and the baseline (initial KPI value, before any improvement measure) for each KPI. By comparing KPIs with these values, it will be possible to monitor the performance of the company in relation to its strategic-based objectives. The detection of the problems and the potential improvements identification is easier when the company performance is mirrored concisely by proper selected and controlled KPIs (Kahn, 2013; Parmenter, 2010). KPIs are also considered a fundamental tool for Lean Manufacturing and Kaizen philosophy implementation, since the final objective in these two approaches is the optimization of processes through a continuous reduction of waste (España, Tsao, & Hauser, 2012; Mourtzis, Fotia, & Vlachou, 2017; Dimitris Mourtzis, Fotia, Vlachou, & Koutoupes, 2017; Sanders, Elangeswaran, & Wulfsberg, 2016). Lastly, for the KPIs to be fully understood and to have an impact on the entire organization, they must be reconciled with visual management. This would make them attractive as well as allow them to be integrated in the company's daily activities and community informal communication (Bhoi, Desai, & Patel, 2014; iF, 2017; Murata & Katayama, 2009).

Despite the referred huge potential of KPIs, its selection (i.e. the set of KPIs used), use and display are not done in the best way by most of the companies. From in-situ observation of the praxis in several companies, the authors of this chapter found a common situation regarding the KPIs selection: the companies have as inspiration to their KPIs, the KPIs in use by other companies. This situation is also confirmed by several experts in KPIs implementation (España et al., 2012; iF, 2017; leanmanufacture, 2017; Marr, 2015; Savkin, 2016). This situation increases the risk of the used KPIs not reflect an assertive and comprehensive performance of the company. So, if the company does not spend enough time properly defining and communicating its strategy, the analysis of results (KPIs) can be compromised and even give incorrect or insufficient indications of necessary changes. Another common mistake in the selection of KPIs comes from collecting of an excessive number of indicators, resulting in an excess of data, and so not useful for decision-making. Lastly, sometimes the KPIs are displayed in an unappealing way and so do not cause the desire impact (iF, 2017; Marr, 2015; Savkin, 2016). Thus, given the difficulty in selecting, showing and exploring KPIs correctly, they are often used inefficiently and ineffectively – they are not selected and used to take full advantage of its potential.

The Shop-Floor KPIs: The Motivation

The above characterization of the use of KPIs was done for KPIs in general. In the particular case of KPIs to be used in the shop-floor, the problems regarding its selection and exploration should be discussed specifically. In the bibliography is possible to find a fair amount of information on the nature of the KPIs and some methodologies about how to select them. But, most of these methodologies are associated with macro performance assessment mainly for top management decision-making, existing only a few publications that refer some aspects about the use of KPIs at shop-floor level (Marr, 2015; Parmenter, 2007; Protzman et al., 2016). In these references is not explicitly described how to select, display and explore KPIs at the shop-floor level. There are also several suggestion of sets of KPIs available in the web, as well as examples of displays to be used at shop-floor level, but they are usually published by consultancy companies resulting in lack of theoretical foundation and comprehensiveness and absence of applicability to each company specificities. This scenario was the base for the motivation to develop the methodology proposed in this chapter.

The Contribution

Having in mind the general misuses of KPIs and the particular gaps on guidance about the selection and use of KPIs at shop-floor level, a methodology is proposed in this chapter. Consequently, the main aim of this methodology is to guide the use of KPIs based on a structured and comprehensive selection, intended to explore the shop-floor level through:

- Overcoming of the difficulty associated with the effective communication of the strategy to all the staff, and its conversion into concrete objectives, to ensure that the chosen KPIs reflect the company strategy as well as its critical points;

- Comprehensive analysis of all the aspects of the production processes, ensuring that any problem or deviation will be identified;
- Ensure a regular review of KPIs, checking that they are a true reflection of the company over a certain period of time;
- Involvement of the staff in the selection of KPIs, so that they feel committed to them and motivated to achieve company's goals;
- Show the KPIs in a way that is engaging, intuitive and easy to assimilate;
- Standardize the way of showing the KPIs;
- Show each employee only the relevant information towards his/her performance increasing and avoiding to spend time with excessive/inadequate information;

As an assumption, the proposed methodology recommends the use of Hoshin Kanri strategy deployment and of the BSC method to assure that all the company collaborators are aware of the strategy of the company. After this fact is assured, the next steps consist in the decision levels identification (i.e. equipment, line/cell, section, business unit) and activities to control (i.e. quality, production, maintenance, etc.) in the company (regarding the production system). A survey is conducted to the several decision levels, with pre-defined universal questions about "what performance to assess" – each question is linked to a KPI, so the methodology includes a KPI database for shop-floor (besides other more general KPIs). The KPIs are selected among the most answered questions for each level of decision and are displayed using intuitive dashboards – visual management boards.

The recurrent use of the proposed methodology allows to take full advantage of the KPIs because it allows their proper selection, design, display and revision. The proposed methodology is also flexible about the type of companies where it can be applied, although it is more adequate for manufacturing and assembling companies.

THE EXISTING MODELS

In general, the set of KPIs selected to monitor the performance of a company are used for the following aims (Kahn, 2013; Marr, 2015; Parmenter, 2010; Savkin, 2016):

- To help in decision making, since they filter the essential information, highlighting it;
- As a mean of communication, allowing managers to transmit objectives and goals to the employees, as a result of the predefined strategy;
- To promote a sense of commitment between all the staff and their work, so they can see just how influential they are in the performance of the company;
- To encourage improvement measures, focused on results, with the intention of improving them;
- To prevent problems, by monitoring the critical values;
- To make comparison with other companies, resulting in the search for better market opportunities, based on the benchmark.

Given the difficulty in selecting "ideal" KPIs, several authors developed different models for choosing them. Table 1 summarizes the three models studied in this chapter. Both Parmenter's model (Parmenter, 2007) and Marr's model (Marr, 2015) are connected with top management, unlike the +QDIP model (Hurley, 2012; Protzman et al., 2016), which is directed at the shop-floor. Parmenter (Parmenter, 2010) and Marr (Marr, 2015) divide the indicators into different groups: for Parmenter there are results indicators (RI) and performance indicators (PI), identifying also the most important (key ones) of each group (Key results indicators – KRIs, Key performance indicators – KPIs); Marr prefers to divide them into strategical and operational. The strategical ones aim to assess the performance regarding the strategic targets and should be measured weekly or monthly; the operational ones aim to monitor the daily performance of the organization.

The way these two authors separate indicators is remarkably similar. In both models the authors also classify the indicators regarding its different time use: i) indicators measured in real time or at most every week, focused on operational objectives and daily actions; ii) indicators measured monthly, focused on strategy and long-term results. This means the first one can be considered as operational indicators, related to the front line and the daily goals of each employee, and the latter can be considered to be used at higher level of management, so more strategic and of greater interest for top management. Nevertheless, there is no defined guidelines to allow the selection of properly adequate KPIs for a specific production system.

Marr (2015) argues that there should be 15 to 20 indicators for each business unit. These should be standardized (making alterations only when strictly necessary) to facilitate their reading, comprehension and contextualization. Parmenter (2010) does not mention this, stating (only) that each company should use 10 KPIs, 80 PIs and RIs, and 10 KRIs. Neither of the authors mention categories of KPIs, even though their books mention the following indicators: financial, client, marketing and sales, project, production, workers and social responsibility.

The +QDIP model (that means Quality, Delivery, Inventory, Production) (Hurley, 2012; Protzman et al., 2016) is a practical model to be used on the shop-floor giving less emphasis to the alignment with the management structure in the choice of KPIs, while still using visual management. This model devises the KPIs in different categories and uses two to four KPIs per category. In some cases an "E" is added to assess environmental performance. Each category is represented by the first letter devised in several sections, one for each day of the month. Red and green colors are daily used to fill the devised sections, depending if the performance on the KPI was below or above the target. This model is used in some companies and recommended by consultant companies, but it does not have a theoretical background.

METHODOLOGY DEVELOPMENT

The methodology developed for the selection, display and exploration of KPIs at the shop-floor level is explained and discussed in this section. The methodology combines two important aspects referred in the KPIs misusing: the harmony between the company strategy and the selected KPIs, and the involvement of the staff in the KPIs selection. It proposes the staff involvement in the KPIs choice to increase the feeling of commitment. It is also intended that the employees feel more stimulated to achieve the defined target. These assumptions are defended by different authors, like Marr (2015) and Parmenter (2010) the main references of this work. It also includes the categorization of KPIs to support its control and organization, and a proposal for its display fostering the visual management. In fact, one of the common difficulties in the use of KPIs is the poor effectiveness of most of the visual boards, so the methodology suggests a way to expose the KPIs to be as clear and intuitive as possible.

Table 1. Comparison between the Parmenter, Marr and + QDIP models, according to their division of indicators, number and update intervals.

Models	Parmenter	Marr	+QDIP
Division of indicators	Results of Indicators (KRIs e RIs) and Performance Indicators (KPIs, PIs)	Strategic indicators and operational indicators	Safety, quality, delivery, cost, inventory, production, environment
Number of indicators	10 KPIs and 10 KRIs 80 Pis and 80 RIs	15 to 20 high level, and a similar number for each business unit	Each category represents some indicators, there being no defined number
Update intervals	KPIs: real time to weekly KRIs: one to four months RIs e PIs: real time to monthly	Strategic indicators: weekly and monthly Operational indicators: real time	Daily in a monthly spectrum

In order to overcome the limitations of the existing methods, the methodology proposed recommends, before its application, the communication of the company strategy to the employees, aiming to ensure that everyone knows their role in the company and to establish a feeling of sharing the same objectives and targets. The methodology can be applied in other circumstances, i.e, without the implementation of this recommendation. However, this communication and sharing is suggested to obtain more realistic results and to be able to choose indicators that truly reflect the company. The authors recommend the use of Hoshin Kanri and BSC methods to define the strategy and allow an easier communication to the company employees and therefore contribute to the alignment of the generated KPIs with the company strategy.

Step 1: Production System Model and Levels

The first difficulty in the methodology development was the fact that KPIs should be explicitly explained to those who are responsible for their performance and for whom they have meaning and relevance. It must be kept in mind that important KPIs for one area can be insignificant for others. Because of this, it is necessary to avoid excess data and/or lack of contextualization.

To solve that problem, a typical production system model was considered to develop the methodology (Figure 1). An organizational pyramid was also used as reference to develop the methodology, taking into consideration different management levels and their responsibilities, as shown in Figure 2. These two models are generic and allow the understanding of the methodology application; they should be adapted for the company real situation when applying the methodology.

The production system model considered (Figure 1) is composed by five distinct production levels. The Production System itself (a macro level) is composed by a set of business units. Each Business Unit, the next level, is made up of a number of Sections, being each Section a family of products. Within the Sections there are various Cells, and each of these consists in a group of different Processes.

This model can be adapted depending on the productive unit being studied. For example, the cell level can be omitted if the product results from only one process (as can be seen in the bottom right corner of (Figure 1). More levels can also be created. For example, when the same company has more than one production system, and wants to monitor all of them in the same way, it is possible to create a sixth level to control that large system.

Figure 1. Production System Model used as reference for the proposed methodology.

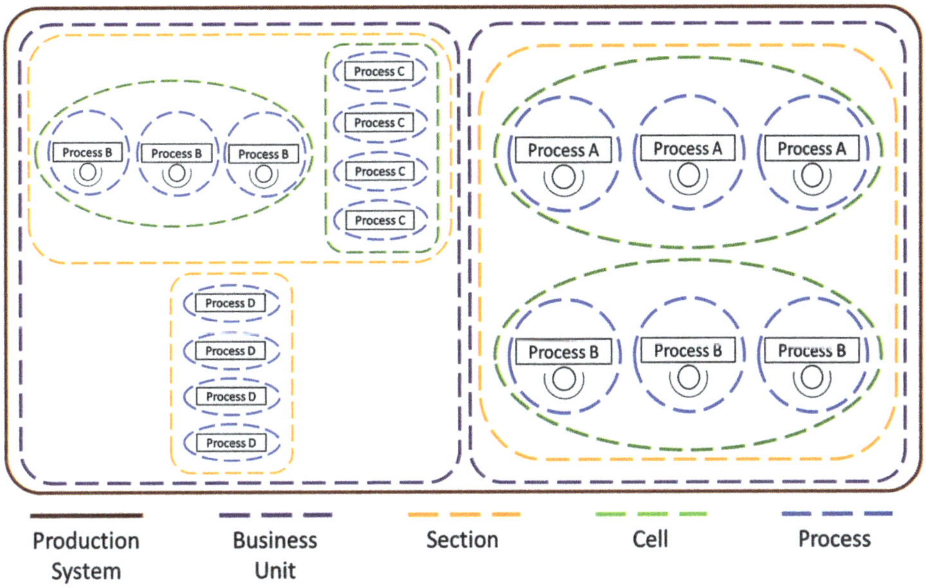

The physical division of the production system of Figure 1, combined with the staff division of responsibilities, results in the organizational pyramid of Figure 2. The relationship proposed in this pyramid was established in order to link the responsibilities of each worker with their place of work: i) the operators are responsible for the process, thus constituting level 1; ii) the cell is the responsibility of the cell leader, forming level 2, etc. Again, both physical and staff divisions can be adapted to each company, depending on its structure and organization. The purpose of this division is to select the appropriated information for each level and guarantee that every worker is involved and tuned with the KPIs of his/her level.

Since every level of the pyramid has different responsibilities, each of these levels needs to check different indicators. This way, they can monitor their work results more effectively. Thus, each group will have different displays, associated with specific KPIs.

Step 2: Classes of KPIs

The KPIs aim to measure different type of performances. At shop-floor level several activities occur simultaneously and a comprehensive set of KPIs must cover all the types of activities. To assure that the final set of KPIs covers all the significant areas of performance in the shop floor, a set of nine Classes is proposed covering all the company value chain. These nine classes were based on the +QDIP model (Hurley, 2012) but proposing a few more classes since that model only consider five or sometimes six classes (Hurley, 2012; Protzman et al., 2016). An acronym will be used to designate each Class in this chapter as indicated in Table 2 (acronym).

A match between the Levels and the Classes is proposed (Table 3). In fact there is no need to assess all the classes of KPIs in all levels, avoiding excessive and redundant information. Some examples explain the need for this matrix. The Mechanical class is relevant for Level 1 (the operators' level). The Suppliers class performance only makes sense to be analyzed in groups of the same production processes

Figure 2. Organizational Pyramid Model used as reference for the proposed methodology, with the color code of Figure 1.

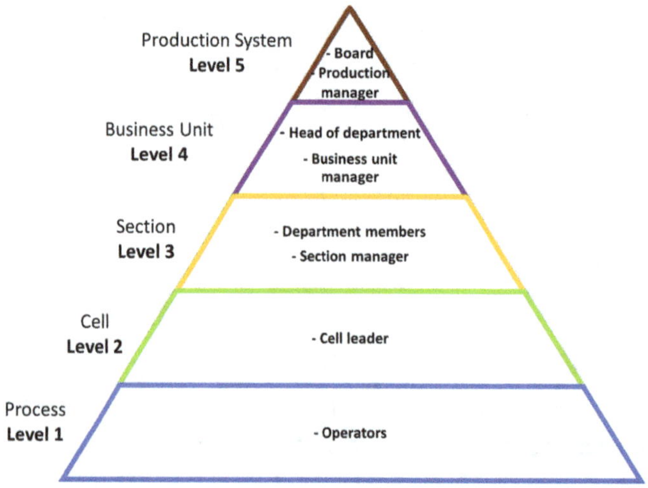

Table 2. List of Classes defined in the proposed methodology

Class Name	Acronym	Content
Suppliers	S	Includes the internal (process to process) and external performance of the suppliers. Evaluate its performance and the performance of the previous process.
Planning	Pl	Evaluate the planning effectiveness; if this is performed correctly and in time. Also, it evaluates the influence of the delays in the performance and the level of resources using according to the planning.
Productivity	Pr	Evaluates the process itself, its performance and levels of productivity.
Quality	Q	Reflects the problems of bad quality and the actions to correct them. Evaluates also the critical parameters of bad quality in each process.
Mechanical	Mec	Evaluates and identify the deviation between the established and the real value of the mechanical parameters of the machines. These parameters are analyzed because they can influence the process and the quality of the parts.
Maintenance	Mai	Referring to all the maintenance (planned or unplanned) points, evaluating its efficiency and cost.
Clients	C	Focused on internal and external clients, studying their satisfaction and targets/ specifications fulfilment.
Environment	E	Evaluates the several types of wastes and environmental related aspects.
Workers	W	Evaluate workers' performance, satisfaction and costs associated to them.

(from level 2 to 5), since the same process has the same type of supplier. About the classes of Clients and Workers, the analysis is only relevant for higher levels of decision (level 3,4 and 5) because it is a very sensitive information.

Following Marr (2015), who suggest the use of 15 to 20 indicators, each class will be represented by three KPIs. Therefore, the KPIs will depend on the company where the methodology is being implemented and the levels which are represented.

The contextualization of the specific characteristics of the company is assured by this and by the latter steps, assuring that all the representative decision levels and types of activities are represented. It is important to mention that, if the company does not need all these classes, it is possible to take them out or add new ones.

Step 3: Key Performance Questions and Surveys

The proposed methodology assures the selection of appropriated and tuned KPIs through the application of surveys to the several levels of the production system. The employees of the several levels will answer to specifically designed surveys (for their level). Depending on the surveys' results three KPIs will be selected for each class in each level. This way, two important aspects are assured by the methodology: KPIs comprehensiveness (by the complete list of classes) and KPIs meaning to its users (they are selected specifically by the ones that will explore them). This is one of the reasons why is so important all the employees are tuned with the company strategy: if they are not aware of it, the selection of KPIs might be not aligned with the mission, objectives and value proposal of the company. Prior to the delivery of the surveys to the employees, and to increase transparency and commitment, a workshop should be held to explain the main aim and advantages of using KPIs as well as the process of KPIs selection.

The way the surveys are designed is now explained. They are based on good practices proposed by other authors. According to Marr (2015) a good understanding of what is intended to measure by the KPI is more important than the selection of KPI. For this reason, he has created the KPQ (Key Performance Question). KPQs guarantee that the information provided by the KPIs is really important and give them a "raison d'être" (Marr, 2015). These questions also help in contextualizing the KPIs and converting them into knowledge.

An intermediary step prior to the construction of the KPQs list, was the creation of a list of possible KPIs to be used at shop-floor level. That list is one important asset of the proposed methodology and was build based on the study and analysis of several published books and on-line KPIs libraries (BI, 2017; Kahn, 2013; Klipfolio, 2017; KPIlibrary, 2017; leanmanufacture, 2017; Marr, 2015; Parmenter, 2010). The most commonly used and recommended KPIs were listed and subsequently divided into the classes. Then, for each KPI, a KPQ was constructed. The KPI will be the operational quantification or measurement of the proposed KPQ. The list of KPIs drawn up leaves room for improvement, since it is not exhaustive and can be modified and/or incremented. In addition, the KPIs listed were translated into formulas, in order to show the data that needs to be collected. In the construction of the KPQs, explicit caution was taken not to make them too explicit in order to avoid biased answers. The intention here was for respondents to focus more on questions they want clarified, rather than on the KPIs that they want to see at the end. The intention was also to achieve impartial choices, as opposed to ones biased in favor

Table 3. Matching of Levels and Classes (refer to acronym of Table 2).

Level	Classes								
	S	Pl	Pr	Q	Mec	Mai	C	E	W
1		X	X	X	X	X		X	
2	X	X	X	X		X		X	
3,4,5	X	X	X	X		X	X	X	X

of KPIs that are easier to measure or that the company generally uses, without taking into consideration their real purpose (Parmenter, 2010). Furthermore, the levels and classes where the each KPQ should be applied were defined (Table 4 and Appendix Table 7).

The KPIs identified as the possible ones for the levels 1 and 2 are the real shop-floor KPIs. Marr (2015) and Parmenter (2010) had defined them but without a clear support for its definition and creation (operational indicators for the first level and KPIs for the latter). In line with the definitions proposed by both authors, these types of indicators should be measured at short intervals, focusing on the operational objectives and daily actions of each employee. For levels 3, 4 and 5, the questions were filtered in such a way that sensitive and/or aggregated information could only reach the highest levels. This is because such information is only important for top management, where it will have significant impact. An example are the costs and aspects related to the macro strategy of the company. If we make a comparison with the models referred above, these can be seen as KRIs (Parmenter, 2010) or strategic indicators (Marr, 2015). So, their purpose is to help monitor the strategy of the company and evaluate the work of various teams working together. The staff at these higher levels have the option of seeing only operational indicators or strategic indicators, or a mixture of both, depending on what they want. This enables them to focus only on the macro strategy, instead of on the operational objectives of each employee/section.

Table 4. Partial list of KPQs and KPIs, with indication of the class and level (full list in Appendix).

	KPIs	KPQs	Levels
S	Quantity of non-compliant	Are we receiving non-compliant parts?	2,3,4,5
	Supplier Quality	Do the supplied parts have the desired quality?	2,3,4,5
	Cost of non-quality	To what extent should we be concerned about suppliers' non-quality costs?	4,5
	Troubleshooting time	Do the suppliers quickly solve the problems we report them?	3,4,5
	Waiting time	Is the supply being done correctly?	2,3,4,5
Pl	Level of inactivity	Is our performance far from planned?	2,3,4,5
	Equipment occupancy fee	Are we using our maximum potential?	1,2,3,4,5
	Boot Time	Is the machine / line starting at the scheduled time?	1,2,3,4,5
	Production in time	Was the work carried out at the planned time?	2,3,4,5
	Set up time	Is the programmed stoppage schedule being fulfilled?	1,2,3,4,5
	Stock rotation	Is our material too long in stock?	3,4,5
	WIP Amount	Is WIP as defined?	2,3,4,5
Pr	Production time	How long does the product take since it starts to be produced until it is finished?	3,4,5
	Set up tasks	Are set up tasks being carried out efficiently?	1,2,3,4
	Cycle time	Is the production cycle time being fulfilled?	1,2,3,4,5
	Unplanned stops	Does this machine / line have unscheduled stops?	1,2,3,4,5
	Performance	Are the parts / assemblies being produced in the established quantity?	1,2,3,4,5
	OEE	Are we operating effectively?	2,3,4,5
	Production cost	Is production being carried out as budgeted?	4,5

continued on following page

Table 4. Continued

	KPIs	KPQs	Levels
Q	First Pass Yield (FPY)	How efficient are our internal operational processes?	2,3,4,5
	Quantity of non-conformities detected at the end of the process	Are we identifying nonconformities at the end of the process?	1,2,3,4,5
	Quantity of non-compliant parts at start-up	Was the production of nonconforming product units excessive during start-up?	1,2,3,4,5
	Non-conformity production cost	To what extent are production costs associated with non-conformities excessive?	4,5
	Time between self-control	Are self-checks performed at the set time?	2,3,4,5
	Recurring problems	Are we encountering recurring quality problems?	1,2,3,4,5
	Cost of rework	To what extent are the cost of rework a problem?	4,5
	Quality troubleshooting time	Are we efficient in solving quality problems?	3,4,5
	Non-quality costs	Is non-quality turning into an overspending for the company?	4,5
(*) Mec	Pressure	Is the machine working under the set pressure conditions?	1
	Temperature	Is the machine working at the set temperature?	1
Mai	MTBF	Are there too many equipment failures?	1,2,3,4,5
	MTTR	Does the equipment take too long to repair?	1,2,3,4,5
	Maintenance time	Is the machine stopped too many times for maintenance?	1,2,3,4,5
	Maintenance Costs	Is the maintenance of equipment a source of high expenses?	4,5
C	Reliability of delivery	Do our customers get what they want when they want?	3,4,5
	Satisfaction	Are our customers satisfied?	3,4,5
	Claims	Are we providing our service as the customer intends?	3,4,5
	Number of orders	Are our orders increasing?	3,4,5
	Order value	Do we have a high monetary value of orders related with this section?	4,5
E	Waste of raw material	Is the raw material being spent in excess?	1,2,3,4,5
	Cost of raw material waste	To what extent should we be concerned about the cost of wasted raw material?	4,5
	Power Consumption	Does energy consumption deviate from the stipulation?	1,2,3,4,5
	Cost of energy consumption	Are energy costs as stipulated?	4,5
	Carbon Emissions	To what extent are carbon emissions a problem?	3,4,5
	Water consumption	To what extent is water consumed a problem?	3,4,5
	Cost of wasted resources	Have the wasted resources a major financial impact?	4,5
W	Immediate problem solving	How autonomous are employees when they meet a problem?	3,4,5
	Satisfaction	Are employees happy with their job?	3,4,5
	5S	Do we have a clean and organized workplace?	3,4,5
	Accidents at work	Is the workplace propitious to accidents?	3,4,5
	Number of training hours	Have the employees been assiduous in the training sessions?	4,5
	Cost of employees	To what extent is the associated cost of the workers from this section a concern?	4,5

(*) These groups / indicators depend on the process, having to be adapted according to the type of process in the company.

It is also possible for each staff member to consult the display of the levels below. Thus, if there are any irregularities, they can trace the source of the problem.

The surveys were based on the outcome of these considerations, with each survey corresponding to one level. So, taking into account the different KPQs and their distribution at different levels, five different surveys were set up with their corresponding KPQs. Figure 3 illustrates an excerpt from one survey and its explanation.

In these surveys employees are requested to rate, on a scale of 1 to 5, how important is to find the answers to these questions (divided into classes). They should answer the survey bearing in mind the strategy of the company and the possibility of improving performance. Workers are also asked to indicate any relevant questions that need answering that is not included in the survey. This would allow the methodology in question to be improved and further developed. Apart from selecting the most important KPQs, the survey also intended to uncover problems that until then had not been recognized or followed.

Figure 3. Outline of a survey with explanation of its content

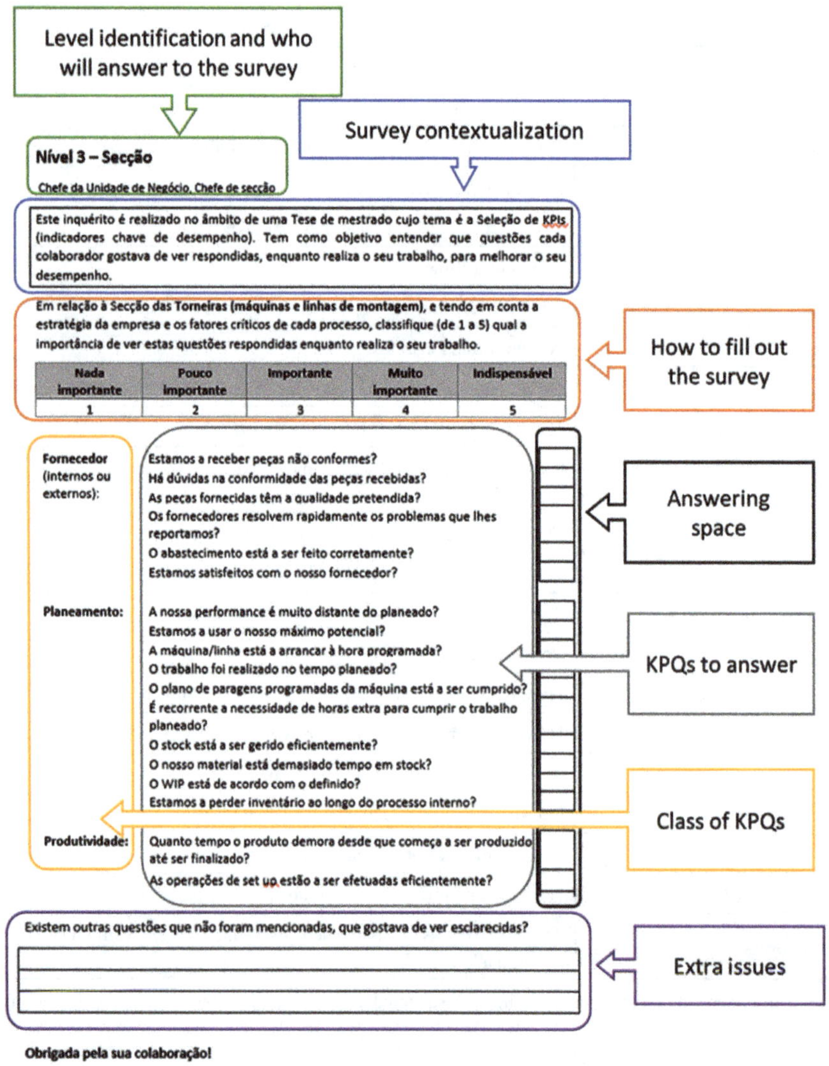

The survey of each level should be handed to all the staff of that level and the level above, as shown in Figure 4. It must be borne in mind that heads and members of each department will only answer questions related to their area of responsibility. For example, maintenance KPIs will be chosen by the maintenance department together with that section manager, as well as the business unit manager and so on.

Step 4: The Selection of the KPIs

The selection of the KPIs is based on the survey's results. The results should be presented in a second workshop and discussed with the employees in each section or business unit, depending on the size company. The KPIs selected are the one with higher voting rate. When even points are obtained by several KPIs, a second level decision criterion should be used, like selecting the one with lower standard deviation. In addition, during the workshop this decision can be shared with the employees and make trade-offs that foster the best performance assessing net of KPIs.

After selecting the KPIs, the target values should be defined and the baseline values should be identified. The target value should be realistic and achievable, taking into account the limitations and constraints of the specific production system. The target value should be set together with the elements involved on its use, aiming to assure peoples empowerment and commitment with the target fulfilment.

Step 5: The Dashboards for KPIs Visualization

KPIs must have visual impact. For this, it is essential that they are presented in a way to be understood immediately, preventing users from getting lost in excess data and/or irrelevant KPIs for their job. KPIs must show the problems and the objectives clearly, in such a way that the user understands them without the need to ask any question (Bell & Davison, 2013; iF, 2017; Parry & Turner, 2006).

Some authors support the use of color coding to quickly understand the state of the process regarding the target values defined by the company (Bell & Davison, 2013; Parry & Turner, 2006). The typical color code is the use of green if the target value is achieved, yellow if the assessed performance is below the target value but higher than the baseline, and the red if is below the baseline. In addition, the numerical

Figure 4. Surveys distribution into different levels

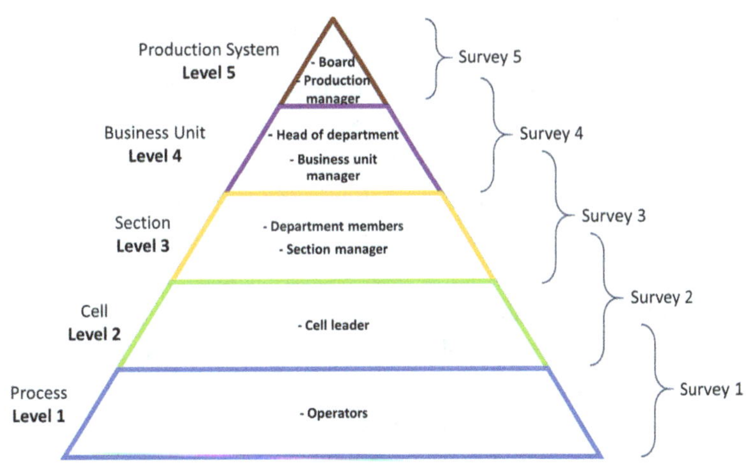

Figure 5. Dashboards types for each level of decision, with indication of the revision period.

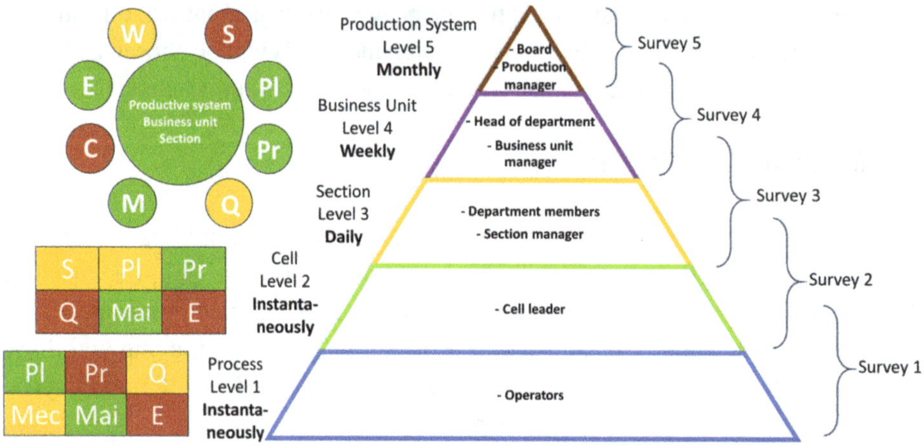

value of the performance should be showed together with the respective color. This type of board should be consistent, easy to understand, visible from a distance and placed near the area in question (Bateman, Philp, & Warrender, 2016; Parry & Turner, 2006).

Given the above, the result of this methodology is a visual management board that summarizes the performance of a particular area, fostering KPIs assertiveness and comprehensibility. So, the present methodology proposes distinct types of displays for each level of decision, as it can be seen in Figure 6.

Dashboard 1 and 2 (level 1 and 2) should be updated in the shorter time considered possible, depending on the limitations of the data collection equipment. A short revision period allows the indicators to be synchronized with the undergoing tasks, which might be useful in these two levels. This also allows to take immediate measures in conformity with the KPIs evolution. If an in-line monitoring is useless because it gives no practical information, a more adequate revision period should be used, being highly advisable that should be lower than 1 day (or 1 shift if the company works in shifts).

For levels 3 to 5, a different dashboard is proposed (similar for these three levels). In the methodology the daily assessment of level 3 KPIs is proposed, for level 4 KPIs the weekly period, and for level 5 monthly. Once again, these revision periods depend on the function that the KPIs must perform for those who view them. Top management is usually more interested understand the performance on a month basis rather than the hourly behavior of a machine, nevertheless the higher levels have access to all the KPIs of the lower levels.

Each of the dashboards proposed (Figure 5) has different divisions and each division has a color depending on the performance of the corresponding KPIs in relation to the target (previously defined by the company). Following the common practice in companies,, the division is green when the performance is better than the target (e.g. the suppliers in Figure 5); yellow when the performance is below the target but above the baseline (see productivity in Figure 5); and red when the performance is below the baseline (e.g. quality in Figure 5).

For levels 1 and 2, data will be updated at short intervals. However, these updates can cover a long period of time. For example, although the number of "ok components" must be updated instantaneously, the target should refer to daily values. Thus, the worker can have an idea of his performance while working. Furthermore, the main point of this is to understand, at the end of the day, if the worker completed all the work that was allocated to him or not. For the company, it is irrelevant if at any point during the day the worker was ahead of or behind schedule. For these reasons, in addition to the KPI value for some KPIs (when justifiable) the history of the KPI is also shown in the same display. Figure 6 represents an example of the level 2 dashboard (a similar one can be used for level 1). This kind of dashboard besides giving information on the color and on the KPIs figures, also allows to visualize the recent evolution of the KPIs.

For the displays referring to levels 3, 4 and 5, it was decided not to show individual KPIs avoiding excess of data, since the aim is to analyze 3 KPIs for each section (Figure 7). The proposed display is formed by 8 small peripheral circles, each one referring to one of the classes. There is also a central circle which reflects the average performance of the 8 classes taken as a whole. In the circles the daily value of each class (numerator) and respective target (denominator) are possible to find, in the case of level 3. When the dashboard is for level 4 the weekly value is used, and the monthly value is used for level 5. This number is the result of the average of the 3 KPIs previously selected by the surveys. Every circle will be painted green, yellow or red, depending on the general performance of each class. The peripheral circles are organized following the order of the process, clockwise. The order is the same for every display. This standardization is intended to make them readily understandable minimizing the possibility of doubt.

In order to have an idea of the history of the events, the last 15 days/12 weeks/12 months of performance (depending on the level in question) is shown in colors around each circle. These small sections also follow a clockwise direction (Figure 7). The first section refers to the day before and the last section before the white one refers to the performance of 15 previously time units. In Figure 7 this evolution is indicated by arrows.

Figure 6. Example of a level 2 dashboard

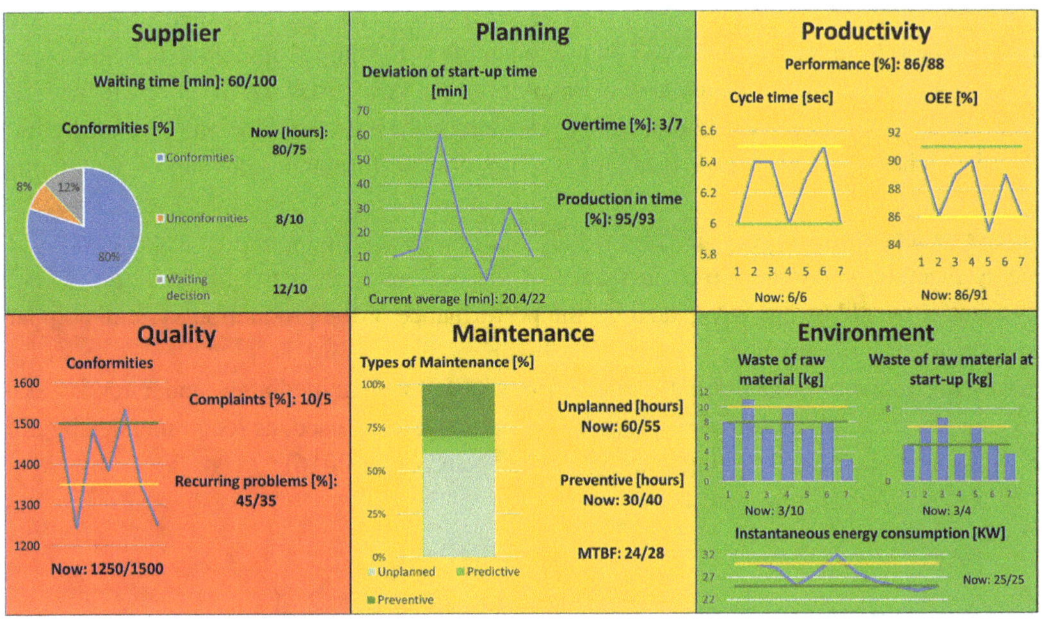

Figure 7. Example of a level 3 dashboard

As already mentioned, each of the small circles in Figure 7 represents one class, and that class is the result of the performance of 3 KPIs. The 3 KPIs of each class must be associated with a group of employees that influence the performance of the KPI with their work and also have the responsibility to react when a KPI value is yellow or red: a procedure must exist to support this action (in the KPI Card explained in the next point of this chapter).

These dashboards should also be used as reference/focus for the recommended periodic meetings in the shop-floor, to assess the good/bad performances of the last period. At the shop-floor level, level 1, 2 and 3, these meetings should be done daily. These meetings also provide an opportunity for ideas brainstorming. The main aim of these meetings is to increase performance and identify where improvements can be made.

Considering levels 4 and 5, the meetings should be done weekly and monthly, respectively. For level 4, a weekly meeting to discuss the performance of the business unit is usually enough. In level 5, a monthly meeting would be enough to analyze the performance of the production system, comparing it with the strategy of the company.

The way these displays are presented enables the KPIs to be immediately understood. Taking all this into consideration, standardized displays can be developed and used, as recommended by different authors (Espana et al., 2012; Henning et al., 2013; Hermann et al., 2016)

Step 6: The KPIs Cards and Surveillance

The methodology proposed in this chapter would not be completed if it doesn't tackle one of the most important aspects for the low effectiveness of KPIs in some companies: the lack of KPI surveillance. An Identity Card per KPI (Figure 8) should be build and maintain by a company employee designated for that job – the KPI Guardian. The KPI guardian has the responsibility to assure that the metrics used and its evolution keeps being understood and being used by the elements that use it (or should use it). Each KPI has different specifications and metrics, so the guardian should also explain the KPI interpretation for new comers in the company/team. The KPI guardian is also responsible to verify if the target value keeps up-to-date or if a performance increasing in the process has to be reflected in the definition of a higher value for the target; briefly, the guardian has to keep the KPI alive and useful and also to propose its extinction or deeper reformulation when becomes useless.

The last component of the proposed methodology, not considered as a step, is its periodic application, aiming to identify KPIs that are not used/useful and to identify novel needs of performance monitoring. The 6 steps proposed foster the alignment of the KPIs with the company strategy, assures the involvement of company elements of different levels in the KPIs generation, assures all dimensions of performance are cover as well all the decision levels, and also that the visualization and usefulness are used to assure their easy interpretation and up-dating. In the next section the results of the application of the methodology to an industrial company are described and discussed.

Figure 8. KPI Identity Card proposed: the fields are listed on the left card and an example is given in the right card.

KPI name	KPQ to which responds
Strategic objective	
Actions associated with its improvement	
Responsible for data collection	Responsible for updating data
Data collection mode	Frequency of data collection
Baseline	Formula and method of quantification
Target	
Benchmark	
Expiration date	Audience
Review date	

Rate of non-conforming parts	Are we identifying nonconformities at the end of the process?
Reduce costs of non-quality - Internal processes	
- Improvement of previous processes - Self-monitoring - Employees training	
Ing. Ana Silva	André Pereira
Through the ERP	Daily
40%	$\frac{Number\ of\ non\text{-}conformities}{Quantity\ produced}$x100
30%	
5%	
1/12/2017	Operators Line manager
First day of each month	

METHODOLOGY APPLICATION: A CASE STUDY

The Company

The methodology was applied in a European producer of products for plastic-based products for toilets, one of the biggest in Europe. The production system is located in Portugal, has around 400 employees, 80 injection molding machines, 1000 molds and 80 production lines (including assembly lines). There are 3 business units in this production site. Each business unit is composed by two or three families of products. The methodology was applied only to a section that produces one of the most complex products. The assembling phase of this section is composed by 8 cells, with 24 operators and 1 cell leader in charge, working 24/7. The injection molding area is formed by 7 machines controlled by 3 technicians and 3 machines operators. This structure is kept unchanged for all the other processes. In addition to the production and assembling areas, there are several parallel departments common to all the business units:

- **Provisioning:** responsible for external supplied parts and for their availability in the company;
- **External Logistics:** responsible for the material storage, transportation and distribution in all the different cells inside the company;
- **Planning:** responsible for planning all the production and dispatch of orders;
- **Maintenance:** responsible for the maintenance of all the equipment and machines, as well as aspects related with energy consumption;
- **Quality:** responsible for the parts inspection and defects, searching their causes and how to avoid it.

The assembly department is responsible for the assembly lines and the injection department for the injection machines. The injection team is composed by machine operators and injection technicians, being the last ones responsible for mold exchanges. The human resources department was also included in this study because its work is associated with all the company's members, and consequently with the production system.

The company uses KPIs being most of them derived from BSC. As a result, they are mainly monetary based KPIs only accessed by top management, and updated monthly and annually. The company also uses a set of shop-floor KPIs that are collected for each shift and exposed in a board named Daily Kaizen for each section (where an every-day meeting of seven minutes occurs with the staff of that section). The KPIs used are: OEE, number of open claims associated, number of actions, number of mold change, stopping times and percentage of non-conformities. These latter KPIs are usually updated daily but they are not visible during the working day, being only used during the Daily Kaizen meeting (Figure 10). In addition to these boards, some screens are exposed near the assembly lines. These are instantaneously updated and give information to the worker about: what product is being produced, how many parts of the model will be produced, how many parts were already produced and if the pace of production established is being fulfilled (Figure 11).

The Model, the Levels and the Surveys

Having in mind the company structure explained above, the identification of the model of the production system for the levels of decision identification was done (Figure 9). It was necessary to divide level 1

Figure 9. Organizational pyramid adapted to the company studied

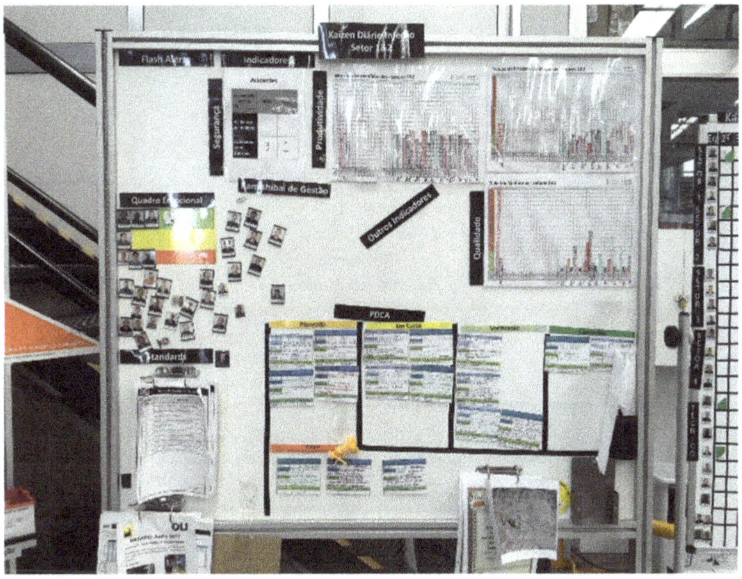

Figure 10. Daily Kaizen board with indicators

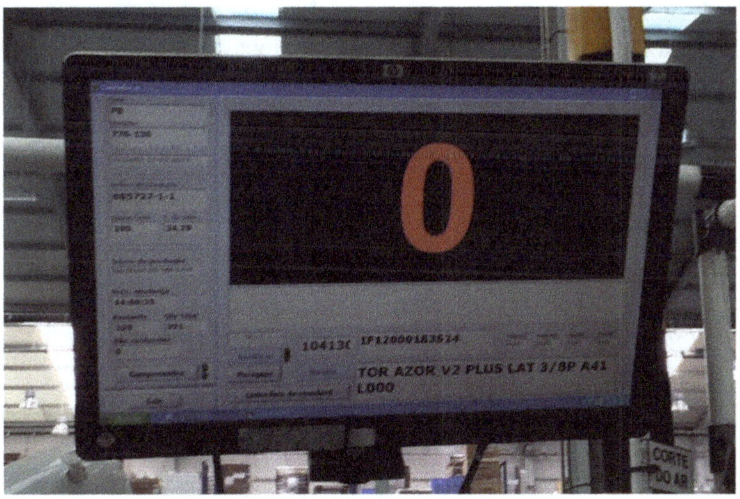

into two different groups, because the two processes studied are quite different and need different KPIs. In accordance with the structure of the company level 2 only relates to assembly. This decision was also made because there is no need to monitor the majority of the KPIs of the machines together, since the information would end up getting lost. Concerning the section (level 3), the company does not have a single person in charge of assembly and injection as stipulated in the model. However, the company has a section manager (responsible for assembly) and a shift master (responsible for injection). So, it was decided to put these two employees together in level 3 with the members of each department. The model was able to be used without any changes for level 5 and 6.

Figure 11. Screen of assembly lines units produced

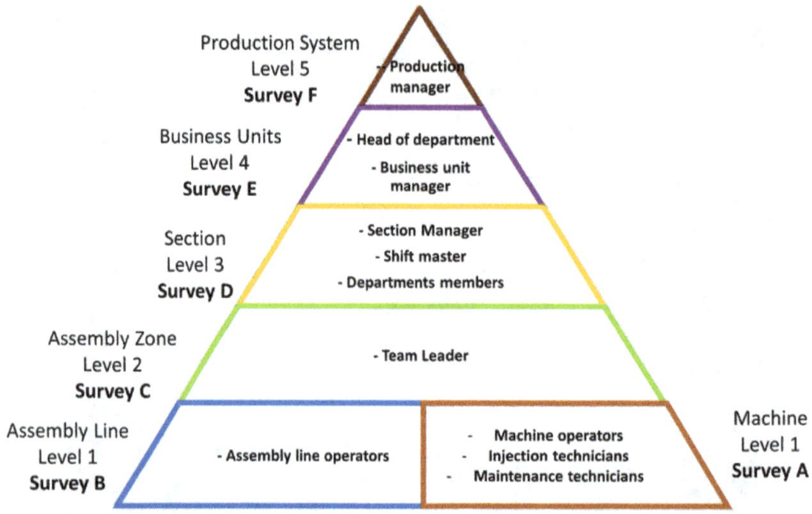

After the identification of the levels of decision, the KPQs were organized and adapted to the different levels, resulting in 6 surveys (A, B, C, D, E, F), one for each division of the pyramid. As stipulated by the methodology, each level answers its corresponding survey and the survey of the level above. Each department should only answer the questions related to their job and their responsibilities.

The choice of classes to be answered by each department was made with the help of the production manager together with the head of the human resources department, with the aim of giving each worker only the information that would help them for improve performance. For this reason, for example, it makes no sense for the logistics department to be involved in the choice of KPIs related to maintenance.

The surveys were handed to the heads of each department, with clear instructions on how to distribute and fill them as described in Table 5.

Table 5. Distribution of the surveys by KPIs classes and company elements

Survey	Regarding to	Classes	Answered by
A	Injection	Pl, Pr, Q, Mec, Mai, E	3 machine operators (Pl, Pr, Q, Mec), injection technicians (Pl, Pr, Q, Mec), 5 maintenance technicians (Mec, Mai, E), 3 shift masters (entire)
B	Assembly line	S, Pl, Pr, Q	24 assembly operators (entire), 1 team leader (entire)
C	Assembly	S, Pl, Pr, Q	1 team leader (entire), 1 section manager (entire), 4 Provisioning members (S), 3 Planning members (Pl, Pr, Q), 6 Quality members (Q), 4 External logistics members (S, Pl), 3 Assembly members (Pr, Q)
D	Tap section	S, Pl, Pr, Q, Mai, C, E, W	1 section manager (entire), 3 shift master (entire), 1 Business unit manager (entire), 5 Provisioning members (S), 4 Planning members (Pl, Pr, Q, C), 7 Quality members (Q), 5 External logistics members (S, Pl), 4 Assembly members (Pr, Q, E), 2 Injection members (Pr, Q, E), 2 Maintenance members (Mai, E), 3 HR members (W).
E	Business unit	S, Pl, Pr, Q, Mai, C, E, W	Heads of: UN (entire), Provisioning (S), Planning (Pl, Pr, Q, C), Quality (Q), External logistics (S, Pl), Assembly (Pr, Q, E), Injection (Pr, Q, E), Maintenance (Mai, E), HR (W). Production manager (entire), management advisor (entire), CEO (entire)
F	Production System	S, Pl, Pr, Q, Mai, C, E, W	Production manager (entire), management advisor (entire), CEO (entire).

KPIs Selection From the Survey's Analysis

After the completed surveys were handed back, the 3 highest rated KPIs for each level and class were selected (Table 6). In cases of an equal rating for two or more KPIs, the company top management decided that the KPI with lower standard deviation was chosen. This reflects greater unanimity regarding the importance of that KPI. The results show the cost related indicators were selected in almost all classes where they were an option (they are only present for levels 5 and 6). This demonstrates that top management is more interested in cost-related KPIs because top management is more focused on aggregated results than on individual processes.

The results also show that there are a group of KPIs which were selected in every survey where they were an option. The explanation for this is that they can summarize effectively the performance of each class:

- **Cycle Time:** important to achieve a good work pace and, consequently, high productivity;
- **Recurring Problems and Complaints Associated With Quality:** representing quality. It reflects the performance of the company when it comes to efficiency in detecting and solving problems;
- **Unplanned Maintenance:** mirrors the performance of maintenance, if there are too many unforeseen breakdowns, as well as if the maintenance is being performed efficiently;
- **Client Satisfaction:** it summarizes all the processes behind the product, from efficient production to delivery on time;
- **Incidents Where the Board Intervened:** it shows, once again, the ability of the company to deal successfully with problems.

Table 6. KPIs selected for each class in each survey (A to F). The KPIs that were selected in all the surveys where they were included are highlighted by a dashed colour.

Niveis		1	1	2	3	4	5
S	Cost of not quality	-	-	-	-		4.0
Pl	Extra hours of labor cost	-	-	-	-		4.0
Pr	Cycle time	4.0	3.96	4.63	4.62	4.5	4.33
	Production cost	-	-	-	-	-	5.0
Q	Recurring problems	4.0	-	4.43	4.55	4.43	4.33
	Non-quality costs	-	-	-	-	4.57	4.33
	Complaints associated with quality	-	-	4.64	4.75	5	5
Mai	Unplanned maintenance	4.25	-	-	4.38	4.5	4.0
	Unplanned maintenance cost	-	-	-			4.33
	Maintenance cost	-	-	-	-		4.33
C	Satisfaction	-	-	-	4.89	5	4.67
	Incidentes em que a administração teve que intervir	-	-	-	4.89	4.5	5.0
	Orders / customer	-	-	-			4.67
E	Cost of wasted resources	-	-	-	-	3.83	4.33
W	Cost of employees	-	-	-	-		4.33

The results also show a significant spread of the votes among several KPIs, meaning each person has a specific idea about what has to be measured. A reason that might contribute to these is the low effectiveness of the communication of the company strategy to the employees that was later recognized by the company top management. It was also decided by the company to repeat the "survey exercise" after 12 months, after a more effective company strategy communication to the employees, and verify if the results are more focus on a few KPIs increasing the relevance of the selected ones. If the same situation persist it means that is a characteristic of the company culture and compromises must be achieved.

Implementing the Selected KPIs

As mentioned before, the company already uses KPIs in the shop-floor and also at the top-management level. Comparing the KPIs selected by the methodology and the KPIs already measured by the company (even though its usefulness was recognized by the company as not effective) there is a match of about 50%. However, most of them are neither updated in the interval of time established by the methodology nor exposed to the right audience. Regarding the KPIs of the Daily Kaizen board of the company related to productivity, all of them were selected by the methodology. Nevertheless, with the dashboards and updating period proposed by the methodology its effectiveness is expected to increase. This means that the way they are being collected may have to change.

For these KPIs and for the remaining KPIs, two kinds of changes were identified for their implementation in the company: i) there are KPIs that are not monitored regularly, being accessed only when there is a non-conformity (i.e. energy consumption), so these kind of KPIs are to be kept but monitored regularly towards the target value; ii) the data collection must be more sensors-based to avoid losing times by the operators and the data collection that cannot be automatized must be supported by devices/ systems to facilitate operator data collecting task (it also increases data accuracy).

The company decided not to implement immediately the dashboards because significant work has to be done regarding the preparation for a more agile data collection that is already an ongoing work. In addition, as referred before the company recognized the low level of employees' knowledge about the company strategy, so the survey's results obtained cannot be considered highly valuable as a way to define the most appropriated KPIs. So, the company decided to give 12 months to project leader to implement what is necessary to be ready to collect process related data in the shop-floor (based on the selected KPIs) and then repeat the survey (but only after the strategy of the company is disseminated among the workers and after some workshops to explain the aims and advantages of the KPIs selection and exploration methodology – the one proposed in this chapter).

CONCLUSION

A novel methodology to select, explore and display KPIs at shop-floor level is proposed in this chapter to overcome several limitations of the current industrial practices and gaps of the existing methods. The limitations and gaps were identified and the steps of the proposed methodology were described. The aim is to answer to the challenges related with generation of a comprehensive and assertive set of indicators yet limited in number to be easy apprehended by the company employees. The comprehensiveness is assured by the definition of seven classes of KPIs covering the several types of activities and by "forcing" the existence of indicators for every classes. The assertiveness is assured by the selection of a specific set of KPIs for each levels of decision in the company (previously identified). The easiness is assured by limiting to three the number of KPIs for each class guaranteeing a final controlled number of KPIs. The use of surveys (based on KPQ) is another contribution for the merging of the company employees with the generated KPIs, by identifying the "most wanted" KPIs in the company. Finally, easy to implement and interpret visual dashboards are proposed for each level of decision, fostering the application of the visual management practices of using the KPIs for monitoring and continuous improving logics. To assure the validity and effectiveness of the KPI an Identity Card is created that includes the nomination of a responsible of verifying the evolution of the use of the KPI, namely the way data is collected, if its variation is taken into account when actions are necessary, if it is used by who should use it and if the target value is updated.

In summary, with the proposed methodology is possible to:

- Chose a group of KPIs which contextualize a process;
- Enable a relation between the strategy of the organization and the selected KPIs;
- Link the KPIs with the workers in order to involve them;
- Achieve a commitment between the workers and selected KPIs;
- Intuitively analyze a KPI performance, in relation to its baseline and target;
- Monitor the selected KPIs;
- Easy and quick display of each process critical points;
- Facilitate the attribution of responsibility to each KPI;
- Periodical review and actualization of KPIs.

The methodology was applied in a company, revealing all the positive points identified. As regards limitations and points to improve, a few conclusions were withdrawn. One of the lessons learned is about the high sensitivity of the methodology effectiveness to the dissemination level the strategy of the company to the employees: weaker their knowledge, weaker are the match between the selected KPIs and company strategy. The other is related with the importance of explaining to the company employees the reasons and potential benefits of having aligned and accurate measures of the company performance, to assure their real commitment. The last one is related with the need to change cultural and technical aspects related with the recurrence and accuracy of data acquisition that tends to delay the full implementation of the methodology. Naturally, the proposed methodology will be easily implemented in a context of Industry 4.0, with the relevant machines, tasks and actions monitored by sensors and other data collecting systems.

REFERENCES

Bateman, N., Philp, L., & Warrender, H. (2016). Visual management and shop floor teams – development, implementation and use. *International Journal of Production Research*, *54*(24), 7345–7358. doi:10.1080/00207543.2016.1184349

Bell, E., & Davison, J. (2013). Visual management studies: Empirical and theoretical approaches. *International Journal of Management Reviews*, *15*(2), 167–184. doi:10.1111/j.1468-2370.2012.00342.x

Bhoi, J. A., Desai, D. A., & Patel, R. M. (2014). The Concept & Methodology of Kaizen. *International Journal of Engineering Development and Research*, *2*(1), 2321–9939.

BI. (2017). *Bilioteca de Indicadores*. Retrieved July 20, 2011, from http://www.bibliotecadeindicadores.com.br/

España, F., Tsao, C., & Hauser, M. (2012). Driving continuous improvement by developing and leveraging lean key performance indicators. *20th Conference of the International Group for Lean Construction*.

Fanning, K. (2016). Big Data and KPIs: A Valuable Connection. *Journal of Corporate Accounting & Finance*, *27*(3), 17–19. doi:10.1002/jcaf.22137

Henning, K., Wolfgang, W., & Johannes, H. (2013). *Recommendations for implementing the strategic initiative INDUSTRIE 4.0*. Final Report of the Industrie 4.0 WG, (April), 82. doi:10.13140/RG.2.1.1205.8966

Hermann, M., Pentek, T., & Otto, B. (2015). Design Principles for Industrie 4.0 Scenarios: A Literature Review. *Technische Universitat Dortmund*, *1*(1), 4–16. doi:10.1109/HICSS.2016.488

Hermann, M., Pentek, T., & Otto, B. (2016). Design principles for industrie 4.0 scenarios. *Proceedings of the Annual Hawaii International Conference on System Sciences*, 3928–3937. 10.1109/HICSS.2016.488

Hurley, B. (2012). Use QDIP Sheets to Identify Environmental Issues. *Lean Six Sigma Environment*. Retrieved from http://leansixsigmaenvironment.org/index.php/use-qdip-sheets-to-identify-environmental-issues/

iF. (2017). *10 ways to make your Visual Management Boards work*. Retrieved from https://www.industryforum.co.uk/resources/blog/10-ways-to-make-your-visual-management-boards-work/

Kahn, S. R. (2013). Key performance indicators: The 75 measures every manager needs to know. *Choice: Current Reviews for Academic Libraries*, *50*(5), 926.

Kaplan, R. S., & Norton, D. P. (2005). The balanced scorecard: Measures That drive performance. *Harvard Business Review*.

Klipfolio. (2017). *KPI Examples*. Retrieved July 20, 2011, from https://www.klipfolio.com/resources/kpi-examples

KPIlibrary. (2017). *KPI library*. Retrieved July 20, 2011, from http://kpilibrary.com/

leanmanufacture. (2017). *Lean KPI's - Key Performance Indicators and performance metrics*. Retrieved from http://www.leanmanufacture.net/kpi.aspx

Marr, B. (2015). Big Data: Using SMART Big Data, Analytics and Metrics To Make Better Decisions and Improve Performance. Wiley.

Marr, B. (2015). *Key Performance Indicators for Dummiers*. Wiley.

Mourtzis, D., Fotia, S., & Vlachou, E. (2017). Lean rules extraction methodology for lean PSS design via key performance indicators monitoring. *Journal of Manufacturing Systems*, *42*, 233–243. doi:10.1016/j.jmsy.2016.12.014

Mourtzis, D., Fotia, S., Vlachou, E., & Koutoupes, A. (2017). A Lean PSS design and evaluation framework supported by KPI monitoring and context sensitivity tools. *International Journal of Advanced Manufacturing Technology*, 1–15. doi:10.100700170-017-0132-5

Murata, K., & Katayama, H. (2009). An evaluation of factory performance utilized KPI/KAI with data envelopment analysis. *Journal of the Operations Research Society of Japan*, *52*(2), 204–220. doi:10.15807/jorsj.52.204

Parmenter, D. (2007). *Key Performance Indicators (KPI): Developing, Implementing, and Using Winning KPIs*. Zhurnal Eksperimental'noi i Teoreticheskoi Fiziki.

Parmenter, D. (2010). *Key Performance Indicators (KPI). Key Performance Indicators (KPI): Developing, Implementing, and Using Winning KPIs* (2nd ed.). Academic Press.

Parry, G., & Turner, C. (2006, January). Application of lean visual process management tools. *Production Planning and Control*, *17*(1), 77–86. doi:10.1080/09537280500414991

Protzman, C., Whiton, F., Kerpchar, J., Lewandowski, C., Stenberg, S., & Grounds, P. (2016). *The Lean Practitioner's Field Book: Proven, Practical, Profitable and Powerful Techniques for Making Lean Really Work*. CRC Press.

Sanders, A., Elangeswaran, C., & Wulfsberg, J. (2016). Industry 4.0 implies lean manufacturing: Research activities in industry 4.0 function as enablers for lean manufacturing. *Journal of Industrial Engineering and Management*, *9*(3), 811–833. doi:10.3926/jiem.1940

Savkin, A. (2016). *KPIs Misuse: Why Does It Happen and How Avoid It*. Retrieved from https://bscdesigner.com/avoiding-kpis-misuse.htm

Tennant, C., & Roberts, P. (2001). Hoshin Kanri: Implementing the catchball process. *Long Range Planning*, *34*(3), 287–308. doi:10.1016/S0024-6301(01)00039-5

Thürer, M. (2013). Hoshin Kanri for the Lean Enterprise: Developing Competitive Capabilities and Managing Profit. *The Quality Management Journal*, *20*(3), 70. doi:10.1080/10686967.2013.11918356

Witcher, B., & Butterworth, R. (1999). Hoshin Kanri: How Xerox manages. *Long Range Planning*, *32*(3), 323–332. doi:10.1016/S0024-6301(99)00036-9

Witcher, B. J., & Chau, V. S. (1993). Balanced scorecard and hoshin kanri : Dynamic capabilities for managing strategic fit. *Management Decision*, *45*(3), 518–538. doi:10.1108/00251740710745115

APPENDIX

Table 7. KPIs list and formulas per class of KPIs.

	KPIs	Formulas
S	Quantity of non-compliant	$\frac{Nr.of\,non-complaint\,units\,provided}{Nr.of\,units\,supplied}$
	Supplier Quality	$\frac{Nr.of\,complaint\,units\,provided}{Nr.of\,units\,supplied}$
	Cost of non-quality	$\frac{Costs\,associated\,with\,non-qualities\,of\,suppliers}{Intended\,costs\,associated\,with\,non-qualities\,of\,suppliers}$
	Troubleshooting time	$\frac{Supplier\,troublesooting\,time}{Intended\,supplier\,troublesooting\,time}$
	Waiting time	$\frac{Unnplaned\,downtime\,due\,to\,suppliers}{Unplanned\,stop\,time}$
Pl	Inactivity Level	$\frac{Actual\,production\,time}{Planned\,production\,time}$
	Equipment occupancy fee	$\frac{Current\,capacity}{Possible\,capacity}$
	Boot Time	$\frac{Production\,starting\,time}{Planned\,production\,starting\,time}$
	Production in time	$\frac{Nr.of\,manufacturing\,orders\,performed\,in\,time}{Nr.of\,manufacturing\,orders}$
	Set up time	$\frac{Set\,up\,time}{Planned\,set\,up\,time}$
	Stock rotation	$\frac{Average\,current\,consumption}{Available\,stock}$
	WIP Amount	$\frac{Nr.of\,WIP\,units}{Nr.stipulated\,units\,for\,WIP}$

continued on following page

Table 7. Continued

	KPIs	Formulas
Pr	Production time	$\frac{Time\ from\ the\ start\ of\ the\ production\ of\ a\ unit\ until\ its\ completion}{Stipulated\ production\ time\ of\ a\ unit}$
	Set up tasks	$\frac{Set\ up\ time}{Stipulated\ set\ up\ time}$
	Cycle time	$\frac{Performed\ cycle\ time}{Intended\ cycle\ time}$
	Unplanned stops	$\frac{Sum\ of\ unscheduled\ stop\ times}{Stoppage\ times}$
	Performance	$\frac{Nr.of\ produced\ units}{Planned\ nr.of\ produced\ units}$
	OEE	$Availability * Performance * Quality$
Q	First Pass Yield (FPY)	$\frac{Nr.of\ units\ without\ defects\ at\ the\ end\ of\ a\ pass}{Nr.of\ units\ at\ the\ beggining\ of\ a\ pass}$
	Quantity of non-conformities detected at the end of the process	$\frac{Nr.of\ non-conformities\ detected\ at\ the\ end\ of\ the\ process}{Nr.of\ produced\ units}$
	Quantity of non-compliant parts at start-up	$\frac{Nr.of\ non-compliant\ parts\ at\ start-up}{Nr.of\ produced\ units}$
	Non-conformity production cost	$\frac{\text{Non-conformity production cost}}{Production\ cost}$
	Time between self-control	$\frac{Time\ for\ self-control}{Planned\ time\ for\ self-control}$
	Recurring problems	$\frac{Nr.of\ recurring\ problems}{Nr.of\ problems}$
	Cost of rework	$\frac{Cost\ of\ rework}{Intended\ Cost\ of\ rework}$
	Quality troubleshooting time	$\frac{Quality\ troubleshooting\ time}{Intended\ quality\ troubleshooting\ time}$
	Non-quality costs	$\frac{Sum\ of\ the\ above\ costs}{Sum\ of\ the\ above\ intended\ costs}$
Mec	Pressure	$\frac{Actual\ value}{Intended\ value}$
	Temperature	

continued on following page

Table 7. Continued

	KPIs	Formulas
Man	MTBF	$\frac{Total\,of\,correct\,runtime}{Nr.of\,failures}$
	MTTR	$\frac{Total\,maintenance\,time}{Nr.of\,failures}$
	Maintenance time	$\frac{Maintenance\,time}{Stoppage\,time}$
	Maintenance Costs	$\frac{Maintenance\,cost}{Stipulated\,maintenance\,cost}$
C	Reliability of delivery	$\frac{Nr.of\,orders\,delivered\,at\,the\,planned\,time}{Nr.of\,orders\,received}$
	Satisfaction	$\frac{Classification\,obtained\,from\,our\,costumers}{Intended\,classification\,obtained\,from\,our\,costumers}$
	Claims	$\frac{Nr.of\,complaints}{Nr.of\,orders}$
	Number of orders	$\frac{Nr.of\,orders\,received}{Intended\,nr.of\,orders\,received}$
	Order value	$\frac{Orders\,value}{Intended\,orders\,value}$
A	Waste of raw material	$\frac{Wasted\,raw\,material}{Total\,raw\,material\,consumed}$
	Cost of raw material waste	$\frac{\text{Cost of raw material wasted}}{\text{Cost of raw material consumed}}$
	Power consumption	$\frac{\text{Power sonsumption}}{Intended\text{ power consumption}}$
	Cost of energy consumption	$\frac{\text{Cost of energy consumption}}{\text{Intended cost of energy consumption}}$
	Carbon Emissions	$\frac{Carbon\,emissions}{Intended\,carbon\,emissions}$
	Water consumption	$\frac{Water\,consumption}{Intended\,water\,consumption}$
	Cost of wasted resources	$\frac{Sum\,of\,the\,above\,costs}{Sum\,of\,the\,above\,intended\,costs}$

continued on following page

Table 7. Continued

	KPIs	Formulas
T	Immediate problem solving	$\frac{Nr.of\ problems\ solved\ immediatly}{Nr.of\ problems}$
	Satisfaction	Evaluation by employees
	5S	$\frac{Workspace\ 5S\ rating}{Intended\ workspace\ 5S\ rating}$
	Accidents at work	$\frac{Nr.of\ work\ accidents * severity(0\ to\ 1)}{Intendcd\ nr.of\ work\ accidents}$
	Number of training hours	$\frac{Nr.of\ training\ hours\ already\ completed}{Intended\ nr.of\ training\ hours}$
	Cost of employees	$\frac{Cost\ of\ employees}{Stipulated\ cost\ of\ employees}$

Chapter 11
Solution Approaches for Reverse Logistics Considering Recovery Options:
A Literature Review

Sevan Katrancioglu
Marmara University, Turkey

Huseyin Selcuk Kilic
Marmara University, Turkey

Cigdem Alabas Uslu
Marmara University, Turkey

ABSTRACT

Reverse logistics stands out as a rapidly gaining concept due to its contribution to both the environment and the economy. There are many problems with reverse logistics. The decision of recovery options is a fundamental issue that serves many purposes. Choosing the right recovery option will also provide the environmental and economic contribution to maximize the benefits. For this purpose, many solution approaches have been produced for different objectives, which are based on the selection of better recovery options. Since solution approaches are directly interacting with problem models and objectives, it is important to determine an appropriate approach to achieve better results. Until now, many different approaches have been implemented, and results are shared. This chapter systematically examines these solution approaches and reveals the achievements in the literature in order to provide directions for future studies.

DOI: 10.4018/978-1-5225-5445-5.ch011

INTRODUCTION

A lot of work is being done on recycling that stands out with its economic and environmental impacts. In order to increase the profitability which is the common denominator, different solution approaches are being produced considering the changes in product and sectoral basis. These solution approaches are specifically developed and improved case by case. Although the solutions are specialized, it can be seen that some specific solution methods are focused when they are being evaluated as solution approaches. Examining these methods will make it easier to find the best method for solving the problem. For this purpose, a general literature review, in which solutions are generally evaluated and put into practice, will provide guidance for further studies.

When the literature is examined in detail by classifying the problems and solution approaches, many papers are seen in this field. Forward and reverse logistics factors are used in models (Giri, Chakraborty, & Maiti, 2017). Five different scenarios were tried to be optimized by combining these two returns: centralized, decentralized (Nash game), and manufacturer-led, retailer-led and third party-led decentralized scenarios. Within the study, the channel-based differences of product returns were also revealed. Another study on closed loop networks was performed (Mehrbod, Tu, Miao, & Wenjing, 2012). They used fuzzy goal programming in multi objective solution design.

Capacity, production and inventory variables were also considered in the studies (Kaya, Bağcı, & Turkay, 2014). It was tried to avoid the ambiguities in the integer programming created. In their two-tiered solution, they have also determined strategic decisions to remove ambiguities. They found that the amount of return of products is much more important than the demand uncertainty. Other researchers also proposed a forward and reverse logistics network model by accepting uncertainties as a risk (El-Sayed, Afia, & El-Kharbotly, 2010). The work progressed through the potential behavior of the inputs.

Another subject focused on green legislation and worked on a very specific solution for cost improvement (Senthil, Srirangacharyulu, & Ramesh, 2014). The solution was made by comparing different methods: a hybrid method using Analytical Hierarchy Process (AHP) and the Fuzzy Technique for Order Preference by Similarity to Ideal Solutions (TOPSIS) is proposed. Similarly, they also used the environment variables as an input variable in their study (Bazan, Jaber, & Zanoni, 2016). Another solution, fuzzy goal programming, in evaluating return options in green approach was used (Subulan, Taşan, & Baykasoğlu, 2015-b).

Another multi objective criteria solution approach was created (Barker & Zabinsky, 2011). In this study, taking into consideration business relationships, it has used a multi objective function in the decision of recycling processes. In the model created with third party companies, cost reduction was used as the basic criterion for optimization.

As a different solution approach, graph theory and matrix were used (Agrawal, Singh, & Murtaza, 2016). The identification of the network that should be followed in the separation of the products has been established on mobile phones in the study which was taken as the target.

Fuzzy is an effective solution method used in some optimization processes (Özceylan & Paksoy, 2013). In this study, a fuzzy multi objective model has been used in the solution step of a multi-criteria approach like other studies. Thanks to Fuzzy, the system administrators have more flexibility and decision making capabilities. Fuzzy approach can also be seen in other papers (Moghaddam, 2015). This study aims at selecting the optimum resources on the network in the ambiguous resource and demand environment.

Network designs are also frequently searched and artificial intelligence solutions are used (Vahdani, Dehbari, & Beni, 2014). Multiple methods have been used together by combining fuzzy possibilistic programming, stochastic programming, and fuzzy multi-objective programming, in a stochastic-possibilistic approach for a design with numerous network elements. As another objective, customer behavior is evaluated by modeling the recycling channels in a coordination (Feng, Govindan, & Li, 2017).

Regarding the existing literature, it can be concluded that there is a requirement of a literature review study which specifically focuses on the solution approaches for selecting better recovery options on reverse logistics. Therefore, in the chapter we propose, our aim is to reveal the articles and analyze them with the perspective of classifying solutions and the effectiveness of better recovery option selection capabilities. Moreover, this study also aims to provide a base for the academics and practitioners that are willing to develop a solution approaches for finding better recovery options on a reverse logistics problem.

PROPOSED METHODOLOGY

A systematic research is followed for the literature review. As shown in Figure 1, the main topic and the sub topics are determined firstly. After definition of topics, key word sets are created. In the selection of papers, up-to-date papers have been given priority. Because trend of studies on this fields are being examined in the paper. The most compatible resources from online libraries are evaluated and analyzed.

Figure 1. Methodology of study

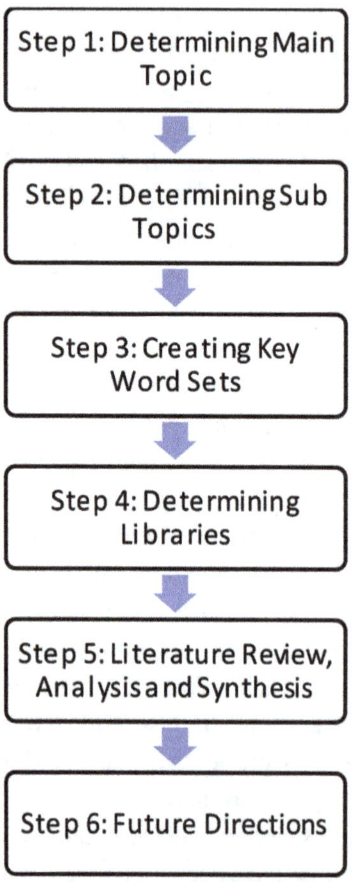

In this literature review, solution approaches for reverse logistics are studied considering different recovery options. Papers and publications are evaluated and categorized in 4 sub topics: problem, subjective, solution approach and case.

Specific key words were selected for finding the source in each sub topic. Care has been taken in selecting keywords in order not to break the sub topic from the main topic integrity. Selected key words are "reverse logistic decisions", "reverse logistic recovery options decision", "reverse logistic risk", "reverse logistic risk management", "solution method for reverse logistic", "environmental recovery options on reverse logistic", "best recovery option solution reverse logistic".

40 top related studies are chosen as a resource of literature review. Research is focused on the recent studies to reveal actual trend of publications in this field. Classification of the papers is shown in Table 1. Moreover, a detailed analysis of the reviewed papers are provided in Appendix (Table 7).

In literature, as shown in Table 1, three literature reviews are published. In first study, literature is reviewed with the perspective of decision making applications using multi-criteria objectives (Rezaei, 2015). Another literature review has focused on modelling techniques of reverse logistics (Bazan, Jaber, & Zanoni, 2016). Mathematical techniques are used to examine the networks and model in real cases. One of common research topic on reverse logistic is network designs. Therefore many studies can be seen in literature. Review with this topic also can show many solution approaches on this field (Govindan & Fattahi, 2017).

LITERATURE REVIEW: SOLUTION APPROACHES FOR REVERSE LOGISTICS CONSIDERING RECOVERY OPTIONS

Reverse logistics is located in the literature as a subdivision of the logistics concept. As there are many problems in this area, many solutions specific to these problems have been put forward. When we deal with reverse logistics, there are many common goals of companies, countries and even individuals. Although corporations generally focus on profits, there are other concerns of countries, such as environmental issues. In many countries, they have assigned responsibilities to companies by legally about recycling. In this framework, recycling in various countries has become mandatory and critical. Each company sets its own recycling strategy by its own products and production structure. In this process, recycling steps are determined and implemented. From the beginning to the end of the process, there are problems that need to be optimized at each point. Factors such as how to recycle products, evaluation

Table 1. Classification of the Reviewed Papers with respect to "Type of Paper"

Type of Papers	# of Publications
Literature Review	3
Solution Approach Based	30
Modelling & Case Study	4
Modelling	3
Total	**40**

of products, costs in process, recycling options, quality and risk have been studied continuously. In this study, a review of the studies focusing on the improvements made at the decision point of recycling, a categorization was made in the following structure shown in Figure 2.

The each categorization shown in Figure 2 has a key function for the studies. Problem definition shows how many different problems are facing with reverse logistic. Also objectives shows the main purpose of the studies and the aim of reverse logistics including recovery options.

The solution approach is the key point that shows how well a problem is solved and how close it is to the satisfying results. For this reason, the problem customized solution approaches are better than the general approaches.

In studies, problems and models are solved with some algorithms. Testing the effectiveness of algorithm is another key point for the studies. Many studies use numerical data set and some others uses real life data from specific products.

With the perspective of recovery options; the most focused recovery options are given below:

- Reuse or resell as new;
- Repair or refurbish and resell;
- Remanufacture and sell;
- Recycle

Figure 2. Classification of studies

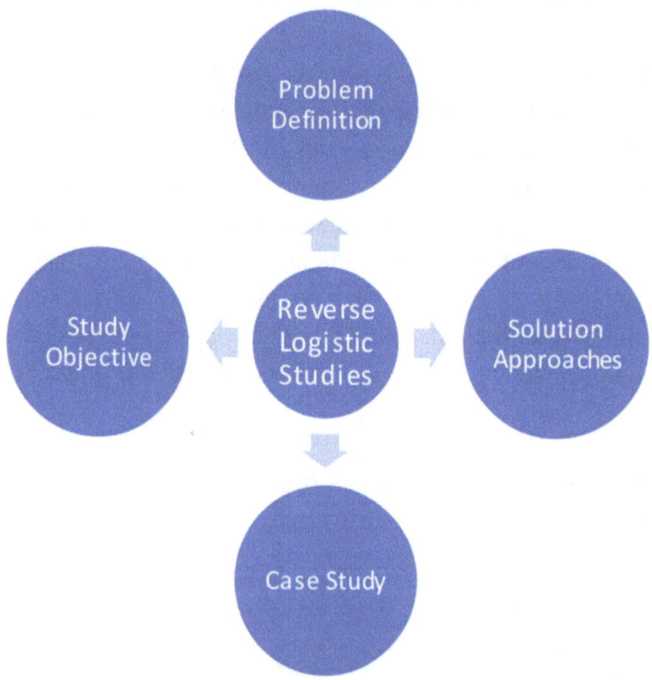

Classification: Problem Definition

Commonly studies have some problems to solve or improve. Problems state a base for models and solution algorithms. Studies may focus on many related problems to solve. Focus of the problems shown in Table 2.

Many studies have more than one problem. But problems are generally divided in two classes: Decision making and Network Design & Implementation. Main reason of this classification is that all studies are searching the best option for the problem by designing networks or making decisions on systems.

Another important point is the tendency of studies to be multipurpose. As shown in Figure 3, studies are focused on two or three problems at once.

Table 2. Problem Focus

Problem Focus	# of Publications Focuses on
Multi Attribute Decision Making	23
Network Design & Implementation	19
Cost	16
Uncertainty	10
Environmental Concerns	8
Design & Modelling	7
Quality	5
Risk Management	3

Figure 3. Studies focused on problems: Distribution of problems considered in studies

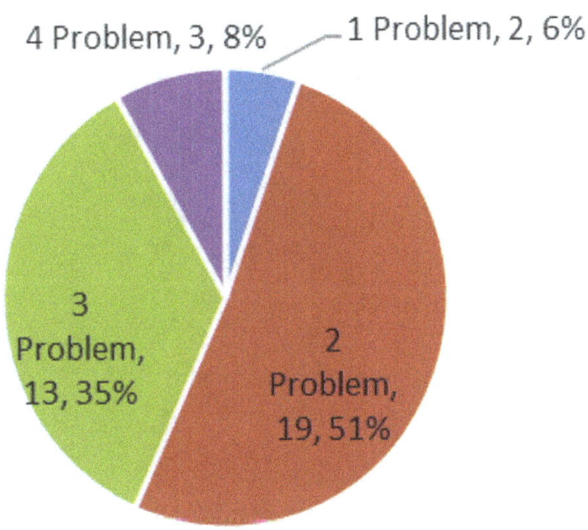

Classification: Study Objective

The pursuit of countries, companies and individuals is based on profitability. Although profitability is the basis of every problem, the way to profitability can be different. The improvements to be made on different variables on the system directly affect the profitability. Therefore studies' objectives differ from optimization of the solutions that would maximize the profit; to selection of the least polluting options for the environment. Also at the complete perspective environmental concerns ultimately make economic sense.

In addition to these, studies aiming to reduce costs or maximizing profit the directly are also examined. These are also based on the improvements on the system. The evaluation of recycling options for reducing environmental pollution, which has recently been intensified, has also been observed in many of these studies. Especially environmental factors have been taken into consideration as a factor in the decisions on the recycling option. Figure 4 shows the distribution of problem objectives.

Figure 4 shows the main objectives of the problems. Many problems have multi-objective oriented solutions. Approximately half of the problems have directly economic objectives and optimization is performed according to this purpose. The other objectives influence profit indirectly and the strategy of the optimization differs from each other.

Process optimization oriented studies are focused on;

- Optimizing manufacturing policy
- Overcoming uncertainties with better organization
- Optimizing network for reducing costs
- Reducing environmental pollutions of hazardous wastes
- Obtaining highest return rates of products

Best decision oriented studies are focused on;

- Best selection of recovery options

Figure 4. Distribution of problem objectives

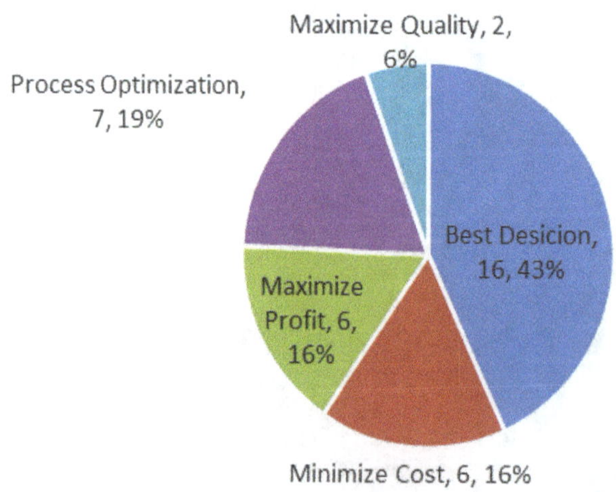

Table 3. Solution Approaches

Solution Approaches	# of Publications
Multi Attribute Decision Making	22
Multi Objective Solution	17
Heuristic Approaches	17
Statistical Approaches	11
Integer-Linear Programming	9
Stochastic Programming	7
Other Mathematical Modelling	6
Model Comparison	2
Graph Theory	1

- Determining appropriate decisions for products
- Uncertainty based recovery options
- Quality based recovery options
- Best selection of third party business partner
- Best model design
- Best selection of inventory strategies

Classification: Solution Approaches

The majority of the studies examined are multipurpose problems. In these problems, different solution methods are continuously applied in search of better solutions. There is a direct relationship between the solution method and the structure of the problem. As each problem differs from the general approach in itself, a single solution is not suitable for every problems. At this stage, it seems that better solutions are found through the combined solution methods. For every problem, minor changes after selection of the base solution method, lead to a better result.

As the objective function, the solution methods are determined to be more than one and to support each other. Table 3 shows the general problem solution approaches in the researched studies.

Statistical methods have been used mainly as supporting methods in modeling and solution approaches. It reveals that it is very useful to eliminate ambiguities and prepare data for the solution method. Many studies have used statistical methods for risk assessment and evaluation since the risk is used as an objective. Another use of statistical methods is quality assessment. One of the variables of the decisions are the quality of the recycled products. For the purpose to evaluate the products statistical methods are used.

In some of the studies, an aggressive objective and solution method was not proposed, so some solutions are based on modeling and comparative results.

In addition to these, more than one solution methods have been used together with optimization-focused studies. Figure 5 shows that approximately half of reviewed studies uses multi solution approaches.

Results of multi solution approaches are more successful than single solution approaches. Approximately half of multi-objective problems uses multi solution approaches as shown in Figure 6.

Figure 5. Number of solution approaches used in problems

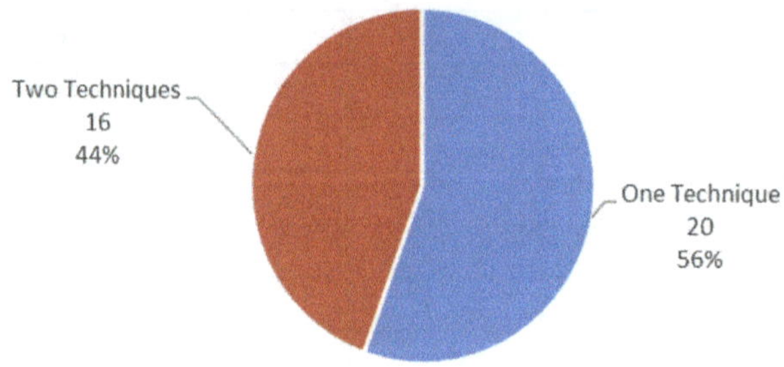

Figure 6. Multi solution approaches used in problems

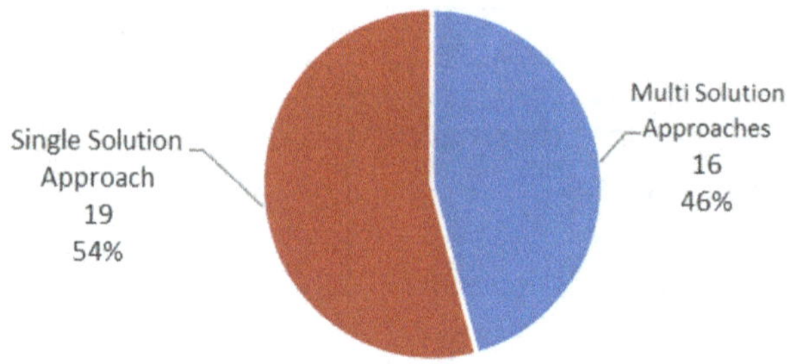

Classification: Case Studies

While solution methods are applied on the problem, different data sets are used. Some have used sectoral data and some have been working with digital data sets. Information about the data used in studies is given in Table 4. In some papers, more than one data set is used.

Sectoral data is also seen in many fields. As many sectors may use recovery options, problems are applied on different products. Table 5 shows studies and sectoral data matrix.

Countries and Publication Counts

In many countries, reverse logistics is being studied. Especially in countries where recycling is legally applied, these studies are predominant. In addition, these studies are carried out in many countries where recycling is important. Table 6 shows the country-based distribution of the authors.

Table 4. Case data

Data Type	# of Publications
Numerical data set	19
Sectoral data set	19

Table 5. Sectors with case data

Sectors	# of Publications
Electronics	4
Automobile	6
Medical Device	1
Textile	1
Home Appliances	2
Domestic Waste	3
Chemicals	1
Hazardous Waste	1

Table 6. Country Based Distribution of Publications

Countries	# of Publications
Turkey, USA	7
China, India	6
Canada, Iran	4
UK	3
Spain, Denmark, The Netherlands, Germany, Italy	2
Tunisia, Egypt, Egypt, Brazil, South Korea, South Korea, Switzerland, New Zealand, Malaysia, Australia	1

CONCLUSION AND FUTURE DIRECTIONS

Reverse logistics is becoming an indispensable process for many companies with its economic and environmental aspects. Each company has to design the recycling process in its own process and specifically for its product. This makes it impossible to establish a common process. This is why recycling is specially planned for many products and production processes.

Recycling is a process that requires planning in itself and involves different risks as it affects revenue with the production process. With well-designed recycling, the need for raw materials will be reduced so that unnecessary resource consumption and cost will come to minimum. Many scientific methods that serve this purpose have been put forward. The most common problem is choosing the best decisions that need to be taken during the recycling process. With the right decisions, maximum profit will be achieved.

One of the most difficult aspects of working in this area is the decision of recycling processes for more than one purpose. Although the main purpose is profitability in general, the ways to achieve this profitability are surrounded by issues such as complex decision making, quality, inventory, procurement, environmental impacts and legal responsibilities. Each of these factors is emerging as a factor that restricts how the process works and blocks each other.

It has been observed that the intersection of many variables has been studied in the studies carried out up to now. When we look at recent studies, especially multi-objective and multi-solution approaches are increasing. Especially environmental effects are seen in the general focus of these studies. This trend shows that work in this framework will continue to be frequent in future work.

REFERENCES

Agrawal, S., Singh, R., & Murtaza, Q. (2016). Disposition decisions in reverse logistics: Graph theory and matrix approach. *Journal of Cleaner Production*, *137*, 93–104. doi:10.1016/j.jclepro.2016.07.045

Aljuneidi, T., & Bulgak, A. (2017). Designing a Cellular Manufacturing System featuring remanufacturing, recycling, and disposal options: A mathematical modeling approach. *CIRP Journal of Manufacturing Science and Technology*, 11.

Amin, S., Zhang, G., & Akhtar, P. (2017). Effects of uncertainty on a tire closed-loop supply chain network. *Expert Systems with Applications*, *73*, 82–91. doi:10.1016/j.eswa.2016.12.024

Ayvaz, B., Bolat, B., & Aydın, N. (2015). Stochastic reverse logistics network design for waste of electrical and electronic equipment. *Resources, Conservation and Recycling*, *104*, 391–404. doi:10.1016/j.resconrec.2015.07.006

Barker, T., & Zabinsky, Z. (2011). A multicriteria decision making model for reverse logistics using analytical hierarchy process. *Omega*, *39*(5), 558–573. doi:10.1016/j.omega.2010.12.002

Bazan, E., Jaber, M., & Zanoni, S. (2016). A review of mathematical inventory models for reverse logistics and the future of its modeling: An environmental perspective. *Applied Mathematical Modelling*, *40*(5-6), 4151–4178. doi:10.1016/j.apm.2015.11.027

Benedito, E., & Corominas, A. (2013). Optimal manufacturing policy in a reverse logistic system with dependent stochastic returns and limited capacities. *International Journal of Production Research*, *51*(1), 189–201. doi:10.1080/00207543.2012.655863

Chen, W., Kucukyazici, B., Verter, V., & Sáenz, M. (2015). Supply chain design for unlocking the value of remanufacturing under uncertainty. *European Journal of Operational Research*, 804-819.

Chinda, T., & Ammarapala, V. (2016). Decision-making on reverse logistics in the construction industry. *Songklanakarin Journal of Science Education and Technology*, 38.

Dhouib, D. (2014). An extension of MACBETH method for a fuzzy environment to analyze alternatives in reverse logistics for automobile tire wastes. *Omega*, *42*(1), 25–32. doi:10.1016/j.omega.2013.02.003

Dutta, P., Das, D., Schultmann, F., & Fröhling, M. (2016). Design and planning of a closed-loop supply chain with three way recovery and buy-back offer. *Journal of Cleaner Production*, *135*, 604–619. doi:10.1016/j.jclepro.2016.06.108

El-Sayed, M., Afia, N., & El-Kharbotly, A. (2010). A stochastic model for forward–reverse logistics network design under risk. *Computers & Industrial Engineering*, *58*(3), 423–431. doi:10.1016/j.cie.2008.09.040

Feng, L., Govindan, K., & Li, C. (2017). Strategic planning: Design and coordination for dual-recycling channel reverse supply chain considering consumer behavior. *European Journal of Operational Research*, *260*(2), 601–612. doi:10.1016/j.ejor.2016.12.050

Giri, B., Chakraborty, A., & Maiti, T. (2017). Pricing and return product collection decisions in a closed-loopsupply chain with dual-channel in both forward and reverse logistics. *Journal of Manufacturing Systems*, *42*, 104–123. doi:10.1016/j.jmsy.2016.11.007

Govindan, K., Fattahi, M., & Keyvanshokooh, E. (2017). Supply chain network design under uncertainty: A comprehensive review and future research directions. *European Journal of Operational Research*, *263*(1), 108–141. doi:10.1016/j.ejor.2017.04.009

Guarnieri, P., Sobreiro, V. A., Nagano, M. S., & Marques Serrano, A. L. (2015). The challenge of selecting and evaluating third-party reverse logistics providers in a multicriteria perspective: A Brazilian case. *Journal of Cleaner Production*, *96*, 209–219. doi:10.1016/j.jclepro.2014.05.040

Hanan, D., & Burnley, S. (2013). A multi-criteria decision analysis assessment of waste paper management options. *Waste Management*, 566-573.

Hatefi, S., & Jolai, F. (2014). Robust and reliable forward–reverse logistics network design under demand uncertainty and facility disruptions. *Applied Mathematical Modelling*, *38*(9-10), 2630–2647. doi:10.1016/j.apm.2013.11.002

John, S., Sridharan, R., Kumar, P., & Krishnamoorthy, M. (2017). Multi-period reverse logistics network design for used refrigerators. *Applied Mathematical Modelling*, *54*, 311–331. doi:10.1016/j.apm.2017.09.053

Jun, H.-B., Lee, D.-H., Kim, J.-G., & Kiritsis, D. (2012). Heuristic algorithms for minimising total recovery cost of end-of-life products under quality constraints. *International Journal of Production Research*, *50*(19), 5330–5347. doi:10.1080/00207543.2011.624562

Kaya, O., Bağcı, F., & Turkay, M. (2014). Planning of capacity, production and inventory decisions in a generic reverse supply chain under uncertain demand and returns. *International Journal of Production Research*, *52*(1), 270–282. doi:10.1080/00207543.2013.838330

Kleber, R., Minner, S., & Kiesmüller, G. (2002). A continuous time inventory model for a product recovery system with multiple options. *International Journal of Production Economics*, *79*(2), 121–141. doi:10.1016/S0925-5273(02)00256-6

Mehrbod, M., Tu, N., Miao, L., & Wenjing, D. (2012). Interactive fuzzy goal programming for a multi-objective closed-loop logistics network. *Annals of Operations Research*, *201*(1), 367–381. doi:10.100710479-012-1192-4

Meng, K., Lou, P., Peng, X., & Prybutok, V. (2016). An improved co-evolutionary algorithm for green manufacturing by integration of recovery option selection and disassembly planning for end-of-life products. *International Journal of Production Research*, *54*(18), 5567–5593. doi:10.1080/00207543.2016.1176263

Moghaddam, K. (2015). Fuzzy multi-objective model for supplier selection and order allocation in reverse logistics systems under supply and demand uncertainty. *Expert Systems with Applications*, *42*(15-16), 6237–6254. doi:10.1016/j.eswa.2015.02.010

Ondemir, O., & Gupta, S. (2014). A multi-criteria decision making model for advanced repair-to-order and disassembly-to-order system. *European Journal of Operational Research*, *233*(2), 408–219. doi:10.1016/j.ejor.2013.09.003

Özceylan, E., & Paksoy, T. (2013). Fuzzy multi-objective linear programming approach for optimising a closed-loop supply chain network. *International Journal of Production Research, 51*(8), 2443–2461. doi:10.1080/00207543.2012.740579

Peña-Fernández, A., Wyke, S., Brooke, N., & Duarte-Davidson, R. (2014). Factors influencing recovery and restoration following a chemical incident. *Environment International, 72*, 98–108. doi:10.1016/j.envint.2014.05.001 PMID:24874002

Ramezani, M., Bashiri, M., & Moghaddam, R. (2013). A new multi-objective stochastic model for a forward/reverse logistic network design with responsiveness and quality level. *Applied Mathematical Modelling, 37*(1-2), 328–344. doi:10.1016/j.apm.2012.02.032

Rezaei, J. (2015). A systematic review of multi-criteria decision-making applications in reverse logistics. *Transportation Research Procedia, 10*, 766–776. doi:10.1016/j.trpro.2015.09.030

Senthil, S., Srirangacharyulu, B., & Ramesh, A. (2014). A robust hybrid multi-criteria decision making methodology for contractor evaluation and selection in third-party reverse logistics. *Expert Systems with Applications, 41*(1), 50–58. doi:10.1016/j.eswa.2013.07.010

Shackleton, M., & Sødal, S. (2010). Harvesting and recovery decisions under uncertainty. *Journal of Economic Dynamics & Control*, 2533-2546.

Sheu, J.-B. (2007). A coordinated reverse logistics system for regional management of multi-source hazardous wastes. *Computers & Operations Research, 34*(5), 1442–1462. doi:10.1016/j.cor.2005.06.009

Subulan, K., Baykasoğlu, A., Özsoydan, F., Taşan, A., & Selim, H. (2015). A case-oriented approach to a lead/acid battery closed-loop supply chain network design under risk and uncertainty. *Journal of Manufacturing Systems, 37*, 340–361. doi:10.1016/j.jmsy.2014.07.013

Subulan, K., Taşan, A., & Baykasoğlu, A. (2015). A fuzzy goal programming model to strategic planning problem of a lead/acid battery closed-loop supply chain. *Journal of Manufacturing Systems, 37*, 243–164. doi:10.1016/j.jmsy.2014.09.001

Subulan, K., Taşan, A., & Baykasoğlu, A. (2015). Designing an environmentally conscious tire closed-loop supply chain network with multiple recovery options using interactive fuzzy goal programming. *Applied Mathematical Modelling, 39*(9), 2661–2702. doi:10.1016/j.apm.2014.11.004

Subulan, K., Taşan, A. S., & Baykasoğlu, A. (2015). Designing an environmentally conscious tire closed-loop supply chain network with multiple recovery options using interactive fuzzy goal programming. *Applied Mathematical Modelling, 39*(9), 2661–2702. doi:10.1016/j.apm.2014.11.004

Trulli, E., Ferronato, N., Torretta, V., Piscitelli, M., Masi, S., & Mancini, I. (2017). *Sustainable mechanical biological treatment of solid waste in urbanized areas with low recycling rates*. Waste Management.

Vahabzadeh, A., Asiaei, A., & Zailani, S. (2015). Reprint of "Green decision-making model in reverse logistics using FUZZY-VIKOR method". *Resources, Conservation and Recycling, 104*, 334–347. doi:10.1016/j.resconrec.2015.10.028

Vahdani, B., Dehbari, S., & Beni, M. (2014). An artificial intelligence approach for fuzzy possibilistic-stochastic multi-objective logistics network design. *Neural Computing & Applications*, *25*(7-8), 1887–1902. doi:10.100700521-014-1679-9

Wadhwa, S., Madaan, J., & Chan, F. (2009). Flexible decision modeling of reverse logistics system: A value adding MCDM approach for alternative selection. *Robotics and Computer-integrated Manufacturing*, *25*(2), 460–469. doi:10.1016/j.rcim.2008.01.006

Xiao, R., Cai, Z., & Zhang, X. (2012). An optimization approach to risk decision-making of closed-loop logistics based on SCOR model. *Optimization*, *61*(10), 1221–1251. doi:10.1080/02331934.2012.688827

Ziout, A., Azab, A., & Atwan, M. (2014). A holistic approach for decision on selection of end-of-life products recovery options. *Journal of Cleaner Production*, *65*, 497–516. doi:10.1016/j.jclepro.2013.10.001

APPENDIX

The studies reviewed are summarized in the relevant headings.

Table 7. Analysis of the Studies Reviewed

Research	Problem	Objective	Solution Approach	Case	Year	Country
(Agrawal, Singh, & Murtaza, 2016)	Four identified alternatives of disposition decisions pair or reuse and resell as new; or repair or refurbish and resell; or remanufacture and sell; or recycle	Selecting best disposition alternative	Graph theory and matrix approach	Returned mobile phones	2016	India
(Aljuneidi & Bulgak, 2017)	Efficient recycling and remanufacturing network for efficient design of sustainable manufacturing enterprise	Designing a network for more efficient and sustainable manufacturing systems.	A mathematical modeling approach, mixed integer linear programming	Numeric data set	2017	Canada
(Amin, Zhang, & Akhtar, 2017)	Designing a network including many partitions with uncertainty.	Maximization of total profit	Decision tree-based methodology	Tire remanufacturing	2015	Canada, UK
(Ayvaz, Bolat, & Aydın, 2015)	Designing a model under uncertain inputs: return quantity, quality, and transportation costs	Maximization of total profit	Sample average approximation method to solve generic multi-echelon, multi-product and capacity constrained two stage stochastic programing model	Electrical and electronic equipment recycling firm in Turkey	2015	Turkey
(Barker & Zabinsky, 2011)	Network design decision making under consideration of business relations vs cost	Cost saving	Sensitivity analysis with multi-criteria decision making model for reverse logistics using analytical hierarchy process	Medical device remanufacturing, residential carpet recycling, commercial carpet recycling	2011	USA
(Bazan, Jaber, & Zanoni, 2016)	Literature review	Research about modeling of reverse logistics inventory systems that are based on EOQ/EPQ and JELS	Literature review	Literature review	2016	Canada, Italy
(Benedito & Corominas, 2013)	Sales dependent product returns model	Optimizing manufacturing policy	Markov decision model	Numeric data set	2013	Spain
(Chen, Kucukyazici, Verter, & Sáenz, 2015)	Overcoming uncertainties of return stream in terms of volume and quality	Assisting original equipment manufacturers about development of their remanufacturing strategy	Two-stage stochastic closed-loop supply chain design model with solution approach integrating sample average approximation with the integer L-shaped method	Case study based on BSH, a leading producer of home appliances in Germany	2015	Spain

continued on following page

Table 7. Continued

Research	Problem	Objective	Solution Approach	Case	Year	Country
(Dhouib, 2014)	Selecting best recovery options of waste tires with ambiguity and imprecise information	Selecting best recovery options of waste tires	Multi-criteria decision analysis	Automobile tire waste	2014	Tunisia
(Dutta, Das, Schultmann, & Fröhling, 2016)	Developing model addresses the possibility of three way recovery options, namely; product remanufacturing, component remanufacturing, and raw material recovery to determine the optimal manufacturing, remanufacturing and recycling quantity	Determine optimal buy-back price for	Modelling	Numeric data set	2016	India, Germany
(El-Sayed, Afia, & El-Kharbotly, 2010)	Network model design under risk factors.	Maximization of the total expected profit	Stochastic mixed integer linear programming	Numeric data set	2010	Egypt
(Feng, Govindan, & Li, 2017)	Designing reverse supply single traditional recycling channel, single online-recycling channel, and a hybrid dual-recycling channel with both centralized and decentralized cases	Selecting more effective channel	Model Comparison	Numeric data set	2017	Denmark, China
(Giri, Chakraborty, & Maiti, 2017)	Making decisions under centralized, decentralized (Nash game),and manufacturer-led, retailer-led and third party-led decentralized scenarios with a model containing a manufacturer, a retailer and a third party	Pricing and return product collection decisions	Numeric analysis	Numeric data set	2017	India
(Govindan & Fattahi, 2017)	Literature review	Comparing supply chain network design under uncertainty	Literature review	Literature review	2017	Denmark, Iran, USA
(Guarnieri, Sobreiro, & Nagano, 2015)	Implementation of The Brazilian National Policy of Solid Waste policies to the companies	Selecting and evaluating third-party reverse logistics providers	Numerical analysis and heuristic approaches	Brazilian case	2015	Brazil
(Hanan & Burnley, 2013)	Waste management decision making considering different options and constraints	Maximizing profit	Multi-criteria decision Analysis	Waste paper	2013	UK
(Hatefi & Jolai, 2014)	Network design considering risk factors: Uncertainty and uncontrollable external factors	Minimizing the nominal cost	Mixed integer linear programing model with augmented p-robust constraints	Numeric data set	2014	Iran
(John, Sridharan, Kumar, & Krishnamoorthy, 2017)	Design of a multi-stage reverse logistics network for product recovery	Profit maximization	Mixed integer linear programming	Refrigerator	2017	India, Australia

continued on following page

Table 7. Continued

Research	Problem	Objective	Solution Approach	Case	Year	Country
(Jun, Lee, Kim, & Kiritsis, 2012)	Optimization recovery cost under quality constraints	Minimizing total recovery cost	Heuristic algorithms	End-of-life products: turbocharger	2012	South Korea, Switzerland
(Kaya, Bağcı, & Turkay, 2014)	Uncertainties in this system, which are the return amounts of the used products and demand for final products	Addressing the disassembly, refurbishing and production operations	Integer programming with two-stage stochastic optimization and robust optimization	Modular products such as computers and mobile phones	2014	Turkey
(Kleber, Minner, & Kiesmüller, 2002)	Inventory management models for reverse logistics considering dynamic demands and returns	Determining the optimal production, remanufacturing, and disposal policy	linear cost model by applying Pontryagin's Maximum Principle	Numeric data set	2012	Germany, The Netherlands
(Mehrbod, Tu, Miao, & Wenjing, 2012)	Considering a multi-product multi-period closed-loop logistics network with different types of facilities	Multi-objective: minimize the total cost, the delivery time of new products, and the collection time of used products	Interactive fuzzy goal programming	Numeric data set	2012	China
(Meng, Lou, Peng, & Prybutok, 2016)	Recovery option selection and disassembly planning considering quality of end of life products	Maximizing profit	An integrated method of multi-target reverse recursion and partial topological sorting with improved co-evolutionary algorithm	Numeric data set	2016	China, USA
(Moghaddam, 2015)	Identify and rank the candidate suppliers and find the optimal number of new and refurbished parts and final products in a reverse logistics network configuration	Find the optimal number of new and refurbished parts and final products	Fuzzy multi-objective mathematical model an a Monte Carlo simulation integrated with fuzzy goal programming	Numeric data set	2015	USA
(Ondemir & Gupta, 2014)	Meeting demands of remaining life based product and component demands as well as recycled material on Sensor-embedded products	determine how to process each and every end-of-life product (EOLP) on hand	linear physical programming (LPP) model	Sensor-embedded products	2014	Turkey, USA
(Özceylan & Paksoy, 2013)	Closed-loop supply chain networks with uncertainty	Network routing optimization	Mixed integer fuzzy mathematical model	Numeric data set	2013	Turkey
(Peña-Fernández, Wyke, Brooke, & Duarte-Davidson, 2014)	Key factors that should be considered when developing a recovery strategy	Case study	Case study	Chemical incident	2014	UK
(Ramezani, Bashiri, & Moghaddam, 2013)	Multi-objectivity of Logistic network design	maximizing the profit, customer responsiveness and quality	Stochastic multi-objective model	Numeric data set	2013	Iran
(Rezaei, 2015)	Literature review	Review of multi-criteria decision-making applications in reverse logistics	Literature review	Literature review	2015	The Netherlands

continued on following page

Table 7. Continued

Research	Problem	Objective	Solution Approach	Case	Year	Country
(Senthil, Srirangacharyulu, & Ramesh, 2014)	Availability of many number of contractors for reverse logistic decisions	Evaluating and selecting the most efficient Reverse Logistics Contractor	Hybrid method using Analytical Hierarchy Process (AHP) and the Fuzzy Technique for Order Preference by Similarity to Ideal Solutions	Numeric data set	2014	India
(Sheu, 2007)	Logistic decisions of hazardous wastes considering cost and environmental issues	Minimizing cost	Linear multi-objective analytical model	Numeric data set – Hazardous waste	2007	China
(Subulan, Taşan, & Baykasoğlu, 2015-a)	Build-up an effective and efficient spent battery collection and recovery systems with significance of different importance and priorities is used to solve the developed model.	maximization of the collection of returned batteries covered by the opened facilities	fuzzy multi-objective, multi-echelon and multi-product mixed integer linear programming model	Lead/acid battery	2015	Turkey
(Subulan, Taşan, & Baykasoğlu, 2015-b)	The degradation of these scrap tires in the environment and the economic benefits of material and energy recovery	Multi-objective, multi-echelon, multi-product, and multi-period logistics network design	Interactive fuzzy goal programming	Used tires	2015	Turkey
(Subulan, Baykasoğlu, Özsoydan, Taşan, & Selim, 2015)	Finding best network design considering financial and collection risks of Lead/acid battery	Multi-Objective: computational efficiency, handling uncertainty and solution quality	Scenario based stochastic and possibilistic mixed integer programming model	Lead/acid battery	2015	Turkey
(Trulli, ve diğerleri, 2017)	Mechanical biological treatment options of waste	Decrease environmental pollution	Comparison of disposal activities and pre-treatment activities	Waste	2017	Italy
(Vahdani, Dehbari, & Beni, 2014)	Green decision-making model in reverse logistics	Finding best recovery options	FUZZY-VIKOR method using interval-valued trapezoidal fuzzy numbers	Numeric data set	2014	New Zealand, Malaysia
(Vahdani, Dehbari, & Beni, 2014)	Considering facility life cycle, designing a cloosed-loop network	Minimizing the total costs	Combination of fuzzy possibilistic programming, stochastic programming, and fuzzy multi-objective programming	Numeric data set	2014	Iran
(Wadhwa, Madaan, & Chan, 2009)	Designing a decision-making model for quantitative and qualitative evaluation based on criteria such as cost/time, legislative factors, environmental impact, quality, market	Determining the decision strategies for best alternative selection for reprocessing	Multiple criteria decision-making (MCDM) model based on fuzzy-set theory	Numeric real-life data set	2009	India, China

continued on following page

Table 7. Continued

Research	Problem	Objective	Solution Approach	Case	Year	Country
(Xiao, Cai, & Zhang, 2012)	Green logistics operation including risk management and strategic planning	Minimization of the risk loss as well as risk management cost and maximization of the profit	Genetic algorithm	Numeric data set – Case study	2012	China, USA
(Ziout, Azab, & Atwan, 2014)	Product recovery option decisions with the perspective of wide spectrum of stakeholders' interests	Best decision on end-of-life product recovery option	Holistic approach	Industrial data example	2014	Canada, USA

Section 4

Green Manufacturing and Sustainable Engineering Concepts

Chapter 12
Multi-Perspective Eco-Efficiency Assessment to Foster Sustainability in Plastic Parts Production:
An Integrated Tool for Industrial Use

Emanuel João Lourenço
Instituto de Ciência e Inovação em Engenharia Mecânica e Engenharia Industrial, Portugal

Nuno Moita
Instituto de Ciência e Inovação em Engenharia Mecânica e Engenharia Industrial, Portugal

Sílvia Esteves
Instituto de Ciência e Inovação em Engenharia Mecânica e Engenharia Industrial, Portugal

Paulo Peças
Universidade de Lisboa, Portugal

Inês Ribeiro
Universidade de Lisboa, Portugal

João Paulo Pereira
Instituto de Ciência e Inovação em Engenharia Mecânica e Engenharia Industrial, Portugal

Luís Miguel Oliveira
Instituto de Ciência e Inovação em Engenharia Mecânica e Engenharia Industrial, Portugal

ABSTRACT

The eco-efficiency assessment is a powerful metric to introduce two components of sustainability assessment in the industrial companies' decisions making: the concurrent consideration of economic and environmental performance. The application of the eco-efficiency concept and of the normative documents is not an easy task, mainly because there are myriad environmental related indicator to consider and acquire. This barrier is higher in the realm of plastic injection molding, where each mold is unique, requiring a recurrent effort of data retrieving for such one-of-a-kind molds. To overcome this barrier, an integrated framework to support the eco-efficiency calculation on a life cycle perspective for a specific type of products, injection molds, is proposed in this chapter. It retrieves a small but representative selected set of eco-efficiency performance indicators. A tool was developed to apply the proposed framework and the results of its application to four real industrial case studies is discussed.

DOI: 10.4018/978-1-5225-5445-5.ch012

INTRODUCTION

Plastic is one of the most common and versatile materials used worldwide, being used to produce a wide variety of products (Ribeiro et al., 2013; Ribeiro et al., 2014) and bringing extensive economic benefits due to their combination of low cost, durability and light weight, among other important characteristics (Ellen MacArthur Foundation, 2016). Most of these plastic products are manufactured through the use of moulds in the injection moulding process, which is a significant cost driver in the plastic industry and simultaneously regulates the injection moulding process efficiency (Ribeiro et al., 2009). Any change in the mould design can affect significantly the injection moulding efficiency and effectiveness, the investment and the production costs and also the resources consumed, emissions, i.e. the environmental impact.

In the injection moulding industry there is an increasing competiveness, even in countries with low production costs. As such, new and innovative technologies and adaptable business models are likely to be developed in order to increase value and decrease cost, without compromising on process efficiency and without increasing environmental impacts (Folgado et al, 2010; Pecas et al, 2009; Ribeiro et al., 2016)]. To accomplish that in the mould's use phase (injection moulding), it is crucial to improve the plastic part's material requirements and the consumed energy because these are considered as the most relevant factors for the economic and environmental performance in this process. The performance in those two aspects, depends on the mould design (which determines the cycle time and the amount of material wasted in the process) and also depends on the injection machine type and power (Thiriez & Gutowski, 2006; Ribeiro et al, 2012, Ribeiro et al., 2013). In the realm of the injection mould industry, moulds with better performance in their use phase require more complex and high embedded engineering, as well as more effort in the manufacturing phase of the mould. As such, this may lead to higher production costs as well as higher energy and material consumption in manufacturing – yet, by assuming that a single mould can produce up to a few million parts, the significance of the economic and environmental impacts of the manufacturing phase in the plastic part overall performance may be considered as residual (Noble et al., 2014; Esteves et al., 2014). As such, a life cycle perspective should be adopted when selecting the most adequate mould design, considering the environmental and cost impacts of mould manufacturing, use and end-of-life phases. So, the evaluation of all life cycle phases is necessary to allow an informed and sustainability conscious decision making process.

A few analyses comprising life cycle of injection moulds have already been developed (Pecas et al., 2009;Ribeiro et al., 2013;Esteves et al., 2014; Pun et al., 20033). These studies focus mainly the comparison of mould design alternatives on a technological scope considering life cycle cost (LCC) and life cycle assessment (LCA) methods. These publications focus on the selection of the most appropriated mould design for a specific application or part geometry and materials. Nevertheless, those analyses do not include the business specificity and the industrial company context. As such, very important aspect to help the industrial decision-making like added-value and profit, as well as significance of the environmental impact to the society are not taken into account in that type of analysis. The inclusion of these aspects within a life cycle analysis (of the alternative mould design) can bring comparison strategy more close to the decision-making processes currently used in industrial companies. There is a concept of the realm of sustainable manufacturing that can help on this analysis: the Eco-Efficiency.

Consequently, the aim of this chapter is to present a tool for industrial use allowing for the operationalization of the practice of Eco-Efficiency in the part and mould design decision-making processes. Broadening the discussion, one can say that understanding the sustainable development definition is the starting point to develop an approach to assess Eco Efficiency for the specificities of the sector of moulds manufacturing and injection moulding. The current definition of sustainable development derives from the one presented in 1983 by the World Commission on Environment and Development, which becomes known later as the Brundtland Commission, which stated that "sustainable development is development that meets the needs of the present without compromising the ability of future generations to meet their own needs" (Paul, 2008). The Eco-Efficiency assessment is one possible approach to evaluate two components of sustainability - it integrates product or service value and environmental influence. Through the quotient of these two performance aspects is possible to assess simultaneously balance between the economic and environmental performance, meaning a higher ratio a more eco-efficient product, system or service. The use of numerical measures of Eco-Efficiency (ratios) allows the use of indicators for continuous improvement of products and services, as well as for internal and external communication and/or reporting of performance along time. When the aim is to apply Eco-Efficiency at product level, there is a recurrent problem related with the need to involve detailed information from several processes (Silalertruksa,2015; Park et al., 2007; Tan et al., 2015) When Eco-Efficiency is applied to a large system or to a company, only the macro performance indicators are used, like the total energy consumed, total amount of materials, etc. as well as value indicator derived from the financial department of the company, so the number of indicators is limited. The assessment of Eco-Efficiency of a product demands for a detailed analysis of the processes and materials used to produce that product, resulting in a large amount of indicators that can be used as numerator (value added, net sales, number of products sold, etc.) and as denominator of the Eco-Efficiency ratio (dozens of categories are possible for each production activity having impact in the environment, e.g. energy, material, water, land and other resources consumed as well as several types of emissions), as referred by several normative documents (ISO, 2012; WBCSD-b, 2000; WBSCD, 2000; Union E.History and Definitions of Eco-Eccificiency, 2010) . The result is a myriad of possible Eco-Efficiency ratios as performance indicators (EEI) (Silalertruksa,2015; Park et al., 2007; Tan et al., 2015) Yet, there is neither approaches published to define one suitable set of EEI for a specific product nor tools to facilitate the choice and analysis of EEI, resulting the selection of indicators for Eco-Efficiency assessment of products in a time-consuming and cloudy analysis.

In addition, there is still no stablished standardized and integrated evaluation of environmental or economic performance for the injection mould sector. The few available assessments methods are based on the aggregation of fragmented information with different definitions and scope (Ellen MacArthur Foundation, 2016 (Hur et al., 2005; Bueno et al., 2016). Hence a significant opportunity is presented for the development of reporting standards and the consolidation of data on a global level (Ellen MacArthur Foundation, 2016).

As such, an integrated tool to support the Eco-Efficiency calculation on a life cycle perspective for a specific type of products, injection moulds, is proposed in this chapter. By the use of this tool a small selected set of environmental performance indicators significant to the injection moulding sector are suggested, based on several case studies analysed and with the real involvement of industrial companies in the study. In fact, if the value indicators are in general only a few, the critical difficulty is usually on the selection of the most appropriated environmental indicators among hundreds of them (ISO, 2012). Besides proposing adequate environmental indicators, the tool also includes a restricted set of EEI that are enough to assess the mould performance and injection moulding process Eco-Efficiency, facilitating the decision-making process and the communication of results. The LCA method is used to analyse the environmental impact throughout the life cycle of the mould, with performance indicators complementing the existing tools that support LCA, chosen specifically to assess mould design and the injection moulding process on a life cycle perspective. These indicators can be used in Eco-Efficiency performance assessing and reporting, contributing to Eco-Efficiency be easily understood by designers and analysts and allow comparisons between different mould design alternatives. To support an agile computing and rigorous analysis of these indicators, the integrated tool developed can translate life cycle data into useful value, environmental and Eco-Efficiency indicators, according to well-established assessment methods and standards.

The chapter begins with this introduction about the motivation and contribution of the study. In the following section the eco-efficiency background is presented to identify the current knowledge, praxis and gaps. The fourth section is dedicated to present the proposed framework to Eco-Efficiency assessment and to discuss its novelty aspects. In the fifth section the application of the framework to real industrial case studies is presented. The chapter finish with a section dedicated do the conclusions.

So, the next section is dedicated to present the present Eco-Efficiency background comprising its definitions and the proposed set of indicators of the most relevant normative documents.

ECO-EFFICIENCY BACKGROUND

Origins and Fundamentals of Eco-Efficiency

World Business Council for Sustainable Development (WBCSD), in the early 1990's, made the following universal interpretation of Eco-efficiency: "Eco-Efficiency is achieved by the delivery of competitively-priced goods and services that satisfy human needs and bring quality of life, while progressively reducing ecological impacts and resource intensity throughout the life-cycle to a level at least in line with the earth's estimated carrying capacity". Concisely, Eco-Efficiency focuses in "creating more value with less impact" (Lehni, 2000).

Nevertheless, 20 years before the WBCSD's proposed definition, Eco-Efficiency was being described as a mean to maintain a balance for the economy using the principle of mass conservation (Huppes et al., 2005). The proposed idea, even if empirical, involved the nature of environmental problems and their implications, keeping in mind that a good solution implies a harmonious relation between economy and environment (Kneese, 2003). This means that Eco-efficiency is an instrument for sustainability assessment that indicates how efficient the relation between the economic activity with nature's goods and services can be (Zhang et al., 2008). In the literature, a disruptive description states Eco-Efficiency as: "(…) Eco-Efficiency should be the extent to which it promotes the recoupling of human perception of environmental issues with human action on the environment, and the concomitant recoupling of collective local organization with locally crafted ecosystem management" (Suh et al., 2005). This description has a very interesting point of view and should not to be ignored, since this definition takes into account the human effects on the environment considering a local perspective, meaning that a generic Eco-Efficiency analysis, which considers generic data and indicators (i.e. generic value aspects and environmental influence), may not be enough to properly evaluate environmental damage. Moreover, Eco-Efficiency must not be seen as simple sustainability assessment metrics, but as an enabler to recouple human perception of environmental issues (Jukkinen, 2001).

Despite different definitions of Eco-Efficiency can be found in the literature, they always embrace the primary objective of Eco-efficiency i.e. "decoupling resource use and pollutant release from economic development" and its strong bond with sustainability (Panel IR, 2014). In general, and in line with the WBCSD, Eco-Efficiency is mentioned by many authors as a tool to help deliver high value products with less environmental impacts (Ng et al., 2015; Sproedt et al., 2015), thus, enabling companies to become more efficient, thus more competitive (Wursthorn et al., 2011).

Eco-Efficiency Assessment

As sated above, Eco-Efficiency quantifies the relationship between economic activity and environmental impacts, which is expressed by the formula, defined by the WBCSD (Lehni, 2000), which should be used to represent Eco-Efficiency as an indicator (Equation 1).

$$\text{Eco-Efficiency Indicators} = \frac{\text{Product or Service Value}}{\text{Environmental Influence(EI)}} \tag{1}$$

It is noteworthy, that in order to define the environmental influence and product service value, the functional unit needs to be defined, enabling to quantify performance of a product system. These functional unit will be the reference unit for environmental and product system value assessment and quantification.

The numerator makes reference to the value, which can be described as monetary indicators that are easy to understand (e.g.: the difference of sales and the cost of all inputs) (Lehni, 2000). Beside the monetary indicators, the product or system value may be measured by a functional characteristic, i.e. a tangible and measurable benefit to the user or a functional value (e.g.: production capacity, parts produced, lifetime) (ISO, 2012; Union E. History and Definitions of Eco-Efficiencyy, 2010).

Regarding the Environmental Influence aspect, the usage of ISO 14040:2006 (International Organization of Standardization, 2006) provides guidelines for LCA that enable a quantification of environmental impacts. These impacts can then be used as the denominator in Eco-Efficiency formula (Equation 1). This methodology is used for assessing environmental impact from "cradle to grave" associated with a production, process, transport or other involved activities by evaluating all energy and resource inputs and outputs. Not being an easy task, due to the necessary amount of data required, this could lead to more accurate results than using just partial life cycle environmental impacts (e.g. only raw material, energy, water, etc.) (Union E. History and Definitions of Eco-Efficiency, 2010; Lehni, 2000). The consideration of a life cycle perspective, by using LCA based indicators, helps identifying and consequently improving cost-benefits over the life cycle (Messagie, 2013) – due to the well-defined stream mapping of the analysis. So, LCA is a very useful tool to quantify environmental influence, and help to understand and to evaluate the magnitude and significance of potential environmental impacts.

In fact, LCA is considered as one of the most advanced tools, to support decision, to improve the environmental performance of products that could play an important role in order to enhance Eco-Efficiency Performance over a product life cycle (Kerr et al., 2001). Several studies, for example: (Wursthorn et al., 2011; Kerr et al., 2001; Korol et al., 2016; Kulak et al., 2016; Jasch, 2008; O'Reilly, 2000), use LCA methodology not only to obtain environmental indicators as input for Eco-Efficiency assessment but also as a support for environmental assessment within EcoDesign approach (Global Reporting Initiative, 2006), that is also a methodology that, if well implemented, could help enhance Eco-Efficiency performance.

Eco-Efficiency assessments can be performed in all types of companies, regardless of their size or area of activity (Alves, 2015), alongside to the products they produce. Unfortunately, the existing myriad of indicators and methodologies found in literature, can lead to an inappropriate use of them. Mostly often results are merely ratios that are analysed as a raw value that as an isolated number, can lead to different interpretations if not under a protocol. It is necessary to have well stablished methodologies, orientated to sectors that transform figures in good understandable results by the community. For this, several Eco-Efficiency assessment methodologies are found. Yet, in terms of implementing Eco-Efficiency as a methods or tool, the literature reveals that there is a lack of an integrated tool, to support Eco-Efficiency assessment (Hukkinen, 2001; Yang et al., 2014; Moller & Schaltegger, 2005; Cerutti et al., 2013; WBCSD-b, 2000).

In the next section, several ways of generating environmental indicators are presented and discussed aiming to describe the complexity of using Eco-Efficiency in an agile and easily understandable way.

Environmental Indicators: Performance and Influence

Several initiatives have proposed different sets of environmental indicators. The National Round Table on the Environment and the Economy (NRTEE), from Canada, has developed environmental indicators to help companies measure, monitor, calculate and report environmental performance indicators in a standardised manner (National Round Table on the Environment and the Economy, 2003; National Round Table on the Environment and the Economy, 2012). The Organization for Economic Co-Operation and Development (OECD), settled a set of indicators to measure environmental performance in order to foster sustainability and simplify reporting(Organisation for Economic Co-operation and Development, 2003); the ISO 14031 proposes a set of environmental indicators to provide management with reliable and verifiable information (Jasch, 2000; O'Reilly, 2000); and The Global Reporting Initiative (GRI) proposes a set of performance indicators based on the environmental, economic and social dimension, to contemplate the company's concerns with sustainability (Global Reporting Initiative, 2014). The proposed indicators are aggregated in Table 1.

In Table 1, each approach has its different criteria. Overall, the current practice of using environmental indicators in business shows a lack of standardisation and some of them do not address the environmental assessment towards the sustainability concept (Olthoorn, 2001). A straight forward critical observation that can be made, with such a wide spectrum of approaches, is that is difficult to understand the effective practical usefulness of some indicators, alongside with the unclear approach to allow its real application. Therefore, there is the need to measure Eco-Efficiency towards a broader but clear life-cycle thinking approach to the sustainability issue – these two, go hand by hand. Still, in Table 1, the different approaches can be distinguished by the different purposes of the indicators. Some very specific, towards a more technical viewpoint (such as NRTEE´s indicators) while others can be very broad, at a sustainability analysis level, touching all the 3 pillars such as the GRI´s approach.

Specifically concerning the plastic injection moulding sector, and despite the use of environmental-oriented indicators for communication, monitoring and evaluation of environmental performance, each mould is unique, which leads to a wide spread of effort in the application of such methodologies (like LCA) for performance assessment and/or comparison. Consequently, the plastic injection sector requires specific analysis in order to translate the mould's life cycle information into useful information, this is, to support the decision making process regarding the improvement of the environmental performance and Eco-Efficiency of the mould, hence promoting the sustainability of the sector. Such specificity promotes intrinsic knowledge throughout all stages of a mould's life cycle, and thus can enable Eco-Efficiency as a decision making approach, in the more complete sustainability assessment pathway.

Table 1. Different approaches to environmental performance indicators definition and purpose

	Indicator	Purpose
NRTEE[47]	Energy intensity indicator	Measures all the direct and indirect fuels used to produce the product(s) or deliver the service(s) per unit of production or service delivered.
	Waste intensity indicator	Quantifies all material that enter the boundaries of the system where the product is produced except the material that ends up as the product and / or as by-products per unit of production or service delivered.
	Water intensity indicator	Measure, track and report water usage, considering the amount of water taken into the project boundary per unit of product or service delivered.
OECD[49]	Tracking environmental progress and performance	Tracking environmental progress and performance, by using the Core Environmental Indicators (CEI)
	Tracking environmental progress and performance	Key Environmental Indicators (KEI) serve wide communication purposes
	Promoting integration	Sectoral Environmental Indicators (SEI) are used to help integrate sectorial issues
	Monitoring progress towards sustainable development, the Decoupling Environmental Indicators (DEI) should be considered since these measure the decoupling of environmental pressure from economic growth.	Decoupling Environmental Indicators (DEI) should be considered since these measure the decoupling of environmental pressure from economic growth
ISO 14031[39,40]	Operational Performance Indicators (OPIs)	Provide information about the environmental performance of the organization's operations.
	Management Performance Indicators (MPIs)	Provide information about management efforts to influence the environmental performance of the organization
	Environmental Condition Indicators (ECIs)	Provide information about the condition of the environment which could be impacted by the organization.
Global Reporting Initiative [50]	Economic	Illustrates the flow of capital among different stakeholders, and the main economic impacts of the organization throughout
	Environmental	Relates inputs and outputs. It covers biodiversity, transport, and product and service-related impacts, as well as environmental compliance and expenditures.
	Social	Concerns the impacts the organization has on the social systems within which it operates.

Given the presented facts, it has become clear that there is a real need for a narrower and deeper analysis of Eco-Efficiency. This analysis must be capable of providing the plastic injection sector with the necessary environmental and Eco-Efficiency performance indicators which will aid mould designers from the very beginning of the designing process. This is in fact the critical decision stage since the decisions taken along it have percussions over the mould's entire life cycle. Thus, an integrated tool should provide straightforward information regarding environmental and Eco-Efficiency performance of the mould throughout its entire life cycle.

THE MOULD MANUFACTURING AND THE PLASTIC PART INJECTION MOULDING

In the manufacturing context of moulds, environmental damage is usually not considered, since the impact of mould manufacturing is marginal in the overall production process (injection moulding), especially for mass production circumstances. In fact, a study by Ribeiro et al. (2008)demonstrated the significance of material and energy consumption for plastic parts production in the overall mould life cycle, contributing more than 95% to the mould life cycle environmental impact. However, most environmental analyses are focused either on the plastic part life cycle (Pun et al., 2003; Lundquist, 2000); on the performance of the injection moulding process (disregarding the mould design influence on it) (Thiriez & Gutowski, 2006; Park & Nguyen, 2014; Madan et al., 2015); or on the processes (milling, electro-discharge machining, grinding, etc.) that are used in mould production (Helu et al., 2012; Newman et al., 2012). None of the analyses focus on i) injection moulding environmental impact (plastic part production; and on the ii) influence of the mould design on the environmental impact of the injection moulding process mould. In fact, the mould is the critical tool that determines the shape, accuracy and the surface finishing of the final plastic parts, and also the material consumption (including material waste) and the energy consumed by influencing the injection moulding cycle time (Rosato et al., 2000). In addition, despite for very large series the environmental impact of the mould production phase can be considered negligible comparing with the injection moulding and plastic parts impacts, it should not be disregarded because it is a life cycle phase of the mould.

Briefly, injection moulds can be manufactured by a combination of different technologies, such as milling and turning, Electro-Discharge Machining (EDM) or wire EDM, or even by more unconventional processes such as laser machining, laser sintering and ultra-sound. The selection of the machining process depends on the mould material, on the geometric characteristic to obtain, on the accuracy required and on the surface finishing. So, the freedom to select among different alternative machining processes is reduced but exists and influences the environmental impact and the economic performance (production costs) of the mould. As regards mould design, it depends directly on the shape to be obtained in plastic and also on the average cycle time and quality required for the part. Several studies present approaches and models to optimize mould design (Oktem et al., 2006; Pandelidis, et al., 1990; Huszar et al., 2015). However, this interaction between part and mould is neither easy nor is always possible to fully model or standardize, as different parts are produced by different moulds which may require different manufacturing processes (Ribeiro et al., 2013; Malloy, 2010). Therefore, for each plastic part a unique mould is required to produce, that will not be produced again. That is why injection moulds manufacturing is considered as "one-of-a-kind" production.

Regarding the plastic part shape and quality requirements constraints, there are usually a few mould design alternatives that can be considered to produce the same plastic part, having influence on the injection moulding process time and resources consumption (Rosato et al., 2000).

Among those alternatives, the most common are the type of feeding system, the type of cooling system, the existence or not of movable components (and its number of movements), among others. For the feeding system there are two main alternatives, i.e. cold runners or hot runners. When using cold runners, the material solidifies in the runner and is ejected with the parts; when using a hot runner system, the material remains in the runners and is used for the next shot. Therefore, and as expected, when cold runners are used the cycle time and material waste in the mould use phase are higher, but the necessary time and resources to produce the mould are lower. The type of feeding system also has influence on energy consumption in the mould use phase, since higher quantities of injected material require higher amount of energy (Folgado et al., 2010; Pecas et al., 2009).

In terms of the cooling system, there are also two major design alternatives: normal cooling or conformal cooling. The latter is mainly used to reduce cycle time and improve the quality of the parts produced, since most defects are a result of inadequate cooling. Another design parameter, the number of movements, influences cycle time of injection moulding: a mould with a higher number of movements will allow a shorter cycle time, but on the other hand moulds with high number of movements are more complex to produce. So, it is clear that the mould design greatly influences the cycle time and material and energy consumption in the use phase, the injection moulding (Rosato et al., 2000).

More engineered moulds, namely with hot runners, conformal cooling and complex moves, require the availability and use of more advanced technologies and materials. These moulds will require a higher investment due to the inherent difficulties of production, which usually may imply higher environmental impacts in the mould production phase. However, despite production costs and environmental impacts, more complex moulds, with better surface finishing and higher number of movements, enable lower cycle times, and more efficient cooling, along with a reduction of material and energy in the injection moulding phase (Ribeiro et al., 2013). Therefore, to fully analyse the influence between mould design alternatives, a comprehensive life cycle analysis must be performed to take into account simultaneously the economic and environmental impact of mould production, together with its performance in the injection moulding process.

There are some publications addressing the subject of mould design influence on the economic and environmental performance, such as, for example, i) Ribeiro et al. (2013; 2011) that discusses the impact of different mould design alternatives and injection machines in the material and energy cost and environmental impact and ii) Gantar et al. (2013)that uses LCA and LCC to study injection moulding designs in their early design stages and conclude that mould design can greatly influence the environmental impact of the injection moulding process. Despite the importance of these studies, the methodologies they proposed are not feasible to be used by companies for each mould since the data gathering is time consuming and the analysis is complex and difficult to incorporate in the industrial decision-making process (mainly in the realm of injection moulding where the moulds manufacturing are one-of-a-kind production). The use of a tool to operationalize the use of Eco-Efficiency as a metrics to compare the performance of different mould designs on a life cycle perspective is one possible answer to fulfil this gap. This tool is proposed in this chapter and is described in the following sections.

THE PROPOSED FRAMEWORK AND THE IMPLEMENTATION TOOL

Description of the Framework

As mentioned in the previous section there is a clear gap regarding eco-efficiency assessment: an agile, complete and accurate approach or framework is required. A framework to fulfil this gap is proposed in this chapter being formalized in a software tool to support decision making. The present section of this chapter describes the proposed framework and a tool specifically developed to assess and compare the eco-efficiency performance of several types of injection moulds. A life cycle perspective is used, so both the injection mould and plastic part production eco-efficiencies are assessed.

The main goal of the developed tool is to support companies' decision making based on economic and environmental indicators, and by the Eco-Efficiency ratios formed by some of them. The tool can be used as an enabler of continuous improvement strategies and mould design improvements towards more competitive products and processes. Therefore, due to the specificity of the injection moulding sector where the tool was applied, its developing process was outlined within a Portuguese project dedicated to the injection moulding sector. The overall aspect and features of the tool are presented in Appendix.

The proposed framework, depicted in Figure 1, comprises five main modules: (i) Scope and boundary definition; (ii) Life Cycle Inventory (LCI); (iii) Computing tool; (iv) Decision indicators; and (v)Report. These modules are described, in detail, later in this section of the chapter.

In order to outline the tool based on the proposed framework (Figure 1), several workshops with various companies from the Portuguese injection mould sector took place. During the workshops, the base ideas and concepts of the framework where presented to the industrials, along with proposed indicators. Subsequently, both, the tool and its indicators where aligned with the real needs of the companies, that were considered being representative of the sector. Such strategy was used in order to have a suitable tool for the sector, namely low effort for: (i) data gathering; (ii) indicators calculation; and (iii) results analysis/reporting. It is worth mentioning that the most important aspect was to define a set of meaningful indicators, which are clearly understood by industrials, and enable companies to compare the performance of different mould designs.

In the following subsections the sequential tasks required to explore the proposed framework are described and discussed regarding its relevance and level of detail. These sequential tasks are in fact modules of the software tool developed for the injection moulding that will be used as a way to better exemplify how the framework concepts and methods should be applied.

Figure 1. Proposed framework for eco-efficiency and environmental performance analysis

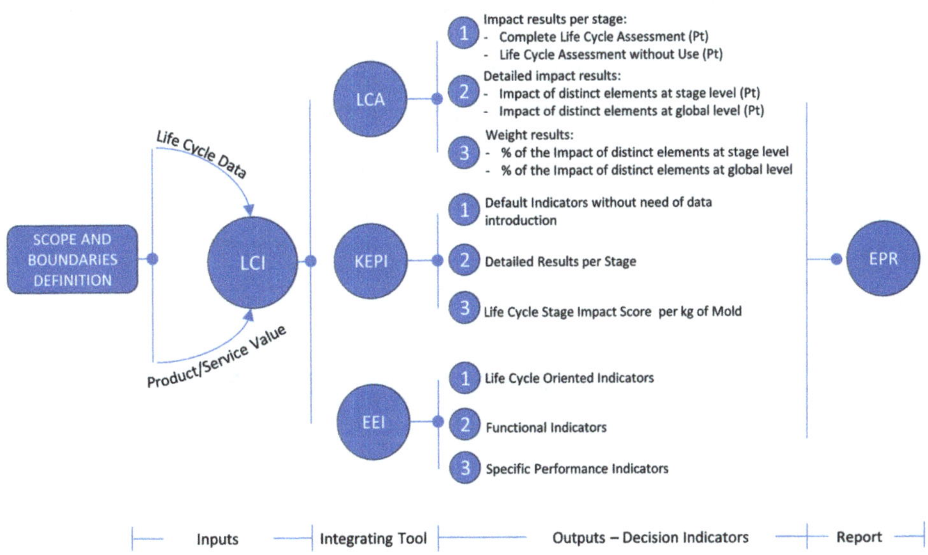

Scope and Boundaries Definition

The first module consists in defining the scope and the boundaries of the assessment. The goal is to establish the purpose of the assessment and to define the functional unit, i.e. reference to which all inputs and outputs are related (Messagie et al., 2013; Oliveira et al., 2015; Weidema et al., 2009)]. The functional unit is an important aspect when performing a LCA, or when assuming a life-cycle thinking approach, since the same comparison must be used to the different production systems and operations considered, and is naturally essential when comparisons are involved.

The boundary definition consists in defining the limits of the assessment, beyond those the flows and process about the functional unit will not be considered, so it identifies the unit processes to be considered in the analysis (Weidema et al., 2004) In the particular case of the developed tool, the functional unit is the injection mould, the boundaries are the mould life cycle phases (Figure 2) that includes the plastic part injection process (the use phase of the mould), and the scope is to compute the eco-efficiency ratios, with special emphasis to the environmental indicators (the value indicators are used as direct inputs).

Life Cycle Inventory

The second module consists in collecting all necessary data regarding the mould's life cycle phases. The LCI data collection stage works as showed in Figure 3. This contains the quantitative data related to the material, energy, emissions and co-products of the different life cycles of the product system under analysis. As mentioned previously, the functional unit and the system boundaries are the delimiting factor to what is included or not and how it is expressed. The data types that are often used are direct and indirect. The first ones relate to data that was actually measured and quantified through a process; the second one refers to data contained in specific databases such as Ecoinvent among others.

Within this module, the major goal is to list all inputs and outputs for each unit process within the limits of the system, and quantify all of the inputs and outputs under analysis, mainly for the calculation of the environmental impacts and KEPIs. It is also necessary to collect value data (e.g. GVA or EBITDA) for the calculation of the eco-efficiency ratios. Yet, as mentioned, the data gathering for the calculations should be agile and simple. With such keen need, the proposed framework was developed on the premise

Figure 2. Injection mould Life Cycle Stages

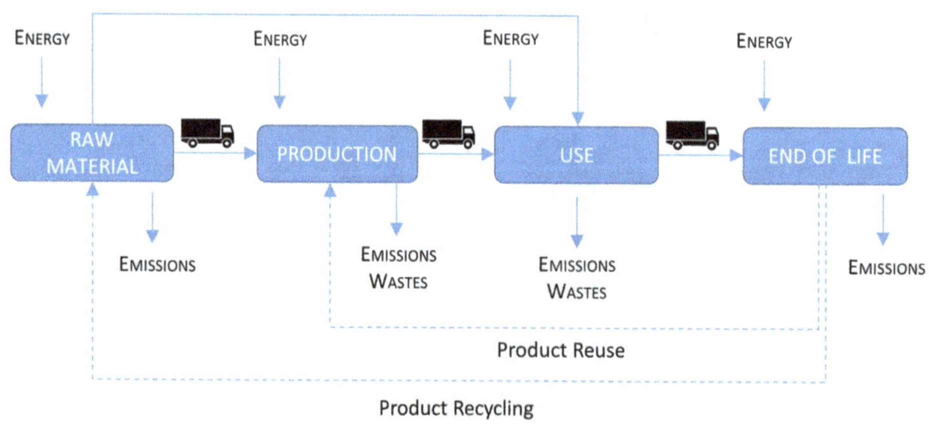

Figure 3. Stages of the creation of an LCI
(Oliveira et al., 2015)

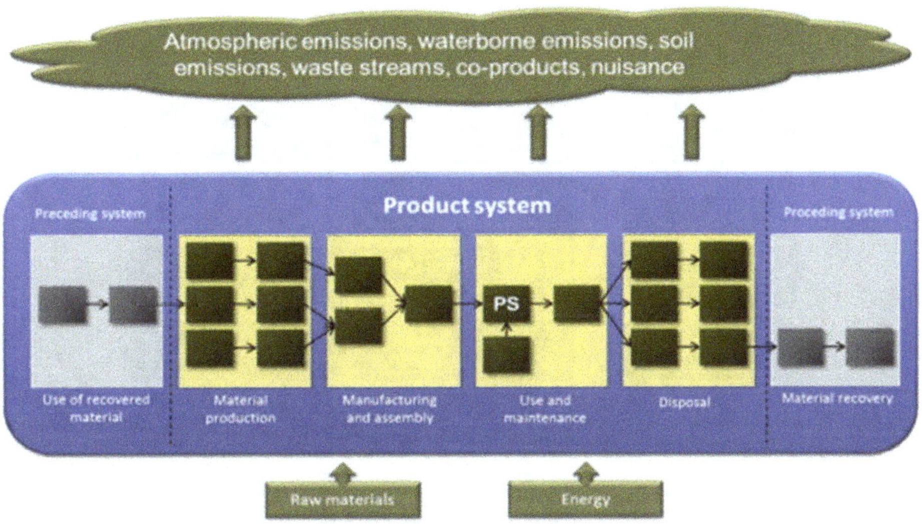

that a dedicated LCI, i.e. sector specific LCI structure, is necessary in order to collect all relevant data for the Eco-Efficiency assessment – this structure depends on the analysis boundaries. For the particular scope and boundaries of this study, the dedicated LCI should enable the characterization of the mould (e.g. size, weight, type of runners, number of cavities, cycle time) and of the produced part (e.g. part weight, material). This results in the dedicated LCI for injection moulding sector.

In this context, a dedicated LCI was developed (as part of the Excel based Integrating tool), based on the typical/usual operations and processes used in each life cycle phase and were then validated by the injection moulding sector companies involved in the study. This logic allows the standardization and simplification of the whole process of gathering LCI information. The required information is organized by modules to simplify the data collection, also to prevent errors from occurring. Figure 4 shows the general panorama of the necessary data for the different lifecycle stages and other relevant information. In Figure 5, the key data for eco-efficiency assessment is presented along with the data necessary to characterize the mould's use stage.

Figure 4. Dedicated LCI structure for eco-efficiency assessment: Data required for consumed resources inventory

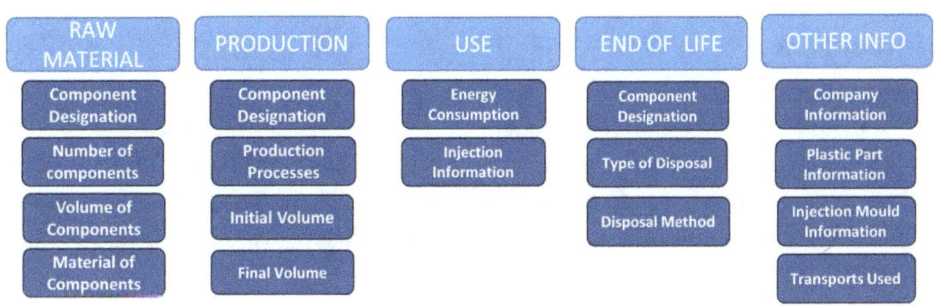

Figure 5. Dedicated LCI structure for eco-efficiency assessment: Inputs required regarding value assessment

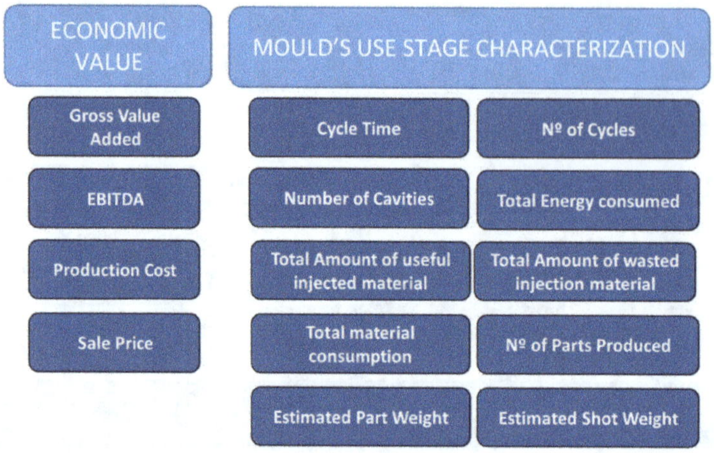

Integrating Tool

The integrating tool module of the proposed framework consists in the calculation of the environmental impacts - LCA, KEPI's and the several proposed eco-efficiency indicators, based on data form LCI and from the value inputs – is the computing engine of the proposed framework. The use of this module allows the reduction of the effort and the time needed to assess Eco-Efficiency performance. Like the dedicated LCI, a dedicated computing engine was developed exclusively for injection moulds, meaning that all the needed information will be dedicated fully oriented to the injection moulding sector.

The simplified scheme of the computing engine of the developed tool for injection moulding sector is shown in Figure 6, that illustrates how information flows towards the data processing and calculations that was developed using MSExcel. This approach was shared with the industrial companies of the sector involved in the study, that show a high preoccupation if a great effort of data inputs is required. After using the computing engine of the developed tool the companies validate it due to the low number as well as explicit and easy to obtains input about mould production and part injection moulding processes. Keeping in mind the proposed approach (Figure 1), after the LCI data is inserted an embedded algorithm will combine the environmental impact factors with the inputs from the inventory and preform the data processing and necessary calculations. Consequently, such computation will generate the several results as indicators - LCA, KEPI and EEI. The logics illustrated in Figure 6 is universal to other kind of products and processes being necessary to adapt to different LCI and inputs.

Decision Indicators

The Decision Indicators are the output component of the framework allowing the assessment of eco-efficiency as well as the environmental performance report. As depicted in Figure 1, there are three sets of indicators, namely the LCA indicators, KEPI and EEI. The developed tool outputs these set of indicators specifically adequate for the injection moulding industry, which are described in the following paragraphs.

Figure 6. Information flows for the data processing and calculations

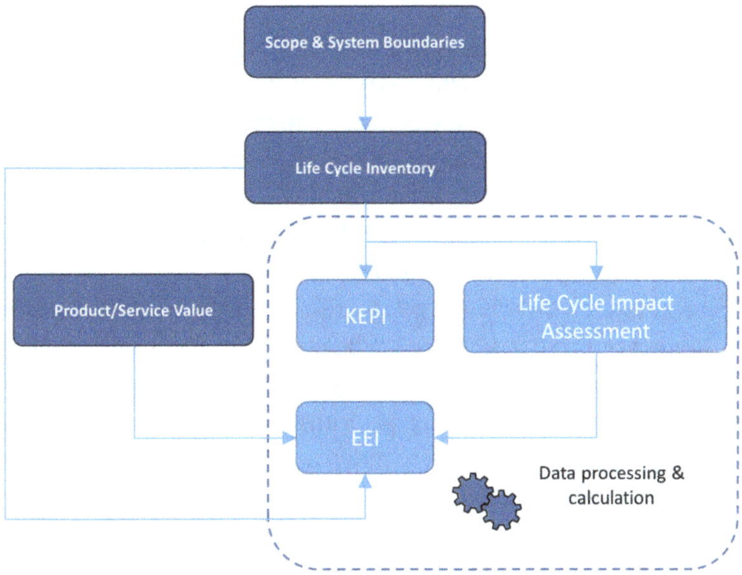

LCA Indicators

The LCA indicators provide an overview of the environmental impact results, which for the injection mould life cycle assessment are the one presented in Figure 7.

Within the proposed framework, the LCA results allows detailed analysis of the environmental impact of the several elements/entities of each life cycle stage (i.e. impact of the material used in a mould component, impact of machining a mould component, etc.). This result is presented for each life cycle stage, yet the influence (%) of each element impact is presented for life cycle stage and at global level. Such weighing (%) enables to identify in a swift manner the major contributors for environmental impact and enables a kind of normalisation that is easily understood by anyone and facilitates comparisons within the life cycle or with other moulds.

Figure 7. Proposed Life Cycle Assessment Indicators for the injection moulding sector

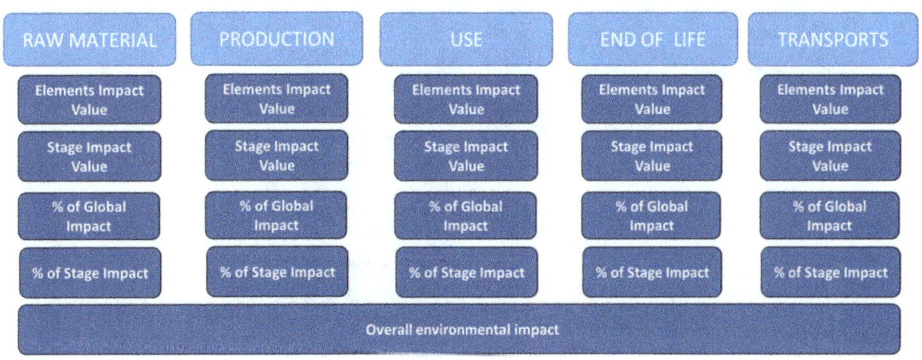

KEPI (Indicators)

Despite the LCA is a cornerstone of the proposed framework, additional calculations are required. KEPI are one of those needs. The KEPIs must be selected for a specific sector or type of product because they depend of the environmental aspects that are technologically related and relevant to the life cycle of a specific type of products. For the studied case of this chapter, the KEPIs were selected having in mind the particularity of the injection moulding industry and the boundary of the analysis that aims to have a life cycle oriented strategy. The KEPI indicators were defined and selected, together with the companies involved in this study fostering its practical usefulness.

The selected indicators are shown in Figure 8. They respect the ISO 14031 and can be used for setting improvement targets. Moreover, these indicators can be useful for: identifying opportunities to improve environmental performance among of mould's its life cycle; providing important information to support the decision-making process in strategic planning, prioritization, project or product reformulation and/or processes. It is important to mention, that all these indicators, except for cycle time and rate of defective products, are normalised by the moulds final weight (kg). Such normalisation is important to enable fair and standard comparisons, as the mould weight was considered a common factor that influences, in a somehow proportional manner, the proposed indicators.

EEI (Indicators)

In general, the EEI proposed by the framework are divided into three sets - life cycle oriented indicators; functional indicators; and performance indicators, are to be used as an ultimate metric of measuring the Eco-Efficiency performance. The EEI present in the developed tool are listed in Figure 9.

The first set of indicators - Life Cycle Oriented, represent clearly the formal definition of eco-efficiency. The Life Cycle Oriented indicators are calculated by the ration between the GVA (the economic value - numerator) and the environmental impact of the different life cycle stages (denominator). It is important to mention that a higher ratio result indicates a higher eco-efficiency performance. These indicators enable to assess and compare the eco-efficiency performance of each life cycle phase and also to compare different moulds' designs or different moulds.

Figure 8. Proposed KEPI for the injection moulding sector

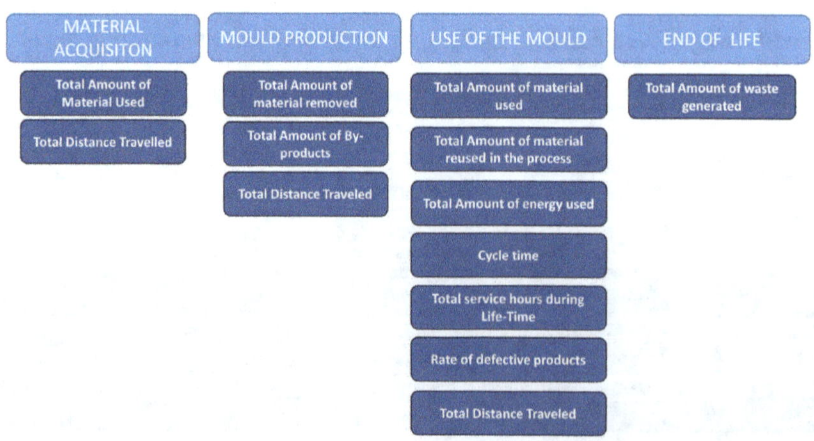

The Functional EEI are primary associated to the functional value of the mould (for instance the number of cycles). Therefore, the objective is to maximize the results, i.e. increase as much as possible the functional value for each unit of impact (Pt). For the third indicator (Tot. WIM/OEI), an exception is made, since the idea is to decrease the amount of wasted injection material per impact (Pt). A performance comparison of different moulds using these indicators, could be a tricky task as some comparisons may give place to misleads, therefore the comparisons should be formal and well thought through.

The last group of suggested EEI - Specific Performance Indicators, target to evaluate the performance of the mould's manufacturing and the injection moulding process. These indicators are sector specific and highly oriented for the moulding industry, as these indicators account for the quantity of energy and material used, and the respective environmental impacts. The first two indicators quantify the specific energy consumption, and the third indicator measures the environmental influence per unit of weight of injected material. As mentioned, eco-efficiency is calculated by the ration between product or service value and environmental influence, therefore these indicators do not follow the formal ratio of eco-efficiency, since the idea is for the EEI to demonstrate the mould's eco-effectiveness during the use phase – which is very important from the industrial point of view..

Report

The final module of the proposed framework consists in an Environmental Performance Report (EPR), which is part of the outputs. The idea of the report, is to comprise information regarding the characterization of the production and use of the product. The report is also intended to include all results namely all LCA indicators, KEPI and EEI, presented and described previously. The EPR is to be used for company internal and external (if there are no confidentiality issues) communication. For the particular case of the injection moulding industry, the developed tool reports the overall performance of the mould production operations as well as of the other life cycle phases, especially the plastic part production (the use phase of the mould).

Figure 9. Proposed EEI for the injection moulding sector

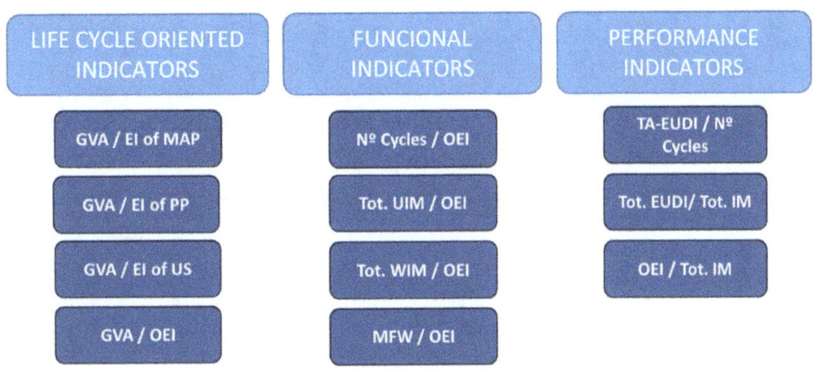

GVA = Gross Value Added | EI of MAP = Environmental influence of materials acquisition stage | EI of PP = Environmental influence of production stage | EI of US = Environmental Influence of use stage | OEI = Overall environmental influence | Tot. UIM = Total amount of useful injected material | Tot. WIM = Total amount of wasted injection material | MFW = Mould's final weight | Tot. EUDI = Total amount of energy used during injection | Tot. IM = Total amount of injected material

APPLICATION CASE

Boundaries of the Analysis

This section focus on the application of the tool developed for the injection moulding sector based on the framework presented in the previous section. The application case involves five plastic injection moulds from different companies. In order to assure confidentiality, the name of the companies was changed to: "Company A"; "Company B"; "Company C"; "Company D"; and "Company E".

After the moulds selection for the tool application, the involved companies started to use the dedicated LCI and then proceed with the specific parametrization and results generation for further analysis. In this way the tool was tested in industrial environment. The tool menus for input and tables and graphs with the outputs can be seen in Appendix.

The scope and boundaries considered for the eco-efficiency assessment in the five moulds were from materials acquisition until mould use phase, including the disposal of the contaminated metallic material. Mould end-of-life, plastic part use phase, disposal and end-of-life were not considered since these aspects were not considered as relevant to the mould making companies.

Life Cycle Inventory

In order to fulfil the eco-efficiency assessment framework, and get all relevant indicators, the LCI should comprise the data presented in the Figure 4 and Figure 5 which for the mould of company A is presented in Table 2 and Table 3. The value related data and specific characterization of the mould use phase are presented for the same mould in Table 4 and Table 5.

In Table 6, Table 7, and Table 8 the LCI data considered for the assessment of the five moulds are presented. The inventory showed is summarized for the considered the key data for this chapter. Analysing LCI data in Table 6 to Table 8 the different nature/features of the moulds involved in the present case study become obvious, for instance: different size, different cycle time, number of injection cycles and type of runners and, of course, different value related data. The developed tool was used to analyse and compute these LCI information in order to obtain the decision indicators.

Calculations Using the Developed Tool

LCA

The LCA was performed using the developed tool, which has embedded the necessary materials and processes (datasets) dedicated to the injection moulding. The datasets consider the Eco-Indicator 99 (H/A) v2.06 method (endpoint-oriented) and use the EcoInvent 2.1 database. All results of the environmental

Table 2. General information of the mould: Example for Company A.

Initial Mass (kg)	Final Mass (kg)	Number of Components (unit)	Removed Mass From Production (kg)
222.77	157.22	153	65.55

Table 3. Detailed information of the mould production and use for all life cycle stages: Example for Company A (N.A.: Data not available).

	Amount	Unit
Materials		
Steel	222.43	kg
Copper	0.34	kg
Production Processes		
Milling - Steel	46.59	kg
Drilling - Steel	15.54	kg
Turning – Steel (CNC)	1.26	kg
Steel Product - Metal Working	2.17	kg
Use		
Electricity -Low Voltage (PT)	4323.41	kWh
Polystyrene	2550.00	kg
Transport to Production		
Transport - Lorry (16-32T)	23.00	km
Transport - Van (<3.5T)	35.50	km
Transport to Use		
N.A.	N.A.	N.A.
Transport to Disposal (P)		
Transport - Lorry (16-32T)	4.00	km
Transport to Disposal		
N.A.	N.A	N.A.
Disposal		
Disposal of Steel - Inert Material Landfill	65.55	kg

Table 4. Economic value data: Example for Company A

Value Data	
Gross Value Added	21 376.00 €
EBITDA	10 036.00 €
Production Cost	11 340.00 €
Sale price	23 850.00 €

impacts are presented using an endpoints method with the unit – Pt (environmental impact at the end of this cause-effect chain) (see example in Table 9). The endpoint was considered in order to simplify comparisons and simplify general understanding.

Regarding the computational aspect of the tool, it enables the parametrization of processes that are transversal and dedicated to the injection moulding sector in order to permit time savings and an agile use of the tool for the calculation of environmental impact. The Excel based tool performs all calcula-

Table 5. Use stage characterization: Example for Company A

Mould's Use Stage Characterization	
Cycle time (sec.)	15.50
Number of Cycles	75000
Total energy consumed (kWh)	4323.42
Total material consumption (kg)	2550.00
Number of Parts Produced	225000
Number of Cavities	3
Total amount of useful injected material (kg)	2250.00
Total amount of wasted injection material (kg)	300.00
Estimated Shot Weight (g)	30.00
Estimated Part Weight (g)	10.00

Table 6. General data of the mould

General Data	Company A	Company B	Company C	Company D	Company E
Initial Mass (kg)	222.77	6623.74	8516.74	396.71	370.66
Final Mass (kg)	157.22	4061.14	7355.66	264.52	307.69
Part Quantity (unit)	153	668	9	310	590
Removed Mass from Production (kg)	65.55	2562.60	1161.07	132.19	62.98

Table 7. Mould use characterization

Mould's Use Phase Characterization	Company A	Company B	Company C	Company D	Company E
Cycle time (sec.)	16	55	38	28	12
Number of Cycles	75000.00	500000.00	50000.00	200000.00	5000000.00
Total energy consumed (kWh)	4323.42	304499.59	409814.48	17886.74	65642.71
Total material consumption (kg)	2550.00	302000.00	209500.00	5000.00	175800.00
Number of Parts Produced	225000.00	1000000.00	100000.00	400000.00	60000000.00
Number of Cavities	3.00	2.00	2.00	2.00	12.00
Total amount of useful injected material (kg)	2250.00	257000.00	209500.00	4000.00	17500.00
Total amount of wasted injection material (kg)	300.00	45000.00	0.00	1000.00	0.00
Estimated Shot Weight (g)	30.00	514.00	419.00	20.00	35.16
Estimated Part Weight (g)	10.00	257.00	209.50	10.00	2.93

Table 8. Economic value (N.A.: Data not available)

Value Data	Company A	Company B	Company C	Company D	Company E
Gross Value Added (€)	21 376.00	55 996.93	80 000.00	19 211.50	N.A.
EBITDA (€)	10 036.00	70 21.62	10 000.00	10 978.00	N.A.
Production Cost (€)	11 340.00	48 975.31	110 000.00	22 205.50	N.A.
Sale price (€)	23 850.00	88 000.00	115 000.00	24 950.00	N.A.

Table 9. Example of the datasets using the Eco-Indicator 99 (H/A) v2.06 method (endpoint-oriented) from the EcoInvent 2.1 database: Example for mould of company A.

	Indicator (Pt/unit)
Material	
Steel, Converter, Unalloyed, At Plant RER	0.104
Copper, At Regional Storage RER	2.013
Production Processes	
Milling, Steel, Average RER	0.275
Drilling, CNC, Steel RER	0.277
Use	
Electricity, Low Voltage, At Grid Pt	0.041
Polystyrene, General Purpose, GPPS, At Plant RER	0.253

tions using Excel Macros. Yet, the accuracy of the calculations, and respective indicators, is always dependent on the quality of the data input.

In Table 10 the detailed results regarding the environmental impacts are presented, as an example of the one of the outcomes from the tool. As mentioned in the previous section, namely depicted in Figure 7, besides the elements impacts (e.g. impact of steel), the contribution (%) of each one of them is also presented, either for stage or global level. The results of the impacts for each life cycle stage are presented in a concise manner in Table 11 and Table 12.

It is noteworthy that the use stage has the highest influence in the total environmental impact, which is caused by the impact of the injection material consumption (Table 10). This is an expected but yet important result for this sector: the mould performance (energy consumption and material waste) has much more relevance than the mould production impact. Such findings relate mainly to the high amount of Polystyrene used to produce plastic parts – 2550 kg (Table 3). Due to such particularity of the injection moulding sector - very high impacts contribution of the use stage, the tool also presents the impacts only related to production of the mould. This is important result, to support decisions related to improve environmental performance related to the mould material, production process and disposal, elsewise, these aspects would be easily neglected. Therefore, the focus for improving the environmental performance must be focused on the use stage.

The summarized set of results for the environmental impacts, for all life cycles stages, of all five moulds are presented in Table 13. The use stage is responsible for more than 96% of the environmental impacts along the defined life cycle, for all five application cases. In order to better understand the high

Table 10. Detailed environmental impacts [Pt - Eco-indicator 99 (H/A)] for each life cycle stage: Example for Company A

	Amount	Unit	% Impact	% Global Impact
Materials				
Steel	16.35	Pt	96%	2%
Copper	0.69	Pt	4%	0%
Production Processes				
Milling - Steel	12.82	Pt	72%	1%
Drilling - Steel	4.30	Pt	24%	0%
Turning – Steel (CNC)	0.39	Pt	2%	0%
Steel Product - Metal Working	0.24	Pt	1%	0%
Use				
Electricity -Low Voltage (PT)	176.26	Pt	18%	17%
Polystyrene	808.83	Pt	82%	79%
Transport to Production				
Transport - Lorry (16-32T)	0.02	Pt	52%	0%
Transport - Van (<3.5T)	0.02	Pt	48%	0%
Transport to Use				
N.A.	N.A.	Pt	N.A.	0%
Transport to Disposal (P)				
Transport - Lorry (16-32T)	0.00	Pt	100%	0%
Transport to Disposal				
N.A.	N.A.	Pt	N.A.	0%
Disposal				
Disposal of Steel - Inert Material Landfill	0.06	Pt	100%	0%
Total	1019.98	PT		

impact obtained in the use stage, the tool enables to look into it - Table 14, which summarizes the use phase results. It has become clear that the injected plastic itself is the principal contributor, for all applications cases, having a much higher impact when compared with the electricity needed for injection.

KEPI

The set of KEPIs selected is presented in Table 15. All the indicators, regardless ir scope, are quantified per kg of Mould, except for operation time and rate of defectives.

For the comparison between different moulds, it is important to mention that the environmental impact per kg of mould is a relevant indicator for the industrial. The aim, of this specific indicator, is to allow the comparison between different typologies of moulds when, at the first sight, seemed as an impossible task because of its different nature. This arises into a new perspective, regarding the mould's environmental performance, suitable to be better explored and understood by the moulds makers and their clients.

Table 11. Mould complete life cycle environmental impact [Pt - Eco-indicator 99 (H/A)]: Example for Company A

Life Cycle Stage	Pt	% Global Impact
Materials Acquisition	17.04	1.67
Transport to Production	0.04	0.00
Production	17.75	1.74
Transport to Disposal (P)	0.00	0.00
Transport to Use	0.00	0.00
Use	985.09	96.58
Transport to Disposal	0.00	0.00
Disposal	0.06	0.01
Total	1 019.99	100.00

Table 12. Mould production and disposal environmental impact [Pt - Eco-indicator 99 (H/A)]: Example for the mould of company A

Life Cycle Stage	Pt	% Global Impact
Materials Acquisition	17.04	48.84
Transport to Production	0.04	0.11
Production	17.75	50.88
Transport to Disposal	0.00	0.01
Disposal	0.06	0.17
Total	34.89	100.00

Table 13. Environmental impacts [Pt - Eco-indicator 99 (H/A)] and the % impacts per stage

Life Cycle Phase	Company A		Company B		Company C		Company D		Company E	
	Pt	%	Pt	%	Pt	%	Pt	%	Pt	%
Materials Acquisition	17.04	1.67	970.18	1.07	1716.37	2,39	29,50	0,95	78,22	0,11
Transport to Production	0.04	0.00	12.96	0.01	0.31	0,00	0,06	0,00	11,49	0,02
Production	17.75	1.74	702.89	0.78	319.50	0,45	36,14	1,16	17,38	0,02
Transport to Disposal (P)	0.00	0.00	0.12	0.00	0.00	0,00	0,00	0,00	0,00	0,00
Transport to Use	0.00	0.00	131.56	0.15	0.00	0,00	10,16	0,33	0,00	0,00
Use	985.09	96.58	8 8731.24	97.99	69 652.16	97,16	3 033,01	97,56	73 922,59	99,86
Transport to Disposal	0.00	0.00	0.00	0.00	0.00	0,00	0,00	0,00	0,00	0,00
Disposal	0.06	0.01	2.26	0.00	1.03	0,00	0,12	0,00	0,00	0,00
Total	1 019.99	100	90 551.22	100	71 689.36	100	3 108,99	100	74 029,68	100

Table 14. Life Cycle Assessment: Use Phase Use Phase panorama for all five moulds

Used Material and Energy		Amount (Pt)	% Use Impact	% Global Impact
Company A	Electricity	176.26	18%	17%
	Polystyrene	808.83	82%	79%
Company B	Electricity.	12409.80	14%	14%
	Polypropylene	76321.44	86%	84%
Company C	Electricity	16707.32	24%	23%
	Polypropylene	52944.84	76%	74%
Company D	Electricity	729.21	24%	24%
	Polycarbonate	2303.80	76%	74%
Company E	Electricity	2676.12	4%	4%
	Nylon	71246.47	96%	96%

The ratios between environmental impacts (Pt) and the kg of mould (Pt/kg mould), for all five mould are presented in Table 16. The higher ratio result, indicates lower environmental performance. Despite mould of company B has the highest overall environmental impact and not the mould with more mass, it does not have the highest impact per kg of mould (22.30 Pt/kg of final Mould). The mould of company E on the other hand, has a very high ratio of impact per each kg of final mould. Such results and findings enable to validate the soundness of this very particular KEPI.

EEI

The last set of indicators are the Eco-Efficiency Indicators. An example is presented in Table 17 for the five moulds. These indicators measure the relation between economic and environmental performance, providing useful information regarding the product/process behaviour.

Regarding the life cycle oriented EEI, is reasonable to state that mould of company A has the highest Eco-efficiency performance (GVA / OEI = 20.96), while the mould of company B has the lowest (GVA / OEI =0.62). These results are highly aligned with the KEPI for Pt/kg of Table 16. Ultimately, in order to simplify the understanding Life Cycle Oriented EEI for the different moulds, an eco-efficiency plot, considering the production costs and the economic value, is presented in Figure 10. It becomes quite clear the moulds from the Companies B and C have the lowest Eco-efficiency performance.

Regarding the functional indicators, it is worth mentioning that the mould from company B and C have a relatively low number of cycle per overall impact (Pt). This information could be used, for instance, to inform the mould designers that, they could design the mould, if possible, for more cycles. Reversely, the mould from company C has the highest relation of used mater per impact and has no waste material.

For the performance indicators, it is possible to state the mould from Company A has a lower energy consumption per cycle and the mould from company C has very high specific energy consumption per cycle. Reversely, mould from Company C has the lowest energy consumption per amount of injected material when compared with the mould of company D.

Table 15. Detailed list of KEPIs specific for the injection moulding industry: Example for the mould of company A

Scope Indicator	Indicator	Amount	Unit
Materials Acquisition			
Materials purchased	Total Amount of material used	1.42	kg /kg of Mould
Transport to production	Total Distance Travelled (Materials Acquisition > Production)	0.37	km/kg of Mould
Production			
Production process	Total Amount of material removed (Mechanical Process)	0.42	kg/kg of Mould
Waste (By-products) generated during production	Total Amount By-products generated within the production process	0.42	kg/kg of Mould
Transport to disposal (P)	Total Distance Travelled (Production > Disposal)	0.03	km/kg of Mould
Use			
Material injected	Total Amount of material used	16.22	kg/kg of Mould
Material injected	Total Amount of material reused in the process	0.00	kg/kg of Mould
Energy used during injection	Total Amount of energy used	27.50	kWh/kg of Mould
Operation time	Cycle time	15.50	Sec/cycle
Operation time	Total service hours during Life-Time	2.05	Hours/kg of Mould
Defective production	Rate of defective products	2.00	%
Disposal			
Waste	Total Amount of waste generated	0.42	kg /kg of Mould

Table 16. Environmental Impact (Pt) per kg of mould per life cycle stage

Life Cycle Stage	Company A	Company B	Company C	Company D	Company E
Materials Acquisition	0.11	0.24	0.23	0.11	0.25
Transport to Production	0.00	0.00	0.00	0.00	0.04
Production	0.11	0.17	0.04	0.14	0.06
Transport to Disposal (P)	0.00	0.00	0.00	0.00	0.00
Transport to Use	0.00	0.03	0.00	0.04	0.00
Use	6.27	21.85	9.47	11.47	240.25
Transport to Disposal	0.00	0.00	0.00	0.00	0.00
Disposal	0.00	0.00	0.00	0.00	0.00
Total	6.49	22.30	9.75	11.75	240.60

All this conclusions, are important and come in handy to support the decision-making process during mould deign, as these indicators do not only relay on environmental or economic performance, they consider the very specific mould functional and performance aspects. Additionally, in Figure 11, an example (Company B) EEI presented to the user is depicted. With such information, a worthy characterization, regarding sustainability performance mould is given in a standardized and understandable manner, to support the decision making process.

Table 17. Eco-efficiency Indicators (N.A.: Data not available)

	Company A	Company B	Company C	Company D	Company E
Life Cycle Oriented Indicators					
GVA / EI of MAP	1254.47	57.72	46.61	651.14	N.A.
GVA / EI of PP	1204.19	79.67	250.39	531.52	N.A.
GVA / EI of US	21.70	0.63	1.15	6.33	N.A.
GVA / OEI	20.96	0.62	1.12	6.18	N.A.
Functional Indicators					
Nº Cycles / OEI	73.53	5.52	0.70	64.33	67.54
Tot. UIM / OEI	2.21	2.84	2.92	1.29	0.24
Tot. WIM / OEI	0.29	0.50	0.00	0.32	0.00
MFW / OEI	0.15	0.04	0.10	0.09	0.00
Performance Indicators					
Tot. EUDI / Nº Cycles	0.0576	0.6090	8.1963	0.0894	0.0131
TA-EUDI / TA-IM	0.0017	0.0010	0.0020	0.0036	0.0004
O-EI / TA-IM	0.0004	0.0003	0.0003	0.0006	0.0004

GVA = Gross Value Added I EI of MAP = Environmental influence of materials acquisition stage I EI of PP = Environmental influence of production stage I EI of US = Environmental Influence of use stage I OEI = Overall environmental influence I Tot. UIM = Total amount of useful injected materialI Tot. WIM = Total amount of wasted injection material I MFW = Mould's final weight I Tot. EUDI = Total amount of energy used during injection I Tot. IM = Total amount of injected material

Figure 10. Example Eco-Efficiency Plot

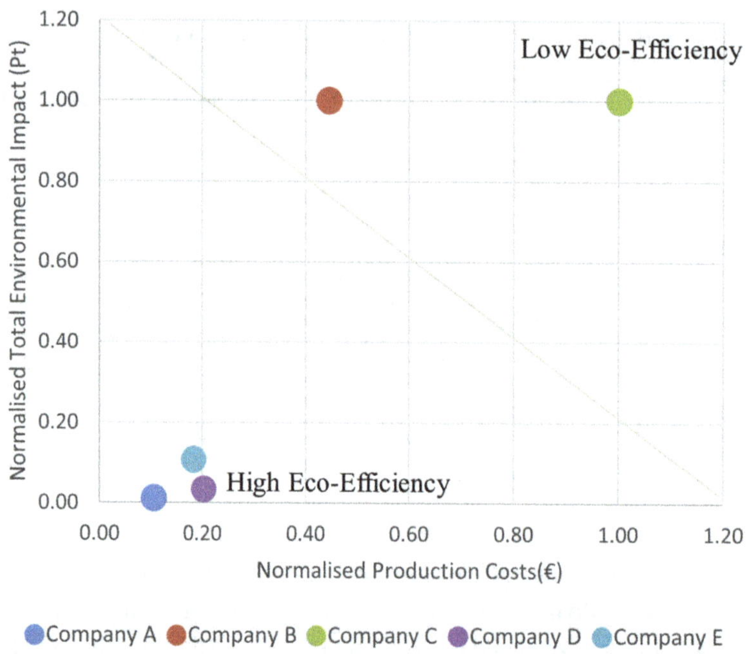

Figure 11. Example Eco-Efficiency from the tool: Company B

Numerator / Denominator		Values	Result
Llife Cycle Oriented Indicators			
Sale Price / Environmental influence of materials acquisition phase	=	88 000,00 € / 970,18 Pt	= 90,70 €/Pt
Sale Price / Environmental influence of production phase	=	88 000,00 € / 702,89 Pt	= 125,20 €/Pt
Sale Price / Environmental Influence of use phase	=	88 000,00 € / 88731,24 Pt	= 0,99 €/Pt
Sale Price / Overall environmental influence	=	88 000,00 € / 90551,22 Pt	= 0,97 €/Pt
Functional Indicators			
Number of cycles / Overall environmental influence	=	500000 Cycles / 90551,22 Pt	= 5,52 Cycles/Pt
Total amount of useful injected material / Overall environmental influence	=	257000 kg / 90551,22 Pt	= 2,84 kg/Pt
Total amount of wasted injection material / Overall environmental influence	=	45000 kg / 90551,22 Pt	= 0,50 kg/Pt
Mould's final weight / Overall environmental influence	=	4061,14 kg / 90551,22 Pt	= 0,04 kg/Pt
Specific Performance Indicators			
Total amount of energy used during injection / Number of cycles	=	304400,59 kWh / 500000 Cycles	= 6,09E-01 kWh/Cycle
Total amount of energy used during injection / Total amount of injected material	=	304400,59 kWh / 302000000 g	= 1,01E-03 kWh/g of Injected material
Overall environmental influence / Total amount of injected material	=	90551,22 Pt / 302000000 g	= 3,00E-04 Pt/g of Injected material

CONCLUSION

The importance of applying Eco-Efficiency assessment are discussed in this chapter and the practical difficulties are identified and discussed. The particular situation of injection moulds manufacturing sector, characterized by one-of-a-kind production, was analysed emphasizing the need of a specific framework to support the calculation of Eco-Efficiency towards a more informed and sustainable decision making. In fact, the several definitions of Eco-efficiency proposed by several normative documents bring a myriad of different environmental indicators. Most of these indicators are not relevant for the specific sector of a product, so they contribute to confuse the user of those documents. This chapter proves that this barriers are even higher for the case of injection moulds, requiring a recurrent need to retrieve specific data for each mould to calculate the Eco-Efficiency. A framework is proposed to overcome this problem, by the simplification of the Eco-Efficiency assessment. This framework is applied to the injection moulds sector, but a similar approach can be used to other sectors.

The benefits of using the proposed framework is a reduced set of Eco-Efficiency related indicators and ratios that allow to assess the products on a life cycle perspective. A specific tool was developed to facilitate the application framework concepts by industrial companies. The tool was applied to four different real industrial moulds from four different companies. The results comparison and analysis allow to identify the potential proposed indicators to support the decision making process related with Eco-Efficiency performance.

ACKNOWLEDGMENT

The authors wish to acknowledge the financial support provided by Project NORTE-01-0145-FEDER-000022 - SciTech - Science and Technology for Competitive and Sustainable Industries, cofinanced by Programa Operacional Regional do Norte (NORTE2020), through Fundo Europeu de Desenvolvimento Regional (FEDER).

The authors wish to acknowledge the financial support provided by Projetos de I&DT Empresas em Co-Promoção – 17637 – ECO2PLAST: Desenvolvimento de sistema de Produção ECOeficiente para produção de componentes PLÁSticos, cofinanced by Programa Operacional Temático Competitividade e Internacionalização, apoiado pelos Fundos Estruturais e de Investimento Europeus no âmbito do Portugal2020.

REFERENCES

Alves, J. L. S., & Dumke De Medeiros, D. (2015). Eco-efficiency in micro-enterprises and small firms: A case study in the automotive services sector. *Journal of Cleaner Production*, *108*, 595–602. doi:10.1016/j.jclepro.2015.07.063

Bueno, C., Hauschild, M. Z., Rossignolo, J. A., Ometto, A. R., & Mendes, N. C. (2016). Sensitivity analysis of the use of Life Cycle Impact Assessment methods: A case study on building materials. *Journal of Cleaner Production*, *112*, 2208–2220. doi:10.1016/j.jclepro.2015.10.006

Cerutti, A. K., Beccaro, G. L., Bagliani, M., Donno, D., & Bounous, G. (2013). Multifunctional Ecological Footprint Analysis for assessing eco-efficiency: A case study of fruit production systems in Northern Italy. *Journal of Cleaner Production*, *40*, 108–117. doi:10.1016/j.jclepro.2012.09.028

Ellen MacArthur Foundation. (2016). *The New Plastics Economy: Rethinking the future of plastics*. Ellen MacArthur Found.

Esteves, S., Lourenço, E. J., Moita, N., Peças, P., Ribeiro, I., & Henriques, E. (2014). Injection Moulding Process Indicators to Foster a More Sustainable Production of Plastic Parts. In 3rd workbook of the cross-sectional group 'Energy-related technical and economic evaluation' of the Cluster of Excellence eniPROD. Wissenschaftliche Scripten.

Folgado, R., Peças, P., & Henriques, E. (2010). Life cycle cost for technology selection: A Case study in the manufacturing of injection moulds. International Journal of Production Economics, 368–78. doi:10.1016/j.ijpe.2010.07.036

Gantar, G., Glojek, A., Mori, M., Nardin, B., & Sekavčnik, M. (2013). Resource efficient injection moulding with low environmental impacts. *Stroj Vestnik/Journal. Mechanical Engineering (New York, N.Y.)*, *59*(3), 193–200.

Global Reporting Initiative. (2006). *Sustainability Reporting Guidelines*. Author.

Global Reporting Initiative. (2014). *G4 Sustainability Reporting Guidelines*. Glob Reporting Initiative. Available from: https://www.globalreporting.org/standards/g4/Pages/default.aspx

Helu, M., Behmann, B., Meier, H., Dornfeld, D., Lanza, G., & Schulze, V. (2012). Impact of green machining strategies on achieved surface quality. *CIRP Annals-Manufacturing Technology.*, *61*(1), 55–58. doi:10.1016/j.cirp.2012.03.092

Hukkinen, J. (2001). Eco-efficiency as abandonment of nature. *Ecological Economics*, *38*(3), 311–315. doi:10.1016/S0921-8009(01)00217-8

Huppes, G., & Ishikawa, M. (2005). A Framework for Quantified Eco-efficiency Analysis. *Journal of Industrial Ecology*, *9*(4), 25–41. Available from: http://doi.wiley.com/10.1162/108819805775247882

Hur, T., Lee, J., Ryu, J., & Kwon, E. (2005). Simplified LCA and matrix methods in identifying the environmental aspects of a product system. *Journal of Environmental Management*, *75*(3), 229–237. doi:10.1016/j.jenvman.2004.11.014 PMID:15829365

Huszar, M., Belblidia, F., Davies, H. M., Arnold, C., Bould, D., & Sienz, J. (2015). Sustainable injection moulding: The impact of materials selection and gate location on part warpage and injection pressure. *Sustainable Materials and Technologies*, *5*, 1–8. doi:10.1016/j.susmat.2015.07.001

International Organization of Standardization. (2006). *UNE-EN ISO 14040 Environmental Management – Life Cycle Assessment –Principles and Framework*. Author.

ISO. (2012). ISO 14045:2012 Preview Environmental management -- Eco-efficiency assessment of product systems -- Principles, requirements and guidelines.

Jasch, C. (2000). Environmental performance evaluation and indicators. *Journal of Cleaner Production*, *8*(1), 79–88. doi:10.1016/S0959-6526(99)00235-8

Kerr, W., & Ryan, C. (2001). Eco-efficiency gains from remanufacturing: A case study of photocopier remanufacturing at Fuji Xerox Australia. *Journal of Cleaner Production*, *9*(1), 75–81. doi:10.1016/S0959-6526(00)00032-9

Kneese, A. (2003). *Environmental Degradation and Institutional Responses: Vol. 1. Handbook of Environmental Economics*. Available from: http://www.sciencedirect.com/science/article/pii/S1574009903010039

Korol, J., Burchart-Korol, D., & Pichlak, M. (2016). Expansion of environmental impact assessment for eco-efficiency evaluation of biocomposites for industrial application. *Journal of Cleaner Production*, *113*, 144–152. doi:10.1016/j.jclepro.2015.11.101

Kulak, M., Nemecek, T., Frossard, E., & Gaillard, G. (2016). Eco-efficiency improvement by using integrative design and life cycle assessment. The case study of alternative bread supply chains in France. *Journal of Cleaner Production*, *112*, 2452–2461. doi:10.1016/j.jclepro.2015.11.002

Lehni, M. (2000). Eco-efficiency. Creating more value with less impact. *World Bus Counc Sustain Dev*, 1–32. Available from: http://www.wbcsd.org

Lundquist, L., Leterrier, Y., Sunderland, P., & Månson, J.-A. E. (2000). Life Cycle Engineering of Plastics. Life Cycle Engineering of Plastics.

Madan, J., Mani, M., Lee, J. H., & Lyons, K. W (2015). Energy performance evaluation and improvement of unit-manufacturing processes: Injection molding case study. *Journal of Cleaner Production, 105*, 157–170. doi:10.1016/j.jclepro.2014.09.060

Malloy, R. A. (2010). Plastic Part Design for Injection Molding. Plastic Part Design. doi:10.3139/9783446433748

Messagie, M., Boureima, F., Mertens, J., Sanfelix, J., Macharis, C., & Mierlo, J. (2013, March). The Influence of Allocation on the Carbon Footprint of Electricity Production from Waste Gas, a Case Study for Blast Furnace Gas. *Energies*, *6*(3), 1217–1232. doi:10.3390/en6031217

Möller, A., & Schaltegger, S. (2005). The Sustainability Balanced Scorecard as a Framework for Eco-efficiency Analysis. *Journal of Industrial Ecology*, *9*(4), 73–83. doi:10.1162/108819805775247927

National Round Table on the Environment and the Economy. (2003). *Environment and Sustainable Development Indicators for Canada*. Available from: papers2://publication/uuid/F0A97390-580F-4EF6-9CED-1EEF3EE889D5

National Round Table on the Environment and the Economy. (2012). *Canada's Opportunity: Adopting Life Cycle Approaches for Sustainable Development*. Author.

Newman, S. T., Nassehi, A., Imani-Asrai, R., & Dhokia, V. (2012). Energy efficient process planning for CNC machining. *CIRP Journal of Manufacturing Science and Technology*, *5*(2), 127–136. doi:10.1016/j.cirpj.2012.03.007

Ng, R., Yeo, Z., Low, J. S. C., & Song, B. (2015). A method for relative eco-efficiency analysis and improvement: Case study of bonding technologies. *Journal of Cleaner Production*, *99*, 320–332. doi:10.1016/j.jclepro.2015.03.004

Noble, J., Walczak, K., & Dornfeld, D. (2014). Rapid tooling injection molded prototypes: A case study in artificial photosynthesis technology. CIRP Procedia, 251–256. doi:10.1016/j.procir.2014.03.035

O'Reilly, M. (2000). ISO 14031: Effective Mechanism to Environmental Performance Evaluation. *Corporate Environmental Strategy*, *7*(3), 267–275. doi:10.1016/S1066-7938(00)80121-9

Oktem, H., Erzurumlu, T., & Erzincanli, F. (2006). Prediction of minimum surface roughness in end milling mold parts using neural network and genetic algorithm. *Materials & Design*, *27*(9), 735–744. doi:10.1016/j.matdes.2005.01.010

Oliveira, L., Messagie, M., Mertens, J., Laget, H., Coosemans, T., & Van Mierlo, J. (2015). Environmental performance of electricity storage systems for grid applications, a life cycle approach. *Energy Conversion and Management*, *101*, 326–335. doi:10.1016/j.enconman.2015.05.063

Olsthoorn, X., Tyteca, D., Wehrmeyer, W., & Wagner, M. (2001). Environmental indicators for business: A review of the literature and standardisation methods. *Journal of Cleaner Production*, *9*(5), 453–463. doi:10.1016/S0959-6526(01)00005-1

Organisation for Economic Co-Operation and Development. (2003). OECD Environmental Indicators: development, measurement and use (Vol. 25). OECD.

Pandelidis, I., & Zou, Q. (1990). Optimization of injection molding design. Part I: Gate location optimization. *Polymer Engineering and Science*, *30*(15), 873–882. doi:10.1002/pen.760301502

Panel IR. (2014). *Decoupling: Natural Resource Use and Environmental*. Author.

Park, H. S., & Nguyen, T. T. (2014). Optimization of injection molding process for car fender in consideration of energy efficiency and product quality. *Journal of Computational Design and Engineering.*, *1*(4), 256–265. doi:10.7315/JCDE.2014.025

Park, P. J., Tahara, K., & Inaba, A. (2007). Product quality-based eco-efficiency applied to digital cameras. *Journal of Environmental Management*, *83*(2), 158–170. doi:10.1016/j.jenvman.2006.02.006 PMID:16697518

Paul, B. D. (2008). A history of the concept of sustainable development: Literature review. Ann Univ Oradea. *Econ Sci Ser.*, *17*(2), 576–580.

Peças, P., Ribeiro, I., Folgado, R., & Henriques, E. (2009). A Life Cycle Engineering model for technology selection: A case study on plastic injection moulds for low production volumes. *Journal of Cleaner Production*, *17*(9), 846–856. doi:10.1016/j.jclepro.2009.01.001

Pun, K.-F., Hui, I.-K., Lewis, W. G., & Lau, H. C. W. (2003). A multiple-criteria environmental impact assessment for the plastic injection molding process: A methodology. *Journal of Cleaner Production*, *11*(1), 41–49. doi:10.1016/S0959-6526(02)00019-7

Ribeiro, I., Kaufmann, J., Schmidt, A., Peças, P., Henriques, E., & Götze, U. (2016). Fostering selection of sustainable manufacturing technologies - A case study involving product design, supply chain and life cycle performance. *Journal of Cleaner Production*, 112.

Ribeiro, I., Peças, P., & Henriques, E. (2013). A life cycle framework to support materials selection for Ecodesign: A case study on biodegradable polymers. *Materials & Design*, 51.

Ribeiro, I., Peças, P., & Henriques, E. (2013). Incorporating tool design into a comprehensive life cycle cost framework using the case of injection molding. *Journal of Cleaner Production*, 53.

Ribeiro, I., Peças, P., & Henriques, E. (2014). Life Cycle Engineering Framework for Technology and Manufacturing Processes Evaluation. In E. Henriques, P. Peças, & A. Silva (Eds.), *Technology and Manufacturing Process Selection: The Product Life Cycle Perspective* (pp. 217–237). Springer. doi:10.1007/978-1-4471-5544-7_11

Ribeiro, I., Peças, P., & Henriques, E. (2012). Assessment of energy consumption in injection moulding process. *Leveraging Technology for a Sustainable World - Proceedings of the 19th CIRP Conference on Life Cycle Engineering*. 10.1007/978-3-642-29069-5_45

Ribeiro, I., Peças, P., & Henriques, E. (2008). Environmental Impact of Plastic Injection Moulds. In *3rd International Conference on Polymers and Moulds Innovations – PMI 2008*. Ghent: University College Ghent.

Ribeiro, I., Peças, P., & Henriques, E. (2011). Life cycle approach to support tooling design decisions. *ICED 11 - 18th International Conference on Engineering Design - Impacting Society Through Engineering Design*.

Ribeiro, I., Pousa, C., Folgado, R., Peças, P., & Henriques, E. (2009). LCC and LCA Simplified Models to Foster the Design of Sustainable Plastic Injection Moulds. *16th CIRP International Conference on Life Cycle Engineering (LCE 2009)*, 99–104.

Rosato, D. V., & Rosato, M. G. (2000). *Injection molding handbook*. Kluwer Academic Publisher. doi:10.1007/978-1-4615-4597-2

Silalertruksa, T., Gheewala, S. H., & Pongpat, P. (2015). Sustainability assessment of sugarcane biorefinery and molasses ethanol production in Thailand using eco-efficiency indicator. *Applied Energy*, *160*, 603–609. doi:10.1016/j.apenergy.2015.08.087

Sproedt, A., Plehn, J., Schönsleben, P., & Herrmann, C. (2015). A simulation-based decision support for eco-efficiency improvements in production systems. *Journal of Cleaner Production*, *105*, 389–405. doi:10.1016/j.jclepro.2014.12.082

Suh, S., & Huppes, G. (2005). Methods for Life Cycle Inventory of a product. *Journal of Cleaner Production*, *13*(7), 687–697. doi:10.1016/j.jclepro.2003.04.001

Tan, H. X., Yeo, Z., Ng, R., Tjandra, T. B., & Song, B. (2015). A sustainability indicator framework for Singapore small and medium-sized manufacturing enterprises. CIRP Procedia, 132–137. doi:10.1016/j.procir.2015.01.028

Thiriez, A., & Gutowski, T. (2006). An Environmental Analysis of Injection Molding. *Proceedings of the 2006 IEEE International Symposium on Electronics and the Environment*, 195–200. 10.1109/ISEE.2006.1650060

Union, E. (2010). *History and Definitions of Eco-Efficiency*. Leonardo da Vinci Program.

WBCSD. (2000a). *Eco-efficiency. Creating more Value with less Impact*. World Bus Counc Sustain Dev.

WBCSD. (2000b). *Measuring ecoefficiency: a guide to reporting company performance*. Geneva, Switzerland. World Business Council for Sustainable Development.

Weidema, B., Wenzel, H., Petersen, C., & Hansen, K. (2004). The product, functional unit and reference flows in LCA. *Environ News, 70*, 46. Available from: http://gfc.force.dk/resources/777.pdf

Weidema, B.P., Ekvall, T., & Heijungs, R. (2009). *Guidelines for application of deepened and broadened LCA*. Guidel Appl Deep broadened LCA Deliv D18 Work Packag 5 CALCAS Proj. 2009, (037075), 49.

Wursthorn, S., Poganietz, W. R., & Schebek, L. (2011). Economic-environmental monitoring indicators for European countries: A disaggregated sector-based approach for monitoring eco-efficiency. *Ecological Economics*, *70*(3), 487–496. doi:10.1016/j.ecolecon.2010.09.033

Yang, Z., Zhou, X., & Xu, L. (2014). Eco-efficiency optimization for municipal solid waste management. *Journal of Cleaner Production*, *104*, 242–249. doi:10.1016/j.jclepro.2014.09.091

Zhang, B., Bi, J., Fan, Z. Y., Yuan, Z. W., & Ge, J. J. (2008). Eco-efficiency analysis of industrial system in China: A data envelopment analysis approach. *Ecological Economics*, *68*(1-2), 306–316. doi:10.1016/j.ecolecon.2008.03.009

APPENDIX

Demonstration Tool

Screens of the developed tool. The goal is to present the overall appearance of the tool.

Figure 12. Company information

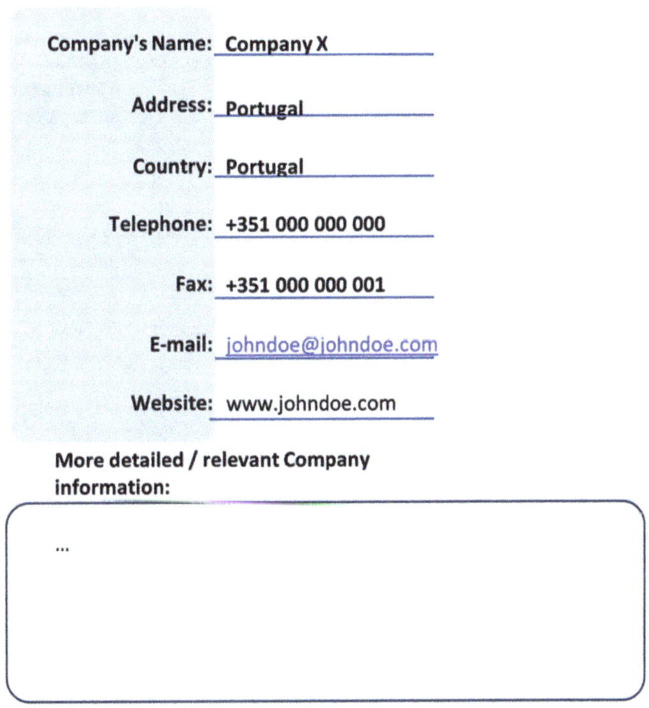

Figure 13. Mould characterization and part description

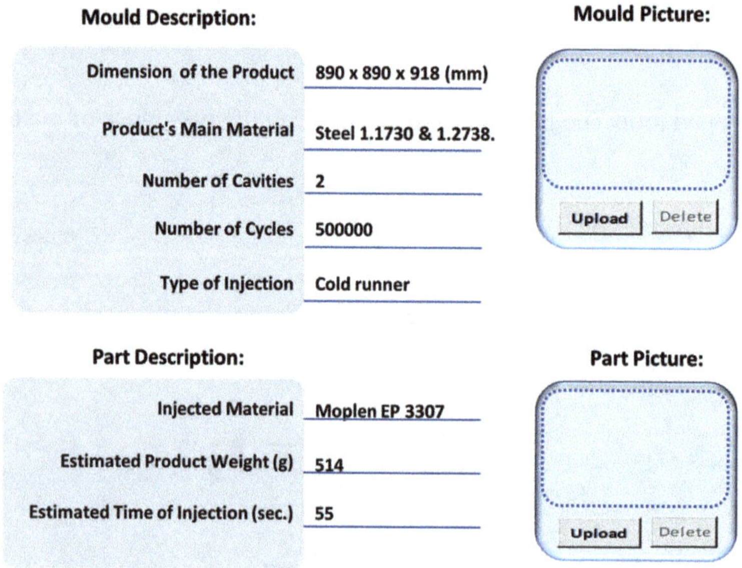

Figure 14. Value Data

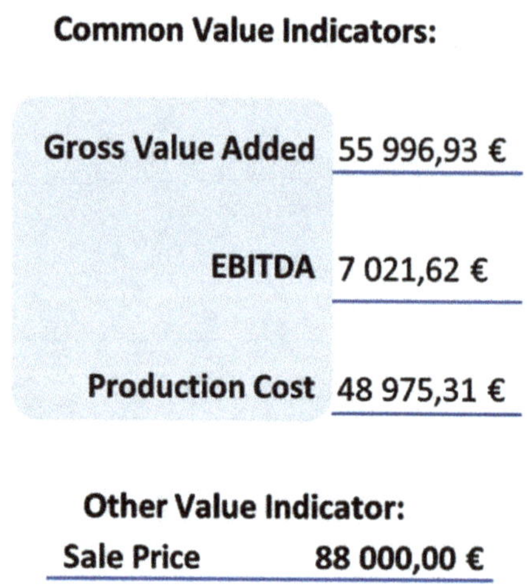

Figure 15. LCI Parameterization

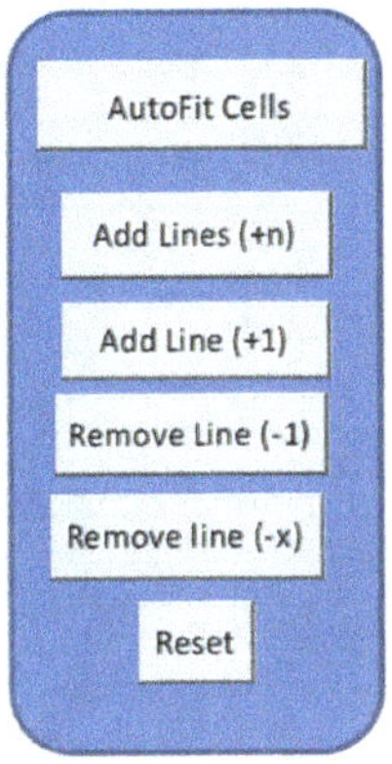

#	Part Reference	Part Name	Part Quantity	Acquisition?	Production?	Disposal?	Process's Number
1	1	Part A	1	Y	Y	Y	3
2	1A	Part B	1	Y	Y	Y	2
3	2	Part C	1	Y	Y	Y	3
4	5	Part D	1	Y	Y	Y	3
5	6	Part E	1	Y	Y	Y	3
6	7	Part F	1	Y	Y	Y	3
7	8	Part G	1	Y	Y	Y	3
8	9	Part H	1	Y	Y	Y	3
9	9A	Part I	1	Y	Y	Y	2
10	12	Part J	16	Y	Y	Y	3
11	13	Part K	2	Y	Y	Y	3
12	14	Part L	10	Y	Y	Y	3
13	16	Part M	4	Y	N	N	0
14	23	Part N	4	Y	N	N	0
15	375	Part O	4	Y	Y	Y	2
16	378	Part P	36	Y	N	N	0

Figure 16. LCI: Materials identification and quantification, and impacts quantification

#	Part Reference	Part Name	Part Quantity	Type of Material	Material Category	Material Subcategory	Material	Unit	Initial Quantity (unitary value)	Final Quantity (unitary value)	Score (Pt)
1	1	Part A	1	Metals	Ferro	Ferro Metals	Steel, Converter, Low-Alloyed	kg	332,11	236,91	55,28
2	1A	Part B	1	Glass	Glass	Glass Materials	Glass Fibre	kg	1,78	1,09	0,24
3	2	Part C	1	Metals	Ferro	Ferro Metals	Steel, Converter, Low-Alloyed	kg	608,15	488,01	113,87
4	5	Part D	1	Metals	Ferro	Ferro Metals	Steel, Converter, Low-Alloyed	kg	106,11	44,19	10,31
5	6	Part E	1	Metals	Ferro	Ferro Metals	Steel, Converter, Low-Alloyed	kg	106,11	44,19	10,31
6	7	Part F	1	Metals	Ferro	Ferro Metals	Steel, Converter, Low-Alloyed	kg	148,32	85,17	19,87
7	8	Part G	1	Metals	Ferro	Ferro Metals	Steel, Converter, Low-Alloyed	kg	188,77	115,21	26,88
8	9	Part H	1	Metals	Ferro	Ferro Metals	Steel, Converter, Low-Alloyed	kg	332,11	234,56	54,73
9	9A	Part I	1	Glass	Glass	Glass Materials	Glass Fibre	kg	20,88	10,84	2,36
10	12	Part J	16	Metals	Ferro	Ferro Metals	Steel, Converter, Low-Alloyed	kg	7,49	3,59	13,39
11	13	Part K	2	Metals	Ferro	Ferro Metals	Steel, Converter, Low-Alloyed	kg	7,49	0,45	0,21
12	14	Part L	10	Metals	Ferro	Ferro Metals	Steel, Converter, Low-Alloyed	kg	7,49	2,24	5,23
13	16	Part M	4	Metals	Ferro	Ferro Metals	Steel, Converter, Low-Alloyed	kg	8,08	8,08	7,55
14	23	Part N	4	Metals	Ferro	Ferro Metals	Steel, Converter, Low-Alloyed	kg	1,55	1,55	1,45
15	375	Part O	4	Metals	Non Ferro	Non Ferro Metals	Aluminium, Production Mix	kg	5,42	3,98	8,83
16	378	Part P	36	Metals	Non Ferro	Non Ferro Metals	Copper	kg	0,03	0,03	2,09

Figure 17. LCI: Production processes identification and quantification, and impacts quantification

#	Part Reference	Part Name	Part Quantity	Type of Process	Process Category	Process	Unit	Quantity (unitary value)	Score (Pt)
1	1	Part A	1	Metalurgic	Chipping	Milling, Steel	kg	42,22	11,62
2	1	Part A	1	Metalurgic	Chipping	Drilling, CNC, Steel	kg	11,13	3,08
3	1	Part A	1	Metalurgic	Chipping	Milling, Steel	kg	41,85	11,52
4	1A	Part B	1	Metalurgic	Chipping	Milling, Steel	kg	0,60	0,17
5	1A	Part B	1	Metalurgic	Chipping	Drilling, CNC Steel	kg	0,08	0,02
6	2	Part C	1	Metalurgic	Chipping	Milling, Steel	kg	38,65	10,64
7	2	Part C	1	Metalurgic	Chipping	Drilling, CNC, Steel	kg	21,58	5,97
8	2	Part C	1	Metalurgic	Chipping	Milling, Steel	kg	59,91	16,49
9	5	Part D	1	Metalurgic	Chipping	Milling, Steel	kg	58,03	15,97
10	5	Part D	1	Metalurgic	Chipping	Drilling, CNC, Steel	kg	3,79	1,05
11	5	Part D	1	Metalurgic	Chipping	Milling, Steel	kg	0,10	0,03
12	6	Part E	1	Metalurgic	Chipping	Milling, Steel	kg	58,03	15,97
13	6	Part E	1	Metalurgic	Chipping	Drilling, CNC, Steel	kg	3,79	1,05
14	6	Part E	1	Metalurgic	Chipping	Milling, Steel	kg	0,10	0,03
15	7	Part F	1	Metalurgic	Chipping	Milling, Steel	kg	28,90	7,95
16	7	Part F	1	Metalurgic	Chipping	Drilling, CNC, Steel	kg	27,40	7,59
17	7	Part F	1	Metalurgic	Chipping	Milling, Steel	kg	6,84	1,88
18	8	Part G	1	Metalurgic	Chipping	Milling, Steel	kg	29,54	8,13
19	8	Part G	1	Metalurgic	Chipping	Drilling, CNC, Steel	kg	37,59	10,41
20	8	Part G	1	Metalurgic	Chipping	Milling, Steel	kg	6,42	1,77
21	9	Part H	1	Metalurgic	Chipping	Milling, Steel	kg	41,67	11,47
22	9	Part H	1	Metalurgic	Chipping	Drilling, CNC, Steel	kg	14,70	4,07
23	9	Part H	1	Metalurgic	Chipping	Milling, Steel	kg	41,19	11,33
24	9A	Part I	1	Metalurgic	Chipping	Milling, Steel	kg	9,11	2,51
25	9A	Part I	1	Metalurgic	Chipping	Drilling, CNC, Steel	kg	0,93	0,26
26	12	Part J	16	Metalurgic	Chipping	Turning, Steel	kg	3,88	19,14
27	12	Part J	16	Metalurgic	Chipping	Drilling, CNC, Steel	kg	0,01	0,06
28	12	Part J	16	Metalurgic	Chipping	Turning, Steel	kg	0,01	0,05
29	13	Part K	2	Metalurgic	Chipping	Turning, Steel	kg	7,04	4,34
30	13	Part K	2	Metalurgic	Chipping	Drilling, CNC, Steel	kg	0,00	0,00
31	13	Part K	2	Metalurgic	Chipping	Turning, Steel	kg	0,00	0,00
32	14	Part L	10	Metalurgic	Chipping	Turning, Steel	kg	5,24	16,13
33	14	Part L	10	Metalurgic	Chipping	Drilling, CNC, Steel	kg	0,01	0,02
34	14	Part L	10	Metalurgic	Chipping	Turning, Steel, CNC	kg	0,01	0,02
60	375	Part O	4	Metalurgic	Chipping	Milling, Aluminium	kg	0,87	2,24
61	375	Part O	4	Metalurgic	Chipping	Drilling, CNC, Aluminium	kg	0,57	1,47

Figure 18. LCI: Transports identification and quantification, and impacts quantification

#	Part Reference	Part Name	Part Quantity	Type of Transport	Transport Category	Transport	Unit	ton.	Unit 1	km	Unit 2	Score (Pt)
1	1	Part A	1	Road	Road Transport	Transport, Lorry 16-32T	tkm	0,33211	t	162	km	0,64552036
2	1A	Part B	1	Road	Road Transport	Transport, Van <3.5T	tkm	0,00178	t	10	km	0,00269121
3	2	Part C	1	Road	Road Transport	Transport, Lorry 16-32T	tkm	0,60815	t	162	km	1,18204103
4	5	Part D	1	Road	Road Transport	Transport, Lorry 16-32T	tkm	0,10611	t	162	km	0,20625131
5	6	Part E	1	Road	Road Transport	Transport, Lorry 16-32T	tkm	0,10611	t	162	km	0,20625131
6	7	Part F	1	Road	Road Transport	Transport, Lorry 16-32T	tkm	0,14832	t	162	km	0,28828121
7	8	Part G	1	Road	Road Transport	Transport, Lorry 16-32T	tkm	0,18877	t	162	km	0,36690336
8	9	Part H	1	Road	Road Transport	Transport, Lorry 16-32T	tkm	0,33211	t	162	km	0,64552036
9	9A	Part I	1	Road	Road Transport	Transport, Van <3.5T	tkm	0,02088	t	10	km	0,03154894
10	12	Part J	16	Road	Road Transport	Transport, Van <3.5T	tkm	0,00749	t	18	km	0,32601042
11	13	Part K	2	Road	Road Transport	Transport, Van <3.5T	tkm	0,00749	t	18	km	0,0407513
12	14	Part L	10	Road	Road Transport	Transport, Van <3.5T	tkm	0,00749	t	18	km	0,20375651
13	16	Part M	4	Road	Road Transport	Transport, Van <3.5T	tkm	0,00808	t	11	km	0,05374274
14	23	Part N	4	Road	Road Transport	Transport, Van <3.5T	tkm	0,00155	t	11	km	0,01032995
15	375	Part O	4	Road	Road Transport	Transport, Van <3.5T	tkm	0,00542	t	18	km	0,05900444
16	378	Part P	36	Road	Road Transport	Transport, Van <3.5T	tkm	0,00003	t	105	km	0,0164911

Figure 19. LCI: Use stage characterization

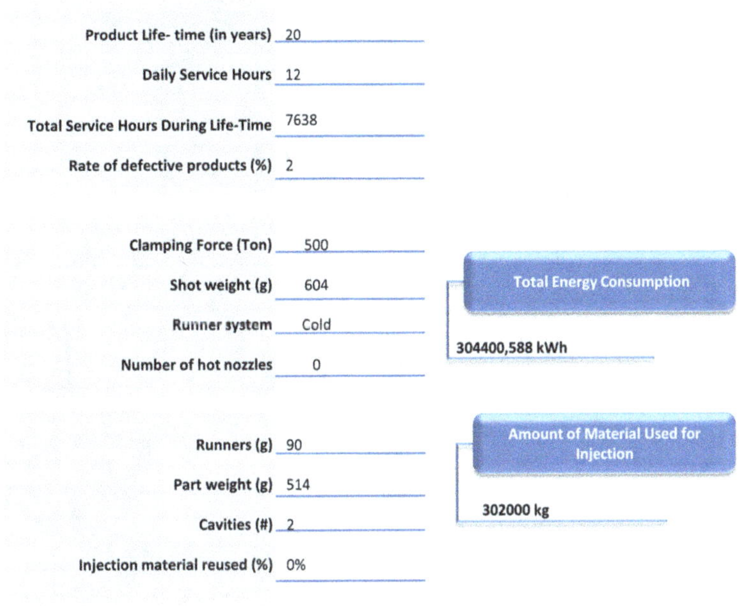

Figure 20. Summary data form LCI

General Information			
Initial Mass (kg)	Final Mass (kg)	Part Quantity (unit)	Removed Mass from Production (kg)
6623,74	4061,14	668	2562,60

Materials Acquisition	Amount	Unit
Steel, Converter, Low-Alloyed, At Plant Rer	6572,63	kg
Glass Fibre, At Plant Rer	22,66	kg
Aluminium, Production Mix, At Plant Rer	21,70	kg
Copper, At Regional Storage Rer	5,02	kg
Electronic Component, Passive, Unspecified, At Plant Glo	1,73	kg

Production Processes	Amount	Unit
Milling, Steel, Average Rer	2037,25	kg
Drilling, Cnc, Steel Rer	312,90	kg
Turning, Steel, Cnc, Average Rer	145,74	kg
Steel Product Manufacturing, Average Metal Working Rer	60,47	kg
Milling, Aluminium, Average Rer	3,47	kg
Drilling, Cnc, Aluminium Rer	2,76	kg

Use	Amount	Unit
Electricity, Low Voltage, At Grid Pt	304400,59	kWh
Polypropylene, Granulate, At Plant Rer	302000,00	kg

Transport to Production	Amount	Unit
Transport, Lorry 16-32T, Euro4 Rer	162,00	km
Transport, Van <3.5T Rer	320,00	km

Transport to Use	Amount	Unit
-	-	-

Transport to Disposal (P)	Amount	Unit
Transport, Lorry 16-32T, Euro4 Rer	4	km

Transport to Disposal	Amount	Unit
-	-	-

Disposal	Amount	Unit
Disposal, Steel, 0% Water, To Inert Material Landfill Ch	2562,60	kg

Figure 21. LCA results for complete life cycle

	Point [Eco-indicator 99 (H/A)]	% of the Global Impact
Materials Acquisition	970,18	1,07
Transport to Production	12,96	0,01
Production	702,89	0,78
Transport to Disposal (P)	0,12	0,00
Transport to Use	131,56	0,15
Use*	88731,24	97,99
Transport to Disposal	0,00	0,00
Disposal	2,26	0,00
Total	90551,22	100

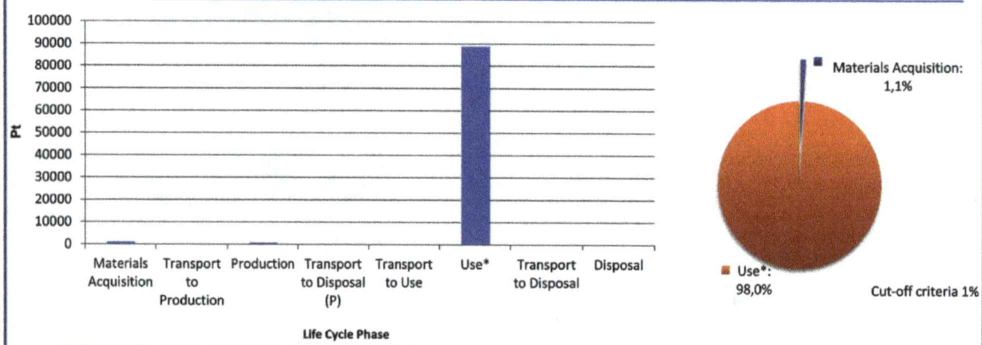

*Cycle time = 55 Sec
Number of Cycles = 500000
Total Energy Consumption = 304400,59 kWh
Total material consumption = 302000 kg

Figure 22. LCA results for Production mould

	Point [Eco-indicator 99 (H/A)]	% of the Global Impact
Materials Acquisition	970,18	57,46
Transport to Production	12,96	0,77
Production	702,89	41,63
Transport to Disposal	0,12	0,01
Disposal	2,26	0,13
Total	1688,42	100

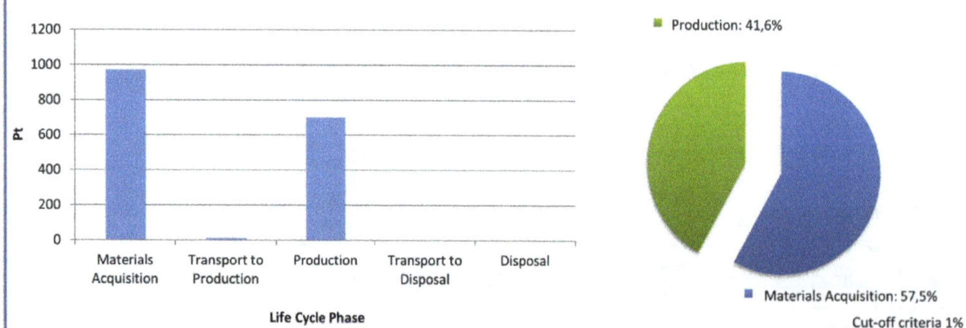

Figure 23. Detailed LCA results

Impact Score

Materials Acquisition	Amount	Unit	% of the Materials Acquisition Impact	% of the Global Impact
Steel, Converter, Low-Alloyed, At Plant Rer	939,55	Pt	96,84%	1,04%
Glass Fibre, At Plant Rer	2,60	Pt	0,27%	0,00%
Aluminium, Production Mix, At Plant Rer	8,83	Pt	0,91%	0,01%
Copper, At Regional Storage Rer	10,11	Pt	1,04%	0,01%
Electronic Component, Passive, Unspecified, At Plant Glo	9,09	Pt	0,94%	0,01%

Production Processes	Amount	Unit	% of the Production Processes Impact	% of the Global Impact
Milling, Steel, Average Rer	560,61	Pt	79,76%	0,62%
Drilling, Cnc, Steel Rer	86,61	Pt	12,32%	0,10%
Turning, Steel, Cnc, Average Rer	44,89	Pt	6,39%	0,05%
Steel Product Manufacturing, Average Metal Working Rer	6,77	Pt	0,96%	0,01%
Milling, Aluminium, Average Rer	2,24	Pt	0,32%	0,00%
Drilling, Cnc, Aluminium Rer	1,77	Pt	0,25%	0,00%

Use	Amount	Unit	% of the Use Impact	% of the Global Impact
Electricity, Low Voltage, At Grid Pt	12409,80	Pt	13,99%	13,72%
Polypropylene, Granulate, At Plant Rer	76321,44	Pt	86,01%	84,41%

Transport to Production	Amount	Unit	% of the Transport to Production Impact	% of the Global Impact
Transport, Lorry 16-32T, Euro4 Rer	12,01	Pt	92,66%	0,01%
Transport, Van <3.5T Rer	0,95	Pt	7,92%	0,00%

Transport to Use	Amount	Unit	% of the Transport to Use Impact	% of the Global Impact
-	-	Pt	-	0,00%

Transport to Disposal (P)	Amount	Unit	% of the Transport to Disposal (P) Impact	% of the Global Impact
Transport, Lorry 16-32T, Euro4 Rer	0,12	Pt	100,00%	0,00%

Transport to Disposal	Amount	Unit	% of the Transport to Disposal Impact	% of the Global Impact
-	-	Pt	-	0,00%

Disposal	Amount	Unit	% of the Transport to Disposal Impact	% of the Global Impact
Disposal, Steel, 0% Water, To Inert Material Landfill Ch	2,3	Pt	100,00%	0,00%

Figure 24. Detailed KEPI results

Materials Acquisition

Scope of the Indicator	Indicator	Amount	Unit
Materials purchased	Total Amount of material used	1,63	kg /kg of mold
Transport to production	Total Distance Traveled (Materials Acquisition > Production)	0,1187	km /kg of mold

Production

Scope of the Indicator	Indicator	Amount	Unit
Production process	Total Amount of material removed (Mechanical Process)	0,63	kg /kg of mold
Waste (By-products) generated during production	Total Amount By-products generated within the production process	0,63	kg /kg of mold
Trasnsport to use	Total Distance Traveled (Production > Use)	0,00	- /kg of mold
Trasnsport to disposal (P)	Total Distance Traveled (Production > Disposal)	0,00	km /kg of mold

Use

Scope of the Indicator	Indicator	Amount	Unit
Material injected	Total Amount of material used	74,36	kg /kg of mold
Material injected	Total Amount of material reused in the process	0,00	kg /kg of mold
Energy used during injection	Total Amount of energy used	74,95	kWh /kg of mold
Operation time	Cycle time	55,00	Sec /cycle
Operation time	Total service hours during Life-Time	1,88	Hours /kg of mold
Defective production	Rate of defective products	2,00	%
Trasnsport to disposal	Total -	-	- /kg of mold

Disposal

Scope of the Indicator	Indicator	Amount	Unit
Waste	Total Amount of waste generated	0,63	kg /kg of mold

Chapter 13
Effective Utilization of Industrial Wastes for Preparing Polymer Matrix Composites:
Usage of Industrial Wastes

Veerasimman Arumugaprabu
Kalasalingam University, India

Deepak Joel Johnson R.
Kalasalingam University, India

Pragatheeswaran R.
Government Polythetic Theni, India

ABSTRACT

The present industry scenario focuses on green manufacturing, in terms of effective reuse and recycling of the industrial wastes generated in enormous amount while preparing the product. The wastes also act as a threat to the society by causing various kinds of pollution. Therefore, the proper safe disposal of the same is a very critical factor. Most of the industries struggled with the enormous disposal of these wastes and finding ways for reuse and disposal. In this chapter, one such way of reuse of these wastes for making composite product is explored. Industrial wastes such as flyash and ricehusk used as fillers of varying weight percentages, 6%, 8%, 30%, 40%, and 50%, wt%, respectively, are reinforced with matrix. The prepared composites were subjected to flexural studies to know the load withstand ability. Results show that the incorporation of both fly ash and rice husk industrial wastes as filler into the polymer matrix increases the flexural strength. In addition, a low-cost product with high strength and good performance is obtained by adding this waste.

DOI: 10.4018/978-1-5225-5445-5.ch013

INTRODUCTION

With the growing needs and the development in technology more number of new products released from the industries to the society. At the same time the manufacturing process involved in preparing such products will be the vital discussion part. The major problem that the industry facing in today scenario is the disposal of the industrial wastes in such a manner that it will not affect the environment. Number of reuse and disposal methods are in practice in the industry, but still the need for much improved disposal of the same is the need for the hour. This chapter emphasis the reuse of two major industrial wastes such as Flyash and Ricehusk for making innovative composite materials. These two wastes are produced in enormous amount from the industry sectors as well as local sectors, for example the industry waste flyash generated not only in industry almost in all other sector where burning of some elements finally result in the flyash. On the other hand almost equivalent to flyash, ricehusk also generated in rapid amount due to the usage of food (rice) throughout the world. The industry is now looking for the potential reuse of these materials and one of the way suggested and discussed in detail is the preparation of composite materials using the same. The attractive features of the composite materials such as low cost, high strength and more resistance which make it suitable for various industrial applications replacing the conventional materials. Different type of composites are prepared by using two phases namely the reinforcement and the matrix phase. Enormous amount even out of any two different materials one we use as reinforcement phase and other use as matrix phase. The different types of matrices are metal, polymer and ceramic is the most commonly used for various composite preparations. The matrix will be chosen based on the better compatibility with the reinforcement for the same the polymer matrix selected for the industrial wastes fly ash and rice husk used as reinforcement.

BACKGROUND

The effective utilization of industrial wastes fly ash and rice husk as potential particle replacement for cement has been done by Satish et al. (2013) in which the compressive strength, flexural strength and tensile strength of the concretes made by mixing fly ash and rice husk in different weight proportions along with cement gives increase in strength. With fly ash as nano filler and micro filler, Al_2O_3, TiO_2 used as reinforcement along with epoxy matrix. Further the mechanical properties studies has been analyzed and noted that the strength decreases with the increasing content of the micro fillers and nano fillers (Ozsoy et al., 2015). In addition to the fly ash filler Nithin Kumar et al. (2015) carried out studies on wood powder as filler on the polymer composites produces promising increase in the mechanical strength. The effect of rice husk silica filler reinforced with high density polyethylene has been studied by Midhun Dominic et al. (2014) in which the surface modification done by using hydrochloric acid and inferred that the mechanical properties of the same higher than that of the pure polymer. The problems related to the disposal of wastes such as fly ash, slags from energy mining and metallurgical industries can be solved by potentially utilizing the waste with geopolymers for producing new composites replacing the existing cementitious materials (Korniejenko et al., 2015).

Sahai and Neha Pawar (2014) performed studies on fly ash filled polyphenylene oxide composites by varying the weight percentages of fly ash using screw extruder machine. The prepared specimens were subjected to mechanical properties and rheological properties analyze and the results were reported. Flexural studies on addition of fly ash along with the geopolymers has been carried out and analyzed using the FEM ABACUS software (Amiri et al., 2016). Michael Ikpi Ofem et al (2011) carried out research on the usage of periwinkle shell, rice husk and cashew nut shell liquid for making an effective hybrid composite. The results show that there is an increase in tensile and flexural strength with the increase in filler content as well as observed decrease in impact strength. Further studies using rice husk (Atan & Awanghe, 2011) waste as mixer along with self compacting concrete for potential cement replacement. The change in phenomenon makes the researchers to turn focus towards the usage of glass fibers filled with rice husk particulates (Routa & Satapathy, 2012), for the same mechanical properties and erosion response studies has been carried out for different compositions and noted an increase in the strength by addition of fillers. The effective means of utilizing the polyester resin (Devi et al., 1998; Sen & Nugay, 2000) along with fly ash has been done with varying weight proportions and treatment conditions using silane.

By using industrial waste cotton cellulose microcrystals (Thambiraj & Shankaran, 2017) were prepared and also from the same cellulose nanocrystals were extracted and the authors reported that from the results obtained the potential use of cotton waste for the crystals preparation makes it suitable for the drug delivery and food packaging applications.

Leandro Wiemes et al (2017) performed studies on the usage of various industrial wastes for the purpose of making bricks. The different types of wastes used were automotive waste, glass waste and wood ash been mixed with clay for the brick formation. The prepared samples were subjected to the flexural studies to know the limit is within allowable standards and also to leaching process. The results suggested the potential use of these three wastes for brick manufacture with potential protection to the environment from this hazardous wastes. Another major potential application is the usage of industrial wastes for the preparation of concretes (Dash et al., 2016), which drastically reduces the demand for the available materials for making concretes. A complete review on the usage of various industrial wastes like waste foundry sand, steel slag, copper slag, imperial smelting furnace slag (ISF slag), blast furnace slag, coal bottom ash, ferrochrome slag and palm oil clinker as coarse particle or fine particles used for preparing the concretes. Physical and mechanical properties studies had been carried out on the prepared concretes and noted that the sustainable standard strength obtained using this wastes for concrete development also durability seems to be good as well as more environment friendly.

Arumuga Prabu et al (2017) carried out studies on the usage of various industrial wastes such as Redmud, Sawdust, Flyash and Rice husk for preparing the composites along with polymer matrices. They noted that the incorporation of these wastes into the matrix increases the mechanical properties to a greater extent. This gives the potentiality of the prepared composites to be used for more industrial applications with the incorporation of natural fibers. The potential usage of redmud industrial waste which is causing severe environmental problems was done by Anu Pande et al (2017) in which they use it as a electromagnetic shielding material. The composite were prepared using redmud and polyaniline. From the testing it was clearly identified that a good shielding effectiveness been noted and makes it more suit for the electromagnetic shield resist application replacing the existing ones.

Wastes such as flyash and molasses (Rao & Birru, 2017) added in different weight percentages with the green sand to obtain modified green sand for the typical molding process to produce different types of parts. A high compression strength and permeability been obtained by using the wastes along with green sand when compare to the without ones. In such a way the wastes were utilized in more quantity reducing the environment hazards. Green metal matrix composites using chicken eggshell waste and aluminium, silicon carbide reinforcement had been studied by Satpal Sharma et al (2017) in which the serious environmental threat produced by the same waste be drastically reduced. From the physical property, strength and thermal behavior studies it was observed that the strength increased and possess better results.

More studies on flyash reinforced polymer matrix composites had been carried out by giving special importance to the automotive applications. Braking application was chosen and the composite prepared using flyash filled and aramid fiber reinforced phenolic composite (Dadkar et al., 2009) had been successfully utilized for the friction reduction with more weight percentage of flyash. The wear performance of the brake pads also increased to a high extent. Further research continued with the enormous waste generated from the society i.e., the plastic bottles and flyash. The Usage of these wastes for potential civil engineering applications had been analyzed and reported by Sushovan Dutta et al. (2016) and strongly recommended that the flyash and plastic bottles effectively used as filler which enhanced the compression property.

The studies continued in flyash with the effect of addition of zirconia (Satapathy, 2000) in different weight percentages and the impact in mechanical properties had been reported. The hardness, MOR and MOE increases with the increase in zirconia content. The abrasion also shows a less volume with the increase in zirconia. Five different polymer concrete mixes were made and compared with the existing Portland cement concrete. One among the five is flyash and observed that the compressive strength will be found to be much higher compared to portland cement (Okoye et al., 2015).

From the thorough literature, it was observed that more works has been done using industrial wastes flyash and ricehusk, this research chapter focuses on the usage of the same along with polyester and epoxy matrix with special reference to the flexural strength.

MAIN FOCUS OF THE CHAPTER

This section deals with the preparation of the composites using the industrial wastes and testing of the same for suitable potential applications. The industrial wastes flyash and rice husk were procured from the local shops in Tamilnadu, India. The unsaturated polyester matrix has been procured from Vasavibala Resins Pvt, Ltd., Chennai, Tamilnadu, India. The following Figure 1 and Figure 2 shows the typical ricehusk and flyash in powder form. The Rice husk filler materials particle size is measured and it ranges from 25-50 microns whereas the fly ash filler material particle size is slightly higher than that of rice husk, it is 40-50 microns.

The step by step procedure for the preparation of the composites are as follows:

Step 1: Two rectangular or square glass plates of size 20X20 cm are used as upper die and lower die. A narrow glass with 3 mm thickness placed at the three adjacent sides of the lower mould.

Step 2: Wax is used as releasing agent applied over the glass plates without any unfilled and then the glass plates are tightly holded by using metal clips at the four corners to keep the dies strongly together.

Step 3: Unsaturated polyester resin of quantity 100 ml taken and is mixed manually well with reinforcement for 2 to 5 min. Then 1 ml of accelerator and catalyst added with polyester matrix to initiate the curing reaction. The same procedure done for epoxy matrix also but the only difference is 50 ml of hardener added to initiate the curing reaction. Then the mixture is allowed to cure for some time.

Step 4: The prepared mould is kept vertically and the prepared resin is poured gently through the opening side of the mould. Then the composition is left without any disturbance allows to cure at room temperature.

Figure 1. Rice husk

Figure 2. Flyash

Step 5: In general the composites will cure throughout the life time but usage purposes for polyester matrix, it is enough to cure for 4- 6 hrs in the mould and then it can be removed. In case of epoxy matrix the specimen removed after 24 hrs. Typical prepared specimens are shown in Figure 3 and 4.

Figure 3. Fly ash-polyester

Figure 4. Rice husk-polyester

To find the flexural property of the fly ash and rice husk reinforced polyester composite, flexural strength of the specimens were tested using three point bending test in UTM as shown in Figure 5 and the values were noted for all the samples

SOLUTIONS AND RECOMMENDATIONS

Solutions

The potential industrial waste disposal problem can be reduced by means of recycling them in a proper way. One of the solutions is by using the wastes for preparing new type composites and the strength af-

Figure 5. Flexural strength using three point bending test

ford by them when comparing with the existing materials. The detail discussion on the flexural strength that is the ability to withstand potential load be given as follows:

Effect of Varying Weight Percentages of Rice Husk on Polyester Resin

The variation in flexural strength of polyester resin by the addition of rice husk industrial waste is shown Figure 6 and 7 shows the effect of low weight percentage of rice husk addition that is 6% and 8% has been shown in Figure 6 it was noted clearly that there is no much difference in the two weight percentages since both has the same values of 50 MPa. Since both the weight percentage the variation in weight is small so there is no change in the flexural strength. At the same time when compared to the pure polyester both weight percentages show a slight decrease in the flexural strength. For pure polyester the flexural strength is 54MPa so a deviation of 4MPa has been observed from the results. It is evident that the mixing of the rice husk with the pure polyester result in decreased strength this is because the nature of rice husk is in the form of particulates which is in semisolid form so the dispersion and the flow pattern of rice husk with the polyester is not in the uniform manner. This leads to the presence of the rice husk here and there gives decreased strength.

The effect of reinforcing high weight percentage rice husk along with polyester in flexural strength is shown in Figure 7 It is strongly observed that the addition of high weight percentages of rice husk reduces the flexural strength. The value reduced from 50 MPa to 31MPa, this values also very low compare to the pure polyester. There is an increase in the flexural strength with the increase in rice husk weight percentage from 30 to 40% but at 50% weight it reduces drastically. This shows a clear note on the dispersion of the rice husk reinforcement with pure polyester is not good which result in poor strength and the variations in values.

Effect of Varying Weight Percentages of Fly Ash on Polyester Resin

The variation in flexural strength of polyester resin by the addition of fly ash industrial waste is shown Figure 8 and 9. The effect of low weight percentage of fly ash addition that is 6% and 8% has been

Figure 6. Flexural strength of 6 wt % and 8 wt% rice husk filled polyester composite

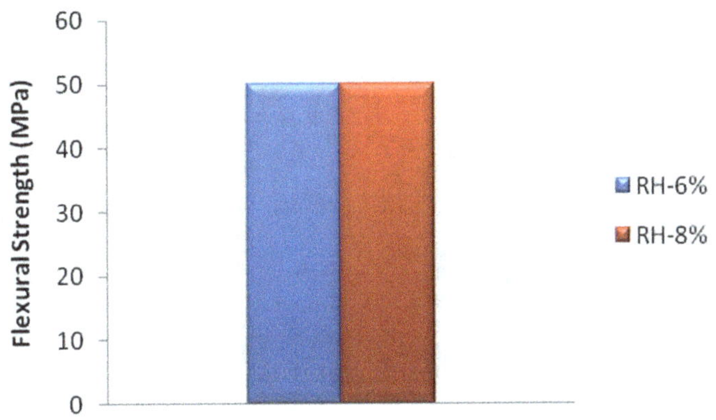

Figure 7. Flexural Strength of 30 wt%, 40 wt% and 50 wt% rice husk filled polyester composite

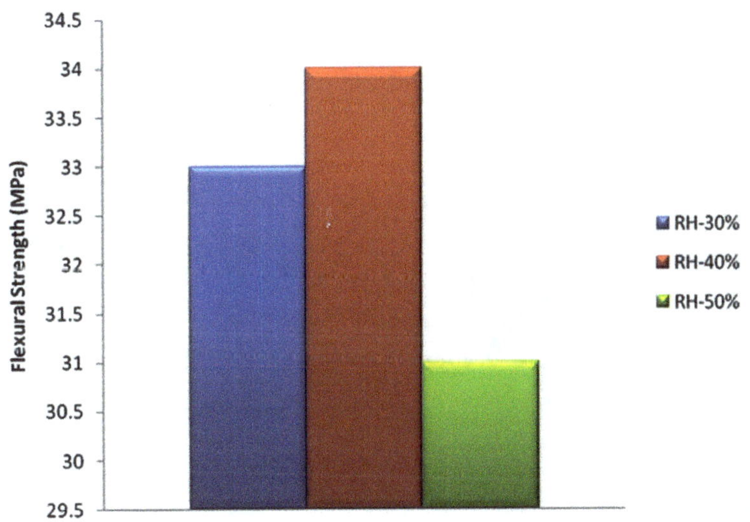

shown in Figure 3. It is noted clearly that there is no much difference in flexural strength between the two weight percentages varying from 6% to 8%Wt, in the case of 6%Wt, the flexural strength is only 42 MPa decreased compare to pure polyester and the most noted thing is the addition of 8%Wt fly ash along with pure polyester gives equal value in the flexural strength. In the case of low weight percentage the flexural strength is 54 MPa compare to rice husk reinforcement. The manual mixing of the fly ash reinforcement along with polyester produce poor results.

The effect of reinforcing high weight percentage flyash along with polyester in flexural strength is shown in Figure 9. It is strongly observed that the addition of high weight percentages of fly ash increases the flexural strength. The value increased from 54 MPa to 60 MPa in the case of 30%Wt. This value also very high compare to the pure polyester. There is an increase in the flexural strength with the increase in fly ash weight percentage from 6 to 30% but at 40% Wt it reduces drastically, and then increases slightly at 50% Wt. This show a clear note on the non uniform distribution of the fly ash reinforcement with pure polyester, results in poor strength and the variations in values. Also in addition the increase in fly ash content decreases the flexural strength compared to pure polyester

Figure 8. Flexural Strength of 6 wt% and 8 wt% fly ash filled polyester composite

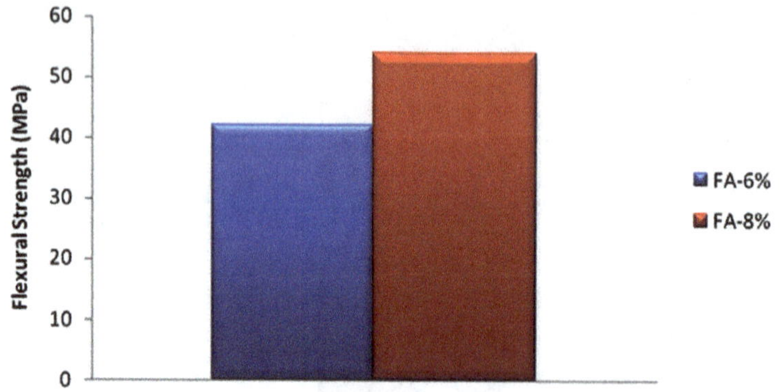

Figure 9. Flexural Strength of 30 wt%, 40 wt% and 50 wt% fly ash filled polyester composite

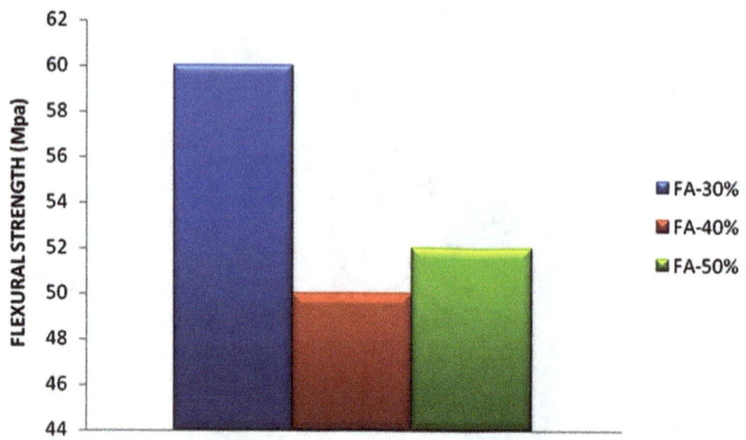

Recommendations

From the present research it was clearly noted that the addition of wastes along with the polymer enhances the strength gradually which makes it suitable for potential filler in polymer for preparing the composites. The mechanical performance of the composites show an increase in the properties which is a positive replica of the potential usage of the industrial wastes. In this way the potential environmental hazard caused by the wastes be drastically reduced.

FUTURE RESEARCH DIRECTIONS

The future research directions focuses on different property studies of the industrial wastes reinforced polymer composites. The studies include wear property, erosion property, thermal property, vibration property and transport studies etc., has been performed to know the performance of the composites and study its suitability for various applications. Also by varying the weight percentage and particle size of the industrial wastes the performance studies has been done.

CONCLUSION

The flexural strength studies on industrial wastes rice husk and fly ash effect on polyester leads to the following conclusions,

- Industrial wastes rice husk and fly ash reinforced polyester composite were fabricated successfully by hand-lay-up technique.
- Flexural properties of the prepared particulate composites are studied and illustrated briefly.
- According to that, among two fillers flyash and rice husk are found to be suitable reinforce materials in polymer composite. In focus polyester is the suitable matrix for both fillers.
- The Flexural strength of Ricehusk reinforced polyester composites shows a drastically decrease in strength when compare to pure polyester whereas the flyash reinforced polyester show increase in flexural strength alone.
- The highest flexural strength value was obtained for the 6% and 8%Wt addition of ricehusk with polyester where this case the strength decreases slightly by 7% due to the improper dispersion.
- On the other side, the highest flexural strength value was obtained for the 30%Wt addition of flyash with polyester where this case the strength increases by 10% due to the better fine dispersion of the particulates. This was found to be the optimal weight percentage.
- From the studies it is revealed that the environmental pollution drastically reduced by reinforcing these wastes with polymer composites as well as along with natural fibers for suitable composite materials production which in term used in automotive sectors.

ACKNOWLEDGMENT

This work was supported by Yeungnam University, Research Grant in 2017 (Postdoctoral Research Associate Program).

REFERENCES

Amiri, A. M., Olfati, A., Najjar, S., & Beiranvand, P. (2016, June). M.H. Naseri Fard, The Effect of Fly Ash on Flexural Capacity Concrete Beams. *Advances in Science and Technology Research Journal*, *10*(30), 89–95. doi:10.12913/22998624/62630

Arumuga Prabu, V., Deepak Joel Johnson, R., Amuthakkannan, P., & Manikandan, V. (2017). Usage of industrial wastes as particulate composite for environment management: Hardness, Tensile and Impact studies. *Journal of Environmental Chemical Engineering*, *5*(1), 1289–1301. doi:10.1016/j.jece.2017.02.007

Atan, M. N., & Awanghe, H. (2011). Compressive and Flexural Strengths of Self-Compacting Concrete using Raw Rice Husk. *Journal of Engineering Science and Technology*, *6*(6), 720–732.

Dadkar, N., Tomar, B. S., & Satapathy, B. K. (2009). Evaluation of flyash-filled and aramid fibre reinforced hybrid polymer matrix composites (PMC) for friction braking applications. *Materials & Design*, *30*(10), 4369–4376. doi:10.1016/j.matdes.2009.04.007

Dash, M. K., Patro, S. K., & Rath, A. K. (2016). Sustainable use of industrial-waste as partial replacement of fine aggregate for preparation of concrete – A review. *International Journal of Sustainable Built Environment*, *5*(2), 484–516. doi:10.1016/j.ijsbe.2016.04.006

Dutta, S., Nadaf, M. B., & Mandal, J. N. (2016). An Overview on the Use of Waste Plastic Bottles and Fly Ash in Civil Engineering Applications. *Procedia Environmental Sciences*, *35*, 681–691. doi:10.1016/j.proenv.2016.07.067

Korniejenko, Mikuła, & Łach. (2015). Fly Ash Based Fiber-Reinforced Geopolymer Composites as the Environmental Friendly Alternative to Cementitious Materials. *Proceedings of 2015 International Conference on Bio-Medical Engineering and Environmental Technology*, 164-171.

Kumar, Chincholi, Hegde, Shivagiri, Revanasiddappa, & Kazi. (2015). Experimental Research on Hardness & Flexural properties of Teak Wood Powder and Fly Ash reinforcement with Epoxy. *International Journal of Current Engineering and Scientific Research, 2*(11), 34-37.

Midhun Dominic, C. D. (2014, March-April). Rice Husk Silica- Efficient Bio Filler in High Density Polyethylene. *International Journal of Advanced Scientific and Technical Research*, *2*(4), 561–569.

Ofem, M. I., Abam, F. I., & Ugot, I. U. (2012). Mechanical Properties of Hybrid Periwinkle and Rice Husk Filled CNSL Composite. *International Journal of Nano and Material Sciences*, *1*(2), 74–80.

Okoye, F. N., Durgaprasad, J., & Singh, N. B. (2015). Mechanical properties of alkali activated flyash/ Kaolin based geopolymer concrete. *Construction & Building Materials*, *98*, 685–691. doi:10.1016/j.conbuildmat.2015.08.009

Ozsoy, Demirkol, Mimaroglu, Unal, & Demir. (2015). The Influence of Micro- and Nano-Filler Content on the Mechanical Properties of Epoxy Composites. *Journal of Mechanical Engineering, 61*(10), 601-609.

Pande, A., Gairola, P., Sambyal, P., Gairola, S. P., Kumar, V., Singh, K., & Dhawan, S. K. (2017). Electromagnetic shielding behavior of polyaniline using Red Mud (industrial waste) as filler in the X e band (8.2e12.4 GHz) frequency range. *Materials Chemistry and Physics*, *189*, 22–27. doi:10.1016/j.matchemphys.2016.12.045

Routa. (2012). Study on mechanical and tribo-performance of rice-husk filled glass–epoxy hybrid composites. *Materials & Design*, *41*(October), 131–141.

Sahai & Pawar. (2014). Studies on Mechanical Properties of Fly Ash Filled PPO Composite with Coupling Agent. International Journal of Chemical, *Environmental & Biological Sciences*, 2(4), 187-192

Saroja Devi, M., Murugesan, V., Rengaraj, K., & Anand, P. (1998). Utilization of flyash as filler for unsaturated polyester resin. *Journal of Applied Polymer Science*, *69*(7), 1385–1391. doi:10.1002/(SICI)1097-4628(19980815)69:7<1385::AID-APP13>3.0.CO;2-T

Satapathy, L. N. (2000). A study on the mechanical, abrasion and microstructural properties of zirconia-flyash material. *Ceramics International*, *26*(1), 39–45. doi:10.1016/S0272-8842(99)00017-6

Satish, H. (2013). Combine Effect of Rice Husk Ash and Fly Ash on Concrete by 30% Cement Replacement. *Procedia Engineering*, *51*, 35–44. doi:10.1016/j.proeng.2013.01.009

Şen, S., & Nugay, N. (2000). Uncured and cured state properties of fly ash filled unsaturated polyester composites. *Journal of Applied Polymer Science*, *77*(5), 1128–1136. doi:10.1002/1097-4628(20000801)77:5<1128::AID-APP21>3.0.CO;2-D

Sharma, S., & Dwivedi, S. P. (2017). Effects of waste eggshells and SiC addition on specific strength and thermal expansion of hybrid green metal matrix composite. *Journal of Hazardous Materials*, *333*, 1–9. doi:10.1016/j.jhazmat.2017.01.002 PMID:28340384

SrinivasaRao, P., & Birru, A. K. (2017). Effect of Mechanical Properties with addition of Molasses and FlyAsh in Green Sand Moulding. *Materials Today: Proceedings*, *4*(2), 1186–1192. doi:10.1016/j.matpr.2017.01.136

Thambiraj, S., & Ravi Shankaran, D. (2017). Preparation and physicochemical characterization of cellulose nanocrystals from industrial waste cotton. *Applied Surface Science*, *412*, 405–416. doi:10.1016/j.apsusc.2017.03.272

Wiemes, L., Pawlowsky, U., & Mymrin, V. (2017). Incorporation of industrial wastes as raw materials in brick's formulation. *Journal of Cleaner Production*, *142*, 69–77. doi:10.1016/j.jclepro.2016.06.174

Chapter 14
Additive Manufacturing Process and Their Applications for Green Technology

Keshavamurthy R.
Dayananda Sagar College of Engineering, India

Vijay Tambrallimath
Dayananda Sagar College of Engineering, India

Prabhakar Kuppahalli
Dayananda Sagar College of Engineering, India

Sekhar N.
Dayananda Sagar College of Engineering, India

ABSTRACT

Growth of nature is an additive process that gives sustainable existence to the structures developed; on the other hand, traditional manufacturing techniques can be wasteful as they are subtractive. Additive manufacturing produces almost nil waste and accordingly preserves raw materials resulting in cost reduction for the procurement of the same. It will also cut down on the carbon emissions that are usually generated from industrial manufacturing. Additive printed objects are lighter as well, making them more efficient, especially when used in the automobile and aerospace industry. Further, the intrinsic characteristics and the promising merits of additive manufacturing process are expected to provide a solution to improve the sustainability of the process. This chapter comprehensively reports on various additive manufacturing processes and their sustainable applications for green technology. The state of the art, opportunities, and future, related to sustainable applications of additive manufacturing have been presented at length.

DOI: 10.4018/978-1-5225-5445-5.ch014

1. INTRODUCTION

Additive Manufacturing (AM) or Rapid manufacturing also called as Rapid prototyping is the innovative way of building up a matter by a process. This follows the nature's principle of constructing a material by adding substances in layered sections depending upon the required condition of the end product to be formed. This is more efficient and better in comparison to the age-old method of building up a material like machining, stamping, cutting etc, Additive Manufacturing creates an end result by adding up the material. The fundamental principle of preparing a model can be classified into two steps, firstly, development of three-dimensional Computer Aided Design and secondly the fabrication of the part. The process is not as simple as it sounds. There is an immense reduction in the difficulty of manufacturing complex 3D parts when looked from a point of conventional manufacturing. The basic understanding required to handle the process requires some basic inputs of dimensional quantities and perceptive knowledge of working of the machine. The material to be used for a particular purpose should possess apt properties.

Manufacturing is all about converting material inputs into products of usage. The sustainability and the environmental assessment can be looked upon, on how efficiently the process can be carried out without detrimental effects on the surroundings. AM proves out of having an advantage of being considered as a sustainable manufacturing process which can be stated with the following reasoning:

1. The material is built up layer wise which supports in the reduction of waste material, complex geometries and lightweight materials can be easily produced which results in lesser material requirement and low energy utilization.
2. On-demand manufacturing of required parts brings down the inventory waste reduction and loss, cuts the transportation related expenses.
3. The health advantages of AM is an added benefit when compared to normal conventional forms of manufacturing, the prolonged exposure of the worker in hazardous environment is drastically reduced. The methods of manufacturing are taking new dimension in the process of developing components and end products when previewed from the point of AM. Numerous researchers are exploring the way of AM for its better sustainability, economic viability and benefits of economic and social impact.

Building of AM product involves certain generic steps to be followed starting from the preparation of CAD model until the printing of final 3D product. The steps can be summarized in the following flow chart (Figure 1) (Gibson et al., 2010).

1.1. Computer Aided Design (CAD)

The basic prerequisite and the first step for the AM process is the understanding and developing a 3D model using CAD. Any professional CAD modeling software can be utilized for the purpose with an output as a solid 3D surface or solid representation.

Figure 1. Flow chart of steps involved in AM

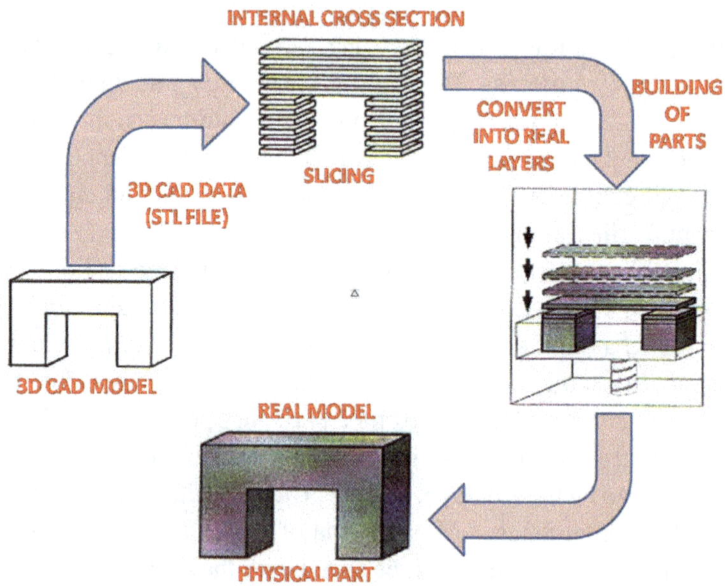

1.2. Stereolithography (STL) Adaptation

The AM machine can understand the input CAD model only when it is in readable format by the machine. Almost every CAD model has the conversion of the format into STL form. The STL file decides the number of layers based on the external closed surface of the original CAD design.

1.3. Manipulation of STL File and Transfer

Once the file is converted to STL it needs to be transferred to the AM machine. Certain manipulations need to be done for the machine to properly fit it into correct size, position, and orientation for building.

1.4. Setup of Machine

Before the build process the AM machine has to be set up in an appropriate condition. These settings would correlate the parameters of build like timings, layer thickness, energy source, material constraint etc.

1.5. Build, Removal, and Post Processing

The whole of AM process is automated and requires no supervision. However external glitches like proper working of software, uninterrupted power supply etc need to be taken care. Once the process of building is over the product needs to be carefully mounted out as there may be some interlock present to hold the material. Temperature to be conditioned should be properly looked after and material needs to taken out when the temperature reaches manageable condition and there should be no any moving parts. The final method of post processing is required for the material which may be weak at this instant and certain manipulations may be required to reach the desired state.

1.6. Application

The product developed may now be used for application and may require some texturing depending on its condition. Painting may also be required depending in where the part is used. The product may be used with other mechanical assemblies too.

2. CLASSIFICATION

The classification of AM is briefly divided into seven categories by ISO/ASTM52900-15. AM includes the building up of material layer by layer. The classification includes binder jetting, direct energy deposition, material extrusion, material jetting, powder bed fusion, sheet lamination and vat polymerization (Figure 2).

2.1 Binder Jetting

Two materials are basically used in this process, one being the material in powder form and the other to be a binder. This acts as an adhesive agent to hold all the powder material in a required dimension. The states of matter of binder will be in liquid form and that of material in powder form. (Figure 3) explains the brief working and constructional process of Binder Jetting. The horizontal movement of the print head takes along the x and y axes of the platform. The alternate layers of the powder material and binder material are being deposited on build platform and after each layer the platform is lowered down for the deposition of the next layer.

Doyle et al. (2015) have studied the influence of layer thickness and part orientation on mechanical characteristics of stainless steel 420 and bronze parts. They have observed that during binder jet printing process, layer thickness had significant impact on mechanical properties of the product by changing its chemical composition whereas part orientation has negligible impact. (Zachary et al., 2017) have fabricated 98% dense parts from ferrous based powder by binder jet 3D printing and infiltration. They have removed stress concentration regions in the inter particle necks and enhanced the strength of fabricated parts by infiltration process. (Gonzalez et al., 2016) have fabricated Aluminum oxide parts using binder jetting technique with 96% part density. Sintered alumina parts exhibited a compressive strength of 131.86 MPa and 9.47 – 5.69 dielectric constant at 20 Hz to 1MHz under optimum process parameters for high

Figure 2. Classification of additive manufacturing processes

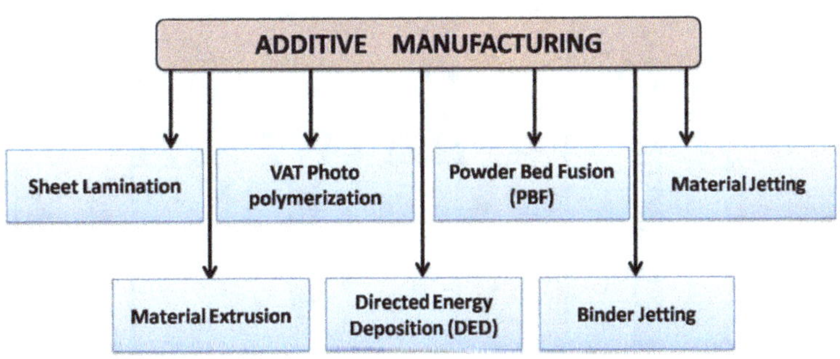

value energy and environmental components. Simon Meteyer et al. (2014) have developed a model for energy consumption and material flow analysis of binder jetting process. The developed mathematical models were useful in predicting energy and material consumption during fabrication. They have concluded that the produced model can be effectively implemented for larger industrial scale binder jetting methods. Further they have highlighted need for estimation of environmental impact and disposal to get full life cycle assessment. Various materials like steel, ceramic, polymer and glass can be processed in this method. (Hadi Miyanaji et. al., 2106) researched on process parameters for dental porcelain ceramic material by binder jetting process. IPS Inline dentine powder was the material used for the experiment which is widely used in dental applications. It was observed that the printing parameters had a great effect on the geometrical properties, which in turn had an effect on final part after sintering. With the increase in the temperature there was a noticeable shrinkage of the final part at the axis away from the base plate of the machine.

2.2 Direct Energy Deposition

It is a procedure that melts metal wire or powder to form an object layer by layer, with utilization of high energy power source such as electron beam, plasma welding torch or a laser. Yuwei Zhai et al. (2014) states this method is basically used to produce metal objects and is a very complex process regularly used for the repair or addition to the existing component. The working principle is same as that of extrusion, multi degree movement arm is used onto which a nozzle is mounted which deposits a melted material on the surface (Figure 4). Quanren Zhang states the classification of DED as LMD, LENS, DMD,DLD. All the mentioned processes which produce very high energy will affect the microstructure, material properties, residual state and thermal distortion of the final part. The material can be melted by various ways such as electron beam, laser, plasma arc etc. Electron beam direct manufacturing is commonly used technique which melts the material when it is deposited on the surface. Morrow et al. (2007) have studied environmental emissions and consumption of energy for manufacturing of molds and dies by direct metal deposition technique. Authors have compared laser-based process with CNC milling and also have

Figure 3. Binder jetting

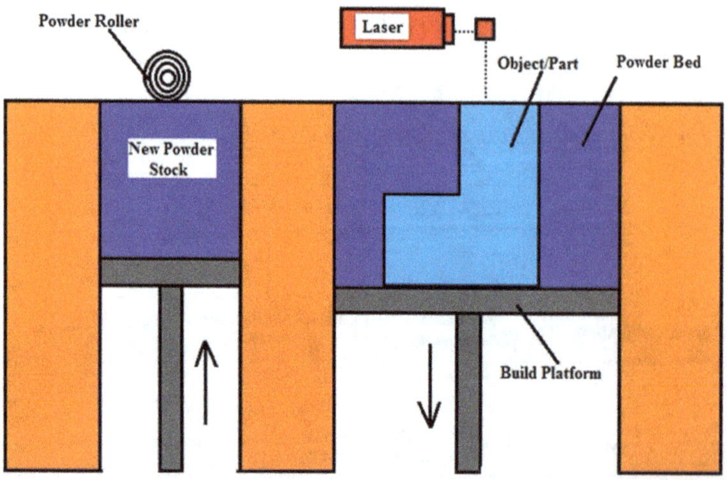

highlighted that consumption of relative energy for direct metal deposition and CNC milling has been determined by the volume ratio of solid to cavity. They have stated that energy consumption and emission are minimum at lower solid cavity volume ratio for direct metal deposition while at higher volume ratios CNC milling process exhibited minimum energy consumption and emissions. They have also indicated that energy intensity for Direct Metal Deposition process offers ample opportunities, remanufacturing and tooling with large reductions in energy consumption, environmental emissions and cost.

2.3 Material Extrusion

A molten material – clay, plastic, cement, ink etc is being extruded from the printer head and becomes solid instantly with greater adhesion between the layers built. The building up of product until a material is complete takes layer by layer. Fuse deposition modeling is one of the efficient methods of manufacturing using the principle of extrusion by using thermoplastic material (*Figure 5)*. Every time a layer is printed, the printer platform is lowered to make space to the for next layer and bonding of the layers take places once the extruded material is out of nozzle. Fuse Deposition Modelling (FDM) is the most efficient and cheapest process available which commonly uses Acrylonitrile Butadiene Styrene (ABS) or Polylactic Acid (PLA) as a thermoplastic material.

Fuda Ning et al. (2015) experimentally investigated the mechanical and flexural properties of ABS thermoplastic with reinforcement of carbon fiber by FDM. ABS pellets and carbon fiber powder were mixed with varying ratios. The filament was extruded at a nozzle temperature of 2200C, with speed of 2m/min and diameter of 2.85mm respectively. ASTM D638-10 standard and ASTM D790-10 standard were followed for tensile and flexural tests. Addition of carbon fiber to ABS showed an increase in tensile strength and Young's modulus but decreased the toughness, yield strength and ductility. 5wt% of carbon fiber to ABS has highest value of flexural properties. (K.G. Jaya Christiyan et.al., 2016) conducted an experiment on ABS reinforced with hydrous magnesium silicate composites. Mechanical and flexural properties were evaluated using ASTM D 638 and ASTM D 790 standard testing methods. The process parameters such as layer thickness and printing speed were analyzed. They concluded that as the printing speed was reduced with the reduction in layer thickness the tensile and flexural properties were maximum.

Figure 4. Direct energy deposition (DED)

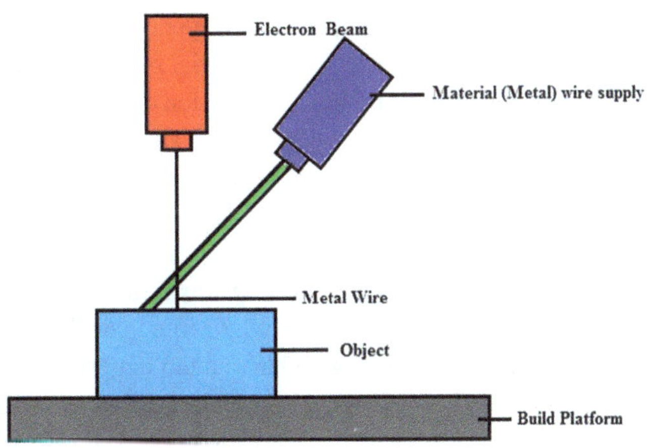

Figure 5. Material extrusion

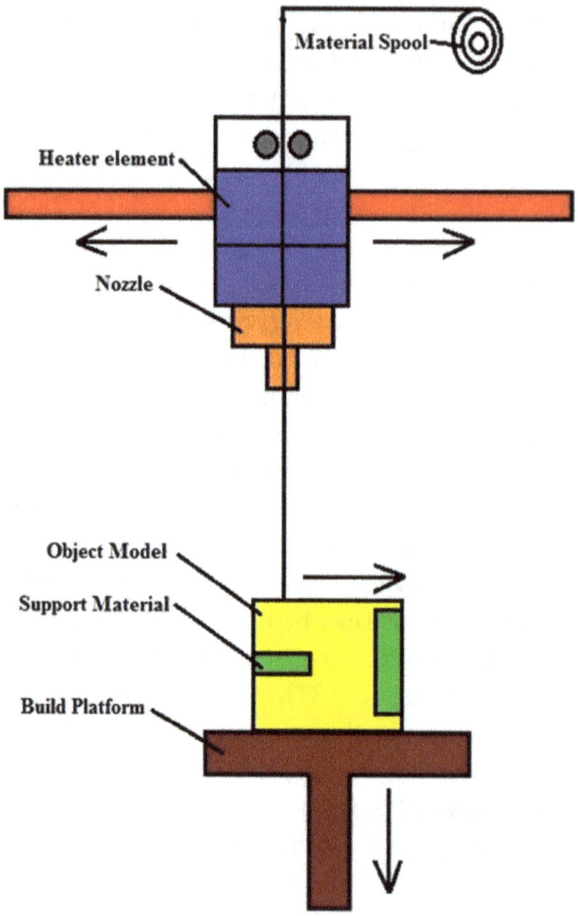

2.4 Material Jetting

This process includes the process of creation of objects in similar pattern with that of 2 dimensional conventional printers. The building up of the product takes in layer by layer form which is dropped onto build platform either continuously or when required. The material moves through the nozzle and which travels in the horizontal path across the platform. The solidification of the material layer is done by passing Ultraviolet (UV) light. Materials like polymer and wax are suitably used in this process. (Kęsy et al., 2010) investigated the parts produced depending on the orientation. The variation in the mechanical properties was observed due to orientation. 16 micro meter super thin photopolymer was jetted on the tray which was quickly hardened by the spreading of UV light. Tensile and hardness tests were conducted in laboratory at 220C and 40% relative humidity. ISO testing was followed for the test. They concluded that the orientation in the internal tray of the part plays a major role in deciding the mechanical properties. Glasschroeder et al. (2015) and Vaezi et al. (2013) have stated material jetting technology exhibits higher capability for an efficient production of parts. Parts can be fabricated by various materials which include ceramics, metals and polymers.

Figure 6. Material jetting

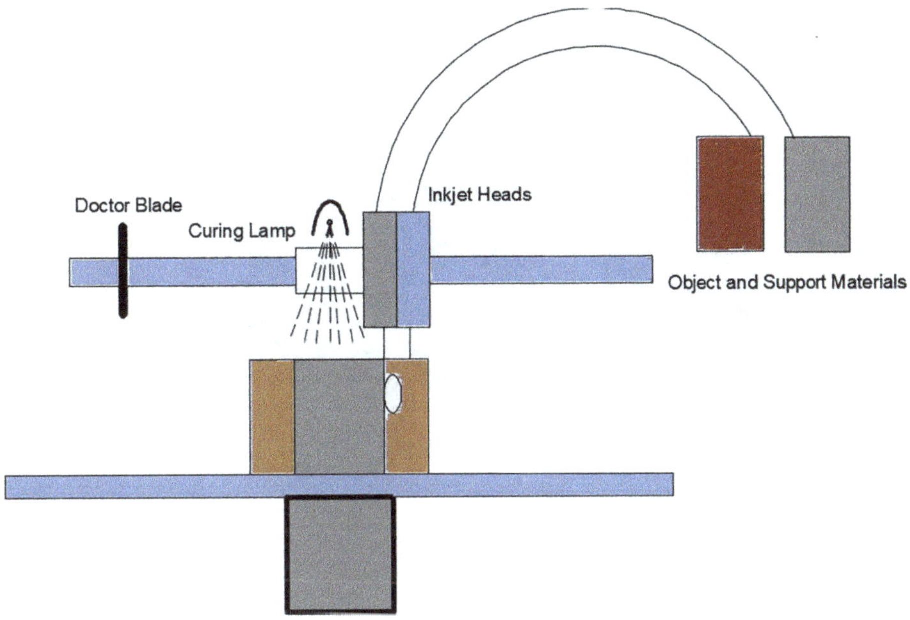

2.5 Powder Bed Fusion

The process of utilizing the high energy laser for fusing tiny particles of small well defined powder layers which may include ceramic, glass, nylon, polystyrene, metals etc. Selective Laser Sintering (SLS) is the most commonly used method of manufacturing using this technology other ways of manufacturing also include direct metal laser sintering, selective heat sintering etc. The platform is lowered down by a layer thickness when each layer is cured. (*Figure 7)* shows the schematic representation of the process. Wide range of material options are available for powder bed fusion technique. One important advantage of powder bed fusion method is, powder itself acts as an internal support structure. Klahn et al. (2016) prototyped lid of a jar by SLS method which provided greater flexibility and opportunistic design feature for snap fitness joint. The higher flexibility and novel design concept of the lid jar was attained by AM where as the conventional method of injection molding would have been a tedious work to carry out. (Bhavar et al., 2104) made a comparative study of SLS and Electron Beam Machining (EBM) methods of powder bed fusion. The building of material is done in vacuum chamber which reduces the thermal gradient and thermal contamination and also for the quality maintenance of the electron beam. High build chamber temperature is maintained (about 700^{0}C – 900^{0}C) which eliminates post heat treatment. The build rate of EBM is 80cm3/hr which is almost four times the value of Selective Laser Melting (SLM). In SLM part manufacturing takes place in argon gas environment to avoid contamination and oxidation. In spite of these facilities, SLM has dominated because of its lower accuracy, high cost and non-availability of higher build volumes of EBM. Powder materials such as polystyrene, ceramics, glass, nylon and metals including steel, aluminum, titanium etc can be used in various processes. This method is relatively inexpensive and suitable for visual models and prototypes and has a large range of options for various materials.

Figure 7. Powder bed fusion

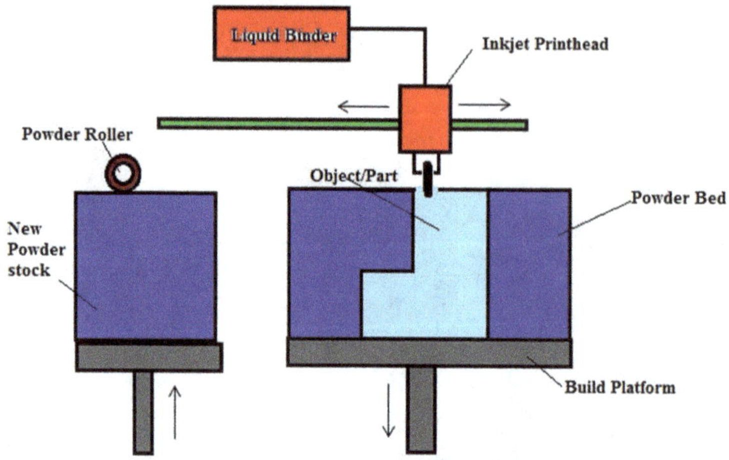

2.6 Sheet Lamination

Basically two methods of manufacturing are evolved much more in this process. Laminated Object Manufacturing (LOM) is one of the processes that use paper as a material and adhesive on par with welding. Cross hatching is the technique that is used in LOM for easier extraction of the material after building *(Figure 8)*. The usage of LOM objects are for models that are used for aesthetic value and are also used for visual models, structural application does not hold good for this process. The second method is Ultrasonic Additive Manufacturing (UAM) which uses ribbons or sheets of metal, which are held together by ultrasonic welding. Removal of additional material and CNC machining is used during welding process. Metals like aluminum, copper, stainless steel, titanium etc are used in UAM. (Kechagias, 2007) investigated the process parameters that influence the surface roughness of the produced by LOM. Numerous process parameters were taken into consideration for the experimental study such as heater temperature, layer thickness, heater speed, laser speed, feeder speed and platform speed. The heater temperature, layer thickness and laser speed were opted to be the parameters of concern which plays an important role in roughness. Optimal process parameters were identified with Taguchi method. Sarahara et al. (2005) have synthesized layered mechanical parts using sheet metal and polymers material with the combination of polymer and sheet metal. High strength parts were manufactured which are suitable for mechanical applications with high damping capacity. A complex thin walled honey comb structure was easily fabricated by this technique which will be extremely difficult by other conventional methods. Quality characteristics of laminated object manufacturing was studied by Kechagias et al. (2007) process parameters such as heater temperature, thickness of layer, platform retract, speed of heater, laser and platform were studied using Taguchi's experimental design. They have concluded that the dimensional accuracy in x-direction was largely influenced by heater speed (89%), heater temperature (5%) and in y-direction heater seed (50%), layer thickness (31%), laser speed (9%), platform speed (6%) and heater temperature (3%). (Weisensel et al., 2004) have employed layer object manufacturing technique to fabricate carbon template of laminate structures using pyrolysed filter papers with the help of pressure less reactive infiltration, laminates were developed into Si-SiC composites. Microstructure and mechanical characteristics of developed parts were largely influenced on porosity, density and distribution of pore size of carbon perform which can be controlled by process parameters during layer object manufacturing.

Figure 8. Sheet lamination

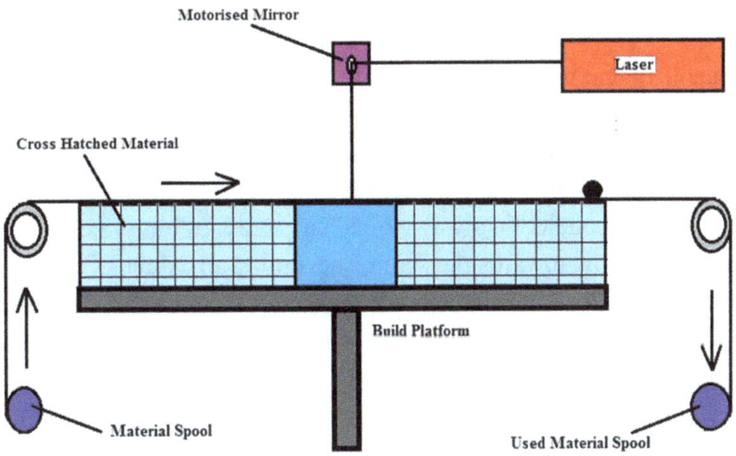

2.7 Vat Polymerization

The process of vat polymerization employs a container filled with a resin, which is a liquid photo-polymer, used to build the model layer wise. Like every other method the platform drops down a layer thickness after one layer is laid upon it, the material is hardened by passing the UV light over the object. Among various processes, STL and polyjet are the two important techniques of manufacturing under vat polymerization. The utilization of the structure is required in this process as the material used is in the form of liquid, unlike powder based where the support is given from unbound material. The curing of the material takes place by the reflection of the UV light or photo polymerization by the method of placing motor controlled mirrors.

Figure 9. VAT polymerization

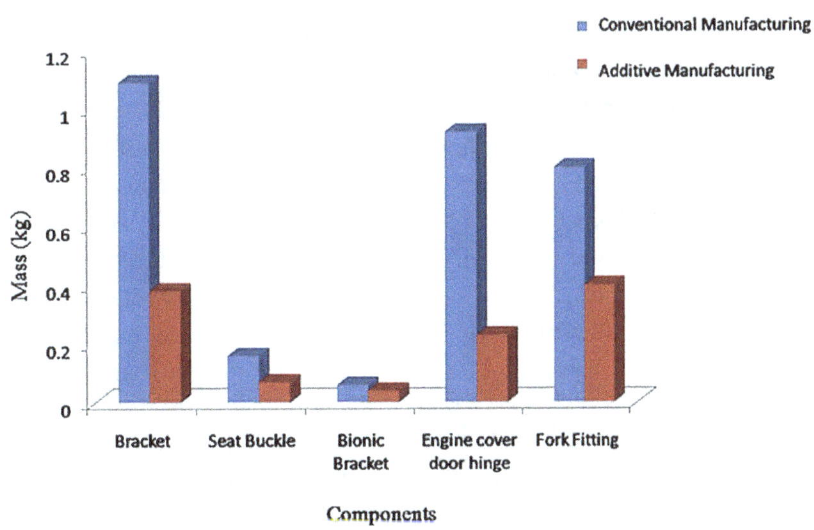

3. SUSTAINABILITY AND GREEN TECHNOLOGY OF ADDTIVE MANUFACTURING

The two terms sustainability and green technology do not mean the same, albeit they can be used together, unless a reader can differentiate properly between the two. Green technology can be adhered to a certain process in manufacturing where it is prepared in an environmental friendly way, where as if the term sustainability is used then the whole of process from procurement of raw material until the delivery of the product, has to be handled in a way which harms less or nil to the environment. The whole of manufacturing sector is shifting towards the same, "becoming sustainable". The governing authority and climate change control bodies of the world has implemented and urged to strictly follow certain rules while manufacturing which includes reducing water and energy, minimizing waste, reduction in emission of harmful gases etc. Additive Manufacturing has paved the path to reach this goal by big giants in manufacturing and also has helped them to embrace the change for the betterment of society as well as the company. This chapter discusses various approaches in line with sustainable technology and recommends implementing and further improvising on the same. Various researchers have worked on the sustainability and green technology of AM. (Tateno et al., 2017) worked on the utilization of AM method for the building of material which are prepared by reusing the products which are already in use. The advantages of adopting this method helps in reduction of transportation cost and also increase in the sustainability of the goods that are remanufactured. The appropriate utilization of the products till it completely losses it strength can be efficiently used with AM. Despeisse et al. (2017), proposed a Sustainable Value Road mapping Tool (SVRT) which combines the road mapping of technical aspects with sustainable criteria which acts as a path for sustainable development towards greener tomorrow. The proposed tool consists of 4 steps which include vision, identification of drivers, identification and prioritization of sustainable business opportunities and finally identifying the enablers to realize these opportunities. (Jackson et al., 2016) have compared the energy utilization of wire based additive- subtractive manufacturing and powder based additive-subtractive manufacturing. The total energy consumption was calculated depending on considering the parameters metal production, deposition and machining. 85% less energy is consumed by wire deposition than powder-based deposition. Thomas Hofstatter et al. (2017), investigated vat polymerization and extrusion techniques to study about the adhesion between the matrix material and the fiber. Fiber arrangement plays a major role in deciding the material strength and Young's modulus. AM injection mould inserts were proven to be more efficient and environmental friendly with flexible prototyping and pilot production whereas FRP helped in improvement of lifetime and reduction in crack propagation velocity. Liang et al. (2010) reviewed the process of AM towards sustainable path which includes the study parameters like internal weight optimization for sustainable product design, improvement in process efficiency, reduction in energy consumption and sustainable production. The utilization of AM process over conventional form is basically adapted in wide industries because of production of optimal topologies of parts with lesser cost and material, no requirement of tooling, jigs and fixtures and the utilization of non-processed material by recycling. AM brings about positive environmental impact on the supply chain, life cycle performance, manufacturing process etc. Dong-Gyu Ahn (2016) examined the process and sustainable approach for greener technology of Direct Metal Additive Manufacturing (DMAM). With sustainability point of view the process of DMAM is approached into 3 categories: (a) product redesign (b) improvement in efficiency of the process (c) remanufacture, repair and life extension of the product. The manufacturing of various products were given preference for light weight structures which was able to be attained by DMAM, it also added to global reduction of CO_2 emissions of the transport vehicles which used to carry the products to the clients from manufacturing site. The DMAM process was utilized here to study the conformal cooling channels and heat sinks were the products exhibited greater efficiency in terms reducing the cooling time of the system and in detection of the faulty parts. Stephen A. Whitmore (2015) gave a detailed report on FDM as a green technology method for the manufacture of hybrid propulsion system used in spacecrafts. Using FDM thermoplastic fuel grains can be fabricated according to port shapes which increases the volumetric efficiencies and enhances the burn properties. Harsha et al. (2015) have studied the sustainable and environmental impacts with a new technique of Additive Manufacturing known as mask-image-projection STL. Life cycle assessment have been studied which projects the damage to resource availability greatly affects the human health. As the build time for any product is increasing the effect on the three parameters by hierarchy approach, human health, ecosystem and resources were noted to increase. However, as the complexity of the product is increasing the time consumed is also increasing which in turn increases the utilization of process energy but yet it is playing negligent role when compared to subtractive manufacturing. However, the author focuses on establishing the AM for mass production as that of conventional form at a faster rate. Simon Ford et. al. (2016) did an exploratory study on challenges and advantages of Additive Manufacturing in a sustainable environment. They explore and throw light on the less understood viable and sustainable options by bringing in AM to the front end of manufacturing. Organizational behavior and present customer base for companies are trying to shift towards AM, with the existing customers. Any manufacturing activity should bring about reduction in the utilization of water, energy, waste and prevent emission of harmful gases to the atmosphere and hence environmental degradation. Additive Manufacturing technique which has become a novel technique of manufacture is adhering to all these cardinal rules of Manufacture the world over and is working constantly towards betterment of society with strict adherence of self-imposed standards.

In order to be sustainable and environmentally friendly, the criteria for any manufacturing activity in the present times is good vision, identification of drivers, identification and prioritization of sustainable business opportunities. Finally, identifying the enablers to realize these opportunities. Also, this includes parameters like internal weight optimization for a good / sustainable product design and improvement in process efficiency. It is gratifying to note that all these are brought about by the AM technique. The features of repair, remanufacturing and life extension of the product are also served very well by the DMAM method. With AM being in the forefront of all manufacturing techniques in the present times, the features of product and process redesign, material input processing, make to order component and product manufacturing have made the process more sustainable and environment friendly at every step of manufacturing. With AM method one can easily achieve sustainable development and a greener tomorrow.

Summarizing the vast base of researchers on their studies for takeover of AM over conventional method holds a greater value for shifting the manufacturing towards new method. Though in initial stages for the mass production yet the advantages have surpassed the subtractive form in terms of energy, efficiency, transportation, life cycle and complexity. The reusability of the material by converting to wire form can be easily done in AM which reduces the procurement of raw materials which in turn saves the ecological sustainability of the environment. Overall AM has and been moving towards the green process and sustainability at every step of manufacturing.

4. CASE STUDIES

AM is a powerful tool that offers competitiveness in vast industrial base, using the process of fabrication direct from 3D CAD model unlocks the opportunity of design potential pushing the limits of manufacturing to higher boundaries with varied applications. The challenges associated with losing the competitiveness between the mass production industries is increasing due to easy availability of goods and parts, every country in the world has almost embraced the conventional forms, hence giving birth to the new method of manufacturing known as AM in which the parts of great quality and design stability in a limited form is manufactured for a specified usage. Numerous researchers have done case studies focusing on the improvement in usability of material with lesser wastage, increasing the efficiency of manufacturing and also reduce the cost involved and also to have lesser impact on environment.

Petrovic et al. (2015) explained industrial application of Additive Manufacturing through case studies. There are numerous advantages of using AM which includes time-to-market reduction due to high speed process which was studied by building a car upright in 28 hours by EBM fabrication method. Reduction in usage of raw material was reduced by 40% in comparison to subtractive manufacturing. Also 95 to 98% of the material can be recycled. The parts were produced without using any tools, moulds, or punches. The parts were produced without residual porosity. The cooling channels in additive techniques can be varied to follow mould cavity surface and provide uniform heat transfer. This also provides a better quality of injected moulds which cannot be varied in subtractive forms. The application of AM in aerospace provides greater advantages like reduced raw materials, reduced buy to fly ratio, freedom of geometrical constraints and reduced energy utilization. By considering the lower price of the vehicle and the lower fuel consumption during its useful life, along with a saving of 100kg of material, resulted in the saving of 4.3 million Pounds per aircraft and 770 Pounds per car. The conformal cooling reduced the cycle time from 6.7 seconds to 5 seconds, this reduction of 1.7 seconds in cycle time translates to improvement of 180 parts which increases the productivity by 33%.

The roundtable summary of Additive Manufacturing: opportunities and constraints held at Royal Academy of Engineering, UK collectively produced the great ideas through case studies by eminent researchers and industrial expertise. Aerospace and automotive are the two major beneficiaries of AM. GE aviation recently bought AM Morris technology, the pioneers in metal Additive Manufacturing and utilized the technology to build up the aero engine fuel nozzle. The conventional method of manufacturing required 20 separate parts to be welded together which is extensively labor intensive and consumes large amount of man hours. The same nozzle can be manufactured more precisely and efficiently by AM which can build up to 20,000 to 25,000 nozzles per year. Graham Tromans, President of AM consultancy GP Tromans associates predicts 50% of jet engine will be additive manufactured in current lifetimes. Neil Mantle from Rolls-Royce says 'AM gives a great opportunity when compared to forging because conventional method of manufacture may take 50 to 60 weeks to manufacture a complete component, while AM can finish it in one month span of time. He again continues, sometimes to prepare a certain component 90% of the material will be machined away which is not the case in AM and stresses to embrace AM technology for Rolls-Royce.

Huang et al. (2015) analyzed the components of aircraft for the reduction of mass by utilizing the Additive Manufacturing technique in comparison to Conventional Manufacturing (CM). Bracket manufactured by the method of machining in conventional form by using Ti alloy had a mass of 1.09

kg whereas the same component manufactured by EBM had a mass of 0.38 kg which reduced the mass by 65%. Similarly seat buckle mass reduction was noted to be 55%, bionic bracket 35%, engine cover door hinge 65% and 50% reduction for fork fitting. CM pathways included forging, turning, milling, machining and casting processes where as AM pathways included SLM, DMLS and EBM processes which clearly indicate the efficient way of utilization of AM *(Figure 9).*

Reduction in mass is largely significant for aerospace parts which have longer service life. Weight reduction for aerospace parts can lead to cost-effective energy savings. The aerospace components developed by Additive Manufacturing process have a significant energy impact compared to conventional techniques. For example, an aircraft bracket which is being used to fix structures in kitchen galleys and lavatories etc. is typically manufactured by milling and machining processes with a very high buy to fly ratio of 8:1. A same bracket with similar functionality and with a topologically optimized geometry can be manufactured by electron beam melting process with a buy to fly ratio of 1.5:1 which is considerably lower than the bracket produced by conventional method. The improved design developed by Additive Manufacturing process was 65% lighter, saving materials used for manufacturing and resulting in use phase energy savings. Further, application of Additive Manufacturing could minimize the energy consumption in the freight and distribution phase. In the case of freight and distribution, energy use is a function of how far the component part has traveled within the supply chain.

AM has played a very significant role in biomedical parts biocompatible and biodegradable parts manufactured with AM were used in coronary bypass implant, guides for maxillofacial surgery also various parts for scaffolds for tissue engineering. Flexible forms of polymers would permit real tissue engineering being base for construction of human organs which causes great difficulty with any conventional forms to attain minute details of human parts.

A stainless steel 316 part was designed for defense applications. Owing to rapid design change and small quantity requirement, customer was waiting for a period of 6 months for part manufacturing. The conventional manufacturing cost was very high. However, the use of

Laser Engineering Net Shaping method (LENS) the proposed part was manufactured in 3 days and the manufacturing cost was 65% lesser than conventional process.

The above discussed case studies clearly demonstrate capability of Additive Manufacturing processes to exhibit sustained performance of the process and design modifications. Further, reduction in consumption of energy and dematerialization can contributes to improvement in the material and energy efficiency. Potential modification in designs can results in fewer components, materials, and interactions. Thus, simple design reduces the amount of material flow and finally, minimizes the environmental impact. When compared with subtractive or conventional manufacturing techniques, it is important to note that Additive Manufacturing techniques produces less waste indicating its high potential green technology applications. It has been reported that 95 to 98% of unused material (powder or resin) in Additive Manufacturing can be reused. Reuse of unused materials significantly contribute to concept of green manufacturing (Petrovoic etal, 2011). Product life cycle analyses of the components synthesized by Additive Manufacturing process have demonstrated that use of Additive Manufacturing techniques helps to reduce production and minimizes the product phases. The major saving will be mainly due to reduction in material usage and handling which reduces the supply chain. In addition, use of phase methods and light weight parts significantly reduces the consumption of energy (Gebler etal, 2014).

5. FUTURE OF ADDITIVE MANUFACTURING AND ITS ROLE IN GREEN TECHNOLOGY

The future of AM looks to be very impressive and exuberant. Every individual will have the ability to grow the required parts with own design and ease. The advantage of using AM would lead to the development of products that is intended towards a specific application instead of building up a complete model. There would be a feasibility to develop a part of a working component alone instead of completely replacing the whole model. The geometrical flexibility actually means the customization of the product according to one's own imaginary design which fits in an individual's unique requirement. Imagine being able to modify the pen depending on the comfort of grip of one's hand.

Envisage the spectacles that fit precisely on your eyes as it was made from an impression with width of face. These things are possible using AM as they have the ability to build any form of complicated products with digital models that may also include biometric data gathered from the specific individual.

(Sunpreet Singh et. al., 2016), reviewed the manufacturing methods of AM or 3D-Printing for medical devices along with the supporting structure for tissue engineering. Few medical devices were permitted to develop the products using AM. Mechanical properties, biocompatibility, in vitro and in vivo performance for various orthopedic applications have been studied. Food and Drug Administration has approved various materials for biomedical applications to be used as implants. However more of concern is to be given to the process parameters to withstand the mechanical and fatigue performance together for load bearing implants. The finalized process parameters for AM are not in par with mechanical standards for varied complex geometries. Very fewer opportunities have been noted so far to get mass production orders especially for screws, nuts etc. Undoubtedly the use of AM technology for prototyping will be carried for the years to come. Time will emerge soon when it is commonplace to manufacture products in low volumes or unique products using AM. Eventually, we may see these machines being used as home fabrication devices, such as photocopying machines.

AM provides opportunities for the industries to experiment on the existing marketable model rather on new designs. The shift of manufacturing to digital form would always keep the required model in a soft file format which would be quickly utilized for repair or remanufacture at any given instant, this kind of approach would provide the necessity incentives for various industrial service oriented business models. This strategy would completely change the base value chains of working environment resulting in sustainable form of work force. However, the changes are not imminent as predicted as it may lead to the huge damages of the present working model of the well established businesses. The shift is going to take place in a designing field which requires a working model with lesser number of parts and this change will subsequently lead to simpler value chains in work force.

6. CONCLUSION

The compensation of reward for AM is being analyzed across industries for products life cycle. The main benefit of AM technology include freedom of new design, material integrated functionalities, enhanced process and product performances, reduced time and customization with cheaper cost. Further growth of AM technologies are required to decrease the duration of the automation process including post processing, enhanced aesthetics, quality and reliability, expanded areas of application and integrate the materials of functionality. Furthermore the AM needs to be analyzed and optimized for greener technology and sustainability and studied for the performance on social impacts.

There arise numerous implications for implementation of AM in the global scenario. The struggle for sustainable development of planet has turned out to be a cold war between the companies and government, industries and consumer activists and in times with government and activists. With the combination of these three it has become a tough step to keep forward with one or the other hindrance from the above mentioned bodies. The steps taken by the organizations to save the environment is claimed to be insufficient by the policy makers and environmentalists. Another idea emerges out claiming to improve the knowledge of the consumers which indirectly makes the organization sustainable. Although, both policies and education are implemented in the required form yet the change is long goal to be achieved. However, there is a realistic and drastic change in the field of manufacturing sector which is moving towards the sustainable form by utilizing AM.

REFERENCES

Ahn, D.-G. (2016). Direct Metal Additive Manufacturing Processes and Their Sustainable Applications for Green Technology. *RE:view*. doi:10.100740684-016-0048-9

Attaran. (2017). *The rise of 3-D printing: The advantages of Additive Manufacturing over traditional manufacturing*. Academic Press. . doi:10.1016/j.bushor.2017.05.011

Bai, Y., & Christopher, B. (2015). An exploration of binder jetting of copper. *Rapid Prototyping Journal*, *21*(2), 177–185. doi:10.1108/RPJ-12-2014-0180

Beese & Clare. (2107). Materials for Additive Manufacturing. *CIRPAnnals - Manufacturing Technology*. doi:10.1016/j.cirp.2017.05.009

Ben Utela, D. S., Anderson, R., & Ganter, M. (n.d.). A review of process development steps for new material systems in three dimensional printing (3DP). *Journal of Manufacturing Processes*. doi:10.1016/j.jmapro.2009.03.002

Bhavar, Kattire, Patil, Khot, Gujar, & Singh. (2014). *A Review on Powder Bed Fusion Technology of Metal Additive Manufacturing*. In 4th International conference and exhibition on Additive Manufacturing Technologies-AM-2014, Bangalore, India.

Bourell, Kruth, Leu, Levy, Rosen, Beese, & Clare. (2017). *Materials for Additive Manufacturing*. Academic Press. . doi:10.1016/j.cirp.2017.05.009

Cedero, Siddel, Amelia, & Elliott. (2017). Strengthening of Ferrous Binder Jet 3D Printed Components through Bronze Infiltration. Additive Manufacturing, 15, 87-92.

Chen, D., Heyer, S., Ibbotson, S., Salonitis, K., Steingrimsson, J. G., & Thiede, S. (2015). Direct Digital Manufacturing: Definition, Evolution and sustainability Implications. *Journal of Cleaner Production*, *107*, 615–625. doi:10.1016/j.jclepro.2015.05.009

Chougrani, L., Pernot, J.-P., Véron, P., & Abed, S. (2017). Lattice structure lightweight triangulation for Additive Manufacturing. *Computer Aided Design*, *90*, 95–104. doi:10.1016/j.cad.2017.05.016

Despeisse, M. (2017). Sustainable Value Roadmapping Framework for Additive Manufacturing. Academic Press. doi:10.1016/j.procir.2016.11.186

Despeisse & Ford. (2015). *Centre for technology management working process*. Academic Press.

Doyle, M., Agarwal, K., Sealy, W., & Schul, K. (2015). Effect of Layer Thickness and Orientation on Mechanical Behaviour of Binder Jet Stainless Steel 420+Bronze Parts. *Procedia Manufacturing, 1*, 251–262. doi:10.1016/j.promfg.2015.09.016

Everton, S. K., Hirsch, M., Stravroulakis, P., Leach, R. K., & Clare, A. T. (2016). Review of in-situ process monitoring and in-situ metrology for metal Additive Manufacturing. *Materials & Design, 95*, 431–445. doi:10.1016/j.matdes.2016.01.099

Ford & Despeisse. (2017). *Additive Manufacturing and sustainability: an exploratory study of the advantages and challenges*. Academic Press. . doi:10.1016/j.jclepro.2016.04.150

Gebler & Visser. (2014). A Global Sustainability Perspective on 3D Printing Technologies. *Energy Policy, 74*(c), 158–167.

Gibson, I., Rosen, D., & Stucker, B. (2010). *Additive Manufacturing Technologies* (2nd ed.). Springer. doi:10.1007/978-1-4419-1120-9

Glasschroeder, J., Prager, E., & Zaeh, M. F. (2015). Powder Bed Based 3D- Printing of Function Integrated Parts. *Rapid Prototyping Journal, 21*(2), 207–215. doi:10.1108/RPJ-12-2014-0172

Gonzalez, J. A., Mireles, J., Lin, Y., & Wicker, R. B. (2016). Characterization of Ceramic Components Fabricated using Binder Jetting Additive Manufacturing Technology. Academic Press.

Hofstatter, Pedersen, Tosello, & Hansen. (2017). *Applications of Fiber-Reinforced Polymers in Additive Manufacturing*. Academic Press. doi:10.1016/j.procir.2017.03.171

Hu & Mahadevan. (2017). *Uncertainty quantification in prediction of material properties during Additive Manufacturing*. Academic Press. . doi:10.1016/j.scriptamat.2016.10.014

Huang, Riddle, Graziano, Warren, Das, Nimbalkar, Cresko, & Masanet. (2015). Energy and emissions saving potential of Additive Manufacturing: the case of lightweight aircraft components. *Journal of Cleaner Production*, 1-12.

Huang, Liu, Mokasdar, & Hou. (2013). *Additive Manufacturing and its societal impact: a literature review*. Academic Press. doi:10.100700170-012-4558-5

Innovating Clean Energy Technologies in Advanced Manufacturing. (2015). *Quadrennial Technology Review 2015, Technology Assessments*, 1-35.

Jackson, Van Asten, Morrow, Min, & Pfefferkorn. (2016). *A Comparison of Energy Consumption in Wire-Based and Powder-Based Additive-Subtractive Manufacturing*. Academic Press. doi:10.1016/j.promfg.2016.08.087

Jaya Christiyan, K. G., Chandrasekhar, U., & Venkateswarlu, K. (2016). A study on the influence of process parameters on the Mechanical Properties of 3D printed ABS composite. *IOP Conf. Series: Materials Science and Engineering, 114*. doi:10.1088/1757-899X/114/1/012109

Kahlin, M., Ansell, H., & Moverare, J. J. (2017). Fatigue behaviour of notched additive manufactured Ti6Al4V with as-built surfaces. *International Journal of Fatigue, 101*, 51–60. doi:10.1016/j.ijfatigue.2017.04.009

Kalsoom, Peristyy, Nesterenko, & Paull. (n.d.). *A 3D printable diamond polymer composite: A novel material for fabrication of low cost thermally conducting devices*. Royal Society of Chemistry.

Kechagias, J. (2007). An experimental investigation of the surface roughness of parts produced by LOM process. *Rapid Prototyping Journal, 13*(1), 17–22. doi:10.1108/13552540710719172

Kellens, Mertens, Paraskevas, Dewulf, & Duflou. (2017). *Environmental Impact of Additive Manufacturing Processes: Does AM contribute to a more sustainable way of part manufacturing?* Academic Press. doi: . doi:10.1016/j.procir.2016.11.153

Kęsy, A., & Kotliński, J. (2010). Mechanical properties of parts produced by using polymer jetting technology. *Archives of Civil and Mechanical Engineering, 10*(3), 37–50. doi:10.1016/S1644-9665(12)60135-6

Klahn, C. (2016). Design Guidelines for Additive Manufactured Snap-Fit Joints. Academic Press. doi:10.1016/j.procir.2016.04.130

Klimek, L., Klein, H. M., Schneider, W., Mosges, R., Schmelzer, B., & Voy, E. D. (1993). Stereolithographic modelling for reconstructive head surgery. *Acta Oto-Rhino-Laryngologica Belgica, 47*(3), 329–334. PMID:8213143

Li, Kucukkoc, & Zhang. (2017). *Production planning in Additive Manufacturing and 3D printing*. Academic Press. . doi:10.1016/j.cor.2017.01.013

Li & Soar. (2009). Characterization of Process for Embedding SiC Fibers in Al 6061 O Matrix Through Ultrasonic Consolidation. *Journal of Engineering Materials and Technology, 131*(2).

Lim, S., Buswell, R. A., Le, T. T., Austin, S. A., Gibb, A. G. F., & Thorpe, T. (2012). Development in Construction Scale Additive Manufacturing Processes. *Automation in Construction, 21*, 262–268. doi:10.1016/j.autcon.2011.06.010

Lindemann, Jahnke, Moi, & Koch. (2012). *Analyzing Product Lifecycle Costs for a Better Understanding of Cost Drivers in Additive Manufacturing*. Academic Press.

Malshe, H., Nagarajan, H., Pan, Y., & Haapala, K. (2015). Profile of Sustainability in Additive Manufacturing and Environmental Assessment of a Novel Stereolithography Process. *ASME 2015 International Manufacturing Science and Engineering Conference Volume 2: Materials; Biomanufacturing; Properties, Applications and Systems; Sustainable Manufacturing*. 10.1115/MSEC2015-9371

Everton, Hirsch, Stravroulakis, Leach, & Clare. (2016). *Review of in-situ process monitoring and in-situ metrology for metal Additive Manufacturing*. Academic Press. doi:10.1016/j.matdes.2016.01.099

Meteyer, Xu, Perry, Fiona, & Zhao. (2014). Energy and Material floe Analysis of Binder Jetting Additive Manufacturing Processess. *Procedia, CIRP15*, 19–25.

Miyanaji, H., Zhang, S., Lassell, A., Zandinejad, A., & Yang, L. (2016). Process Development of Porcelain Ceramic Material with Binder Jetting Process for Dental Applications. *JOM*, *68*(3), 831–841. doi:10.100711837-015-1771-3

Morrow, W. R., Qi, H., Kim, I., Mazumder, J., & Skerlos, S. J. (2007). Environmental Aspects of Laser Based and Conventional Tool and Die Manufacturing. *Journal of Cleaner Production*, *15*(10), 932–943. doi:10.1016/j.jclepro.2005.11.030

Murr & Johnson. (2017). 3D metal droplet printing development and advanced materials Additive Manufacturing. *Journal of Materials Research and Technology*. doi:10.1016/j.jmrt.2016.11.002

Nikzad, M., Masood, S. H., & Sbarski, I. (2011). Thermo-mechanical properties of a highly filled polymeric composites for Fused Deposition Modeling. *Materials & Design*, *32*(6), 3448–3456. doi:10.1016/j.matdes.2011.01.056

Nimbalkar, S., Cox, D., Visconti, K., & Cresko, J. (2014). Life Cycle Energy Assessment Methodology and Additive Manufacturing Energy Impacts Assessment Tool. *Proceedings from the LCA XIV International Conference*, 130-141.

Ning, Cong, & Qiu, Wei, & Wang. (2015). Additive Manufacturing of carbon fiber reinforced thermoplastic composites using fused deposition modeling. *Composites. Part B, Engineering*, *80*, 369–378.

Nukman, Farooqi, Al-Sultan, Rahman, Alnasser, & Bhuiyan. (2017). *A Strategic Development of Green Manufacturing Index (GMI) Topology Concerning the Environmental Impacts*. Academic Press. doi: . doi:10.1016/j.proeng.2017.04.107

Paranthaman, M. P., Shafer, C. S., Elliott, A. M., Siddel, D. H., McGuire, M. A., Springfield, R. M., ... Ormerod, J. (2016). Binder Jetting: A Novel NdFeB Bonded Magnet Fabrication Process. *JOM*, *68*(7), 1978–1982. doi:10.100711837-016-1883-4

Petrovic, V., Juan, V. H. G., Ferrando, O. J., Gordillo, J. D., Jose, R. B. P., & Griñan, L. P. (2011). Additive layered manufacturing: Sectors of industrial application shown through case studies. *International Journal of Production Research*, *49*(4), 1061–1079. doi:10.1080/00207540903479786

Petrovoic, Gonzale, Ferrando, Gordillo, Purchades, & Grinan. (2011). Additive Layer Manufacturing: Sectors of Industrial Applications shown through case studies. *International Journal of Production Research*, *49*(4), 1061-1079.

Rengier, Mehndiratta, von Tengg-Kobligk, Zechmann, Unterhinninghofen, Kauczor, & Giesel. (2010). *3D printing based on imaging data: review of medical applications*. Academic Press. doi:10.100711548-010-0476-x

Sarahara, H., Tsutsumi, M., & Chino, M. (2005). Development of Layered Manufacturing System using Sheet Metal-Polymer Lamination for Mechanical Parts. *International Journal of Advanced Manufacturing Technology*, *27*(3-4), 268–273. doi:10.100700170-004-2163-y

Schmidt, Merklein, Bourell, Dimitrov, Hausotte, Wegener, … Levy. (2017). Laser based Additive Manufacturing in industry and academia. *CIRP Annals - Manufacturing Technology*. doi:10.1016/j.cirp.2017.05.011

Singh, Haverinen, Dhagat, & Jabbour. (2010). Inkjet Printing – Process and Applications. *Advance Materials, 22*(6), 673–685.

Singh, S., & Ramakrishna, S. (2017). *Biomedical applications of Additive Manufacturing: present and future*. Current Opinion in Biomedical Engineering. doi:10.1016/j.cobme.2017.05.006

Tammas-Williams & Todd. (2017). *Design for Additive Manufacturing with site-specific properties in metals and alloys*. Academic Press. . doi:10.1016/j.scriptamat.2016.10.030

Tateno & Kondoh. (2017). *Environmental load reduction by customization for reuse with Additive Manufacturing*. Academic Press. doi: . doi:10.1016/j.procir.2016.11.219

Vaezi, M., Chianrabutra, S., Mellor, B., & Yang, S. (2013). Multiple Material Additive Manufacturing Part 1: A Review. *Virtual and Physical Prototyping, 8*(1), 19–50. doi:10.1080/17452759.2013.778175

van Noort, R. (2012). The future of dental devices is digital. *Dental Materials, 28*(1), 3–12. doi:10.1016/j.dental.2011.10.014 PMID:22119539

Wang, Jiang, Zho, Gou, & Hui. (2017). *3D printing of polymer matrix composites: A review and prospective*. Academic Press. doi:10.1016/j.compositesb.2016.11.034

Weisenel, L., Travitzky, N., Seiber, H., & Grell, P. (2004).. *Laminated Object Manufacturing of SiSiC Composite*. Academic Press.

Whitmore. (2015). Additive Manufacturing as an enabling technology for "green" hybrid spacecraft propulsion. *Recent Advances in Space Technologies (RAST), 2015 7th International Conference*. 10.1109/RAST.2015.7208305

Wong & Hernandez. (2012). *International Scholarly Research Network ISRN Mechanical Engineering Volume*. doi:10.5402/2012/208760

Zeng, Q., Xu, Z., Tian, Y., & Qin, Y. (2016). Advancement in Additive Manufacturing & numerical modelling considerations of direct energy deposition process. In *Proceeding of the 14th International Conference on Manufacturing Research*. IOS Press.

Zeng, Q., Xu, Z., Tian, Y., & Qin, Y. (2016). Advancement in Additive Manufacturing & numerical modelling considerations of direct energy deposition process. In *Proceeding of the 14th International Conference on Manufacturing Research*. IOS Press.

Zhai, Y., Lados, D. A., & LaGoy, J. L. (2014). Additive Manufacturing: Making Imagination the Major Limitation. *JOM, 66*(5), 808–816. doi:10.100711837-014-0886-2

Zhang, Liu, & To. (2017). *Role of anisotropic properties on topology optimization of additive manufactured load bearing structures*. Academic Press. . doi:10.1016/j.scriptamat.2016.10.021

Chapter 15
Spark Plasma Sintering of Mg-Zn-Mn-Si-HA Alloy for Bone Fixation Devices:
Fabrication of Biodegradable Low Elastic Porous Mg-Zn-Mn-Si-HA Alloy

Chander Prakash
Lovely Professional University, India

Sunpreet Singh
Lovely Professional University, India

Ahmad Majdi Abdul-Rani
Universiti Teknologi PETRONAS, Malaysia

M. S. Uddin
University of South Australia, Australia

B. S. Pabla
National Institute of Technical Teachers' Training and Research, India

Sanjeev Puri
Panjab University, India

ABSTRACT

In this chapter, low elastic modulus porous Mg-Zn-Mn-(Si, HA) alloy was fabricated by mechanical alloying and spark plasma sintering technique. The microstructure, topography, elemental, and chemical composition of the as-sintered bio-composite were characterized by optical microscope, FE-SEM, EDS, and XRD technique. The mechanical properties such as hardness and elastic modulus were determined by nanoindentation technique. The as-sintered bio-composites show low ductility due to the presence of Si, Ca, and Zn elements. The presence of Mg matrix was observed as primary grain and the presence of coarse Mg2Si, Zn, and CaMg as a secondary grain boundary. EDS spectrum and XRD pattern confirms the formation of intermetallic biocompatible phases in the sintered compact, which is beneficial to form apatite and improved the bioactivity of the alloy for osseointegration. The lowest elastic modulus of 28 GPa was measured. Moreover, the as-sintered bio-composites has high corrosion resistance and corrosion rate of the Mg was decreased by the addition of HA and Si element.

DOI: 10.4018/978-1-5225-5445-5.ch015

INTRODUCTION

Describe Metallic biomaterials like stainless steel, cobalt chromium, titanium and its alloys were used for hard tissue replacement for the decade (Niinomi et. al., 2012; Geetha et. al., 2009; Bartolo et. al., 2012; Vaccaro et. al., 2003; Bartolo et. al., 2009). The main drawback with these materials are used as permanent implants, thus after bone healing the implants were taken off from the body by additional surgery causes an increase in costs to the healthcare, as well as anxiety to the patient (Prakash et. al., 2016; Prakash et. al., 2015; Prakash et. al., 2017). In the current, the magnesium base alloys and bio-composite have gained much attention of researchers for the design and development orthopedic implants due to their high biodegradability and unique combination of mechanical properties such as low elastic modulus and high strength (Staiger et. al., 2016). Till date no perfect magnesium alloy available that possessed the adequate putrescible integrity to corrosion for a certain period of bone healing (Vahidgolpayegani et. al., 2017). The major drawback of Mg alloy, which hampers their eminent utility, is their rapid corrosion rate in the human body (Uddin et. al. 2015). Through decades past, numerous procedures and techniques were practiced to retard the corrosion rate, predominantly, to delay the degradation of the Mg alloy at a pace that matches bone healing (Cui et. al., 2008; Hassel et. al., 2007; Salahshoor et. al., 2013). The mechanical alloying of the element is the most promising technique yet to control the corrosion rate and to improve the mechanical properties (Uddin et. al. 2017). Till date various alloying elements such as zinc (Zn), aluminium (Al), silver (Ag), yttrium (Y), zirconium (Zr), neodymium (Nd), silicon (Si), and manganese (Mn) have been used as alloying elements to enhance the mechanical properties and corrosion behavior of Mg alloy. In this regard, various Mg-based alloys such as Mg–CA binary alloy (Staiger et. al., 2006), Mg–Zn (Xu et. al., 2007), Mg–Sr (Li et. al., 2008), Mg-Mn-Zn (Xu et. al., 2008; Zhang et. al., 2009), Mg-Zn-Y (Zhang et. al., 2008) and Mg-Zn-Mn-CA alloys (Zhang et. al., 2008) and much more had been developed. Among them, the implant made by the alloying of Ca, Mn, Z, and Si elements improved the mechanical properties and corrosion resistance (Witte et. al., 2007). In reality, there are only a few elements including Ca, Mn, Zn, and Si that can be accommodated in the human body and can also retard the biodegradation of Mg alloys. The element Ca and Zn are the abundant minerals necessary for the bone ingrowth in human and one of the most favoring alloying elements used for magnesium alloys (Khanra et. al., 2010, Gu et. al., 2010; Ye et. al., 2010; Zhao et. al., 2011; Viswanathan et. al., 2013). The manganese (Mn) decreases the corrosion rate of the Mg-alloys.

Recently, silicon (Si) element has been accounted as a necessary mineral to facilitate the healing and assist to strengthen the immune system of human body (Zhang et. al., 2010). Moreover, it also plays an important role in development and growth of tissues (Hamu et. al., 2008). However, Mg-Si-based alloys showed low ductility, high strength, and high corrosion resistance because of the presence of large Mg_2Si particle and eutectic phases. The addition of Ca element in the Mg-Mn-Zn alloy not only refined the grain size but also improves the mechanical properties and corrosion resistance. Moreover, element Ca is a main and prime part of the human bone that can stimulate the bone ingrowth process. The alloying of Zn and Mn elements in the Mg matrix improved both tensile strength and corrosion resistance (Liu et. al,. 2016; Zhang et. al., 2008). The previous studies showed that Mg-Zn-Mn alloy is biodegradable in a human biological environment without any dregs in hosted area and surrounding (Fu et. al., 2017; Bakhsheshi-Rad et. al., 2014). In this contrast, various fabrication routes and methodologies such as hot isostatic pressing (HIP), hot sintering, electrical resistance furnace, high-frequency induction furnace, and high-frequency induction heat sintering has been utilized for the sintering of magnesium-based alloys (Khalil & Almajid, 2012; Sunil et al., 2014; zheng et al., 2011). These techniques produce a nanoporous structure, merely they're long sintering time, temperature weakens the mechanical and electrochemical properties, which failed the implant at cyclic load condition and long-term performances (Ye et. al., 2010; Zhao et. al., 2011; Viswanatha et. al., 2013; Sun et. al., 2013, Khalil et. al., 2012; Suni et. al., 2014; Zheng et. al., 2011). On the other hand, the application of Spark Plasma Sintering (SPS) technique using powder metallurgy route consisting of mechanical alloying powders for the manufacture of porous Mg-alloys has been reported a novel and potent technique (Lala et. al., 2011). As a novel powder sintering process with distinct characteristics compared with conventional sintering techniques, the spark plasma sintering (SPS) technique enables alloys with good properties to be constructed in a shorter holding time and a high pressure at rapid heating and cooling rates (>100 °C/min) and relatively lower sintering temperature.

From the literature, it is seen that hydroxyapatite (HA) and silicon (Si) has been used for the mechanical alloying in Mg-based alloys to enhance the corrosion resistance and mechanical properties. Very limited research studies are available reported the effect of Si and HA addition in the Mg-Zn-Mn alloy on the microstructure, mechanical properties, and corrosion resistance. However, the effect of Si and HA addition on the elastic modulus, which is an essential property required during stress shielding of Mg-based alloy, has not been clearly brought out. Hence, the objective of the present work is to fabricate porous biodegradable, low elastic Mg-based alloy by high-energy ball milling and then sintering the milled product using the spark plasma technique. This paper presents the fabrication of fabrication of porous Mg-Zn-Mn-(Si-HA) alloy by spark plasma sintering process using the powder metallurgical route, which is urgently needed for the next generation of implants. A comprehensive and critical investigation on the effect of Si and HA addition on the microstructure, morphology, phase composition, microhardness, elastic modulus, and corrosion resistance of the as-prepared alloys were conducted to evaluate the efficacy of the SPS technique as a porous implant fabrication for preparing materials for biomedical applications.

MATERIALS AND METHODS

In the current study, Mg-Zn-Mn-Si, Mg-Zn-Mn-HA, and Mg-Zn-Mn-Si-HA biodegradable alloys were in-situ fabricated by mechanical alloying of powders such as Magnesium (Mg) powder along with Manganese (Mn), Zinc (Zn), Silicon (Si), and hydroxyapatite (HA) and consolidation using spark plasma sintering (SPS) technique. The Elemental powders of Mg (Purity 99.9%, 25μm), Mn (Purity 99.9%, 25μm), Zn (Purity 99.9%, 25μm), Si (Purity 99.9%, 25μm), and HA (Purity 99.9%, 0.5 μm) were taken as alloying elements. In the Mg-Zn-Mn-Si biocomposite, the powders were mixed in an atomic weight percentage of 5% Mn, 1% Zn, and 10% Si, and the rest is Mg (84%). In the Mg-Zn-Mn-HA biocomposite, the powders were mixed in an atomic weight percentage of 5% Mn, 1% Zn, and 10% HA, and rest is Mg (84%). In the Mg-Zn-Mn-Si-HA biocomposite, the powders were mixed in an atomic weight percentage of 5% Mn, 1% Zn, 5% Si and 5% HA, and the rest is Mg (84%). The desired amount of powder was weighed and loaded in stainless steel vial for alloying. The stainless steel balls were used for homogenization and alloying of powder. The ball to powder weight ratio of 5:1 was selected. The powder mixture was mechanically alloyed in a high energy planetary ball mill (Fritsch, Pulverisette 7) at a rotational speed of 300 RPM for 8 hrs. The as-blended powder mixture was first pre-heated at 200 °C for 2 h in an argon atmosphere (1 l/min) to evaporate the moisture and then consolidated via spark plasma sintering (SPS) method in a graphite die. The consolidation of powder mixture was carried out at sintering temperatures of 350, 400, 450 K with a heating rate of 50 K/min (holding time 5 min) under vacuum conditions at a uniaxial pressure of 30, 40, and 50 MPa, as per the procedure adopted elesweher (Lala et. al., 2011). The circular compacts of diameter 20 mm and thickness 4 mm were sintered. Figure 1 shows the photograph of high-energy planetary ball mill, spark plasma sintering (SPS-5000) machine, blended sample of powder in a graphite die, the red-hot sample under vacuum condition during the process, and as-consolidated sample.chapter.

Samples for microstructure examination were cut precisely from the as-sintered compact with the low speed diamond cutter. The samples were grounded by SiC papers and polished to an average surface roughness R_a of 0.55 μm by adequate polishing methods. Then the polished samples were etched with Acetic-Picral (5 ml Acetic Acid, 6 g Picric Acid, 10 ml Water, 100 ml Ethanol) for 10s. The microstructure examination was conducted optical microscope and field emission scanning electron microscopy (FE-SEM; JEOL 7600F), respectively. The elemental and phase composition the compact was accessed by energy dispersive spectroscopy (EDS) and X-ray diffraction (XRD; X'pert-PRO) using CuKα radiation at 45 kV, 40 mA at incident angle 2θ. The elastic modulus and hardness are the prime and main desirable characteristics of the biomaterial. The mechanical properties such as hardness and young's modulus were determined, basically, the main requirements for orthopedic implants are considered as mechano-biological antibacterial and corrosion properties. The nanoindentation is the only approach for calculating both elastic modulus and hardness of the material in a single test. The nanoindentation test was conducted on the Hyistron TI-950 indentation system using Berkovich indenter. The hardness and elastic modulus of the composite were determined using the Oliver-Pharr method. The surface microhardness of the HA deposited surface was measured by Vickers hardness tester (HMV-G21ST, SHIMADZU, Japan). The cross-sectional surface was utilized to test the surface hardness and prepared using appropriate methods, as referred to previous research study (Khanra et al., 2010). An indentation load of 0.2 N was applied at a dwell time of 15 s. The observations were replicated 3-times to reduce the error.

Figure 1. (a) High energy planetary ball mill and (b) spark plasma sintering (SPS) machine

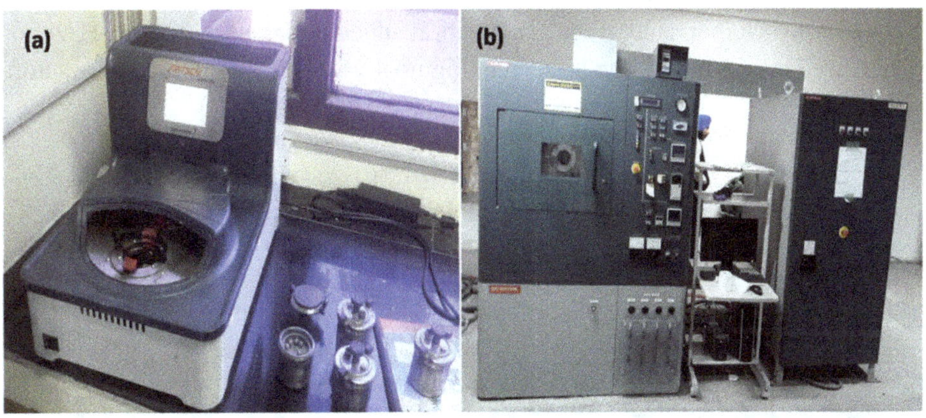

RESULTS AND DISCUSSIONS

It was noted that the net weight of starting powder mixture before mechanical alloying was 10 gm, and approximately 10 gm powder was recovered after mechanical alloying. It can be clearly ascertained that the powder mixture was homogenized properly and grain size reduced significantly during milling due to plastic deformation and fragmentation. It can be also noted that the powder particles were defused with each other due to heat generation during milling. Figure 2 shows the photograph of the sintered alloy.

The Porous, semi-porous, and fully dense structure was received at 350 °C, 400 °C, and 450 °C under 40 MPa applied pressure. The variance of density and porosities of sintered bio-composites with regard to applied pressure and sintering temperature was measured, as seen in Figure 3. The relative compactness of the sintered alloys was greatly increased with applied pressure, from 30 MPa to 50 MPa under 400 °C, as can be found out in Figure 3 (a). The application of pressure during sintering is helpful for getting rid of pores from powder compact and offer additional driving force for compaction. It can be clearly considered that as the applied force per unit area increased the porosity in the sintered compact decreases, as can be found out in Figure 3 (b). The relative density of the as-sintered alloy increases with increases in sintering temperature. As the sintering temperature increases from 350 to 450°C, relative density varies directly and increases from 96.2 to 99.56% and porosity starts to decrease from 25% at 350 OC and reaches the lowest level of 5% at a temperature of 450°C, as can be found out in Figure 3 (c). Increased sintering temperature helps the materials to fuse easily which in turn reduces the porosity and makes it denser (Zhang et al., 2010; Liu et al., 2016; Hamu et al., 2007; Zhang et al., 2008; Fu et al., 2017). The main aim of the present work is to fabricate porous alloy, so we consider the 40 MPa is the optimal condition of applied pressure for the fabrication of porous alloy with favorable density. It can be clearly seen that at 40 MPa applied pressure the porosity in the sintered compact decreases, as the sintering temperature increases from 350°C to 450°C, as can be seen in Figure 3 (d).

Figure 4 shows the optical microstructure of as-sintered Mg-Zn-Mn-Si, Mg-Zn-Mn-HA, and Mg-Zn-Mn-(Si, HA) alloys at a sintering temperature of 400 °C under 40 MPa applied pressure. The size of grains is large ~157 μm in the as-sintered Mg-Zn-Mn-Si alloy, as can be seen in the microstructure image Figure 4(a). It can be clearly seen that the grain size decreased with the addition of HA content. The grain size was about ~55 μm observed in microstructure image of Mg-Zn-Mn-HA alloy, as can be seen in Figure 4(b). On the other hand, if the Si element is added into the Mg-Zn-Mn-HA alloy, the grain size is again increased to ~68 μm, as can be seen in Figure 4(c). In all the sintered alloys, the secondary

Figure 2. Photograph of sintered alloy (a) Mg-Zn-Mn-HA, (b) Mg-Zn-Mn-Si, and (c) Mg-Zn-Mn-Si-HA at 40 MPa applied pressure and 400 °C sintering temperature

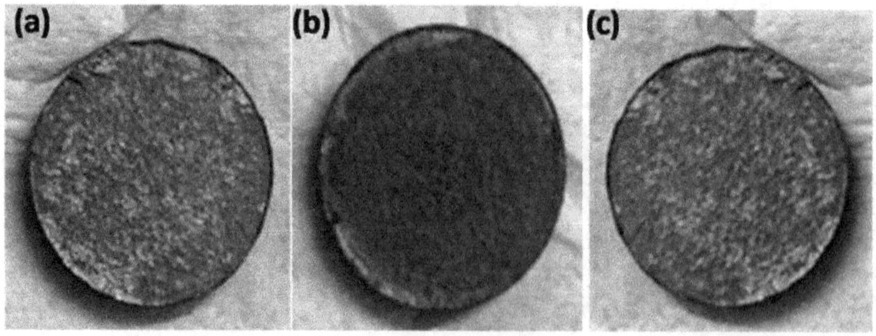

Figure 3. Variation of porosity and density with respect to applied pressure at constant temperature of 400°C

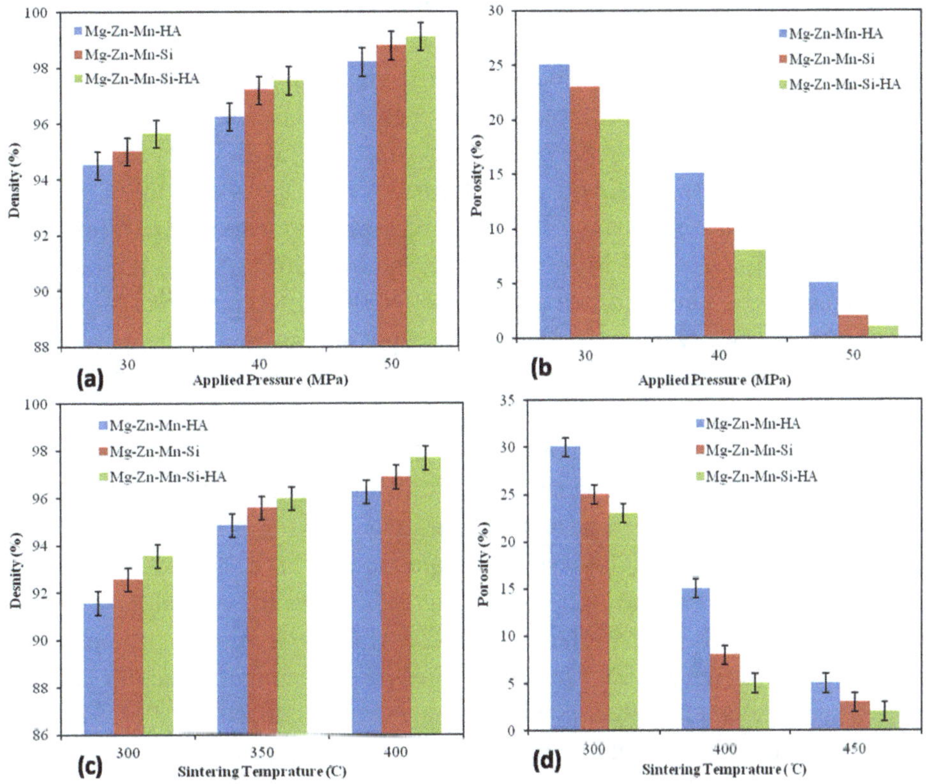

phases mainly distributed along grain boundary and some at interdendritic inner grain. Figure 5 shows the SEM micrograph and EDS spectrum of the as-fabricated Mg-Zn-Mn-Si, Mg-Zn-Mn-Ha, and Mg-Zn-Mn-Si-HA alloy sintered at 40 MPa applied pressure and 400 °C sintering temperatures. The results show that formation of porous structure. It has also reported that large amounts of gases were released due to a chemical reaction between the alloying elements that lead to the creation of porosity in the structure of the sintered composite. The Mg-Zn-Mn-Si structure typically consists of primary α-Mg dendrites and eutectoid with a large lamellar structure which is made up of α–Mg, $MgZn_2$, MnSi, and Mg_2Si phase as shown in Figs. 5 (a-b). The alloying element Zn existed with hexagonal close-packed structure and $MgZn_2$ as the secondary phase in the Mg matrix. The $MgZn_2$ phase appears as clusters of small particles which are aligned. The intermetallics Mn_5Si_3 and Mg_2Si phase often appear in a polygonal shape. The associated EDS spectrum confirmed the presence of Mg, Zn, Mn, and Si elements predominantly existed in eutectic structures. Under the high magnification, dark and grey phases were observed on Mg-matrix, as shown in Figure 5 (b). The dark phase was identified by EDS to be a $MgZn_2$ phase and the gray phase to be a Mg_2Si phase. As the hydroxyapatite (HA) was used as an alloying element instead of silicon (Si), the typical eutectic structure disappeared and needle-like MgCaO phases formed and tended to distribute along the grain boundary as shown in Figure 5 (c-d). The hydroxyapatite (HA) decomposed and formed β-tricalcium phosphate (β-TCP) with pores. Under higher magnification, the needle-like was clearly identified which covers the whole matrix. The associated EDS spectrum of the sintered alloy confirmed the presence of Mg, Zn, Mn, Ca, P, and O elements. The presence of Ca, P, and O elements confirmed that the hydroxyapatite decomposed due to the sintering temperature and formed various oxides in the structure. When Si and HA were added in the Mg-Zn-Mn, the microstructure shows different morphology. Some dark phases, grey phases, and bright phases were observed in magnesium matrix. The dark phases were identified to be Mg matrix, gray phases to be CaMgSi, and bright phase to be a Mg_2Si phase. The typical eutectic structure disappeared and needle-like MgCaO phases formed. The associate EDS spectrum confirmed the presence of Mg, Zn, Mn, Si, Ca, P, and O element. During sintering, these elements react with each other and formed various types of oxides.

The XRD pattern confirms the formation of various oxide phases on the sintered alloy. Figure 6 shows the XRD pattern of the Mg-Zn-Mn-Si, Mg-Zn-Mn-HA, and Mg-Zn-Mn-Si-HA bio-composite sintered at 40 MPa applied pressure and sintering temperature 400 °C. The XRD pattern of Mg-Zn-Mn-Si shows the formation of Mg_2Si, $Mg_{0.97}$ $Zn_{0.03}$, Mn_5Si_3, and SiO_2 possible oxides in the as-sintered alloy. The formation of Mg_2Si and SiO_2 enhances the corrosion resistance of the alloy. The XRD pattern of Mg-Zn-Mn-HA shows the formation of MgCaO, β-TCP, Mn-CaO, and CaMgZn phases. The XRD pattern of Mg-Zn-Mn-Si-HA shows the formation of CaMgSi, Mg_2Si, Mn_5Si_3, β-TCP, Mn-CaO, and CaMgZn phases. The formation of these phases is beneficial to form apatite, thus improved the bioactivity of the alloy for osseointegration between alloy and natural bone tissues (Zhang et. al., 2010; (Hamu et. al., 2008).

The elastic modulus and hardness of each sintered alloy sample have been determined from the load-displacement curves with a maximum load of 1 N. Figure 7 shows the typical loading and unloading curve of the Mg-Zn-Mn-Si, Mg-Zn-Mn-HA, and Mg-Zn-Mn-Si-HA samples. The minimum value of elastic modulus is found for the Mg-Zn-Mn-HA, whereas the maximum value of elastic modulus for Mg-Zn-Mn-Si-HA sample is relatively very high compared to the doped samples. The Mg-Zn-Mn-Si sample has an intermediate range of elastic modulus between Mg-Zn-Mn-HA and Mg-Zn-Mn-Si-HA

Figure 4. Microstructure of as-sintered alloys: (a) Mg-Zn-Mn-Si, (b) Mg-Zn-Mn-HA, and (c) Mg-Zn-Mn-Si-HA alloy

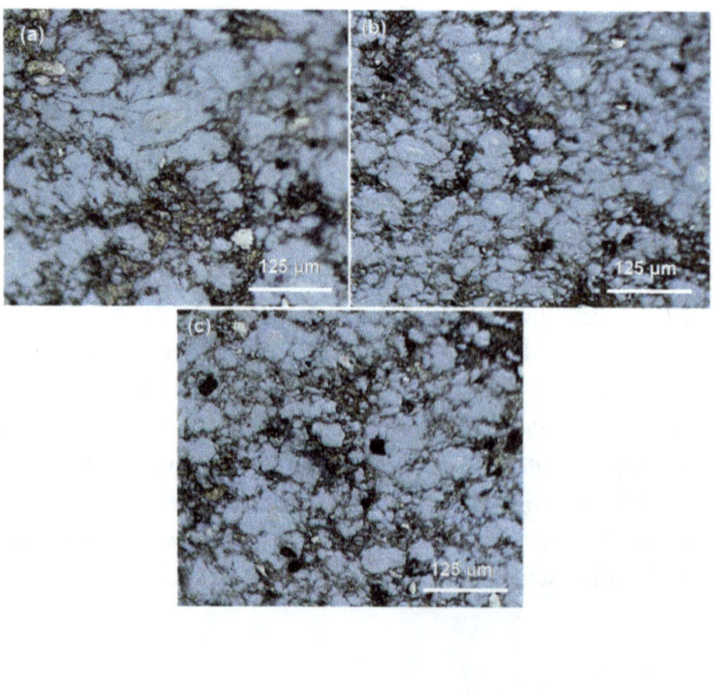

Figure 5. SEM micrograph and associated EDS spectrum of as sintered (a) Mg-Zn-Mn-Si, (b) Mg-Zn-Mn-HA, and (c) Mg-Zn-Mn-Si-HA alloy

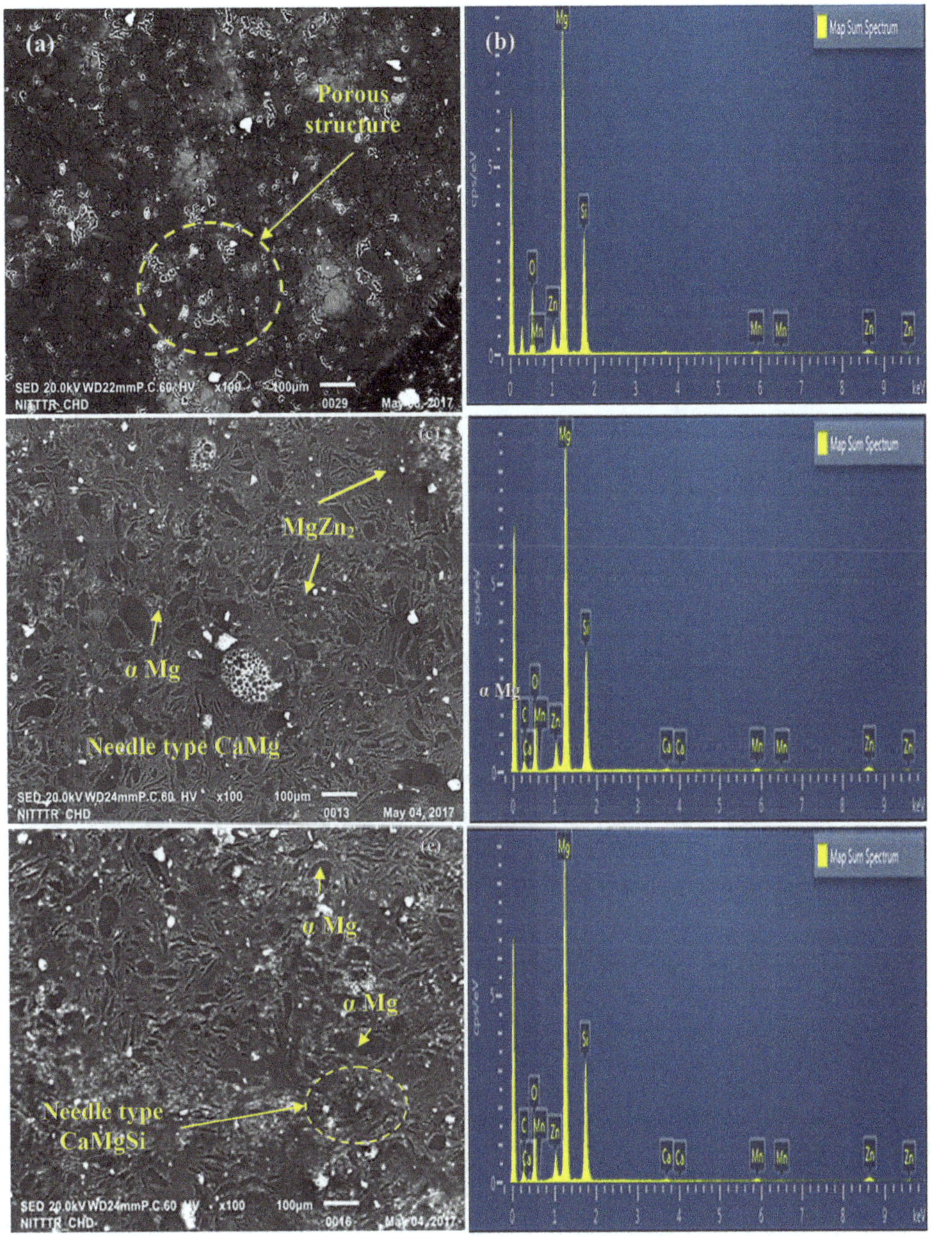

sample. The elastic modulus measured for the Mg-Zn-Mn-HA, Mg-Zn-Mn-Si, and Mg-Zn-Mn-Si-HA samples are 30 GPa, 38 GPa, and 46 GPa respectively. Table 1 shows the observed value of parameters used to determine the elastic modulus and hardness. As depicted from the plots that the slope of the unloading curve of Mg-Zn-Mn-Si-HA samples is less than the Mg-Zn-Mn-HA and Mg-Zn-Mn-Si samples. The slope of the unloading curve of Mg-Zn-Mn-HA sample is highest among all the samples. Similar

Figure 6. XRD pattern of Mg-Zn-Mn-HA, Mg-Zn-Mn-Si, Mg-Zn-Mn-Si-HA alloy sintered at 40 MPa applied pressure and sintering temperature 400 °C

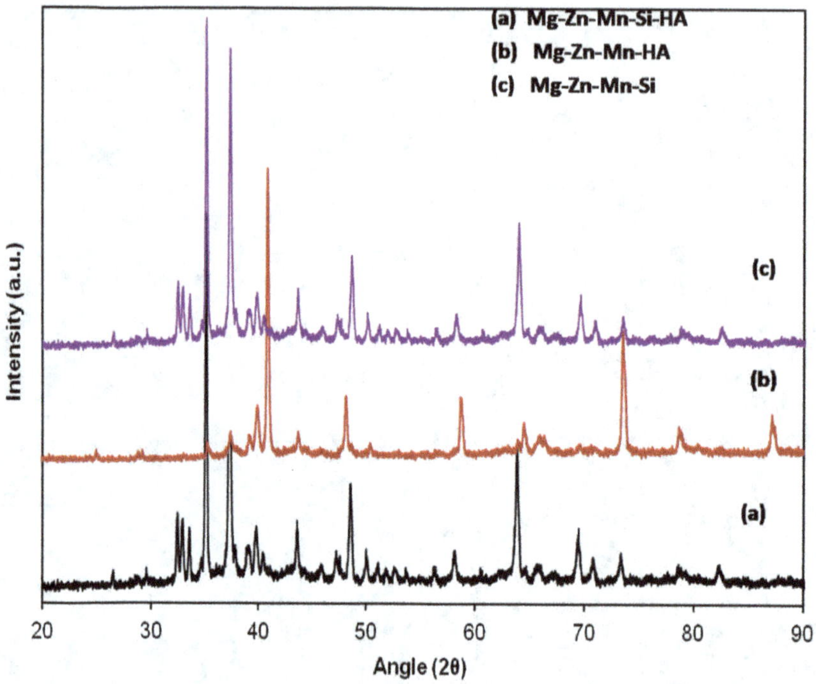

to elastic modulus, the hardness of the Mg-Zn-Mn-Si-HA samples is exceedingly high compared to the Mg-Zn-Mn-HA and Mg-Zn-Mn-Si samples. The minimum hardness value is obtained for the Mg-Zn-Mn-HA sample. It is evident from the indented image that the Mg-Zn-Mn-HA sample has a low hardness as the size of the indent is compared to the indent size of the Mg-Zn-Mn-Si and Mg-Zn-Mn-Si-HA.

CONCLUSION

Porous Mg-Zn-Mn-Si, Mg-Zn-Mn-HA, and Mg-Zn-Mn-Si-HA alloys with interconnected pore characteristics were fabricated by mechanical alloying and Spark plasma sintering technique. Results showed that >50 μm average pore size and 30%–60% porosity could be achieved by adding HA and Si with 10 weight%. The Porous Mg-Zn-Mn-(Si-HA) alloy possessed not only the low elastic modulus of 29-45 GPa (very close to that of human bone) but also high hardness (86–200 HV). Although the amount of HA and Si partially decomposed and formed secondary phases, the SPS porous Mg-Zn-Mn-(Si-HA) alloys still showed good bioactivity. The EDS spectrum and XRD analysis of sintered Mg-Zn-Mn-(Si-HA) alloys confirmed the presence of calcium (Ca) and phosphate (P) along with base alloy elements (Mg, Zn, Mn, Si, and HA) elements. The presence of Ca, P and O element confirmed and conferred the bioactivity of the alloy prone to afford bone tissue. The XRD phase composition confirms the formation of biomimetic oxide phases such as CaMg, MgSi2, Mg-Zn, Mn-Si, SiO2, Mn-CaO, Mn-P, Ca-Mn-O,

Figure 7. Nano-indentation plot and Micro-hardness measurements of Mg-Zn-Mn-HA, Mg-Zn-Mn-Si, Mg-Zn-Mn-Si-HA alloy sintered at 40 MPa applied pressure and sintering temperature 400 °C

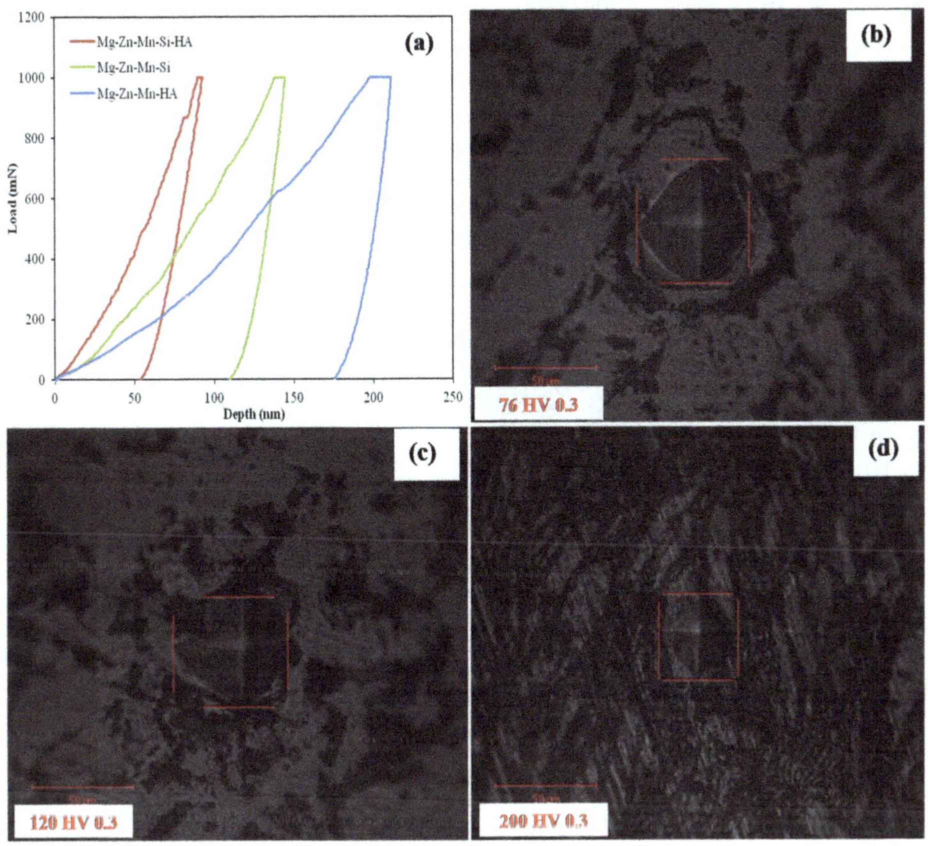

Table 1. Elastic modulus and hardness of the sintered alloy

Parameters	Mg Alloy Consolidated by Spark Plasma Sintering		
	Mg-Zn-Mn-Si	Mg-Zn-Mn-HA	Mg-Zn-Mn-Si-HA
Elastic Modulus, E (GPa)	39.75	32.42	45.87
Hardness, H (GPa)	1.18	0.75	1.97
Hardness, H (HV)	120	76	200

ZnO2, and CaMgSi in the porous layer. This is attributed due to complex reactions between HA and alloy elements occurred during the sintering process. These biocompatible phases conferred bioactivity and improved the biocompatibility of the implant. Moreover, improved the improved the corrosion resistance. However, the combination of interconnected pore characteristics, low elastic modulus, high corrosion resistance and enhanced bioactivity might make porous Mg-Zn-Mn-(Si-HA) alloys prepared by SPS a promising candidate for Orthopaedic applications.

REFERENCES

Bakhsheshi-Rad, H. R., Idris, M. H., Abdul-Kadir, M. R., Ourdjini, A., Medraj, M., Daroonparvar, M., & Hamza, E. (2014). Mechanical and bio-corrosion properties of quaternary Mg–Ca–Mn–Zn alloys compared with binary Mg–Ca alloys. *Materials & Design*, *53*, 283–292. doi:10.1016/j.matdes.2013.06.055

Bartolo, P., Kruth, J.P., Silva, J., Levy, G., Malshe, A., Rajurkar, K., Mitsuishi, M., Ciurana, J., & Leu, M. (2012). Biomedical production of implants by additive electro-chemical and physical processes. *CIRP Annals - Manufacturing Technology*, *61*, 635-655.

Bartolo, P. J., Chua, C. K., Almeida, H. A., Chou, S. M., & Lim, A. S. C. (2009). Biomanufacturing for tissue engineering: Present and future trends. *Virtual and Physical Prototyping*, *4*(4), 203–216. doi:10.1080/17452750903476288

Cui, F., Yang, J., Jiao, Y., Yin, Q., Zhang, Y., & Lee, I. S. (2008). Calcium phosphate coating on magnesium alloy for modification of degradation behaviour. *Frontiers of Materials Science in China*, *2*(2), 143–148. doi:10.100711706-008-0024-6

Fu, J., Liu, K., Duc, W., Wang, Z., Li, S., & Du, X. (2017). Microstructure and mechanical properties of the as-cast Mg-Zn-Mn-Ca alloys. *IOP Conf. Series: Materials Science and Engineering*, 182.

Geetha, M., Singh, A. K., Asokamani, R., & Gogia, A. K. (2009). Ti based biomaterials, the ultimate choice for orthopaedic implants – A review. *Progress in Materials Science*, *54*(3), 397–425. doi:10.1016/j.pmatsci.2008.06.004

Gu, X., Zhou, W., Zheng, Y., Dong, L., Xi, Y., & Chai, D. (2010). Microstructure, mechanical property, bio-corrosion and cytotoxicity evaluations of Mg/HA composites. *Materials Science and Engineering C*, *30*(6), 827–832. doi:10.1016/j.msec.2010.03.016

Hamu, G. B., Eliezer, D., & Shin, K. S. (2007). The role of Si and Ca on new wrought Mg–Zn–Mn based alloy. *Materials Science and Engineering A*, *447*(1-2), 35–43. doi:10.1016/j.msea.2006.10.059

Hassel, T., Bach, F. W., & Krause, C. (2007). Influence of alloy composition on the mechanical and electrochemical properties of binary Mg-Ca alloys and its corrosion behaviour in solutions at different chloride concentrations. *Proc. 7th Int. Conf. Magnesium Alloys and Their Applications*, 789–95.

Khalil, K. A., & Almajid, A. (2012). Effect of high-frequency induction heat sintering conditions on the microstructure and mechanical properties of nanostructured magnesium/ hydroxyapatite nanocomposites. *Materials & Design*, *36*, 58–68. doi:10.1016/j.matdes.2011.11.008

Khanra, A. K., Jung, H. C., Yu, S. H., Hong, K. S., & Shin, K. S. (2010). Microstructure and mechanical properties of Mg–HAP composites. *Bulletin of Materials Science*, *33*(1), 43–47. doi:10.100712034-010-0006-z

Lala, S., Maity, T. N., Singh, M., Biswas, K., & Pradhan, S. K. (2017). Effect of doping (Mg,Mn,Zn) on the microstructure and mechanical properties of spark plasma sintered hydroxyapatites synthesized by mechanical alloying. *Ceramics International*, *43*(2), 2389–2397. doi:10.1016/j.ceramint.2016.11.027

Li, Z., Gu, X., Lou, S., & Zheng, Y. Z. (2008). The development of binary Mg–Ca alloys for use as biodegradable materials within bone. *Biomaterials, 29*(10), 1329–1344. doi:10.1016/j.biomaterials.2007.12.021 PMID:18191191

Liu, X., Sun, J., Zhou, F., Yang, Y. H., Chang, R., Qiu, K., ... Zheng, Y. (2016). Micro-alloying with Mn in Zn–Mg alloy for future biodegradable metals application. *Materials & Design, 94*, 95–104. doi:10.1016/j.matdes.2015.12.128

Niinomi, M., Nakai, M., & Hieda, A. (2012). Development of new metallic alloys for biomedical applications. *Acta Biomaterialia, 8*(11), 3888–3903. doi:10.1016/j.actbio.2012.06.037 PMID:22765961

Prakash, C., Kansal, H. K., Pabla, B. S., & Puri, S. (2015). Processing and characterization of novel biomimetic nanoporous bioceramic surface on β-Ti implant by powder mixed electric discharge machining. *Journal of Materials Engineering and Performance, 24*(9), 3622–3633. doi:10.100711665-015-1619-6

Prakash, C., Kansal, H. K., Pabla, B. S., & Puri, S. (2016). Powder Mixed Electric Discharge Machining an Innovative Surface Modification Technique to Enhance Fatigue Performance and Bioactivity of β-Ti Implant for Orthopaedics Application. *Journal of Computing and Information Science in Engineering, 14*, 1–9.

Prakash, C., Kansal, H. K., Pabla, B. S., Puri, S., & Aggarwal, A. (2016). Electric discharge machining-a potential choice for surface modification of metallic implants for orthopedics applications: A review. *Proceedings of the Institution of Mechanical Engineers. Part B, Journal of Engineering Manufacture, 230*(2), 231–253. doi:10.1177/0954405415579113

Prakash, C., & Uddin, M.S. (2017). Surface modification of β-phase Ti implant by hydroaxyapatite mixed electric discharge machining to enhance the corrosion resistance and in-vitro bioactivity. *Surface and Coatings Technology, 236*(Part A), 134-145.

Salahshoor, M., & Guo, Y. B. (2013). Process mechanics in ball burnishing biomedical magnesium–calcium alloy. *International Journal of Advanced Manufacturing Technology, 64*(1-4), 133–144. doi:10.100700170-012-4024-4

Staiger, M. P., Pietak, A. M., Huadmai, J., & Dias, G. (2006). Magnesium and its alloys as orthopedic biomaterials: A review. *Biomaterials, 27*(9), 1728–1734. doi:10.1016/j.biomaterials.2005.10.003 PMID:16246414

Staiger, M. P., Pietak, A. M., Huadmai, J., & Dias, G. (2006). Magnesium and its alloys as orthopedic biomaterials: A review. *Biomaterials, 27*(9), 1728–1734. doi:10.1016/j.biomaterials.2005.10.003 PMID:16246414

Sun, J., Chen, M., Cao, G., Bi, Y., Liu, D., & Wei, J. (2013). The effect of nano-hydroxyapatite on the microstructure and properties of Mg–3Zn–0.5Zr alloy. *Journal of Composite Materials, 48*(7), 825–834. doi:10.1177/0021998313478259

Sunil, B. R., Ganapathy, C., Kumarn, T. S. S., & Chakkingal, U. (2014). Processing and mechanical behavior of lamellar structured degradable magnesium hydroxyapatite implants. *Journal of the Mechanical Behavior of Biomedical Materials, 40*, 178–189. doi:10.1016/j.jmbbm.2014.08.016 PMID:25241282

Uddin, M. S., Hall, C., & Murphy, P. (2015). Surface treatments for controlling corrosion rate of biodegradable Mg and Mg-based alloy implants. *Science and Technology of Advanced Materials*, *16*(5), 053501. doi:10.1088/1468-6996/16/5/053501 PMID:27877829

Uddin, M. S., Rosman, H., Hall, C., & Murphy, P. (2017). Enhancing the corrosion resistance of biodegradable Mg-based alloy by machining-induced surface integrity: Influence of machining parameters on surface roughness and hardness. *International Journal of Advanced Manufacturing Technology*, *90*(5-8), 2095–2108. doi:10.100700170-016-9536-x

Vaccaro, A. R., Singh, K., Haid, R., Kitchel, S., Wuisman, P., Taylor, W., ... Garfin, S. (2003). The use of bioabsorbable implants in the spine. *The Spine Journal*, *3*(3), 227–237. doi:10.1016/S1529-9430(02)00412-6 PMID:14589204

Vahidgolpayegani, Wen, Hodgson, & Li. (2017). *Production methods and characterization of porous Mg and Mg alloys for biomedical applications*. Woodhead Publishing.

Viswanathan, R., Rameshbabu, N., Kennedy, S., Sreekanth, D., Venkateswarlu, K., Sandhya, R. M., & Muthupandi, V. (2013). Plasma electrolytic oxidation and characterization of spark plasma sintered magnesium/hydroxyapatite composites. *Materials Science Forum*, *765*, 827–831. doi:10.4028/www.scientific.net/MSF.765.827

Witte, F., Feyerabend, F., Maier, P., Fischer, J., Stormer, M., Blawert, C., ... Hort, N. (2007). Biodegradable magnesium–hydroxyapatite metal matrix composites. *Biomaterials*, *28*(13), 2163–2174. doi:10.1016/j.biomaterials.2006.12.027 PMID:17276507

Xu, L. P., Yu, G. N., Zhang, E. L., Pan, F., & Yang, K. (2007). In vivo corrosion behavior of Mg–Mn–Zn alloy for bone implant application. *Journal of Biomedical Materials Research*, *83A*(3), 703–711. doi:10.1002/jbm.a.31273 PMID:17549695

Xu, L. P., Yu, G. N., Zhang, E. L., Pan, F., & Yang, K. (2008). In vitro corrosion behavior of Mg alloys in a phosphate buffered solution for bone implant application. *Journal of Materials Science. Materials in Medicine*, *19*(3), 1017–1025. doi:10.100710856-007-3219-y PMID:17665099

Ye, X., Chen, M., Yang, M., Wei, J., & Liu, D. (2010). In vitro corrosion resistance and cytocompatibility of nano-hydroxyapatite reinforced Mg–Zn–Zr composites. *Journal of Materials Science. Materials in Medicine*, *21*(4), 1321–1328. doi:10.100710856-009-3954-3 PMID:20012772

Zhang, E., & Yang, L. (2008). Microstructure, mechanical properties and bio-corrosion properties of Mg–Zn–Mn–Ca alloy for biomedical application. *Materials Science and Engineering A*, *497*(1-2), 111–118. doi:10.1016/j.msea.2008.06.019

Zhang, E., Yang, L., Xu, J., & Chen, H. (2010). Microstructure, mechanical properties and bio-corrosion properties of Mg–Si(–Ca, Zn) alloy for biomedical application. *Acta Biomaterialia*, *6*(5), 1756–1762. doi:10.1016/j.actbio.2009.11.024 PMID:19941979

Zhang, E. L., He, W. W., Du, H., & Yang, K. (2008). Microstructure, mechanical properties and corrosion properties of Mg–Zn–Y alloys with low Zn content. *Materials Science and Engineering A*, *488*(1-2), 102–111. doi:10.1016/j.msea.2007.10.056

Zhang, E. L., & Yang, L. (2008). Microstructure, mechanical properties and bio-corrosion properties of Mg–Zn–Mn–Ca alloy for biomedical application. *Materials Science and Engineering A*, *497*(1-2), 111–118. doi:10.1016/j.msea.2008.06.019

Zhang, E. L., Yin, D. S., Xu, L. P., Yang, L., & Yang, K. (2009). Microstructure, mechanical and corrosion properties and biocompatibility of Mg–Zn–Mn alloys for biomedical application. *Materials Science and Engineering C*, *29*(3), 987–993. doi:10.1016/j.msec.2008.08.024

Zhao, J., Yu, Z., Yu, K., & Chen, L. (2011). Biodegradable behaviors of Mg–6%Zn–5%Hydroxyapatite biomaterial. *Advanced Materials Research*, *239–242*, 1287–1291. doi:10.4028/www.scientific.net/AMR.239-242.1287

Zheng, B., Ertorer, O., Li, Y., Zhou, Y., Mathaudhu, S. N., Tsao, C. Y. A., & Laverni, J. E. (2011). High strength, nano-structured Mg–Al–Zn alloy. *Materials Science and Engineering A*, *528*(4-5), 2180–2191. doi:10.1016/j.msea.2010.11.073

Chapter 16
Assessment of Remanufacturability Index for an Automotive Product: A Case Study

Jayakrishna Kandasamy
VIT University, India

Shama M. S.
Trinity College of Engineering, India

Aravind Raj Sakthivel
VIT University, India

Sultan M. T. H.
University of Putra Malaysia, Malaysia

Vimal K. E. K.
National Institute of Technology Patna, India

V. Sharath Kumar Reddy
VIT University, India

Jyoteesh Gutha
VIT University, India

ABSTRACT

The increasing competition among the manufacturing organizations and stringent government regulation forces the manufacturing organizations to implement sustainability principles in manufacturing. Sustainability focuses on material, product development, and manufacturing process orientations. End of life (EoL) disposal of the product is very much important in the modern scenario. The remanufacturing is a vital strategy for attaining sustainability in manufacturing. The assessment of remanufacturability of products needs to be done during the design stage so as to provide the manufacturers the guidelines for sustainable product development. In this context, this chapter presents the insights on remanufacturability index assessment for a typical automotive product. The practical implications of the study are also being discussed.

DOI: 10.4018/978-1-5225-5445-5.ch016

INTRODUCTION

Modern manufacturing industries strive to attain the goals of sustainable development. Remanufacturing is one of the best strategies to attain sustainability because the requirement of energy and materials are less when compared to new manufacture and it can be transformed into cost savings (Amaya et al. 2010 ; Heese et al. 2005). Remanufacturing can be defined as ''the process of rebuilding a product, during which the product is cleaned, inspected and disassembled; defective components are replaced; and the product is reassembled, tested and inspected again to ensure it meets or exceeds newly manufactured product standards'' (Sundin and Bras 2005). The assessment of remanufacturability of products in the design stage gives information about the ability of the product to be remanufactured and guidelines for improving the product design features. In this study, the Remanufacturability Index (RI) of an automotive product is assessed based on Disassembly Index, Cleaning and Inspection Index (ICI) and the State Index (SI) which is determined based on the state of the components after usage period (Bras and Hammond 1996) and it helps to decide whether to go for remanufacturing of a particular used product or not. This study extended the principles of Dixit (2006) to evaluate the remanufacturing ability of an automotive product. The study identified cost consumed for remanufacturability operations of the selected automotive product and various factors influencing these operations. The unique aspect of the study is that it performs component analysis and classification of components into different categories to determine the necessary remanufacturing operations and resource consumption.

LITERATURE REVIEW

The literature review was reviewed from the perspective of remanufacturing and its assessment.

Review on Remanufacturing

Kaustov et al. (2017), study concentrated to recognize the outline design criteria of an item that can upgrade its remanufacturability and hence, utilizing these outline design criteria a hierarchical model is created to assess the remanufacturability between the alternate products. The weights of the design criteria and the sub criteria is computed by the Analytical Hierarchy process (AHP) integrated with fuzzy analysis is used then axiomatic design approach (AD) is used to evaluate all the alternatives.

Qin et al. (2016), presented a new method for finding the optimum active remanufacturing time which involves the both economic and environmental indicators. And they derived the optimum active remanufacturing time for oil cylinder by considering the cost index and environmental index it came around 6.25 years of successful operation.

Mahdi et al. (2014), this study explains that the returned products can be processed in two ways through remanufacturing, first way uses the dedicated capacity in which returns are remanufactured and other by using merged capacity returns from two different markets are remanufactured. Analytical queueing models are used for the admission decision, which conclude on acceptance. The above model explains the concluded threshold value in order to increase the total profit of remanufacturing system. It also allows to know the collaboration between overall utilization and the arrival process.

Farazee et al. (2016), the study reveals the introduction of new tool multi method simulation which is used to measure the economic and environmental performance of circular product systems. This is the first ever tool for measuring the economic and environmental performance of circular product systems which is also useful in designing simulation dynamic model. It forecast's system's behaviour and also shows the effect on economic and environmental performance of circular product systems.

Ardeshir et al. (2017), study gives new sorting method on both product external (market trends) and internal factors (product identity data, future reusability of components, product health status). Main purpose of this method is to better the decision making operations in remanufacturing of a product by integrating end of life, end of use and optimal sorting policies.

Seitz (2007) studied about the driving forces of product recovery and remanufacturing in automotive industry. The author discussed that the primary motive for remanufacturing included ethical and moral responsibility, product take-back and recovery legislation, as well as the economic benefits of remanufacturing. The author found that the remanufactured products provided a valuable source of spare parts and under-warranty engines for the vehicle manufacturers. The market share, brand protection and customer orientation were found as more dominant reasons for undertaking the recovery operation than the primary motives for remanufacturing with respect to a Vehicle Manufacturer.

White et al. (2003) suggested that the product recovery was an environmentally desirable substitute to disposal, and illustrated the recovery of products to frame an environmental research agenda for recovery management. They pointed out that remanufacturing will reduce the environmental impacts as well as cost of manufacturing processes.

Ijomah et al. (2007) discussed about the influence of product features in remanufacturability of the products. They conducted workshops and categorized the products into repairable, reconditionable and remanufacturable states. They identified that the major issue for design for remanufacture is that neglecting the features that prevent the product brought back.

Subramoniam et al. (2011) developed a Remanufacturing Decision Making Framework (RDMF) and it was validated for an automotive industry. They identified the important factors to make valid remanufacturing decisions in the conceptual stage of product development. They concluded that the RDMF would help to incorporate environmentally conscious thinking in the remanufacturing process in automotive industries.

Rathore et al. (2011) studied about the remanufacturing activities prevailed in Indian markets to attain sustainability. They conducted a study in an Indian mobile phone market to investigate the acceptance of remanufactured products among the customers. According to them, the markets lack a clear strategy for implementing the remanufacturing procedure.

Review on Assessment of Remanufacturing

Bras and Hammond (1996) established metrics based on the product design characteristics to effectively evaluate the remanufacturability of the products in the design stage. The metrics can only be effectively used during or after the embodiment phase of product design.

Ferrer (2001) suggested a new framework for remanufacturing procedure for products based on two design metrics like disassemblability and reusability. The developed framework aids to balance the motives of product recovery and the difficulties to obtain the relevant information about components to be remanufactured.

Liu et al. (2002) pointed out the principles and technologies for design for recyclability and they introduced a method to assess the recyclability in the design stage using artificial neural network (ANN) technique. They have stressed that the easiness of recyclability greatly depend upon the design characteristics of the products. The assessment of recyclability would direct towards green and environmentally conscious design.

Desai and Mital (2003) studied about disassemblability of a product and they proposed methodology to improve the disassemblability. They identified the factors relevant for the non destructive process of disassembling; numerical values were assigned to the factors according to the varying levels of difficulties to provide quantitative design information. They remarked that the scoring system can be used for any kind of disassembling operations and tools.

Villalba et al. (2004) pointed out that the potential for recycling the material should be included in the assessment and optimization of disassembly and design for disassembly. They concluded that the profit to loss margin of recycling determines the economical feasibility to disassemble the product to recover the materials that can be recycled.

Dixit (2006) proposed a methodology to numerically calculate the Remanufacturability Index of the products by considering all possible after life scenario of the products. They performed a case study to find the remanufacturing potential of the electric staple gun and they found the factors which influence the remanufacturability index of the particular product.

Mathieux et al. (2008) introduced a new method by analyzing the strength and weakness of earlier methods to improve product design based on the recovery aspects. Multi criteria and multi scenario of recoverability were assessed during the product development stage. A case study was conducted to evaluate the multi criteria recoverability of Television set and concluded that the recoverability of the products decidedly depends upon the scenarios which were considered.

Du et al. (2012) introduced an integrated method for evaluating the remanufacturability of used machine tool to analyze the technology feasibility, economic feasibility and environmental benefits of machine tool remanufacturing. In this method, the weight of each index is determined using Analytical Hierarchy Process. They conducted a case study and applied the method to the remanufacturing process of used planer B2025 to examine its feasibility and validity.

METHODOLOGY TO FIND RE-MANUFACTURABILITY INDEX

The methodology followed is shown in figure 1. Involves selection of candidate product, understanding the product features and splitting into appropriate modules and components. The state of the component is identified and Remanufacturability Index of components is calculated using relevant data. The RI of each module is determined and the overall product RI is calculated. This is followed by derivation of end of life disposal activities. Though remanufacturing is a well-researched topic many authors have concentrated on various elements of remanufacturing individually. But there exist a need for comprehensive remanufacturability assessment and also the assessment during the design stage. This study has been done to fulfill these aspects.

Figure 1.

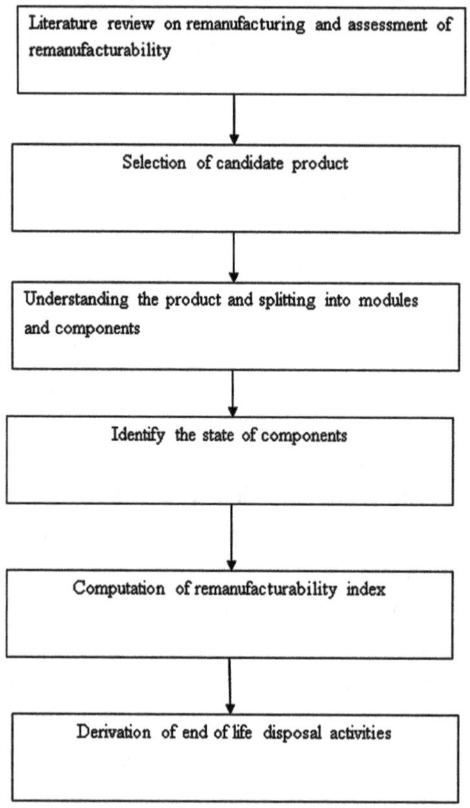

CASE STUDY

The case study was conducted in an Indian automotive industry. The case organization is the manufacturer of hydraulic power rack and pinion steering gear assembly, integral power steering and power steering pump assembly. The company aspires to enhance environmentally friendliness in their product design and development practices. The details of the case study are presented in the following subsections.

Understanding the Automotive Component

The selected automotive product is disassembled into four modules and the modules are splitted into components. The first module M1 (Nylone sleeve assembly) consists of Nylone sleeve (C1) and Head screw (C2). The second module M2 (Wheel end assembly) includes Wheel end (C3), Bush (C4), Rubber bush (C5), external circlip (C6), Spacer (C7) and lock spacer (C8). The accessories like protector serrated end (C9), protector wheel end (C10) and Nut (C11) form the third module (M3). The final module (M4) is outer tube assembly. It consists of outer tube (C12) and protector outer tube (C13).

Figure 2.

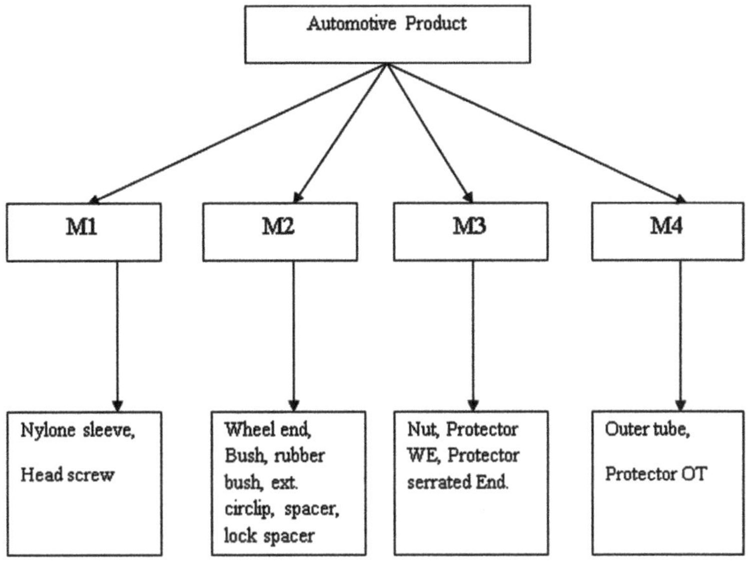

The case study extended the principles of Dixit (2006) to evaluate the remanufacturability. Remanufacturability Index (RI) of components consists of three indices namely Disassembly Index (DI), Inspection and Cleaning Index (ICI) and State Index (SI). Out of which, the first two indices are base indices. It is necessary for all the components. The SI is determined by state of the components which is classified into reusable, repairable, reconditionable, recyclable, and scrap (environmental). After disassembling, cleaning and inspection the state of the components is determined. It is found that the wheel end is repairable, the bush is reconditionable and the rubber bush is scrap. All other components fall under reusable category.

Calculation of RI of Components

DI and ICI are determined based on the cost of disassembly (DC), cost of cleaning and inspection (CIC) and original cost of the components (OCC). The Repairability Index (RPI) and Reconditionability Index (RCI) are calculated using cost of repair and cost of recondition. Environmental Index (EI) is determined by cost of dumping the component. Reusability index (RUI) depends on estimated worth of the component (EWC) and the OCC. The input data are collected and are shown in Table 1.

Cost of repair for C3, cost of recondition for C4, cost of dumping for C5 are 28.571, 5.229, and 0.333 respectively. DI, ICI, SI are calculated using the following equations (Dixit 2006 and Ferror 2001).

$$DI = \frac{DC}{OCC} \tag{1}$$

$$ICI = \frac{CIC}{OCC} \tag{2}$$

Table 1. Input data

Component	Quantity	OCC	DC	CIC	State of Component	EWC
C1	1	3.5	0.7	0.175	Reuse	3.02
C2	1	0.5	0.05	0.012	Reuse	0.431
C3	1	50	15	3.75	repair	43.143
C4	2	9.15	0.458	0.114	recondition	7.895
C5	2	3.325	0.332	0.083	environmental	2.869
C6	1	2.15	0.215	0.054	reuse	1.855
C7	1	5.75	0.575	0.144	reuse	4.962
C8	1	2	0.1	0.025	reuse	1.726
C9	1	3.75	0.375	0.094	reuse	3.236
C10	1	2	0.1	0.025	reuse	1.73
C11	1	1	0.05	0.012	reuse	0.863
C12	1	60	18	4.5	reuse	51.771
C13	1	2	0.3	0.075	reuse	1.726

$$RPI = \frac{RPC}{OCC} \tag{3}$$

$$RCI = \frac{RCC}{OCC} \tag{4}$$

$$EI = \frac{EC}{OCC} \tag{5}$$

$$RUI = \frac{EWC}{OCC} \tag{6}$$

All indices except reusability index should be as minimum as possible. To combine the individual indices, effective DI, ICI, RPI, RCI, EI are calculated by subtracting from one. All the indices do not influence the remanufacturability of the components equally. Weight of each index is determined based on influence factors, which depends upon the individual components and the base factors. The base factors considered for the study are cost, time, other resource used. An excerpt of the factors considered for determining the weight and the rating against base factors of Nylon sleeve and Wheel end are shown in Tables 2-7.

The weight of each index is taken as the total score obtained. The relative weight (RW) of each index is then determined using the following formula (Dixit, 2006).

Table 2. Determination of DI weight for Nylone sleeve

	Cost	Time	Other Resources	Total
Component disconnection	2	2	2	6
Design complexity	2	1	2	5
Functional complexity	2	2	2	6
Total	6	5	6	17

Table 3. Determination of ICI weight for Nylone sleeve

	Cost	Time	Other Resources	Total
Inspection method	3	3	3	9
Cleaning method	2	2	2	6
safety of cleaning	3	3	3	9
Total	8	8	8	24

Table 4. Determination of RUI weight for Nylon sleeve

	Cost	Time	Other Resources	Total
Technology cycle	2	2	2	6
Wear rate	2	2	2	6
Obsolescence rate	2	2	2	6
Total	6	6	6	18

Table 5. Determination of DI weight for Wheel end

	Cost	Time	Other Resources	Total
Component disconnection	2	2	2	6
Design complexity	2	2	2	6
Functional complexity	1	1	1	3
Total	5	5	5	15

Table 6. Determination of ICI weight for Wheel end

	Cost	Time	Other Resources	Total
Inspection method	1	1	1	3
Cleaning method	2	2	2	6
safety of cleaning	2	2	2	6
Total	5	5	5	15

Table 7. Determination of RPI weight for Wheel end

	Cost	Time	Other Resources	Total
Special set up required	1	1	1	3
Design complexity	2	2	2	6
Functional complexity	1	1	1	3
Total	4	4	4	12

$$RWDI = \frac{Weight\,DI}{Weight\,DI + Weight\,ICI + Weight\,SI} \tag{7}$$

$$RWICI = \frac{Weight\,DI}{Weight\,DI + Weight\,ICI + Weight\,SI} \tag{8}$$

$$RWSI = \frac{Weight\,SI}{Weight\,DI + Weight\,ICI + Weight\,SI} \tag{9}$$

The RI of the components is calculated using inverse weight addition method (Bras and Hammond 1996).

$$RI = \frac{1}{\frac{RWDI}{EDI} + \frac{RWICI}{EICI} + \frac{RWSI}{ESI}} \tag{10}$$

The results are shown in Table 8.

The module RI is determined by taking the average of component RI. The contribution of each module in the product is different. The weight of each module is obtained from the design experts of the case organization.

RESULTS AND DISCUSSIONS

The RI of the automotive product is calculated using the following formula.

$$RI = \frac{1}{\frac{weight\,of\,M1}{RI\,of\,M1} + \frac{weignt\,of\,M2}{RI\,of\,M2} + .. + \frac{weight\,of\,Mn}{RI\,of\,Mn}} \tag{11}$$

Table 8. RI of components

	DI	ICI	EDI	EICI	ESI	RWDI	RWICI	RWSI	RI
c1	0.20	0.050	0.80	0.950	0.862	0.29	0.41	0.30	0.87584
c2	0.10	0.025	0.90	0.975	0.862	0.33	0.33	0.33	0.919152
C3	0.30	0.075	0.70	0.925	0.571	0.36	0.36	0.28	0.717627
C4	0.05	0.012	0.95	0.987	0.570	0.37	0.37	0.26	0.820264
C5	0.10	0.025	0.90	0.975	0.899	0.37	0.37	0.26	0.926324
C6	0.10	0.025	0.90	0.975	0.863	0.30	0.40	0.30	0.91641
C7	0.10	0.025	0.90	0.975	0.863	0.35	0.30	0.35	0.907323
C8	0.05	0.012	0.95	0.988	0.863	0.35	0.30	0.35	0.927833
C9	0.10	0.025	0.90	0.975	0.862	0.29	0.38	0.33	0.913757
C10	0.05	0.012	0.95	0.987	0.863	0.28	0.36	0.36	0.928985
C11	0.05	0.012	0.95	0.987	0.863	0.28	0.36	0.36	0.928985
C12	0.30	0.075	0.70	0.925	0.862	0.31	0.38	0.31	0.824441
C13	0.15	0.038	0.85	0.963	0.863	0.30	0.35	0.35	0.891155

Where n is the number of modules in the selected assembly. It was found that the RI of the candidate product is 8.75. The potential for remanufacture of the selected automotive product is very high. The module RI and product RI are shown in Table 9.

RI of wheel end, phosphor bronze bush, and outer tube is less compared to that of other components. The wheel end needs repair that consumes additional cost and energy. The main damage found in wheel end is bending of the wheel due to the applied torque during worst condition. The strategy for repair is straightening the wheel end after inspection. The effective DI is also low for the component because the disassembling cost is high. Recondition is needed for bush before reusing because wear and tear is the main problem for phosphor bronze bushes. Usage of special protection covering is one of the strategies to protect these bushes from damage. For outer tube, the disassembly cost is high that reduces the overall RI. Most of the components of the product fall under reusable category. After disassembly, detailed inspection and testing, they can be reused for new manufacture or as a source of spare parts.

CONCLUSION

Modern manufacturing industries are trying to implement the principles of sustainability to ensure social, environmental benefits along with economic profit. Remanufacturing of old products and components will give less cost substitute for the new products. The assessment of potential for remanufacturing of the products provides guidance to the manufacturers to take decision about remanufacturing. It also gives ideas about necessary changes required for remanufacturing in the design stage of products. In this study, remanufacturing assessment starts with disassembly and detailed inspection of the automotive product and its components. The components are classified into reusable, repairable, reconditionable and environmental based on the state of the components. RI of the product is determined based on the cost requirement of these remanufacturing operations. The weight of the indices is determined based on the factors affecting these remanufacturing processes. The RI of the product is found to be very high.

Remanufacturing of the products helps to generate more economic profit because the products consume less resources and cost. Also it reduces environmental impact. The limitation of the study is that the method attaches importance to cost requirements of remanufacturing operations than the other influencing factors. The remanufactured products are always cheaper than the new manufacture but customer satisfaction may be less due to lack of product differentiation. In future, more number of studies could be performed for wide range of automotive products; also newer indices for remanufacturing also could be developed.

Table 9. RI of automotive product

Module	Module RI	Weight of Module	Product RI
M1	0.897	0.15	0.875
M2	0.870	0.45	
M3	0.924	0.1	
M4	0.858	0.3	

REFERENCES

Amaya, J., Zwolinski, P., & Brissaud, D. (2010). Environmental Benefits of Parts Remanufacturing: The Truck Injector Case. *17th CIRP International Conference on Lifecycle Engineering*.

Asif, F. M., Lieder, M., & Rashid, A. (2016). Multi-method simulation based tool to evaluate economic and environmental performance of circular product systems. *Journal of Cleaner Production, 139*, 1261–1281. doi:10.1016/j.jclepro.2016.08.122

Bras, B., & Hammond, R. (1996). Towards Design for Remanufacturing –Metrics for Assessing Remanufacturability. *Proceedings of 1st international workshop on reuse*, 5-22.

Chakraborty, K., Mondal, S., & Mukherjee, K. (2017). Analysis of product design characteristics for remanufacturing using Fuzzy AHP and Axiomatic Design. *Journal of Engineering Design, 28*(5), 338–368. doi:10.1080/09544828.2017.1316014

Desai, A., & Mital, A. (2003). Evaluation of disassemblability to enable design for disassembly in mass production. *International Journal of Industrial Ergonomics, 32*(4), 265–281. doi:10.1016/S0169-8141(03)00067-2

Dixit, S.B. (2006). *Product design: A conceptual development of product remanufacturability index.* Academic Press.

Du, Y., Cao, H., Liu, F., Li, C., & Chen, X. (2012). An integrated method for evaluating the remanufacturability of used machine tool. *Journal of Cleaner Production, 20*(1), 82–91. doi:10.1016/j.jclepro.2011.08.016

Fathi, M., Zandi, F., & Jouini, O. (2015). Modeling the merging capacity for two streams of product returns in remanufacturing systems. *Journal of Manufacturing Systems, 37*, 265–276. doi:10.1016/j.jmsy.2014.08.006

Ferrer, G. (2001). Theory and Methodology On the widget of remanufacturing operation. *European Journal of Operational Research, 135*(2), 373–393. doi:10.1016/S0377-2217(00)00318-0

Heese, H., Cattani, K., Ferrer, G., Gilland, W., & Roth, A. (2005). Competitive advantage through take-back of used products. *European Journal of Operational Research, 164*(1), 143–157. doi:10.1016/j.ejor.2003.11.008

Ijomah, W., McMahon, C., Hammond, G., & Newman, S. (2007). Development of robust design-for-remanufacturing guidelines to further the aims of sustainable development. *International Journal of Production Research, 45*(18), 4513–4536. doi:10.1080/00207540701450138

Ijomah, W. L., McMahon, C. A., Hammond, G. P., & Newman, S. T. (2007). Development of design for remanufacturing guidelines to support sustainable manufacturing. *Robotics and Computer-integrated Manufacturing, 23*(6), 712–719. doi:10.1016/j.rcim.2007.02.017

Liu, Z. F., Liu, X. P., Wang, S. W., & Liu, G. F. (2002). Recycling strategy and a recyclability assessment model based on an artificial neural network. *Journal of Materials Processing Technology, 129*(1-3), 500–506. doi:10.1016/S0924-0136(02)00625-8

Mashhadi, A. R., & Behdad, S. (2017). Optimal sorting policies in remanufacturing systems: Application of product life-cycle data in quality grading and end-of-use recovery. *Journal of Manufacturing Systems*, *43*, 15–24. doi:10.1016/j.jmsy.2017.02.006

Mathieux, F., Froelich, D., & Moszkowicz, P. (2008). ReSICLED: A new recovery-conscious design method for complex products based on a multicriteria assessment of the recoverability. *Journal of Cleaner Production*, *16*(3), 277–298. doi:10.1016/j.jclepro.2006.07.026

Rathore, P., Kota, S., & Chakrabarti, A. (2011). Sustainability through remanufacturing in India: A case study on mobile handsets. *Journal of Cleaner Production*, *19*(15), 1709–1722. doi:10.1016/j.jclepro.2011.06.016

Seitz, M. A. (2007). A critical assessment of motives for product recovery: The case of engine remanufacturing. *Journal of Cleaner Production*, *15*(11-12), 1147–1157. doi:10.1016/j.jclepro.2006.05.029

Subramoniam, R., Huisingha, D., Chinnam, R. B., & Subramoniam, S. (2011). Remanufacturing Decision-Making Framework (RDMF): Research validation using the analytical hierarchical process. *Journal of Cleaner Production*, 1–9.

Sundin, E., & Bras, B. (2005). Making functional sales environmentally and economically beneficial through product remanufacturing. *Journal of Cleaner Production*, *13*(9), 913–925. doi:10.1016/j.jclepro.2004.04.006

Villalba, G., Segarra, M., Chimenos, J. M., & Espiell, F. (2004). Using the recyclability index of materials as a tool for design for disassembly. *Ecological Economics*, *50*(3-4), 195–200. doi:10.1016/j.ecolecon.2004.03.026

White, C. D., Masanet, E., & Meisner, C. (2003). Product recovery with some byte: An overview of management challenges and environmental consequences in reverse manufacturing for the computer industry. *Journal of Cleaner Production*, *11*(4), 445–458. doi:10.1016/S0959-6526(02)00066-5

Xiang, Q., Zhang, H., Jiang, Z., Zhu, S., & Yan, W. (2016). A Decision-making Method for Active Remanufacturing Time Based on Environmental and Economic Indicators. *International Journal of Online Engineering*, *12*(12).

KEY TERMS AND DEFINITIONS

CIC: Cleaning and inspection cost.
DC: Disassembly cost.
DI: Disassembly index.
EDI: Effective disassembly index.
EI: Environmental index.
EICI: Effective inspection and cleaning index.
ESI: Effective state index.
EWC: Estimated worth of the component.
ICI: Inspection and cleaning index.

OCC: Original cost of components.
RCI: Reconditionability index.
RI: Remanufacturability index.
RPI: Repairability index.
RUI: Reusability index.
RWDI: Relative weight of disassembly index.
RWICI: Relative weight of inspection and cleaning index.
RWSI: Relative weight of state index.
SI: State index.

Chapter 17
Application of Cluster Analysis for Identifying Potential Automotive Organizations Towards the Conduct of Green Manufacturing Sustainability Studies

Jayakrishna Kandasamy
VIT University, India

Vimal K. E. K.
National Institute of Technology Patna, India

Aravind Raj Sakthivel
VIT University, India

V. Sharath Kumar Reddy
VIT University, India

Babulal K. S.
Dire Dawa University, Ethiopia

ABSTRACT

Increasing legislative concerns and rapidly transforming technologies pressurizes the global competitive landscape to deploy smart, safe, and sustainable green manufacturing. This chapter scrutinizes organizational sustainability of the automobile components manufacturing organizations located in the state of Tamil Nadu, India using hierarchy cluster analysis towards setting up a benchmark on sustainability of organizations. Along with the triple bottom line (TBL) of sustainable development, the organizational responsibility and government legislation in achieving sustainability were selected as the five major governing variables during the conduct of this case study. As a result, 25 automotive components manufacturing organizations chosen from for this study were classified into three clusters, confirming a particular organization as the most suitable one for the conduct of green manufacturing sustainability studies. According to the distinctiveness of the assorted clusters, suggestions were also proposed for improving the organizational sustainability further.

DOI: 10.4018/978-1-5225-5445-5.ch017

1. INTRODUCTION

The World Commission on Environment and Development declaration defines sustainable development as a process of change in which the exploitation of resources, the direction of investments, the orientation of technological development, and institutional change are made consistent with the future as well as present needs (World Commission on Environment and Development, 1987). Sustainable development was articulated by Brundtland commission as development that "meets the needs of the present without compromising the ability of future generations to meet their own needs" (UN, 19987). Sustainable approaches are comprehensive and holistic in nature for creating environmentally benign, socially equitable and economically viable products and systems. In practical terms, triple bottom line accounting means expanding the traditional economic reporting framework to take into account ecological and social performance in addition to financial performance, as shown in Figure 1.

Business sustainability requires the inclusion of the objectives of sustainable development, social equity, economic efficiency and environmental performance, into organization's operational practices. Organizations that compete globally are progressively more obligatory to consign and report on the overall sustainability performance of operational proposals. The term industrial sustainability as coined by the Institute for Manufacturing at the University of Cambridge and it defines industrial sustainability as conceptualization, design and manufacture of goods and services that meet the needs of the present generation while not diminishing economic, social and environmental opportunity in the long term (Jansson, et al., 2000). Zeng et al. (2008) defined industrial sustainability as the development that meets the needs of economic growth, social development, environmental protection and results in industrial advantage for short and long term future of their region located. These definitions continue to evolve, but still the benchmark to call an organization as sustainable is found not to be available, realizing the need to formulate a measuring scale; sustainability studies are to be conducted but selecting a suitable organization for the conduct of sustainability studies is not an easy task. For an automotive organization

Figure 1.

that is characterized by very long engineering cycles and multiple iterations to align technology, quality, and customer demand, incorporating sustainability into its business model and its products will be posing a significant challenge to respond to the growing pressure to capture their product's environmental footprint. This case study formulates the basis towards identifying most suitable automotive organizations for carrying out sustainability studies using cluster analysis, a multivariate data analytic technique. For this study, ISO 14000:2004 certified organizations have been selected and responsibility and involvement of the organization towards sustainability studies have been ensured for the selected organization. The results concluded that the 'Organization 16' as most suitable automotive organization for the conduct of sustainability studies. After selecting the best organization from the cluster, steps will be taken towards the conduct of sustainability studies. The uniqueness of the study is the application of clustering mechanism for grouping organizations in automotive sector towards the conduct of sustainability studies. The novelty of this study is proved by its substantial state of practical implications it could bring through the conduct of sustainability studies, in the process of making the organization more sustainable. The structure of the chapter includes the literature review on sustainability, Corporate Social Responsibility (CSR) and cluster analysis followed by the methodology adopted in this study. The consecutive section includes the background details of the case study, identification of suitable sector for the conduct of case study, selection of functionality variables for sustainable development and data analysis. After analysis, the results are interpreted and the conclusions were drawn.

2. LITERATURE REVIEW

The literature was reviewed from three perspectives namely sustainability, CSR and application of cluster analysis.

2.1. Literature Review on Sustainability

Deloitte and Touche (1992) defines business sustainability as ''adopting business strategies and activities that meet the needs of the enterprise and its stakeholders today while protecting, sustaining and enhancing the human and natural resources that will be needed in the future''. A company that embarks on the path of sustainability needs to carefully examine its mission, vision and values (Sheate et al., 2003). The principles of sustainability help businesses to reduce unnecessary risks, avoid waste generation, increase material and energy efficiency, innovate new environmentally friendlier products and services and obtain operating permits from local communities (Goldin & Winters, 1995). Sustainable organization also reduces production costs and prevents environmental problems for maintaining green and clean atmosphere (Bevilacqua, Ciarapica, & Giacchetta, 2007). Lopez et al. (2007) refers sustainability to an organization's activities, typically considered voluntary, that demonstrate the inclusion of social and environmental concerns in business operations and interactions with stakeholders.

2.2. Literature Review on CSR

CSR is an issue that has been receiving greater attention in discussions on business and sustainability (Sweeney & Coughlan, 2008). Rochon et al. (2010) have presented the best management practices for corporate, academic and governmental transfer of sustainable technologies to developing countries. The authors also presented some examples of recognized modes of sustainable technology transmission and develop best management practices for future sustainable technology transfer. CSR requires accountability by all leaders, individuals, organizations, stakeholders, customers, and community members, and yet accountability is complex. The factors which influence the effectiveness of corporate accountability are multiple and tightly interconnected. Dolan (2004) presented a study on this interconnectedness and its relationship to accountability. Lozano (2008) commented that CSR is becoming a leading principle of top management and of entrepreneurs. The number of observations in research in this field clearly delineated models, leadership competencies, accountability, and structure of partnerships as well as organizational challenges, limitations and ethics. Organizations can reexamine their pattern of behavior in the TBL framework and begin their journey towards a sustainable approach that is integrated into their business strategy (Vosburgh, 2007).

2.3 Literature Review on Applications of Cluster Analysis

The literature was reviewed from the perspective of utilization of cluster analysis in identifying the highly sustainable cluster. Sneath (1957) described a methodology to handle large quantity of taxonomic data to yield the outline of classification based on equally weighted features. This enables the similarity to be expressed numerically, and to measure taxonomic rank comparing with the conventional classification method based on a bacteria study. Lee et al. (1998) proposed a non-parametric approach that integrates neural network method with cluster analysis to estimate the development cost for software project planning, control and management increasing the training efficacy of the network. Cowgill and Harvey (1999) proposed Genetic Algorithm (GA) for performing cluster analysis. GA clustering technique maximizes a variance-ratio (VR) based goodness-of-fit criterion defined in terms of external cluster isolation and internal cluster homogeneity. But this methodology does not guarantee to recover the cluster solution which exhibits the global maximum of this fitness function. Cuevas et al. (2001) proposed a cluster methodology with density estimation with the idea of estimating the population clusters. The results of the study proved a good performance in stipulation with efficiency and robustness, compared with two classical cluster algorithms k- means and single linkage. Liang et al. (2005) proposed a cluster analysis method based on fuzzy equivalence relation. The fuzzy equivalence based on the fuzzy compatibility relation with various rating attitudes was taken into account in the aggregation process to assure more convincing and accurate cluster analysis by the authors. Wang and Yang (2009) proposed a fuzzy modeling method using Enhanced Objective Cluster Analysis (EOCA) to obtain the compact and robust approximate Takagi–Sugeno–Kang (TSK) fuzzy model. The Objective Cluster Analysis (OCA) algorithm was employed to obtain the fuzzy rule prototypes, enhanced by introducing the Relative Dissimilarity Measure (RDM) and a new consistency criterion to represent the similarity degree between the clusters, reducing the time and error over computation to a great extent. The authors also identified the consequence parameters adopting Stable Kalman Filter (SKF) algorithm. Gelbard et al. (2009) used a single algorithm to overcome the shortcomings in the process of determining a particular cluster by considering cross-cultural research employing Multi-Algorithm Voting (MAV) methodology. The contribution of the authors enables the researchers in avoiding arbitrary decisions in determining the number of clusters. Che and Wang (2016) in order to differentiate part suppliers effectively, proposed a hybrid approach based on K-means, simulated annealing algorithm (SA), convergence factor particle swarm optimization (CPSO), and the Taguchi method abbreviated as KSACPSO. To prove that KSACPSO approach has good clustering performance, the case study of notebook computer was adopted to carry out the clustering procedures on parts suppliers, and compared the differences between the proposed approach and other hybrid methods.

Fu Gu et.al (2016) introduced a novel strategy to adjust reused plastic for demanding industrial applications by outlining reasonable formulae which can fulfill the specialized prerequisites of the applications. The proposed technique consolidates the principal component estimate and hierarchical clustering analysis of some selected automobile parts were procured and benchmarked in this study.

Neri et.al (2017) in their study, Evaluated the dynamics of national economies through cluster analysis within the input-state-output sustainability framework .This framework was valuable and comprehensive apparatus for surveying nation about their performances over time and improving guidelines for the characterization of nations under a sustainability point of view. In order to observe system behavior and pattern of development this analysis was conducted over 83 countries in between 2000-2008.

Molin et.al (2016), used cluster analysis to recognize multi model travel groups and to detail the effects of socio – demographic, perception and work related and attitudinal variables on the likelihood of having a place with each of the five distinguished classes such as youthful people, high training, small family units, and availability of, which gives trust in the outcomes.

Miguel Barros and Susana Azevedo (2016) proposed a framework to evaluate and screen the level of sustainability in an automotive supply chain network, by contextual analysis, considering a few organizations from United Kingdom (UK).

Karen Novia et al. (2016), examined a 142 cities dataset which includes annual/capita water use (m^3/year/capita). As an index of hydro climatic water supply, it adds 0.5 grid water annual budget value (P-PET/yr). From this findings, urban water supply and demand, were regulated using hierarchical cluster analysis.

From the literature review, it was evident that cluster analysis could be used as a tool in identifying the suitable organization towards the conduct of sustainability studies for any given geographical location and the application of cluster analysis in sustainability field were also found to be minimal.

3. METHODOLOGY

The methodology adopted during the conduct of this study in shown in Figure 2.

A detailed literature survey has been conducted on selecting the major functionality variables in deciding the sustainability of an organization, followed by the identification of suitable sector for immediate conduct of the study. Next step is to find out the most suitable organization using cluster analysis. Interpretations were derived from the identified clusters and defining the differences between the clusters for studying the sustainability levels forms the consecutive steps of this case study.

4. BACKGROUND DETAILS OF THE CASE STUDY

There always exists a question over setting up a benchmark, to identify the sustainability of an organization. The lack of a benchmark to confirm an organization as sustainable requires more elaborate sustainability studies. To conduct preliminary studies, an organization needs to be selected first to understand the factors facilitating and hindering the organization from being sustainable. This case study concentrates more on the process of identifying one such suitable organization for the conduct of sustainability studies.

Figure 2.

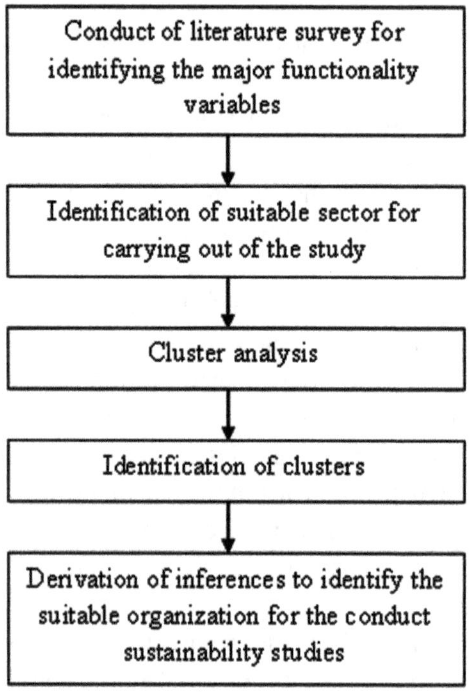

4.1 Identification of Suitable Sector for the Conduct of Case Study

Tamilnadu is one of the well developed states of India, in terms of industrial development. Tamilnadu is also being popularly hailed as "Detroit" of India as it has large automobile and ancillary sector. Automobile industry plays a crucial role in the state economy and has been one of the key driving factors, contributing 8% to State Gross State Domestic Product (GDP) and giving direct employment to 2,20,000 people. More than 100 companies in the automotive and auto ancillary industry are located in this state, maintaining highest production norms by implementing internationally recognized quality standards. Tamilnadu accounts for 21% of passenger cars, 33% of the Commercial vehicles and 35% of automobile components produced in India. The size of the auto industry in Tamilnadu is estimated to grow to Indian National rupees (INR) 67500 - INR.90000 crores (US $15-20 billion) by 2015. Major automobile manufacturers like Ford, Hyundai, Ashok Leyland, etc. have their manufacturing base in Tamilnadu.

Considering the value of automobile sectors contribution towards the GDP, 25 potential automotive organizations were selected for the conduct of this case study in the state of Tamilnadu, India. The selected organizations were explained about this study through an orientation meeting conducted at the top management level. The workforce level of the selected organization was in the range of 100 – 1100 employees per shift and the sales range in the year 2010 is 100 to 15000 million in INR. The organizations also assured to conduct sustainability studies to the possible extent.

4.2 Selection of Functionality Variables for Sustainable Development

The assortment of variables is a significant task in performing cluster analysis. Based on the detailed literature survey conducted on industrial sustainability, five critical variables for identifying the suitable automotive organization were selected. The functionality variables for selecting the suitable automotive organization are,

- Economic credibility (V_1)
- Societal contribution (V_2)
- Environmental performance (V_3)
- Organizational commitment towards sustainable development (V_4) and
- Government Legislations (V_5)

The above five are the comprehensive variables of the sustainability. These variables were identified based on the literature survey (Jain & Dubes, 1998) and after discussing with the stakeholders of the organizations. The three variables Economic credibility, societal contribution and Environmental performance were selected because these three are the main perspectives of sustainability. Organizational commitment towards sustainable development variable is selected to ensure the involvement and commitment of the management towards sustainability studies. Economic credibility (V_1) includes financial health, potential trading opportunities, economic performance and other financial benefits of that organization in comparison to the other organizations of automotive sector in Tamilnadu. The values of V_1 for the organizations can be easily obtained from the market share values of the organizations under study. Societal contribution (V_2) is defined as the organizational engagement with its external and internal human resources, stake holder participation and macro social performance. The values of V_2 are obtained based upon a general survey conducted in the organizations. The environmental performance (V_3) of the organizations indicates the difference between the quantities of environmental preservation activities carried out against the environmental impact created by that organization. V_3 values were obtained from the pollution control board of that region where the organization is located. Organizational commitment (V_4) is the promise made by that organization towards sustainable development in making their products, processes and locations more sustainable in comparison with the other organizations under study. Government legislations (V_5) are the rules and regulations enforcing the organizations to be sustainable in nature with more responsibility to the society they are located. 25 organizations considered under this study are designated as A_1, A_2, A_3...and A_{25}. The normalized values of each variable were obtained using equation (1), where V_i^s is the normalized value of the ith variable V_i (i=1, 2, 3, 4). The data collected against the variables V_i (i=1, 2, 3, 4) and their corresponding normalized values are presented in Table 1. The normalized values computed were used for conducting cluster analysis in this case study and are computed using Equation 1.

$$V_i^s = \frac{V_i - \min V_i}{\max V_i - \min V_i}. \tag{1}$$

Table 1. Functionality and normalized values of variables

Organization	Functionality Variable Values					Normalized Variable Values				
	V_1	V_2	V_3	V_4	V_5	V_1	V_2	V_3	V_4	V_5
A1	1.072	0.62	1.5	1.058	1.063	0.004	0.001	0.065	0	0.000
A2	1.692	2.28	2.8	1.527	2.075	0.197	0.057	0.403	0.007	0.038
A3	2.367	0.96	2	1.146	1.618	0.406	0.013	0.195	0.001	0.021
A4	2.912	2.49	2.4	1.558	2.340	0.575	0.064	0.299	0.007	0.048
A5	2.516	3.03	4.3	1.693	2.885	0.452	0.083	0.792	0.009	0.068
A6	1.859	2.52	4.8	1.345	2.631	0.248	0.066	0.922	0.004	0.059
A7	3.218	8.44	1.8	3.56	4.255	0.67	0.266	0.143	0.035	0.119
A8	2.659	2.34	1.25	2.645	2.224	0.497	0.059	0	0.022	0.043
A9	1.116	0.93	1.25	1.243	1.135	0.018	0.012	0	0.003	0.003
A10	2.674	4	1.25	2.112	2.509	0.502	0.116	0	0.015	0.054
A11	1.391	3.5	1.5	1.553	1.986	0.103	0.099	0.065	0.007	0.034
A12	1.612	3.3	4.2	2.9	3.003	0.172	0.092	0.766	0.026	0.072
A13	2.886	6.28	3.8	4.639	4.401	0.567	0.193	0.662	0.051	0.125
A14	1.442	4.63	3.5	3.682	3.314	0.119	0.137	0.584	0.037	0.084
A15	2.291	8.88	4.1	4.358	4.907	0.383	0.281	0.74	0.047	0.144
A16	4.279	30.05	5	72.079	27.852	1	1	0.974	1	1.000
A17	2.217	0.78	2.1	1.154	1.563	0.36	0.006	0.221	0.001	0.019
A18	3.151	10.68	2.34	3.086	4.814	0.65	0.342	0.283	0.029	0.140
A19	1.059	0.59	2.6	1.052	1.325	0	0	0.351	0	0.010
A20	1.171	2.09	1.9	1.209	1.593	0.035	0.051	0.169	0.002	0.020
A21	1.25	2.94	3.1	1.234	2.131	0.059	0.08	0.481	0.003	0.040
A22	2.253	4.49	4	2.039	3.196	0.371	0.132	0.714	0.014	0.080
A23	1.629	2.34	1.6	1.271	1.710	0.177	0.059	0.091	0.003	0.024
A24	1.69	4.05	5.1	6.903	4.436	0.196	0.117	1	0.082	0.126
A25	2.217	0.78	2.1	1.204	1.575	0.36	0.006	0.221	0.002	0.019

4.3 Data Analysis

Cluster analysis groups objects based only on information found in the data that describes the objects and their relationships. The goal is that the objects within a group be similar to one another and different from the objects in other groups. The greater the similarity within a group and greater the difference between groups, the better or more distinct is the clustering. The types of clustering include hierarchical (Nested) versus partitioned (Un-nested), exclusive versus overlapping versus fuzzy, and complete versus partial are the types of clustering's. There are three major important techniques in cluster analysis known as, K-means, Agglomerative Hierarchical Clustering (AHC) approach and Density Based Clustering Algorithm (DBSCAN). AHC approach used in this study refers to a collection of closely related clustering techniques that produce a homogeneous hierarchical clustering by starting with each point as a singleton cluster and then repeatedly merging the two closest clusters until a single, all-encompassing cluster remains. The clustering method uses the dissimilarities or distances between objects while forming the clusters.

5. ANALYSIS

The Statistical Package for the Social Sciences (SPSS) version 18 was used in for statistical analysis in this case study using which 'distances' between data points were calculated in terms of the specified variables. A hierarchical tree diagram, called a dendrogram on SPSS, was produced to show the linkage points. The clusters are linked at increasing levels of dissimilarity. The actual measure of dissimilarity depends on the measure used. The fusion values or linkage distances were calculated by SPSS. The AHC approach was used in this case study to form groups from the normalized data computed from the chosen 25 organizations by the five functionality variables selected. There are more than five variables in this case study so it was preferred to use squared Euclidean distance as the correlation measures are influenced by differences in scale, to calculate the distance between each pair of organizations.

6. RESULTS

The dendrogram, a graphical representation of the cluster trees, begins with each object or case in a class by itself. In the dendrogram plots (shown in Figure 3) more and more objects are linked together and larger and larger clusters of increasingly dissimilar elements are amalgamated until one cluster is left. The horizontal axis denotes the fusion or linkage distance. At each node in the graph, the criterion distance at which the respective elements were linked together, a new single cluster is formed. The vertical direction shows the organizations identified by organizational number (A_1, A_2, A_3...and A_{25}).

Table 2 shows the agglomeration schedule explaining the average linkage between groups showing two clear clusters and a minor one.

Organization A_{16} is the only organization that comes under cluster I as the functionality variables economic credibility (V_1), societal contribution (V_2), environmental performance (V_3), organizational commitment (V_4) and government legislations (V_5) with values 4.279, 30.050, 5.000, 72.079 and 63.089 respectively for this organization is comparatively very high when compared with other organizations. The cluster II consists of seven organizations namely A_6, A_{24}, A_{12}, A_5, A_{22}, A_{15} and A_{13} with higher relative societal contribution (V_2), environmental performance (V_3) and government legislations (V_5) values in contrast with other organizations excluding organization A_{16}. To be explicit, cluster III includes 17 organizations A_1, A_9, A_{11}, A_{23}, A_{20}, A_{14}, A_{21}, A_2, A_{19}, A_{17}, A_{25}, A_3, A_4, A_8, A_{10}, A_7 and A_{18}. The organizations in cluster III have low economic credibility (V_1), societal contribution (V_2) and have low environmental performance (V_3) value rating as compared with other organizations. But nevertheless the contributions of each and every organization in all the three clusters are not negligible as they have their own contribution towards the sustainable GDP of the automobile clusters located in the Tamilnadu state of India.

6.2. Results Interpretation

The conduct of the study enables the grouping of organizations into three clusters. This kind of grouping will enable the identification of commonalities among the organizations in a cluster thereby the

Figure 3.

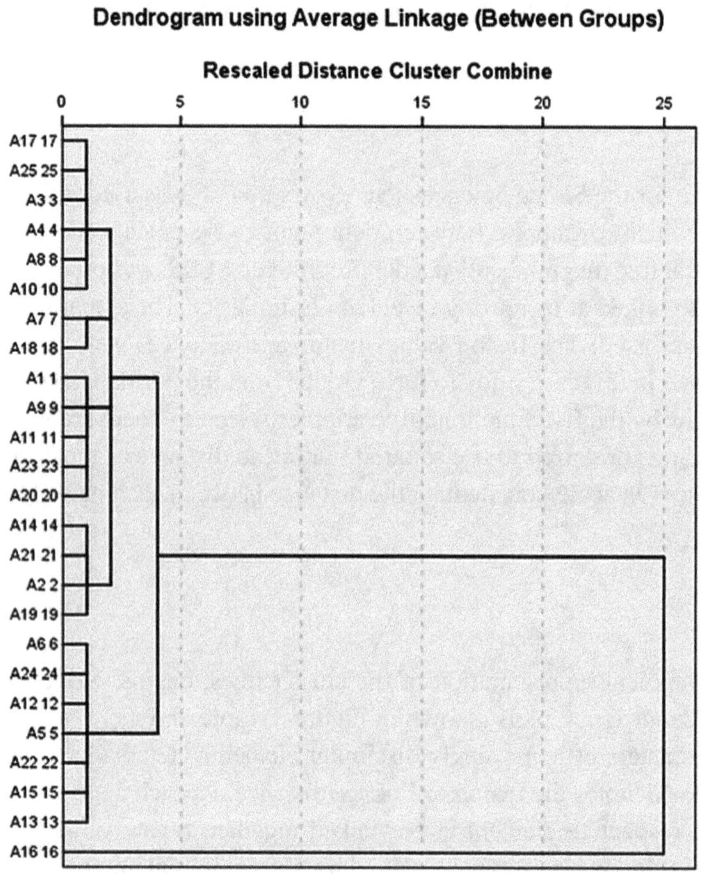

implementation of sustainable concepts become easier. During the conduct of the study, the importance of management commitment was understood. The activities in this study were carried out by the cross functional team formulated which involves education and training to the team members involved in this study. The practitioners working in the field also expressed the usefulness of the approach towards conducting further studies.

7. CONCLUSION

The novelty of the study is to apply clustering mechanism for grouping organizations in automotive sector towards the conduct of sustainability studies. Adding sustainability considerations will challenge automotive organizations, especially eco-concepts incorporated into already prevailing complex design decisions that strive to balance cost, time to market, safety, and quality. The usage of cluster analysis as a tool to setup a benchmark, to identify the sustainability level of the organizations proves to be effective in the process of identifying one such best organization for the conduct of sustainability studies. Cluster analysis was used to examine the 25 automotive components manufacturing organizations in the state of

Table 2. Agglomeration schedule

Agglomeration Schedule						
Stage	Cluster Combined		Coefficients	Stage Cluster First Appears		Next Stage
	Cluster 1	Cluster 2		Cluster 1	Cluster 2	
1	17	25	0	0	0	2
2	3	17	0.003	0	1	16
3	8	10	0.003	0	0	18
4	1	9	0.005	0	0	10
5	11	23	0.008	0	0	9
6	5	22	0.015	0	0	13
7	14	21	0.021	0	0	12
8	6	24	0.022	0	0	14
9	11	20	0.022	5	0	10
10	1	11	0.026	4	9	21
11	7	18	0.026	0	0	20
12	2	14	0.037	0	7	17
13	5	15	0.041	6	0	15
14	6	12	0.047	8	0	19
15	5	13	0.048	13	0	19
16	3	4	0.052	2	0	18
17	2	19	0.056	12	0	21
18	3	8	0.076	16	3	20
19	5	6	0.12	15	14	23
20	3	7	0.153	18	11	22
21	1	2	0.169	10	17	22
22	1	3	0.277	21	20	23
23	1	5	0.486	22	19	24
24	1	16	3.664	23	0	0

Tamilnadu, India, by selecting five functionality variables purely based on the sustainable development aspects. The result analysis clustered 25 organizations into four clusters and inferred that the organization A_{16} as the most suitable organization for the conduct of sustainability studies in the process of setting up a benchmark for sustainable organization whose functionality variable values were comparatively high compared to other organization in this case study. The conduct of sustainability studies in the selected organization (A_{16}) will facilitate implementation of sustainability concepts and tools such as Life Cycle Assessment (LCA), Environmentally Conscious Quality Function Deployment (ECQFD), Environmental Impact Assessment (EIA), and other activities enabling the organization in turning out to be more sustainable and patronizing it to be a benchmark organization. The conduct of such studies will enable the organizations to improve their ability in designing products with synchronized functionality, safety, and sustainability in a cost-effective manner in comparison with their competitive organizations. As a future direction of research, this study can be further extended to other sectors contributing majorly to the GDP to identify the suitable organization among them and an overall comparative sustainability check has to be performed as a process of validation.

The sustainability studies will further need to be expanded in product/process perspectives in the identified organization. Product studies will be based upon selecting alternate materials as a surrogate to existing materials ensuring their system performance and life. Design studies will be based on 6R concepts (Reduce, Reuse, Recycle, Repair, Recondition, and Remanufacture). Process studies will be based on finding out the most suitable alternate manufacturing processes as compared with the existing processes followed in the organization based upon each product. The best alternative will be identified and implemented in the organization in order to improve their sustainability level in comparison with the sustainability benchmark set by the organization (A_{16}) identified in this case study. The present study focuses on the organizations in automotive sector towards the conduct of sustainability studies. Also, the approach could be extended for similar advanced manufacturing paradigms like leanness, agility and leagility.

ACKNOWLEDGMENT

The authors are grateful to Department of Science and Technology (DST), New Delhi, India for funding towards the implementation of project titled "Development of a model for ensuring sustainable product design in automotive organizations" (Ref No.SR/S3/MERC-0102/2009). This research study forms a part of this chief research project.

REFERENCES

Barros, M. M., & Azevedo, S. G. (2016). Supply chain sustainability assessment: the case of the automotive industry. *Proceedings of International Conference on Electrical, Mechanical, and Industrial Engineering*. 10.2991/icemie-16.2016.30

Bevilacqua, M., Ciarapica, F. E., & Giacchetta, G. (2007). Development of a sustainable product lifecycle in manufacturing firms: A case study. *International Journal of Production Research*, *45*(18–19), 4073–4098. doi:10.1080/00207540701439941

Che, Z. H., & Wang, H. S. (2010). Hybrid approach for supplier cluster analysis. *Computers & Mathematics with Applications (Oxford, England)*, *59*(2), 745–763. doi:10.1016/j.camwa.2009.10.018

Cowgill, M. C., Harvey, R. J., & Watson, L. T. (1999). A genetic algorithm approach to Cluster analysis. *Computers & Mathematics with Applications (Oxford, England)*, *37*(7), 99–108. doi:10.1016/S0898-1221(99)00090-5

Cuevas, A., Febrero, M., & Fraiman, R. (2001). Cluster analysis: A further approach based on density estimation. *Computational Statistics & Data Analysis*, *36*(4), 441–459. doi:10.1016/S0167-9473(00)00052-9

Deloitte & Touche. (1992). *ISSD. Business strategy for sustainable development: leadership and accountability for the 90s*. Author.

Dolan, P. (2004). Sustainable leadership. *Leader to Leader*, *33*(33), 8–12. doi:10.1002/ltl.82

Gelbard, R., Carmeli, A., Bittmann, R. M., & Ronen, S. (2009). Cluster analysis using multi-algorithm voting in cross-cultural studies. *Expert Systems with Applications*, *36*(7), 10438–10446. doi:10.1016/j.eswa.2009.01.071

Goldin, I., & Winters, L. A. (1995). *The Economics of Sustainable Development*. Cambridge, UK: University of Cambridge Press. doi:10.1017/CBO9780511751905

Gu, F., Hall, P., & Miles, N. J. (2016). Development of composites based on recycled polypropylene for injection moulding automobile parts using hierarchical clustering analysis and principal component estimate. *Journal of Cleaner Production*, *137*, 632–643. doi:10.1016/j.jclepro.2016.07.028

Hollenstein, H. (2003). Innovation modes in the Swiss service sector: A cluster analysis based on firm-level data. *Research Policy*, *32*(5), 845–863. doi:10.1016/S0048-7333(02)00091-4

Jain, A. K., & Dubes, R. C. (1998). *Algorithms for clustering data*. Prentice Hall.

Jansson, P. M., Gregory, M. J., Barlow, C., Phaal, R., Farrukh, C. J. P., & Probert, D. R. (2000). *Industrial sustainability a review of UK and international research and capabilities*. Cambridge, UK: University of Cambridge.

Lee, A., Cheng, C. H., & Balakrishnan, J. (1998). Software development cost estimation: Integrating neural network with cluster analysis. *Information & Management*, *34*(1), 1–9. doi:10.1016/S0378-7206(98)00041-X

Liang, G. S., Chou, T. Y., & Chen Han, T. C. (2005). Cluster analysis based on fuzzy equivalence relation. *European Journal of Operational Research*, *166*(1), 160–171. doi:10.1016/j.ejor.2004.03.018

Lopez, M., Garcia, A., & Rodriguez, L. (2007). Sustainable development and corporate performance: A study based on the Dow Jones Sustainability Index. *Journal of Business Ethics*, *75*(3), 285–300. doi:10.100710551-006-9253-8

Lozano, R. (2008). Envisioning sustainability three-dimensionally. *Journal of Cleaner Production*, *16*(17), 1838–1846. doi:10.1016/j.jclepro.2008.02.008

Molin, E., Mokhtarian, P., & Kroesen, M. (2016). Multimodal travel groups and attitudes: A latent class cluster analysis of Dutch travelers. *Transportation Research Part A, Policy and Practice*, *83*, 14–29. doi:10.1016/j.tra.2015.11.001

Neri, L., D'Agostino, A., Regoli, A., Pulselli, F. M., & Coscieme, L. (2017). Evaluating dynamics of national economies through cluster analysis within the input-state-output sustainability framework. *Ecological Indicators*, *72*, 77–90. doi:10.1016/j.ecolind.2016.08.016

Noiva, K., Fernández, J. E., & Wescoat, J. L. Jr. (2016). Cluster analysis of urban water supply and demand: Toward large-scale comparative sustainability planning. *Sustainable Cities and Society*, *27*, 484–496. doi:10.1016/j.scs.2016.06.003

Rochon, G. L., Niyogi, D., Fall, S., Quansah, J. E., Biehl, L., Araya, B., ... Thiam, T. (2010). Best management practices for corporate, academic and governmental transfer of sustainable technologies to developing countries. *Clean Technologies and Environmental Policy*, *12*(1), 19–30. doi:10.100710098-009-0218-3

Sheate, W., Dagg, S., Richardson, J., Aschemann, R., Palerm, J., & Steen, U. (2003). Integrating the environment into strategic decision-making: Conceptualizing policy SEA. *European Environment*, *13*(1), 1–18. doi:10.1002/eet.305

Sneath, P. H. A. (1957). The application of computers to taxonomy. *Journal of General Microbiology*, *17*(1), 201–226. PMID:13475686

Sweeney, L., & Coughlan, J. (2008). Do different industries report corporate social responsibility differently? An investigation through the lens of stakeholder theory. *Journal of Marketing Communications*, *14*(2), 113–124. doi:10.1080/13527260701856657

UN. (1987). World Commission on Environment and Development, Report of The World Commission on Environment and Development: Our Common Future, Annex To General Assembly Document A/42/427.

Vinodh, S. (2010). *Assessment of sustainability using multi-grade fuzzy approach*. Clean Tech Environ Policy. doi:10.100710098-010-0333-1

Vosburgh, R. M. (2007). Special edition articles. Human Resource Planning, 30(1), 7–8.

Wang, N., & Yang, Y. (2009). A fuzzy modeling method via Enhanced Objective Cluster Analysis for designing TSK model. *Expert Systems with Applications*, *36*(10), 12375–12382. doi:10.1016/j.eswa.2009.04.048

World Commission on Environment and Development. (1987). *From one earth to one world: an over view*. Retrieved from http://mom.gov.af/Content/files/Bruntland_Report.pdf

Zeng, S. X., Liu, H. C., Tam, C. M., & Shao, Y. K. (2008). Cluster analysis for studying industrial sustainability: An empirical study in Shanghai. *Journal of Cleaner Production*, *16*(10), 1090–1097. doi:10.1016/j.jclepro.2007.06.004

Compilation of References

Abdalla, I. I., Ibrahim, T., & Mohd Nor, N. (2014). Analysis of a Tubular Linear Permanent Magnet Motor for Reciprocating Compressor Applications. *Applied Mechanics and Materials*, *448*, 2114–2119.

Abdul Majeed, M., Vijayaraghavan, L., Malhotra, S. K., & Krishnamurthy, R. (2008). Ultrasonic machining of Al_2O_3/$LaPO_4$ composites. *Journal of Machine Tools & Manufacture*, *48*(1), 40–46. doi:10.1016/j.ijmachtools.2007.07.012

Abedin, B., & Sohrabi, B. (2009). Graph theory application and web page ranking for website link structure improvement. *Behaviour & Information Technology*, *28*(1), 63–72. doi:10.1080/01449290701840948

Acharya, B. G., Jain, V. K., & Batra, J. L. (1986). Multi objective optimization of the ECM process. *Precision Engineering*, *8*(2), 88–96. doi:10.1016/0141-6359(86)90091-7

Adya, N., Alam, M., Ravindranath, T., Mubeen, A., & Saluja, B. (2005). Corrosion in titanium dental implants: Literature review. *Journal of Indian Prosthodontic Society*, *5*(3), 126–131. doi:10.4103/0972-4052.17104

Agarwala, M. K., Jamalabad, V. R., Langrana, N. A., Safari, A., Whalen, P. J., & Danforth, S. C. (1996). Structural quality of parts processed by fused deposition. *Rapid Prototyping Journal*, *2*(4), 4–19. doi:10.1108/13552549610732034

Agrawal, S., Singh, R., & Murtaza, Q. (2016). Disposition decisions in reverse logistics: Graph theory and matrix approach. *Journal of Cleaner Production*, *137*, 93–104. doi:10.1016/j.jclepro.2016.07.045

Ahn, D.-G. (2016). Direct Metal Additive Manufacturing Processes and Their Sustainable Applications for Green Technology. *RE:view*. doi:10.100740684-016-0048-9

Ahn, D., Kweon, J. H., Choi, J., & Lee, S. (2012). Quantification of surface roughness of parts processed by laminated object manufacturing. *Journal of Materials Processing Technology*, *212*(2), 339–346. doi:10.1016/j.jmatprotec.2011.08.013

Al Mashagbeh, M., & Khamesee, M. B. (2015). Virtual performance evaluation of an industrial SCARA robot prior to real-world task. *Microsystem Technologies*, *21*(12), 2605–2609. doi:10.100700542-015-2502-y

Alangi, N., Mukherjee, J., Anupama, P., Verma, M. K., Chakravarthy, Y., Padmanabhan, P. V. A., ... Gantayet, L. M. (2011). Liquid uranium corrosion studies of protective yttria coatings on tantalum substrate. *Journal of Nuclear Materials*, *10*(1-3), 39–45. doi:10.1016/j.jnucmat.2010.12.307

Alberdi, A., Suarez, A., Artaza, T., Escobar-Palafox, G. A., & Ridgway, K. (2013). Composite Cutting with Abrasive Water Jet. *Procedia Engineering*, *63*, 421–429. doi:10.1016/j.proeng.2013.08.217

Alexander, P., Allen, S., & Dutta, D. (1998). Part orientation and build cost determination in layered manufacturing. *Computer Aided Design*, *30*(5), 343–356. doi:10.1016/S0010-4485(97)00083-3

Aliyu, A. A. A., Abdul-Rani, A. M., Ginta, T. L., Prakash, C., Axinte, E., Razak, M. A., & Ali, S. (2017). A review of additive mixed-electric discharge machining: Current status and future perspectives for surface modification of biomedical implants. *Advances in Materials Science and Engineering*, *2017*, 1–23. doi:10.1155/2017/8723239

Aljuneidi, T., & Bulgak, A. (2017). Designing a Cellular Manufacturing System featuring remanufacturing, recycling, and disposal options: A mathematical modeling approach. *CIRP Journal of Manufacturing Science and Technology*, 11.

Al-Otaibi, Z., & Jack, A. (2008). *Utilising SMC in single phase permanent magnet linear motors for compressor applications*. Academic Press.

Alshamasin, M. S., Ionesco, F., & Al-Kasasbeh, R. T. (2012). Kinematic modelling and Simulation of a SCARA Robot by Using Solid Dynamics and Verification by MATLAB/Simulink.International Journal of Modelling. *Identification and Control*, *15*(1), 28–38. doi:10.1504/IJMIC.2012.043938

Alves, J. L. S., & Dumke De Medeiros, D. (2015). Eco-efficiency in micro-enterprises and small firms: A case study in the automotive services sector. *Journal of Cleaner Production*, *108*, 595–602. doi:10.1016/j.jclepro.2015.07.063

Amaya, J., Zwolinski, P., & Brissaud, D. (2010). Environmental Benefits of Parts Remanufacturing: The Truck Injector Case. *17th CIRP International Conference on Lifecycle Engineering*.

Amin, S., Zhang, G., & Akhtar, P. (2017). Effects of uncertainty on a tire closed-loop supply chain network. *Expert Systems with Applications*, *73*, 82–91. doi:10.1016/j.eswa.2016.12.024

Amiri, A. M., Olfati, A., Najjar, S., & Beiranvand, P. (2016, June). M.H. Naseri Fard, The Effect of Fly Ash on Flexural Capacity Concrete Beams. *Advances in Science and Technology Research Journal*, *10*(30), 89–95. doi:10.12913/22998624/62630

Arachchige & Abderrahmane. (2013). Design of reconfigurable joints for the advanced robotic systems. In *Proceedings of the National Conference on Undergraduate Research (NCUR)*. La Crosse, WI: University of Wisconsin.

Arjun, N. R., Ganiger, M., & Idris, M. (2012). *Studies on optimization of Selective Laser Sintering process to manufacture Fuel tanks*. Available at http://amsi.org.in

Arumuga Prabu, V., Deepak Joel Johnson, R., Amuthakkannan, P., & Manikandan, V. (2017). Usage of industrial wastes as particulate composite for environment management: Hardness, Tensile and Impact studies. *Journal of Environmental Chemical Engineering*, *5*(1), 1289–1301. doi:10.1016/j.jece.2017.02.007

Ashley, S. (1991). Rapid prototyping systems. *Mechanical Engineering (New York, N.Y.)*, *113*(4), 34.

Ashman, A. (1971). Acrylic resin tooth implant: A progress report. *The Journal of Prosthetic Dentistry*, *25*(3), 342–347. doi:10.1016/0022-3913(71)90197-1 PMID:5276857

Asif, F. M., Lieder, M., & Rashid, A. (2016). Multi-method simulation based tool to evaluate economic and environmental performance of circular product systems. *Journal of Cleaner Production*, *139*, 1261–1281. doi:10.1016/j.jclepro.2016.08.122

Asokan, P., Kumar, R. R., Jeyapaul, R., & Santhi, M. (2008). Development of multi-objective optimization models for electrochemical machining process. *International Journal of Advanced Manufacturing Technology*, *39*(1-2), 55–63. doi:10.100700170-007-1204-8

Atan, M. N., & Awanghe, H. (2011). Compressive and Flexural Strengths of Self-Compacting Concrete using Raw Rice Husk. *Journal of Engineering Science and Technology*, *6*(6), 720–732.

Attaran. (2017). *The rise of 3-D printing: The advantages of Additive Manufacturing over traditional manufacturing*. Academic Press. . doi:10.1016/j.bushor.2017.05.011

Ayvaz, B., Bolat, B., & Aydın, N. (2015). Stochastic reverse logistics network design for waste of electrical and electronic equipment. *Resources, Conservation and Recycling*, *104*, 391–404. doi:10.1016/j.resconrec.2015.07.006

Bacchewar, P. B., Singhal, S. K., & Pandey, P. M. (2007). Statistical modelling and optimization of surface roughness in the selective laser sintering process. *Proceedings of the Institution of Mechanical Engineers. Part B, Journal of Engineering Manufacture*, *221*(1), 35–52. doi:10.1243/09544054JEM670

Bai, Y., & Christopher, B. (2015). An exploration of binder jetting of copper. *Rapid Prototyping Journal*, *21*(2), 177–185. doi:10.1108/RPJ-12-2014-0180

Bakhsheshi-Rad, H. R., Idris, M. H., Abdul-Kadir, M. R., Ourdjini, A., Medraj, M., Daroonparvar, M., & Hamza, E. (2014). Mechanical and bio-corrosion properties of quaternary Mg–Ca–Mn–Zn alloys compared with binary Mg–Ca alloys. *Materials & Design*, *53*, 283–292. doi:10.1016/j.matdes.2013.06.055

Balasundaram, G., Sato, M., & Webster, T. J. (2006). Using hydroxyapatite nanoparticles and decreased crystallinity to promote osteoblast adhesion similar to functionalizing with RGD. *Biomaterials*, *27*(14), 2798–2805. doi:10.1016/j.biomaterials.2005.12.008 PMID:16430957

Balasundaram, G., & Webster, T. J. (2006). A perspective on nanophase materials for orthopedic implant applications. *Journal of Materials Chemistry*, *16*(38), 3737–3745. doi:10.1039/b604966b

Balasundarm, G., & Webster, T. J. (2007). An overview of Nano-Polymers for Orthopaedic Applications. *Macromolecular Bioscience Journal*, *7*(5), 635–642. doi:10.1002/mabi.200600270

Barbieri, N., Manco, G., & Ritacco, E. (2011, April). A probabilistic hierarchical approach for pattern discovery in collaborative filtering data. In *Proceedings of the 2011 SIAM International Conference on Data Mining* (pp. 630-641). Society for Industrial and Applied Mathematics. 10.1137/1.9781611972818.54

Barker, T., & Zabinsky, Z. (2011). A multicriteria decision making model for reverse logistics using analytical hierarchy process. *Omega*, *39*(5), 558–573. doi:10.1016/j.omega.2010.12.002

Barnett, S.O. (n.d.). Investment casting - the multi-process technology. *Foundry Trade Journal International*, 1-11.

Barros, M. M., & Azevedo, S. G. (2016). Supply chain sustainability assessment: the case of the automotive industry. *Proceedings of International Conference on Electrical, Mechanical, and Industrial Engineering*. 10.2991/icemie-16.2016.30

Bártolo, P. J., & Gibson, I. (2011). History of stereolithographic processes. In Stereolithography (pp. 37-56). Springer US.

Bartolo, P., Kruth, J.P., Silva, J., Levy, G., Malshe, A., Rajurkar, K., Mitsuishi, M., Ciurana, J., & Leu, M. (2012). Biomedical production of implants by additive electro-chemical and physical processes. *CIRP Annals - Manufacturing Technology*, *61*, 635-655.

Bartolo, P. J., Chua, C. K., Almeida, H. A., Chou, S. M., & Lim, A. S. C. (2009). Biomanufacturing for tissue engineering: Present and future trends. *Virtual and Physical Prototyping*, *4*(4), 203–216. doi:10.1080/17452750903476288

Bateman, N., Philp, L., & Warrender, H. (2016). Visual management and shop floor teams – development, implementation and use. *International Journal of Production Research*, *54*(24), 7345–7358. doi:10.1080/00207543.2016.1184349

Bauer, S., Schmuki, P., von der Mark, K., & Park, J. (2013, April). Engineering biocompatible implant surfaces part 1: Materials and surfaces. *Progress in Materials Science*, *58*(3), 261–326. doi:10.1016/j.pmatsci.2012.09.001

Bazan, E., Jaber, M., & Zanoni, S. (2016). A review of mathematical inventory models for reverse logistics and the future of its modeling: An environmental perspective. *Applied Mathematical Modelling*, *40*(5-6), 4151–4178. doi:10.1016/j.apm.2015.11.027

Beese & Clare. (2107). Materials for Additive Manufacturing. *CIRPAnnals - Manufacturing Technology*. doi:10.1016/j.cirp.2017.05.009

Bell, E., & Davison, J. (2013). Visual management studies: Empirical and theoretical approaches. *International Journal of Management Reviews*, *15*(2), 167–184. doi:10.1111/j.1468-2370.2012.00342.x

Ben Utela, D. S., Anderson, R., & Ganter, M. (n.d.). A review of process development steps for new material systems in three dimensional printing (3DP). *Journal of Manufacturing Processes*. doi:10.1016/j.jmapro.2009.03.002

Benedito, E., & Corominas, A. (2013). Optimal manufacturing policy in a reverse logistic system with dependent stochastic returns and limited capacities. *International Journal of Production Research*, *51*(1), 189–201. doi:10.1080/00207543.2012.655863

Bevilacqua, M., Ciarapica, F. E., & Giacchetta, G. (2007). Development of a sustainable product lifecycle in manufacturing firms: A case study. *International Journal of Production Research*, *45*(18–19), 4073–4098. doi:10.1080/00207540701439941

Bhattacharyya, A., Sur, B., & Sorkhel, S. K. (1973). Analysis of optimum parametric combination in electro-chemical machining. *Ann. CIRP*, *22*(1), 59–60.

Bhavar, Kattire, Patil, Khot, Gujar, & Singh. (2014). *A Review on Powder Bed Fusion Technology of Metal Additive Manufacturing*. In 4th International conference and exhibition on Additive Manufacturing Technologies-AM-2014, Bangalore, India.

Bhoi, J. A., Desai, D. A., & Patel, R. M. (2014). The Concept & Methodology of Kaizen. *International Journal of Engineering Development and Research*, *2*(1), 2321–9939.

BI. (2017). *Bilioteca de Indicadores*. Retrieved July 20, 2011, from http://www.bibliotecadeindicadores.com.br/

Bidanda, B., & Bártolo, P. J. (Eds.). (2007). *Virtual prototyping & bio manufacturing in medical applications*. Springer Science & Business Media.

Bing, Q., & Sun, C. T. (2008). Specimen size effect in off-axis compression tests of fiber composites. *Composites. Part B, Engineering*, *39*(1), 20–26. doi:10.1016/j.compositesb.2007.02.010

Boschetto, A., & Bottini, L. (2015). Roughness prediction in coupled operations of fused deposition modeling and barrel finishing. *Journal of Materials Processing Technology*, *219*, 181–192. doi:10.1016/j.jmatprotec.2014.12.021

Boschetto, A., Bottini, L., & Veniali, F. (2016). Finishing of fused deposition modeling parts by CNC machining. *Robotics and Computer-integrated Manufacturing*, *41*, 92–101. doi:10.1016/j.rcim.2016.03.004

Bras, B., & Hammond, R. (1996). Towards Design for Remanufacturing –Metrics for Assessing Remanufacturability. *Proceedings of 1st international workshop on reuse*, 5-22.

Brin, S., & Page, L. (2012). Reprint of: The anatomy of a large-scale hypertextual web search engine. *Computer Networks*, *56*(18), 3825–3833. doi:10.1016/j.comnet.2012.10.007

Bueno, C., Hauschild, M. Z., Rossignolo, J. A., Ometto, A. R., & Mendes, N. C. (2016). Sensitivity analysis of the use of Life Cycle Impact Assessment methods: A case study on building materials. *Journal of Cleaner Production*, *112*, 2208–2220. doi:10.1016/j.jclepro.2015.10.006

Burstein, G. T., Liu, C., & Souto, R. (2005). The effect of temperature on the nucleation of corrosion pits on titanium in Ringer's physiological solution. *Biomaterials*, *26*(3), 245–256. doi:10.1016/j.biomaterials.2004.02.023 PMID:15262467

Carach, J., Hloch, S., Hlavacek, P., Gombar, M., Klichova, D., Botko, F., ... Lehocka, D. (2016). Hydro-abrasive disintegration of alloy Monel K-500- the influence of technological and abrasive factors on the surface quality. *Procedia Engineering*, *149*, 17–23. doi:10.1016/j.proeng.2016.06.633

Carter, D. R., & Beaupre, G. S. (1998). Mechanobiology of skeletal regeneration. *Clinical Orthopaedics and Related Research*, *335*, 541–555. PMID:9917625

Cedero, Siddel, Amelia, & Elliott. (2017). Strengthening of Ferrous Binder Jet 3D Printed Components through Bronze Infiltration. Additive Manufacturing, 15, 87-92.

Cerutti, A. K., Beccaro, G. L., Bagliani, M., Donno, D., & Bounous, G. (2013). Multifunctional Ecological Footprint Analysis for assessing eco-efficiency: A case study of fruit production systems in Northern Italy. *Journal of Cleaner Production*, *40*, 108–117. doi:10.1016/j.jclepro.2012.09.028

Chakraborty, K., Mondal, S., & Mukherjee, K. (2017). Analysis of product design characteristics for remanufacturing using Fuzzy AHP and Axiomatic Design. *Journal of Engineering Design*, *28*(5), 338–368. doi:10.1080/09544828.2017.1316014

Chakradhar, D., & Gopal, A. V. (2011). Multi-objective optimization of electrochemical machining of EN31 steel by grey relational analysis. *International Journal of Modeling and Optimization, 1*(2), 113.

Chalmers, R. E. (2001). Rapid tooling technology from Ford country. *Manufacturing Engineering*, *127*(5), 36.

Chandra, G. (2014). *Bio-implants Market by Product Types- Global Opportunity Analysis and Industry Forecast, 2013 – 2020*. Retrieved from https://www.alliedmarketresearch.com/bio-implants-market

Chaturvedi, T. P. (2009). An overview of the corrosion aspect of dental implants (titanium and its alloys). *Indian Journal of Dental Research*, *20*(1), 91–98. doi:10.4103/0970-9290.49068 PMID:19336868

Cheah, C. M., Chua, C. K., Lee, C. W., Feng, C., & Totong, K. (2015). Rapid prototyping and tooling techniques: A review of applications for rapid investment casting. *International Journal of Advanced Manufacturing Technology*, *25*(3-4), 308–320. doi:10.100700170-003-1840-6

Chen, W., Kucukyazici, B., Verter, V., & Sáenz, M. (2015). Supply chain design for unlocking the value of remanufacturing under uncertainty. *European Journal of Operational Research*, 804-819.

Chen, D., Heyer, S., Ibbotson, S., Salonitis, K., Steingrimsson, J. G., & Thiede, S. (2015). Direct Digital Manufacturing: Definition, Evolution and sustainability Implications. *Journal of Cleaner Production*, *107*, 615–625. doi:10.1016/j.jclepro.2015.05.009

Chenghao, L., Shusen, W., Naibao, H., Zhihong, Z., Shuchun, Z., & Jing, R. (2015). Effects of Lanthanum and Cerium Mixed Rare Earth Metal on Abrasion and Corrosion Resistance of AM60 Magnesium Alloy. *Rare Metal Materials and Engineering*, *44*(3), 521–526. doi:10.1016/S1875-5372(15)30031-X

Chen, L., Siorest, T. E., & Wong, W. C. K. (1996). Kerf Characteristics in Abrasive Waterjet Cutting of Ceramic Materials. *Journal of Machine Tools & Manufacture*, *36*(11), 1201–1206. doi:10.1016/0890-6955(95)00108-5

Chen, X., Li, D., Wu, H., Tang, Y., & Zhao, L. (2011). Analysis of ceramic shell cracking in stereolithography-based rapid casting of turbine blade. *International Journal of Advanced Manufacturing Technology*, *55*(5), 447–455. doi:10.100700170-010-3064-x

Chen, Y.-M., Fan, S.-Y., & Lu, W.-S. (2008). Performance analysis of linear permanent-magnet motors with finite-element analysis. *IEEE Transactions on Magnetics*, *44*(3), 377–385. doi:10.1109/TMAG.2008.915618

Che, Z. H., & Wang, H. S. (2010). Hybrid approach for supplier cluster analysis. *Computers & Mathematics with Applications (Oxford, England)*, *59*(2), 745–763. doi:10.1016/j.camwa.2009.10.018

Chia, H. N., & Wu, B. M. (2015). Recent advances in 3D printing of biomaterials. *Journal of Biological Engineering*, *9*(1), 4. doi:10.118613036-015-0001-4 PMID:25866560

Chinda, T., & Ammarapala, V. (2016). Decision-making on reverse logistics in the construction industry. *Songklanakarin Journal of Science Education and Technology*, 38.

Chiu, Y. Y., Liao, Y. S., & Hou, C. C. (2003). Automatic fabrication for bridged laminated object manufacturing (LOM) process. *Journal of Materials Processing Technology*, *140*(1), 179–184. doi:10.1016/S0924-0136(03)00710-6

Chockalingam, K., Jawahar, N., Ramanathan, K. N., & Banerjee, P. S. (2006). Optimization of stereolithography process parameters for part strength using design of experiments. *International Journal of Advanced Manufacturing Technology*, *29*(1), 79–88. doi:10.100700170-004-2307-0

Choobineh, F., & Jain, V. K. (1993). A fuzzy sets approach for selecting optimum parameters of an ECM process. *Processing of Advanced Materials*, *3*, 225–232.

Chougrani, L., Pernot, J.-P., Véron, P., & Abed, S. (2017). Lattice structure lightweight triangulation for Additive Manufacturing. *Computer Aided Design*, *90*, 95–104. doi:10.1016/j.cad.2017.05.016

Chua, C. K., Chou, S. M., Lin, S. C., Eu, K. H., & Lew, K. F. (1998). Rapid prototyping assisted surgery planning. *International Journal of Advanced Manufacturing Technology*, *14*(9), 624–630. doi:10.1007/BF01192281

Chua, C. K., Hong, K. H., & Ho, S. L. (1999). Rapid tooling technology. Part 2. A case study using arc spray metal tooling. *International Journal of Advanced Manufacturing Technology*, *15*(8), 609–614. doi:10.1007001700050109

Cooper, K. (2001). *Rapid prototyping technology: selection and application*. CRC Press. doi:10.1201/9780203910795

Core Tech System Inc. (n.d.). *Engineering Talk From Core Tech Systems*. Core Tech Systems.

Cowgill, M. C., Harvey, R. J., & Watson, L. T. (1999). A genetic algorithm approach to Cluster analysis. *Computers & Mathematics with Applications (Oxford, England)*, *37*(7), 99–108. doi:10.1016/S0898-1221(99)00090-5

Crasto, A. S., & Kim, R. Y. (1994). The Effects of Constituent Properties on the Compression Strength of Advanced Composites. In S. E. Groves & A. L. Highsmith (Eds.), *Compression Response of Composite Structures, ASTM STP 1185* (pp. 177–192). Philadelphia: American Society for Testing and Materials. doi:10.1520/STP24338S

Cuevas, A., Febrero, M., & Fraiman, R. (2001). Cluster analysis: A further approach based on density estimation. *Computational Statistics & Data Analysis*, *36*(4), 441–459. doi:10.1016/S0167-9473(00)00052-9

Cui, F., Yang, J., Jiao, Y., Yin, Q., Zhang, Y., & Lee, I. S. (2008). Calcium phosphate coating on magnesium alloy for modification of degradation behaviour. *Frontiers of Materials Science in China*, *2*(2), 143–148. doi:10.100711706-008-0024-6

Dadkar, N., Tomar, B. S., & Satapathy, B. K. (2009). Evaluation of flyash-filled and aramid fibre reinforced hybrid polymer matrix composites (PMC) for friction braking applications. *Materials & Design*, *30*(10), 4369–4376. doi:10.1016/j.matdes.2009.04.007

Das, A., Madras, G., Dasgupta, N., & Umarji, A. M. (2003). Binder removal studies in ceramic thick shapes made by laminated object manufacturing. *Journal of the European Ceramic Society*, *23*(7), 1013–1017. doi:10.1016/S0955-2219(02)00266-2

Dash, M. K., Patro, S. K., & Rath, A. K. (2016). Sustainable use of industrial-waste as partial replacement of fine aggregate for preparation of concrete – A review. *International Journal of Sustainable Built Environment*, *5*(2), 484–516. doi:10.1016/j.ijsbe.2016.04.006

Das, M. K., Kumar, K., Barman, T. K., & Sahoo, P. (2014). Optimization of surface roughness and MRR in electrochemical machining of EN31 tool steel using grey-Taguchi approach. *Procedia Materials Science*, *6*, 729–740. doi:10.1016/j.mspro.2014.07.089

Das, M. T., & Dulger, L. C. (2004). Mathematical modelling, simulation and experimental verification of a scara robot. *Simulation Modelling Practice and Theory*, *13*(3), 257–271. doi:10.1016/j.simpat.2004.11.004

Dedoussis, V., Canellidis, V., & Mathioudakis, K. (2008). Aerodynamic experimental investigation using stereolithography fabricated test models: The case of a linear compressor blading cascade. *Virtual and Physical Prototyping*, *3*(3), 151–157. doi:10.1080/17452750802120201

Dedoussis, V., & Giannatsis, J. (2004). Stereolithography assisted redesign and optimisation of a dishwasher spraying arm. *Rapid Prototyping Journal*, *10*(4), 255–260. doi:10.1108/13552540410551388

Delfs, P., Li, Z., & Schmid, H. J. (2015). Mass finishing of laser sintered parts. *Proceedings of the 26th Annual International Solid Freeform Fabrication Symposium*, 514-526.

Deloitte & Touche. (1992). *ISSD. Business strategy for sustainable development: leadership and accountability for the 90s*. Author.

Derzija, B. H., Ahmet, C., Muhamed, M., & Almina, D. (2015). Experimental study on surface roughness in abrasive water jet cutting. *Procedia Engineering*, *100*, 394–399. doi:10.1016/j.proeng.2015.01.383

Desai, A., & Mital, A. (2003). Evaluation of disassemblability to enable design for disassembly in mass production. *International Journal of Industrial Ergonomics*, *32*(4), 265–281. doi:10.1016/S0169-8141(03)00067-2

Despeisse & Ford. (2015). *Centre for technology management working process*. Academic Press.

Despeisse, M. (2017). Sustainable Value Roadmapping Framework for Additive Manufacturing. Academic Press. doi:10.1016/j.procir.2016.11.186

Dhouib, D. (2014). An extension of MACBETH method for a fuzzy environment to analyze alternatives in reverse logistics for automobile tire wastes. *Omega*, *42*(1), 25–32. doi:10.1016/j.omega.2013.02.003

Dixit, S.B. (2006). *Product design: A conceptual development of product remanufacturability index*. Academic Press.

Dolan, P. (2004). Sustainable leadership. *Leader to Leader*, *33*(33), 8–12. doi:10.1002/ltl.82

Dong, Y., Tang, J., Xu, B., & Wang, D. (2005). An application of swarm optimization to nonlinear programming. *Computers & Mathematics with Applications (Oxford, England)*, *49*(11-12), 1655–1668. doi:10.1016/j.camwa.2005.02.006

Dotchev, K., & Soe, S. (2006). Rapid manufacturing of patterns for investment casting: Improvement of quality and success rate. *Rapid Prototyping Journal*, *12*(3), 156–164. doi:10.1108/13552540610670735

Doyle, M., Agarwal, K., Sealy, W., & Schul, K. (2015). Effect of Layer Thickness and Orientation on Mechanical Behaviour of Binder Jet Stainless Steel 420+Bronze Parts. *Procedia Manufacturing*, *1*, 251–262. doi:10.1016/j.promfg.2015.09.016

Durrieu, M. C., Pallu, S., Guillemot, F., Bareille, R., Amédée, J., Baquey, C., ... Dard, M. (2004). Grafting RGD containing peptides onto hydroxyapatite to promote osteoblastic cells adhesion. *Journal of Materials Science. Materials in Medicine*, *15*(7), 779–786. doi:10.1023/B:JMSM.0000032818.09569.d9 PMID:15446238

Dutta, P., Das, D., Schultmann, F., & Fröhling, M. (2016). Design and planning of a closed-loop supply chain with three way recovery and buy-back offer. *Journal of Cleaner Production*, *135*, 604–619. doi:10.1016/j.jclepro.2016.06.108

Dutta, S., Nadaf, M. B., & Mandal, J. N. (2016). An Overview on the Use of Waste Plastic Bottles and Fly Ash in Civil Engineering Applications. *Procedia Environmental Sciences*, *35*, 681–691. doi:10.1016/j.proenv.2016.07.067

Du, Y., Cao, H., Liu, F., Li, C., & Chen, X. (2012). An integrated method for evaluating the remanufacturability of used machine tool. *Journal of Cleaner Production*, *20*(1), 82–91. doi:10.1016/j.jclepro.2011.08.016

El-Dardery, M. A. (1982). Economic study of electrochemical machining. *International Journal of Machine Tool Design and Research*, *22*(3), 147–158. doi:10.1016/0020-7357(82)90023-3

Ellen MacArthur Foundation. (2016). *The New Plastics Economy: Rethinking the future of plastics.* Ellen MacArthur Found.

El-Sayed, M., Afia, N., & El-Kharbotly, A. (2010). A stochastic model for forward–reverse logistics network design under risk. *Computers & Industrial Engineering*, *58*(3), 423–431. doi:10.1016/j.cie.2008.09.040

Ermi, A. M., & James, L. A. (1986). Miniature Center-Cracked-Tension Specimen for Fatigue Crack Growth Testing. In W. R. Corwin & G. E. Lucas (Eds.), *The Use of Small-Scale Specimens for Testing Irradiated Material, ASTM STP 888* (pp. 261–275). Philadelphia: American Society for Testing and Materials. doi:10.1520/STP33009S

España, F., Tsao, C., & Hauser, M. (2012). Driving continuous improvement by developing and leveraging lean key performance indicators. *20th Conference of the International Group for Lean Construction.*

Esteves, S., Lourenço, E. J., Moita, N., Peças, P., Ribeiro, I., & Henriques, E. (2014). Injection Moulding Process Indicators to Foster a More Sustainable Production of Plastic Parts. In 3rd workbook of the cross-sectional group 'Energy-related technical and economic evaluation' of the Cluster of Excellence eniPROD. Wissenschaftliche Scripten.

Everton, S. K., Hirsch, M., Stravroulakis, P., Leach, R. K., & Clare, A. T. (2016). Review of in-situ process monitoring and in-situ metrology for metal Additive Manufacturing. *Materials & Design*, *95*, 431–445. doi:10.1016/j.matdes.2016.01.099

Fang & Li. (2013). Four degrees of freedom SCARA robot kinematics modelling and simulation analysis. *International Journal of Computer, Consumer and Control*, *2*(4), 20-27.

Fang, L., Kong, X. L., Su, J. Y., & Zhou, Q. D. (1993). Movement patterns of abrasive particles in three-body abrasion. *Wear*, *162/164*, 782–789. doi:10.1016/0043-1648(93)90079-2

Fang, L., Kong, X. L., & Zhou, Q. D. (1992). A wear tester capable of monitoring and evaluating the movement pattern of abrasive particles in three-body abrasion. *Wear*, *159*(1), 115–120. doi:10.1016/0043-1648(92)90292-G

Fanning, K. (2016). Big Data and KPIs: A Valuable Connection. *Journal of Corporate Accounting & Finance*, *27*(3), 17–19. doi:10.1002/jcaf.22137

Fathi, M., Zandi, F., & Jouini, O. (2015). Modeling the merging capacity for two streams of product returns in remanufacturing systems. *Journal of Manufacturing Systems*, *37*, 265–276. doi:10.1016/j.jmsy.2014.08.006

Fedak, Durovsky, & Uveges. (2014). Analysis of Robotic System Motion in SimMechanics and MATLAB GUI Environment. *MATLAB Applications for the Practical Engineer*, 565-581. doi:10.5772/58371

Feng, H., Song, Y., Zuo, Z., Shang, J., Wang, Y., & Roskilly, A. P. (2015). Stable Operation and Electricity Generating Characteristics of a Single-Cylinder Free Piston Engine Linear Generator: Simulation and Experiments. *Energies*, *8*(2), 765–785. doi:10.3390/en8020765

Feng, L., Govindan, K., & Li, C. (2017). Strategic planning: Design and coordination for dual-recycling channel reverse supply chain considering consumer behavior. *European Journal of Operational Research*, *260*(2), 601–612. doi:10.1016/j.ejor.2016.12.050

Ferrer, G. (2001). Theory and Methodology On the widget of remanufacturing operation. *European Journal of Operational Research*, *135*(2), 373–393. doi:10.1016/S0377-2217(00)00318-0

Fleury, E., & Ha, J. S. (1998). Small punch tests to estimate the mechanical properties of steels for steam power plant: I. Mechanical strength. *International Journal of Pressure Vessels and Piping*, *75*(9), 699–706. doi:10.1016/S0308-0161(98)00074-X

Flowers, J., & Moniz, M. (2002). Rapid prototyping in technology education. *Technology Teacher*, *62*(3), 7.

Folgado, R., Peças, P., & Henriques, E. (2010). Life cycle cost for technology selection: A Case study in the manufacturing of injection moulds. International Journal of Production Economics, 368–78. doi:10.1016/j.ijpe.2010.07.036

Ford & Despeisse. (2017). *Additive Manufacturing and sustainability: an exploratory study of the advantages and challenges*. Academic Press. . doi:10.1016/j.jclepro.2016.04.150

Fowler, G., Pashby, I. R., & Shipway, P. H. (2009). The effect of particle hardness and shape when abrasive water jet milling titanium alloy Ti6Al4V. *Wear*, *266*(7-8), 613–620. doi:10.1016/j.wear.2008.06.013

Fu, J., Liu, K., Duc, W., Wang, Z., Li, S., & Du, X. (2017). Microstructure and mechanical properties of the as-cast Mg-Zn-Mn-Ca alloys. *IOP Conf. Series: Materials Science and Engineering*, 182.

Funk, J. G., & Sykes, G. F. (1986, Fall). The Effects of Radiation on the Interlaminar Fracture Toughness of a Graphite/Epoxy Composite. *Journal of Composites Technology & Research*, *8*(3), 92–97. doi:10.1520/CTR10328J

Gantar, G., Glojek, A., Mori, M., Nardin, B., & Sekavčnik, M. (2013). Resource efficient injection moulding with low environmental impacts. *Stroj Vestnik/Journal. Mechanical Engineering (New York, N.Y.)*, *59*(3), 193–200.

Gao, M., Liu, K., & Wu, Z. (2010). Personalisation in web computing and informatics: Theories, techniques, applications, and future research. *Information Systems Frontiers, 12*(5), 607-629.

Gao, M., & Wu, Z. (2010). Incorporating Personalized Contextual Information in Item-based Collaborative Filtering Recommendation. *JSW*, *5*(7), 729–736. doi:10.4304/jsw.5.7.729-736

Gao, M., Wu, Z., & Jiang, F. (2011). Userrank for item-based collaborative filtering recommendation. *Information Processing Letters*, *111*(9), 440–446. doi:10.1016/j.ipl.2011.02.003

Gebler & Visser. (2014). A Global Sustainability Perspective on 3D Printing Technologies. *Energy Policy*, *74*(c), 158–167.

Geetha, M., Singh, A. K., Asokamani, R., & Gogia, A. K. (2009). Ti based biomaterials, the ultimate choice for orthopaedic implants – A review. *Progress in Materials Science*, *54*(3), 397–425. doi:10.1016/j.pmatsci.2008.06.004

Gelbard, R., Carmeli, A., Bittmann, R. M., & Ronen, S. (2009). Cluster analysis using multi-algorithm voting in cross-cultural studies. *Expert Systems with Applications*, *36*(7), 10438–10446. doi:10.1016/j.eswa.2009.01.071

Gettleman, L., Nathanson, D., & Myerson, R. L. (1977). Effect of rapid curing procedures on polymer implant materials. *The Journal of Prosthetic Dentistry*, *37*(1), 74–82. doi:10.1016/0022-3913(77)90195-0 PMID:264323

Ghosh, D., Doloi, B., & Das, P. K. (2015). Parametric analysis and optimization on abrasive water jet cutting of silicon nitride ceramics. *Journal of Precision Technology*, *5*(3/4), 294–311. doi:10.1504/IJPTECH.2015.073833

Gibson, I., Rosen, D. W., & Stucker, B. (2010). *Additive Manufacturing Technologies Rapid Prototyping to Direct Digital Manufacturing*. Springer-Verlag New York.

Gibson, I., Rosen, D., & Stucker, B. (2010). *Additive Manufacturing Technologies* (2nd ed.). Springer. doi:10.1007/978-1-4419-1120-9

Gieras, J. F. (2008). *Advancements in electric machines*. Springer.

Gieras, J. F., Piech, Z. J., & Tomczuk, B. (2011). *Linear synchronous motors: transportation and automation systems*. CRC Press.

Gieras, J. F., Wang, R.-J., & Kamper, M. J. (2008). *Axial flux permanent magnet brushless machines* (Vol. 1). Springer. doi:10.1007/978-1-4020-8227-6

Giri, B., Chakraborty, A., & Maiti, T. (2017). Pricing and return product collection decisions in a closed-loopsupply chain with dual-channel in both forward and reverse logistics. *Journal of Manufacturing Systems*, *42*, 104–123. doi:10.1016/j.jmsy.2016.11.007

Glasschroeder, J., Prager, E., & Zaeh, M. F. (2015). Powder Bed Based 3D- Printing of Function Integrated Parts. *Rapid Prototyping Journal*, *21*(2), 207–215. doi:10.1108/RPJ-12-2014-0172

Global Reporting Initiative. (2006). *Sustainability Reporting Guidelines*. Author.

Global Reporting Initiative. (2014). *G4 Sustainability Reporting Guidelines*. Glob Reporting Initiative. Available from: https://www.globalreporting.org/standards/g4/Pages/default.aspx

Goldin, I., & Winters, L. A. (1995). *The Economics of Sustainable Development*. Cambridge, UK: University of Cambridge Press. doi:10.1017/CBO9780511751905

Gong, G., Zhang, B., Zhang, H., & Li, W. (2006). Pressure less sintering of machinable $Al_2O_3/LaPO_4$ composites in N_2 atmosphere. *International Journal of Ceramics International*, *32*(3), 349–352. doi:10.1016/j.ceramint.2005.03.002

Gonzalez, J. A., Mireles, J., Lin, Y., & Wicker, R. B. (2016). Characterization of Ceramic Components Fabricated using Binder Jetting Additive Manufacturing Technology. Academic Press.

Gori, M., Pucci, A., Roma, V., & Siena, I. (2007, January). ItemRank: A Random-Walk Based Scoring Algorithm for Recommender Engines. *IJCAI (United States)*, *7*, 2766–2771.

Gouasmi, M., Ouali, M., Fernini, B., & Meghatria, M. (2012). Kinematic Modelling and Simulation of a 2-R Robot Using SolidWorks and Verification by MATLAB/Simulink. *International Journal of Advanced Robotic Systems*, *9*(6), 245. doi:10.5772/50203

Govindan, K., Fattahi, M., & Keyvanshokooh, E. (2017). Supply chain network design under uncertainty: A comprehensive review and future research directions. *European Journal of Operational Research*, *263*(1), 108–141. doi:10.1016/j.ejor.2017.04.009

Greer, S. E. (2009). A comparison of the ancient metal casting materials and processes to modern metal casting materials and processes. Hartford, CT: Rensselaer Polytechnic Institute.

Guarnieri, P., Sobreiro, V. A., Nagano, M. S., & Marques Serrano, A. L. (2015). The challenge of selecting and evaluating third-party reverse logistics providers in a multicriteria perspective: A Brazilian case. *Journal of Cleaner Production*, *96*, 209–219. doi:10.1016/j.jclepro.2014.05.040

Gudimetla, P., Wang, J., & Wong, W. (2002). Kerf formation analysis in the abrasive waterjet cutting of industrial ceramics. *Journal of Materials Processing Technology*, *128*(1-3), 123–129. doi:10.1016/S0924-0136(02)00437-5

Gu, F., Hall, P., & Miles, N. J. (2016). Development of composites based on recycled polypropylene for injection moulding automobile parts using hierarchical clustering analysis and principal component estimate. *Journal of Cleaner Production, 137*, 632–643. doi:10.1016/j.jclepro.2016.07.028

Gunduz, O., Daglilar, S., Salman, S., Ekren, N., Agathopoulos, S., & Oktar, F. N. (2008). Effect of Yttria-doping on Mechanical Properties of Bovine Hydroxyapatite (BHA). *Journal of Composite Materials, 42*(13), 1281–1287. doi:10.1177/0021998308092196

Guo, N., & Leu, M. C. (2013, September 1). Additive manufacturing: Technology, applications and research needs. *Frontiers of Mechanical Engineering, 8*(3), 215–243. doi:10.100711465-013-0248-8

Gu, X., Zhou, W., Zheng, Y., Dong, L., Xi, Y., & Chai, D. (2010). Microstructure, mechanical property, bio-corrosion and cytotoxicity evaluations of Mg/HA composites. *Materials Science and Engineering C, 30*(6), 827–832. doi:10.1016/j.msec.2010.03.016

Hague, R., Mansour, S., & Saleh, N. (2004). Material and design considerations for rapid manufacturing. *International Journal of Production Research, 42*(22), 4691–4708. doi:10.1080/00207840410001733940

Halidi, S. N., & Abdullah, J. (2012). Moisture effects on the ABS used for fused deposition modeling rapid prototyping machine. In *Humanities, Science and Engineering Research (SHUSER), IEEE Symposium* (pp. 839-843). IEEE. 10.1109/SHUSER.2012.6268999

Halloran, J. W., Tomeckova, V., Gentry, S., Das, S., Cilino, P., Yuan, D., ... Alabi, T. R. (2011). Photopolymerization of powder suspensions for shaping ceramics. *Journal of the European Ceramic Society, 31*(14), 2613–2619. doi:10.1016/j.jeurceramsoc.2010.12.003

Hamu, G. B., Eliezer, D., & Shin, K. S. (2007). The role of Si and Ca on new wrought Mg–Zn–Mn based alloy. *Materials Science and Engineering A, 447*(1-2), 35–43. doi:10.1016/j.msea.2006.10.059

Hanan, D., & Burnley, S. (2013). A multi-criteria decision analysis assessment of waste paper management options. *Waste Management*, 566-573.

Harlan, N. R., Bourell, D. L., Beaman, J. J., & Reyes, R. (2001). Titanium castings using laser-scanned data and selective laser-sintered zirconia molds. *Journal of Materials Engineering and Performance, 10*(4), 410–413. doi:10.1361/105994901770344818

Harry Asada, H. (2016, March 12). *Introduction to Robotics*. Retrieved from https://ocw.mit.edu/courses/mechanical-engineering/2-12-introduction-to-robotics-fall-2005/lecture-notes/chapter4.pdf

Harun, W.S., Safian, S., & Idris, M.H. (2009). Evaluation of ABS patterns produced from FDM for investment casting process. *Computational Method and Experiments in Materials Characterisation IV, 1*, 319-28.

Hascalik, A., Çaydaş, U., & Gürün, H. (2007). Effect of traverse speed on abrasive waterjet machining of Ti–6Al–4V alloy. *Materials & Design, 28*(6), 1953–1957. doi:10.1016/j.matdes.2006.04.020

Hassel, T., Bach, F. W., & Krause, C. (2007). Influence of alloy composition on the mechanical and electrochemical properties of binary Mg-Ca alloys and its corrosion behaviour in solutions at different chloride concentrations. *Proc. 7th Int. Conf. Magnesium Alloys and Their Applications*, 789–95.

Hatefi, S., & Jolai, F. (2014). Robust and reliable forward–reverse logistics network design under demand uncertainty and facility disruptions. *Applied Mathematical Modelling, 38*(9-10), 2630–2647. doi:10.1016/j.apm.2013.11.002

Heese, H., Cattani, K., Ferrer, G., Gilland, W., & Roth, A. (2005). Competitive advantage through take-back of used products. *European Journal of Operational Research, 164*(1), 143–157. doi:10.1016/j.ejor.2003.11.008

Heine, R. W., Loper, C. R., & Rosenthal, P. C. (1955). *Principles of metal casting*. Tata McGraw-Hill Education.

Helu, M., Behmann, B., Meier, H., Dornfeld, D., Lanza, G., & Schulze, V. (2012). Impact of green machining strategies on achieved surface quality. *CIRP Annals-Manufacturing Technology.*, *61*(1), 55–58. doi:10.1016/j.cirp.2012.03.092

Henning, K., Wolfgang, W., & Johannes, H. (2013). *Recommendations for implementing the strategic initiative INDUSTRIE 4.0.* Final Report of the Industrie 4.0 WG, (April), 82. doi:10.13140/RG.2.1.1205.8966

Hermann, M., Pentek, T., & Otto, B. (2015). Design Principles for Industrie 4.0 Scenarios: A Literature Review. *Technische Universitat Dortmund, 1*(1), 4–16. doi:10.1109/HICSS.2016.488

Hernandez, V., Bravo, G., Jose de Jesus, R., & Pacheco, J. (2011). Kinematics for the SCARA and the Cylindrical Manipulators. *ICIC Express Letters. Part B, Applications*, *2*(2), 421–425.

Hloch, S., Valicek, J., & Simkulet, V. (2009). Estimation of the smooth zone maximal depth at surfaces created by Abrasive Waterjet. *Journal of Surface Science and Engineering*, *3*, 347–359.

Hofstatter, Pedersen, Tosello, & Hansen. (2017). *Applications of Fiber-Reinforced Polymers in Additive Manufacturing*. Academic Press. doi:10.1016/j.procir.2017.03.171

Hollenstein, H. (2003). Innovation modes in the Swiss service sector: A cluster analysis based on firm-level data. *Research Policy*, *32*(5), 845–863. doi:10.1016/S0048-7333(02)00091-4

Hu & Mahadevan. (2017). *Uncertainty quantification in prediction of material properties during Additive Manufacturing*. Academic Press. . doi:10.1016/j.scriptamat.2016.10.014

Hu, X., & Eberhart, R. (2002). Multiobjective optimization using dynamic neighborhood particle swarm optimization. In *Evolutionary Computation, 2002. CEC'02. Proceedings of the 2002 Congress on* (*Vol. 2*, pp. 1677-1681). IEEE.

Huang, Liu, Mokasdar, & Hou. (2013). *Additive Manufacturing and its societal impact: a literature review*. Academic Press. doi:10.100700170-012-4558-5

Huang, Riddle, Graziano, Warren, Das, Nimbalkar, Cresko, & Masanet. (2015). Energy and emissions saving potential of Additive Manufacturing: the case of lightweight aircraft components. *Journal of Cleaner Production*, 1-12.

Hukkinen, J. (2001). Eco-efficiency as abandonment of nature. *Ecological Economics*, *38*(3), 311–315. doi:10.1016/S0921-8009(01)00217-8

Huppes, G., & Ishikawa, M. (2005). A Framework for Quantified Eco-efficiency Analysis. *Journal of Industrial Ecology*, *9*(4), 25–41. Available from: http://doi.wiley.com/10.1162/108819805775247882

Hurley, B. (2012). Use QDIP Sheets to Identify Environmental Issues. *Lean Six Sigma Environment*. Retrieved from http://leansixsigmaenvironment.org/index.php/use-qdip-sheets-to-identify-environmental-issues/

Hur, T., Lee, J., Ryu, J., & Kwon, E. (2005). Simplified LCA and matrix methods in identifying the environmental aspects of a product system. *Journal of Environmental Management*, *75*(3), 229–237. doi:10.1016/j.jenvman.2004.11.014 PMID:15829365

Huszar, M., Belblidia, F., Davies, H. M., Arnold, C., Bould, D., & Sienz, J. (2015). Sustainable injection moulding: The impact of materials selection and gate location on part warpage and injection pressure. *Sustainable Materials and Technologies*, *5*, 1–8. doi:10.1016/j.susmat.2015.07.001

Hutyrova, Z., Scuska, J., Hloch, S., Hlava, P., & Zele, M. (2016). Turning of wood plastic composites by water jet and abrasive water jet. *Journal of Advance Manufacturing Technology*, *84*, 1615–1624.

Hwa-Hsing, T., Ming-Lu, C., & Hsiao-Chuan, Y. (2011). Slurry based selective laser sintering of polymer-coated ceramic powders to fabricate high strength alumina parts. *Journal of the European Ceramic Society*, *31*(8), 1383–1388. doi:10.1016/j.jeurceramsoc.2011.02.020

Ibrahim, T. (2009). *Short-stroke, single-phase tubular permanent magnet motors for refrigeration applications*. The University of Sheffield.

iF. (2017). *10 ways to make your Visual Management Boards work*. Retrieved from https://www.industryforum.co.uk/resources/blog/10-ways-to-make-your-visual-management-boards-work/

Ijomah, W. L., McMahon, C. A., Hammond, G. P., & Newman, S. T. (2007). Development of design for remanufacturing guidelines to support sustainable manufacturing. *Robotics and Computer-integrated Manufacturing*, *23*(6), 712–719. doi:10.1016/j.rcim.2007.02.017

Ijomah, W., McMahon, C., Hammond, G., & Newman, S. (2007). Development of robust design-for-remanufacturing guidelines to further the aims of sustainable development. *International Journal of Production Research*, *45*(18), 4513–4536. doi:10.1080/00207540701450138

Innovating Clean Energy Technologies in Advanced Manufacturing. (2015). *Quadrennial Technology Review 2015, Technology Assessments*, 1-35.

International Organization of Standardization. (2006). *UNE-EN ISO 14040 Environmental Management – Life Cycle Assessment –Principles and Framework*. Author.

Ionescu, F. (2007). Modelling and Simulation in Mechatronics. In IFAS inter.confer.MCPL2007 (pp. 26-29). Sibiu, Romania: Academic Press.

Ionescu, F., Chojnowski, F., & Constantin, G. (2002). Virtual Reality in Mechanical Engineering: Modelling and Simulation with Solid Dynamics. *ARA-Journal*, *1*, 27.

ISO. (2012). ISO 14045:2012 Preview Environmental management -- Eco efficiency assessment of product systems -- Principles, requirements and guidelines.

Jackson, Van Asten, Morrow, Min, & Pfefferkorn. (2016). *A Comparison of Energy Consumption in Wire-Based and Powder-Based Additive-Subtractive Manufacturing*. Academic Press. doi:10.1016/j.promfg.2016.08.087

Jain, A. K., & Dubes, R. C. (1998). *Algorithms for clustering data*. Prentice Hall.

Jain, N. K., & Jain, V. K. (2007). Optimization of electro-chemical machining process parameters using genetic algorithms. *Machining Science and Technology*, *11*(2), 235–258. doi:10.1080/10910340701350108

Jang, S.-M., Choi, J.-Y., Cho, H.-W., & Lee, S.-H. (2005). Dynamic characteristic analysis and experiments of moving-magnet linear actuator with cylindrical Halbach array. *IEEE Transactions on Magnetics*, *41*(10), 3814–3816. doi:10.1109/TMAG.2005.854931

Jansson, P. M., Gregory, M. J., Barlow, C., Phaal, R., Farrukh, C. J. P., & Probert, D. R. (2000). *Industrial sustainability a review of UK and international research and capabilities*. Cambridge, UK: University of Cambridge.

Jasch, C. (2000). Environmental performance evaluation and indicators. *Journal of Cleaner Production*, *8*(1), 79–88. doi:10.1016/S0959-6526(99)00235-8

Jaya Christiyan, K. G., Chandrasekhar, U., & Venkateswarlu, K. (2016). A study on the influence of process parameters on the Mechanical Properties of 3D printed ABS composite. *IOP Conf. Series: Materials Science and Engineering, 114*. doi:10.1088/1757-899X/114/1/012109

Jazar, R. N. (2009). *Theory of applied robotics*. New York: Springer.

Jha, Dutta, & Saha. (2014). Analysis of Dynamics of SCORA-ER14 Robot in MATLAB. *International Journal of Innovative Research in Advanced Engineering*, *1*(4), 145–150.

Jiang, F., & Gao, M. (2009). Collaborative filtering approach based on item and personalized contextual information. *Proceedings of the International Symposium on Intelligent Information Systems and Applications Qingdao*, 63-66.

John, S., Sridharan, R., Kumar, P., & Krishnamoorthy, M. (2017). Multi-period reverse logistics network design for used refrigerators. *Applied Mathematical Modelling*, *54*, 311–331. doi:10.1016/j.apm.2017.09.053

Jones, S., & Yuan, C. (2003). Advances in shell moulding for investment casting. *Journal of Materials Processing Technology*, *135*(2), 258–265. doi:10.1016/S0924-0136(02)00907-X

Jun, H.-B., Lee, D.-H., Kim, J.-G., & Kiritsis, D. (2012). Heuristic algorithms for minimising total recovery cost of end-of-life products under quality constraints. *International Journal of Production Research*, *50*(19), 5330–5347. doi:10.1080/00207543.2011.624562

Kahlin, M., Ansell, H., & Moverare, J. J. (2017). Fatigue behaviour of notched additive manufactured Ti6Al4V with as-built surfaces. *International Journal of Fatigue*, *101*, 51–60. doi:10.1016/j.ijfatigue.2017.04.009

Kahn, S. R. (2013). Key performance indicators: The 75 measures every manager needs to know. *Choice: Current Reviews for Academic Libraries*, *50*(5), 926.

Kalpakjian, S. (1984). *Manufacturing processes for engineering materials*. Pearson Education India.

Kalsoom, Peristyy, Nesterenko, & Paull. (n.d.). *A 3D printable diamond polymer composite: A novel material for fabrication of low cost thermally conducting devices*. Royal Society of Chemistry.

Kamachimudali, U., & Sridhar, T. (2003). Corrosion of bio implants. *Sadhana Academy Proceedings in Engineering Sciences*, *28*, 601–637.

Kaplan, R. S., & Norton, D. P. (2005). The balanced scorecard: Measures That drive performance. *Harvard Business Review*.

Kasada, R., Ono, H., & Kimura, A. (2006). Small specimen test technique for evaluating fracture toughness of blanket structural materials. *Fusion Engineering and Design*, *81*(8-14), 981–986. doi:10.1016/j.fusengdes.2005.08.088

Kaya, O., Bağcı, F., & Turkay, M. (2014). Planning of capacity, production and inventory decisions in a generic reverse supply chain under uncertain demand and returns. *International Journal of Production Research*, *52*(1), 270–282. doi:10.1080/00207543.2013.838330

Kechagias, J. (2007). An experimental investigation of the surface roughness of parts produced by LOM process. *Rapid Prototyping Journal*, *13*(1), 17–22. doi:10.1108/13552540710719172

Kellens, Mertens, Paraskevas, Dewulf, & Duflou. (2017). *Environmental Impact of Additive Manufacturing Processes: Does AM contribute to a more sustainable way of part manufacturing?* Academic Press. doi:. doi:10.1016/j.procir.2016.11.153

Kennedy, R. (1995, November). J. and Eberhart, Particle swarm optimization. *Proceedings of IEEE International Conference on Neural Networks IV*, 1000. 10.1109/ICNN.1995.488968

Kerr, W., & Ryan, C. (2001). Eco-efficiency gains from remanufacturing: A case study of photocopier remanufacturing at Fuji Xerox Australia. *Journal of Cleaner Production*, *9*(1), 75–81. doi:10.1016/S0959-6526(00)00032-9

Kęsy, A., & Kotliński, J. (2010). Mechanical properties of parts produced by using polymer jetting technology. *Archives of Civil and Mechanical Engineering, 10*(3), 37–50. doi:10.1016/S1644-9665(12)60135-6

Khalil, K. A., & Almajid, A. (2012). Effect of high-frequency induction heat sintering conditions on the microstructure and mechanical properties of nanostructured magnesium/ hydroxyapatite nanocomposites. *Materials & Design, 36*, 58–68. doi:10.1016/j.matdes.2011.11.008

Khan, A. A., & Haque, M. M. (2007). Performance of different abrasive materials during abrasive water jet machining of glass. *Journal of Materials Processing Technology, 191*(1-3), 404–407. doi:10.1016/j.jmatprotec.2007.03.071

Khanra, A. K., Jung, H. C., Yu, S. H., Hong, K. S., & Shin, K. S. (2010). Microstructure and mechanical properties of Mg–HAP composites. *Bulletin of Materials Science, 33*(1), 43–47. doi:10.100712034-010-0006-z

Kienapfel, H., Sprey, C., Wilke, A., & Griss, P. (1999). Implant fixation by bone ingrowth. *The Journal of Arthroplasty, 14*(3), 355–368. doi:10.1016/S0883-5403(99)90063-3 PMID:10220191

Kima, B. J., Kasadaa, R., Kimuraa, A., & Tanigawab, H. (2011). Effects of specimen size on fracture toughness of phosphorous added F82H steels. *Fusion Engineering and Design, 86*(9-11), 2403–2408. doi:10.1016/j.fusengdes.2011.04.001

Kim, H., Choi, J. W., & Wicker, R. (2010). Scheduling and process planning for multiple material stereolithography. *Rapid Prototyping Journal, 16*(4), 232–240. doi:10.1108/13552541011049243

Kim, R. Y., Crasto, A. S., & Yum, Y. J. (1994). Analysis of a Miniature Sandwich Compression Specimen. In S. E. Groves & A. L. Highsmith (Eds.), *Compression Response of Composite Structures, ASTM STP 1185* (pp. 338–350). Philadelphia: American Society for Testing and Materials. doi:10.1520/STP24347S

Kim, Y., Hong, M.-H., Lee, S. H., Kim, E.-P., Lee, S., & Noh, J.-W. (2006). The Effect of Yttrium Oxide on the Sintering Behavior and Hardness of Tungsten. *Metals and Materials International, 12*(3), 245–248. doi:10.1007/BF03027538

Klahn, C. (2016). Design Guidelines for Additive Manufactured Snap-Fit Joints. Academic Press. doi:10.1016/j.procir.2016.04.130

Kleber, R., Minner, S., & Kiesmüller, G. (2002). A continuous time inventory model for a product recovery system with multiple options. *International Journal of Production Economics, 79*(2), 121–141. doi:10.1016/S0925-5273(02)00256-6

Klimek, L., Klein, H. M., Schneider, W., Mosges, R., Schmelzer, B., & Voy, E. D. (1993). Stereolithographic modelling for reconstructive head surgery. *Acta Oto-Rhino-Laryngologica Belgica, 47*(3), 329–334. PMID:8213143

Klipfolio. (2017). *KPI Examples*. Retrieved July 20, 2011, from https://www.klipfolio.com/resources/kpi-examples

Kneese, A. (2003). *Environmental Degradation and Institutional Responses: Vol. 1. Handbook of Environmental Economics*. Available from: http://www.sciencedirect.com/science/article/pii/S1574009903010039

Knights, M. (2001). Rapid tooling is ready for prime time. *Plastics Technology, 47*(1), 46–53.

Kopac, J., & Krajnik, P. (2007). Robust design of flank milling parameters based on grey-taguchi method. *Journal of Materials Processing Technology, 191*(1-3), 400–403. doi:10.1016/j.jmatprotec.2007.03.051

Korniejenko, Mikuła, & Łach. (2015). Fly Ash Based Fiber-Reinforced Geopolymer Composites as the Environmental Friendly Alternative to Cementitious Materials. *Proceedings of 2015 International Conference on Bio-Medical Engineering and Environmental Technology*, 164-171.

Korol, J., Burchart-Korol, D., & Pichlak, M. (2016). Expansion of environmental impact assessment for eco-efficiency evaluation of biocomposites for industrial application. *Journal of Cleaner Production, 113*, 144–152. doi:10.1016/j.jclepro.2015.11.101

KPIlibrary. (2017). *KPI library*. Retrieved July 20, 2011, from http://kpilibrary.com/

Kruth, J. P. (1991). Material incress manufacturing by rapid prototyping techniques. *CIRP Annals-Manufacturing Technology*, *40*(2), 603–614. doi:10.1016/S0007-8506(07)61136-6

Kruth, J. P., Levy, G., Klocke, F., & Childs, T. H. (2007). Consolidation phenomena in laser and powder-bed based layered manufacturing. *CIRP Annals-Manufacturing Technology*, *56*(2), 730–759. doi:10.1016/j.cirp.2007.10.004

Kruth, J. P., Mercelis, P., Van Vaerenbergh, J., Froyen, L., & Rombouts, M. (2005). Binding mechanisms in selective laser sintering and selective laser melting. *Rapid Prototyping Journal*, *11*(1), 26–36. doi:10.1108/13552540510573365

Kruth, J. P., Wang, X., Laoui, T., & Froyen, L. (2003). Lasers and materials in selective laser sintering. *Assembly Automation*, *23*(4), 357–371. doi:10.1108/01445150310698652

Kulak, M., Nemecek, T., Frossard, E., & Gaillard, G. (2016). Eco-efficiency improvement by using integrative design and life cycle assessment. The case study of alternative bread supply chains in France. *Journal of Cleaner Production*, *112*, 2452–2461. doi:10.1016/j.jclepro.2015.11.002

Kumar, Chincholi, Hegde, Shivagiri, Revanasiddappa, & Kazi. (2015). Experimental Research on Hardness & Flexural properties of Teak Wood Powder and Fly Ash reinforcement with Epoxy. *International Journal of Current Engineering and Scientific Research, 2*(11), 34-37.

Kumar, Madhusoodanan, & Rupani. (2006). Miniature specimen technique as an NDT tool for estimation of service life of operating pressure equipment. *International Conference & Exhibition on Pressure Vessel and Piping*.

Kumar, S., & Kruth, J. P. (2010). Composites by rapid prototyping technology. *Materials & Design*, *31*(2), 850–856. doi:10.1016/j.matdes.2009.07.045

Kunjan, C., Jawahar, N., Chandrasekhar, U., Praveen, J., & Karthic, M. (2016). Development of Process Model for Optimal Selection of Process Parameters for Geometric Tolerances and Surface Roughness in Stereolithography. *International Journal of Advanced Design and Manufacturing Technology*, *9*(3), 103–107.

Kurtza, S. M., Jewetta, C. W., & Bergstr, J. S. (2002). Miniature specimen shear punch test for UHMWPE used in total joint replacements. *Biomaterials*, *23*(9), 1907–1919. doi:10.1016/S0142-9612(01)00316-7 PMID:11996031

Kütük, M. E., Halicioglu, R., & Dulger, L. C. (2015). Kinematics and Simulation of a Hybrid Mechanism: MATLAB/SimMechanics. *Journal of Physics: Conference Series*, *574*, 451–458. doi:10.1088/1742-6596/574/1/012016

Lala, S., Maity, T. N., Singh, M., Biswas, K., & Pradhan, S. K. (2017). Effect of doping (Mg,Mn,Zn) on the microstructure and mechanical properties of spark plasma sintered hydroxyapatites synthesized by mechanical alloying. *Ceramics International*, *43*(2), 2389–2397. doi:10.1016/j.ceramint.2016.11.027

Lan, P. T., Chou, S. Y., Chen, L. L., & Gemmill, D. (1997). Determining fabrication orientations for rapid prototyping with stereolithography apparatus. *Computer Aided Design*, *29*(1), 53–62. doi:10.1016/S0010-4485(96)00049-8

leanmanufacture. (2017). *Lean KPI's - Key Performance Indicators and performance metrics*. Retrieved from http://www.leanmanufacture.net/kpi.aspx

Lee, E. L., Kuo, T. T., & Lin, S. D. (2016). *A Collaborative Filtering-Based Two Stage Model with Item Dependency for Course Recommendation*. Academic Press.

Lee, J. H., Kim, H. Y., Jun, M. J., & Lee, S. C. (2011). *Optimum shape design of single-sided linear induction motors using response surface methodology and finite element method.* Paper presented at the Electrical Machines and Systems (ICEMS), 2011 International Conference on. 10.1109/ICEMS.2011.6073769

Lee, A., Cheng, C. H., & Balakrishnan, J. (1998). Software development cost estimation: Integrating neural network with cluster analysis. *Information & Management*, *34*(1), 1–9. doi:10.1016/S0378-7206(98)00041-X

Lee, C. W., Chua, C. K., Cheah, C. M., Tan, L. H., & Feng, C. (2004). Rapid investment casting: Direct and indirect approaches via fused deposition modelling. *International Journal of Advanced Manufacturing Technology*, *23*(1-2), 93–101. doi:10.100700170-003-1694-y

Lee, J., & Soutis, C. (2008). Measuring the notched compressive strength of composite laminates: Specimen size effects. *Composites Science and Technology*, *68*(12), 2359–2366. doi:10.1016/j.compscitech.2007.09.003

Lee, M. G., Choi, Y.-M., Lee, S. Q., Lim, D., & Gweon, D.-G. (2006). Design of high precision linear stage with double-sided multi-segmented trapezoidal magnet array and its compensations for force ripples. *Mechatronics*, *16*(6), 331–340. doi:10.1016/j.mechatronics.2006.01.006

Lehni, M. (2000). Eco-efficiency. Creating more value with less impact. *World Bus Counc Sustain Dev*, 1–32. Available from: http://www.wbcsd.org

Li & Soar. (2009). Characterization of Process for Embedding SiC Fibers in Al 6061 O Matrix Through Ultrasonic Consolidation. *Journal of Engineering Materials and Technology, 131*(2).

Li, Kucukkoc, & Zhang. (2017). *Production planning in Additive Manufacturing and 3D printing*. Academic Press. . doi:10.1016/j.cor.2017.01.013

Liang, G. S., Chou, T. Y., & Chen Han, T. C. (2005). Cluster analysis based on fuzzy equivalence relation. *European Journal of Operational Research*, *166*(1), 160–171. doi:10.1016/j.ejor.2004.03.018

Lim, S., Buswell, R. A., Le, T. T., Austin, S. A., Gibb, A. G. F., & Thorpe, T. (2012). Development in Construction Scale Additive Manufacturing Processes. *Automation in Construction*, *21*, 262–268. doi:10.1016/j.autcon.2011.06.010

Lindemann, Jahnke, Moi, & Koch. (2012). *Analyzing Product Lifecycle Costs for a Better Understanding of Cost Drivers in Additive Manufacturing*. Academic Press.

Linden, G., Smith, B., & York, J. (2003). Amazon. com recommendations: Item-to-item collaborative filtering. *IEEE Internet Computing*, *7*(1), 76–80. doi:10.1109/MIC.2003.1167344

Liu, B. (2007). *Web data mining: exploring hyperlinks, contents, and usage data*. Springer Science & Business Media.

Liu, X., & Niebur, G. L. (2008). Bone ingrowth into a porous coated implant predicted by mechano-regulatory tissue differentiation algorithm. *Biomechanics and Modeling in Mechanobiology*, *7*(4), 335–344. doi:10.100710237-007-0100-3 PMID:17701434

Liu, X., Sun, J., Zhou, F., Yang, Y. H., Chang, R., Qiu, K., ... Zheng, Y. (2016). Micro-alloying with Mn in Zn–Mg alloy for future biodegradable metals application. *Materials & Design*, *94*, 95–104. doi:10.1016/j.matdes.2015.12.128

Liu, Z. F., Liu, X. P., Wang, S. W., & Liu, G. F. (2002). Recycling strategy and a recyclability assessment model based on an artificial neural network. *Journal of Materials Processing Technology*, *129*(1-3), 500–506. doi:10.1016/S0924-0136(02)00625-8

Li, W., Chau, K., & Jiang, J. (2011). Application of linear magnetic gears for pseudo-direct-drive oceanic wave energy harvesting. *IEEE Transactions on Magnetics*, *47*(10), 2624–2627. doi:10.1109/TMAG.2011.2146233

Li, Z., Gu, X., Lou, S., & Zheng, Y. Z. (2008). The development of binary Mg–Ca alloys for use as biodegradable materials within bone. *Biomaterials*, *29*(10), 1329–1344. doi:10.1016/j.biomaterials.2007.12.021 PMID:18191191

Long, M., & Rack, H. J. (1998). Titanium alloys in total joint replacement – a materials science perspective. *Biomaterials, 19*(18), 1621–1639. doi:10.1016/S0142-9612(97)00146-4 PMID:9839998

Lopez, M., Garcia, A., & Rodriguez, L. (2007). Sustainable development and corporate performance: A study based on the Dow Jones Sustainability Index. *Journal of Business Ethics, 75*(3), 285–300. doi:10.100710551-006-9253-8

Lost Core Injection Moulding. (n.d.). Copyright. Retrieved from http://LostCoreInjectionMoulding.htm

Lozano, R. (2008). Envisioning sustainability three-dimensionally. *Journal of Cleaner Production, 16*(17), 1838–1846. doi:10.1016/j.jclepro.2008.02.008

Lucon, E., Bicego, V., D'Angelo, D., & Fossati, C. (1993). Evaluating a Service- Exposed Component's Mechanical Properties by Means of Subsized and Miniature Specimens. In Small Specimen Test Techniques Applied to Nuclear Reactor Vessel Thermal Annealing and Plant Life Extension. American Society for Testing and Materials. doi:10.1520/STP12738S

Lundquist, L., Leterrier, Y., Sunderland, P., & Månson, J.-A. E. (2000). Life Cycle Engineering of Plastics. Life Cycle Engineering of Plastics.

Ma, Yu, Cao, Zheng, & Liu. (2014). The Kinematic Analysis and Trajectory Planning Study of High-Speed SCARA Robot Handling Operation. *Applied Mechanics and Materials, 687-691*, 294-299. Retrieved from www.scientific.net/AMM.687-691.294

Madan, J., Mani, M., Lee, J. H., & Lyons, K. W. (2015). Energy performance evaluation and improvement of unit-manufacturing processes: Injection molding case study. *Journal of Cleaner Production, 105*, 157–170. doi:10.1016/j.jclepro.2014.09.060

Madrazo, I., Zamorano, C., Magallón, E., Valenzuela, T., Ibarra, A., Salgado-Ceballos, H., ... Guízar-Sahagún, G. (2009). Stereolithography in spine pathology: A 2-case report. *Surgical Neurology, 72*(3), 272–275. doi:10.1016/j.surneu.2008.04.034 PMID:18614210

Mahesh, M., Wong, Y. S., Fuh, J. Y., & Loh, H. T. (2004). Benchmarking for comparative evaluation of RP systems and processes. *Rapid Prototyping Journal, 10*(2), 123–135. doi:10.1108/13552540410526999

Maiti, K., & Sil, A. (2006). Preparation of Rare earth oxide doped alumina ceramics, their hardness and fracture toughness determinations. *Indian Journal of Engineering and Materials Sciences, 13*, 443–450.

Make Parts Fast. (2018). Retrieved from www.makepartsfast.com

Malloy, R. A. (2010). Plastic Part Design for Injection Molding. Plastic Part Design. doi:10.3139/9783446433748

Malshe, H., Nagarajan, H., Pan, Y., & Haapala, K. (2015). Profile of Sustainability in Additive Manufacturing and Environmental Assessment of a Novel Stereolithography Process. *ASME 2015 International Manufacturing Science and Engineering Conference Volume 2: Materials; Biomanufacturing; Properties, Applications and Systems; Sustainable Manufacturing*. 10.1115/MSEC2015-9371

Manahan, M. P., Sr., Martin, F. J., & Stonesifer, R. B. (1999). Results of the ASTM Instrumented/Miniaturized Round Robin Test Program. In Pendulum Impact Testing: A Century of Progress. American Society for Testing and Materials.

Manahan, M. P., Stonesifer, R. B., Soong, Y., & Burger, J. M. (1995). Miniaturized Notch Test Specimen and Test Machine Design. In Pendulum Impact Machines: Procedures and Specimens for Verification. American Society for Testing and Materials. doi:10.1520/STP14656S

Marchelli, G., Prabhakar, R., Storti, D., & Ganter, M. (2011). The guide to glass 3D printing: Developments, methods, diagnostics and results. *Rapid Prototyping Journal*, *17*(3), 187–194. doi:10.1108/13552541111124761

Marr, B. (2015). Big Data: Using SMART Big Data, Analytics and Metrics To Make Better Decisions and Improve Performance. Wiley.

Marr, B. (2015). *Key Performance Indicators for Dummiers*. Wiley.

Mashhadi, A. R., & Behdad, S. (2017). Optimal sorting policies in remanufacturing systems: Application of product life-cycle data in quality grading and end-of-use recovery. *Journal of Manufacturing Systems*, *43*, 15–24. doi:10.1016/j.jmsy.2017.02.006

Mathieux, F., Froelich, D., & Moszkowicz, P. (2008). ReSICLED: A new recovery-conscious design method for complex products based on a multicriteria assessment of the recoverability. *Journal of Cleaner Production*, *16*(3), 277–298. doi:10.1016/j.jclepro.2006.07.026

McHale, G., Shirtcliffe, N. J., Aqil, S., Perry, C. C., & Newton, M. I. (2004). Topography driven spreading. *Physical Review Letters*, *93*(3), 36–102. doi:10.1103/PhysRevLett.93.036102 PMID:15323838

McLoughlin, J. R. (1972). The Preparation and Testing of Miniature Carbon Fiber Reinforced Composites. In *Composite Materials: Testing and Design*. American Society for Testing and Materials. 10.1520/STP27752S

MCP. (n.d.). Retrieved from http://www.mcp-group.de

Meessen, K. J., Gysen, B., Paulides, J., & Lomonova, E. A. (2008). Halbach permanent magnet shape selection for slotless tubular actuators. *IEEE Transactions on Magnetics*, *44*(11), 4305–4308. doi:10.1109/TMAG.2008.2001536

Mehrbod, M., Tu, N., Miao, L., & Wenjing, D. (2012). Interactive fuzzy goal programming for a multi-objective closed-loop logistics network. *Annals of Operations Research*, *201*(1), 367–381. doi:10.1007/10479-012-1192-4

Mekonnen, B. G., Bright, G., & Walker, A. (2016). *A Study on State of the Art Technology of Laminated Object Manufacturing (LOM). In CAD/CAM, Robotics and Factories of the Future* (pp. 207–216). New Delhi: Springer.

Meng, K., Lou, P., Peng, X., & Prybutok, V. (2016). An improved co-evolutionary algorithm for green manufacturing by integration of recovery option selection and disassembly planning for end-of-life products. *International Journal of Production Research*, *54*(18), 5567–5593. doi:10.1080/00207543.2016.1176263

Messagie, M., Boureima, F., Mertens, J., Sanfelix, J., Macharis, C., & Mierlo, J. (2013, March). The Influence of Allocation on the Carbon Footprint of Electricity Production from Waste Gas, a Case Study for Blast Furnace Gas. *Energies*, *6*(3), 1217–1232. doi:10.3390/en6031217

Meteyer, Xu, Perry, Fiona, & Zhao. (2014). Energy and Material floe Analysis of Binder Jetting Additive Manufacturing Processess. *Procedia, CIRP15*, 19–25.

Michel, O. (2004). Professional Mobile Robot Simulation. *International Journal of Advanced Robotic Systems*, *1*(1), 39–42. doi:10.5772/5618

Midhun Dominic, C. D. (2014, March-April). Rice Husk Silica- Efficient Bio Filler in High Density Polyethylene. *International Journal of Advanced Scientific and Technical Research*, *2*(4), 561–569.

Min, W., Miyahara, D., Yokoi, K., Yamaguchi, T., Daimon, K., Hikichi, Y., ... Ota, T. (2001). Thermal and Mechanical Properties of Sintered $LaPO_4$–Al_2O_3 Composites. *Materials Research Bulletin*, *36*(5-6), 939–945. doi:10.1016/S0025-5408(01)00555-4

Misch, C. E. (1998). *Contemporary implant dentistry*. St. Louis, MO: Mosby Elsevier.

Miyanaji, H., Zhang, S., Lassell, A., Zandinejad, A., & Yang, L. (2016). Process Development of Porcelain Ceramic Material with Binder Jetting Process for Dental Applications. *JOM*, *68*(3), 831–841. doi:10.100711837-015-1771-3

Moghaddam, K. (2015). Fuzzy multi-objective model for supplier selection and order allocation in reverse logistics systems under supply and demand uncertainty. *Expert Systems with Applications*, *42*(15-16), 6237–6254. doi:10.1016/j.eswa.2015.02.010

Mogilevsky, P., Boakye, E. E., & Hay, R. S. (2007). Solid solubility and thermal expansion in a $LaPO_4$–YPO_4 System. *Journal of the American Ceramic Society*, *90*(6), 1899–1907. doi:10.1111/j.1551-2916.2007.01653.x

Molin, E., Mokhtarian, P., & Kroesen, M. (2016). Multimodal travel groups and attitudes: A latent class cluster analysis of Dutch travelers. *Transportation Research Part A, Policy and Practice*, *83*, 14–29. doi:10.1016/j.tra.2015.11.001

Möller, A., & Schaltegger, S. (2005). The Sustainability Balanced Scorecard as a Framework for Eco-efficiency Analysis. *Journal of Industrial Ecology*, *9*(4), 73–83. doi:10.1162/108819805775247927

Morrow, W. R., Qi, H., Kim, I., Mazumder, J., & Skerlos, S. J. (2007). Environmental Aspects of Laser Based and Conventional Tool and Die Manufacturing. *Journal of Cleaner Production*, *15*(10), 932–943. doi:10.1016/j.jclepro.2005.11.030

Morvan, S., Hochsmann, R., & Sakamoto, M. (2005). ProMetal RCT (TM) process for fabrication of complex sand molds and sand cores. *Rapid Prototyping*, *11*(2), 1–7.

Mourtzis, D., Fotia, S., & Vlachou, E. (2017). Lean rules extraction methodology for lean PSS design via key performance indicators monitoring. *Journal of Manufacturing Systems*, *42*, 233–243. doi:10.1016/j.jmsy.2016.12.014

Mourtzis, D., Fotia, S., Vlachou, E., & Koutoupes, A. (2017). A Lean PSS design and evaluation framework supported by KPI monitoring and context sensitivity tools. *International Journal of Advanced Manufacturing Technology*, 1–15. doi:10.100700170-017-0132-5

Mueller, B., & Kochan, D. (1999). Laminated object manufacturing for rapid tooling and patternmaking in foundry industry. *Computers in Industry*, *39*(1), 47–53. doi:10.1016/S0166-3615(98)00127-4

Murata, K., & Katayama, H. (2009). An evaluation of factory performance utilized KPI/KAI with data envelopment analysis. *Journal of the Operations Research Society of Japan*, *52*(2), 204–220. doi:10.15807/jorsj.52.204

Murphy, S. V., & Atala, A. (2014). 3D bioprinting of tissues and organs. *Nature Biotechnology*, *32*(8), 773–785. doi:10.1038/nbt.2958 PMID:25093879

Murr & Johnson. (2017). 3D metal droplet printing development and advanced materials Additive Manufacturing. *Journal of Materials Research and Technology*. doi:10.1016/j.jmrt.2016.11.002

Murr, L. E., Gaytan, S. M., Ramirez, D. A., Martinez, E., Hernandez, J., Amato, K. N., ... Wicker, R. B. (2012). Metal fabrication by additive manufacturing using laser and electron beam melting technologies. *Journal of Materials Science and Technology*, *28*(1), 1–14. doi:10.1016/S1005-0302(12)60016-4

Narayan, R. (Ed.). (2012). Materials for medical devices: ASM Handbook, Vol 23. ASM International.

National Round Table on the Environment and the Economy. (2003). *Environment and Sustainable Development Indicators for Canada*. Available from: papers2://publication/uuid/F0A97390-580F-4EF6-9CED-1EEF3EE889D5

National Round Table on the Environment and the Economy. (2012). *Canada's Opportunity: Adopting Life Cycle Approaches for Sustainable Development*. Author.

Neri, L., D'Agostino, A., Regoli, A., Pulselli, F. M., & Coscieme, L. (2017). Evaluating dynamics of national economies through cluster analysis within the input-state-output sustainability framework. *Ecological Indicators*, *72*, 77–90. doi:10.1016/j.ecolind.2016.08.016

Newman, S. T., Nassehi, A., Imani-Asrai, R., & Dhokia, V. (2012). Energy efficient process planning for CNC machining. *CIRP Journal of Manufacturing Science and Technology*, *5*(2), 127–136. doi:10.1016/j.cirpj.2012.03.007

Ng, R., Yeo, Z., Low, J. S. C., & Song, B. (2015). A method for relative eco-efficiency analysis and improvement: Case study of bonding technologies. *Journal of Cleaner Production*, *99*, 320–332. doi:10.1016/j.jclepro.2015.03.004

Niinomi, M. (2003). Fatigue performance and cyto-toxicity of low rigidity titanium alloy, Ti–29Nb–13Ta–4.6Zr. *Biomaterials*, *24*(16), 2673–2683. doi:10.1016/S0142-9612(03)00069-3 PMID:12711513

Niinomi, M., & Nakai, M. (2011). Titanium-Based Biomaterials for Preventing Stress Shielding between Implant Devices and Bone. *International Journal of Biomaterials*, *2011*, 1–10. doi:10.1155/2011/836587 PMID:21765831

Niinomi, M., Nakai, M., & Hieda, A. (2012). Development of new metallic alloys for biomedical applications. *Acta Biomaterialia*, *8*(11), 3888–3903. doi:10.1016/j.actbio.2012.06.037 PMID:22765961

Nikolakopoulos, A. N., Kouneli, M. A., & Garofalakis, J. D. (2015). Hierarchical itemspace rank: Exploiting hierarchy to alleviate sparsity in ranking-based recommendation. *Neurocomputing*, *163*, 126–136. doi:10.1016/j.neucom.2014.09.082

Nikzad, M., Masood, S. H., & Sbarski, I. (2011). Thermo-mechanical properties of a highly filled polymeric composites for Fused Deposition Modeling. *Materials & Design*, *32*(6), 3448–3456. doi:10.1016/j.matdes.2011.01.056

Nimbalkar, S., Cox, D., Visconti, K., & Cresko, J. (2014). Life Cycle Energy Assessment Methodology and Additive Manufacturing Energy Impacts Assessment Tool. *Proceedings from the LCA XIV International Conference*, 130-141.

Ning, Cong, & Qiu, Wei, & Wang. (2015). Additive Manufacturing of carbon fiber reinforced thermoplastic composites using fused deposition modeling. *Composites. Part B, Engineering*, *80*, 369–378.

Niu, S., Ho, S., & Fu, W. (2011). Performance Analysis of a Novel Magnetic-Geared Tubular Linear Permanent Magnet Machine. *IEEE Transactions on Magnetics*, *47*(10), 3598–3601. doi:10.1109/TMAG.2011.2148167

Noble, J., Walczak, K., & Dornfeld, D. (2014). Rapid tooling injection molded prototypes: A case study in artificial photosynthesis technology. CIRP Procedia, 251–256. doi:10.1016/j.procir.2014.03.035

Noiva, K., Fernández, J. E., & Wescoat, J. L. Jr. (2016). Cluster analysis of urban water supply and demand: Toward large-scale comparative sustainability planning. *Sustainable Cities and Society*, *27*, 484–496. doi:10.1016/j.scs.2016.06.003

Noorani, R. (2006). *Rapid prototyping: principles and applications*. John Wiley & Sons Incorporated.

Norouzi, Y., Rahmati, S., & Hojjat, Y. (2009). A novel lattice structure for SL investment casting patterns. *Rapid Prototyping Journal*, *15*(4), 255–263. doi:10.1108/13552540910979776

Novakova-Marcincinova, L. (2012). Application of fused deposition modeling technology in 3D printing rapid prototyping area. *Manuf. and Ind. Engg.*, *11*(4), 35–37.

Nozawa, Katoh, & Kohyama. (2005). Evaluation of Tensile Properties of SiC/SiC Composites with Miniaturized Specimens. *Materials Transactions, 46*(3), 543-551.

Nozawa, Ozawa, Kondo, Hinoki, Katoh, Snead, & Kohyama. (2005). Tensile, Flexural, and Shear Properties of Neutron Irradiated SiC/SiC Composites with Different Fiber-Matrix Interfaces. *Journal of ASTM International, 2*(3).

Nozawa, T., Hinoki, T., Katoh, Y., Kohyama, A., & Lara-Curzio, E. (2002). Specimen Size Effects on Tensile Properties of 2D/3D SiC/SiC Composites. In M. A. Sokolov, J. D. Landes, & G. E. Lucas (Eds.), *Small Specimens Test Techniques: Fourth Volume, ASTM STP 1418*. West Conshohocken, PA: ASTM International. doi:10.1520/STP10828S

Nukman, Farooqi, Al-Sultan, Rahman, Alnasser, & Bhuiyan. (2017). *A Strategic Development of Green Manufacturing Index (GMI) Topology Concerning the Environmental Impacts*. Academic Press. doi: . doi:10.1016/j.proeng.2017.04.107

O'Reilly, M. (2000). ISO 14031: Effective Mechanism to Environmental Performance Evaluation. *Corporate Environmental Strategy*, *7*(3), 267–275. doi:10.1016/S1066-7938(00)80121-9

Ofem, M. I., Abam, F. I., & Ugot, I. U. (2012). Mechanical Properties of Hybrid Periwinkle and Rice Husk Filled CNSL Composite. *International Journal of Nano and Material Sciences*, *1*(2), 74–80.

Okonkwo, R. (2006). Design and performance of permanent-magnet DC linear motors. *IEEE Transactions on Magnetics*, *42*(9), 2179–2183. doi:10.1109/TMAG.2006.880397

Okoye, F. N., Durgaprasad, J., & Singh, N. B. (2015). Mechanical properties of alkali activated flyash/Kaolin based geopolymer concrete. *Construction & Building Materials*, *98*, 685–691. doi:10.1016/j.conbuildmat.2015.08.009

Oktem, H., Erzurumlu, T., & Erzincanli, F. (2006). Prediction of minimum surface roughness in end milling mold parts using neural network and genetic algorithm. *Materials & Design*, *27*(9), 735–744. doi:10.1016/j.matdes.2005.01.010

Oliveira, L., Messagie, M., Mertens, J., Laget, H., Coosemans, T., & Van Mierlo, J. (2015). Environmental performance of electricity storage systems for grid applications, a life cycle approach. *Energy Conversion and Management*, *101*, 326–335. doi:10.1016/j.enconman.2015.05.063

Olsthoorn, X., Tyteca, D., Wehrmeyer, W., & Wagner, M. (2001). Environmental indicators for business: A review of the literature and standardisation methods. *Journal of Cleaner Production*, *9*(5), 453–463. doi:10.1016/S0959-6526(01)00005-1

Omar, M. F., Sharif, S., Ibrahim, M., Hehsan, H., Busari, M. N., & Hafsa, M. N. (2012). Evaluation of direct rapid prototyping pattern for investment casting. *Advanced Materials Research*, *463*, 226–233. doi:10.4028/www.scientific.net/AMR.463-464.226

Ondemir, O., & Gupta, S. (2014). A multi-criteria decision making model for advanced repair-to-order and disassembly-to-order system. *European Journal of Operational Research*, *233*(2), 408–219. doi:10.1016/j.ejor.2013.09.003

Onuh, S. O., & Hon, K. K. (1998). Optimising build parameters for improved surface finish in stereolithography. *International Journal of Machine Tools & Manufacture*, *38*(4), 329–342. doi:10.1016/S0890-6955(97)00068-0

Organisation for Economic Co-Operation and Development. (2003). OECD Environmental Indicators: development, measurement and use (Vol. 25). OECD.

Özceylan, E., & Paksoy, T. (2013). Fuzzy multi-objective linear programming approach for optimising a closed-loop supply chain network. *International Journal of Production Research*, *51*(8), 2443–2461. doi:10.1080/00207543.2012.740579

Ozsoy, Demirkol, Mimaroglu, Unal, & Demir. (2015). The Influence of Micro- and Nano-Filler Content on the Mechanical Properties of Epoxy Composites. *Journal of Mechanical Engineering, 61*(10), 601-609.

Pal, D. K., Ravi, B., & Bhargava, L. S. (2002). E-manufacturing one-off intricate castings using rapid prototyping technology. In *International Conference on e-Manufacturing, Bhopal* (pp. 259-63). American Foundry Society.

Pande, A., Gairola, P., Sambyal, P., Gairola, S. P., Kumar, V., Singh, K., & Dhawan, S. K. (2017). Electromagnetic shielding behavior of polyaniline using Red Mud (industrial waste) as filler in the X e band (8.2e12.4 GHz) frequency range. *Materials Chemistry and Physics*, *189*, 22–27. doi:10.1016/j.matchemphys.2016.12.045

Pandelidis, I., & Zou, Q. (1990). Optimization of injection molding design. Part I: Gate location optimization. *Polymer Engineering and Science*, *30*(15), 873–882. doi:10.1002/pen.760301502

Pandey, P. M., Thrimurthulu, K., & Reddy, N. V. (2004). Optimal part deposition orientation in FDM by using a multicriteria genetic algorithm. *International Journal of Production Research*, *42*(19), 4069–4089. doi:10.1080/00207540410001708470

Panel IR. (2014). *Decoupling: Natural Resource Use and Environmental*. Author.

Pan, W., Zhong, H., Xu, C., & Ming, Z. (2015). Adaptive Bayesian personalized ranking for heterogeneous implicit feedbacks. *Knowledge-Based Systems*, *73*, 173–180. doi:10.1016/j.knosys.2014.09.013

Paranthaman, M. P., Shafer, C. S., Elliott, A. M., Siddel, D. H., McGuire, M. A., Springfield, R. M., ... Ormerod, J. (2016). Binder Jetting: A Novel NdFeB Bonded Magnet Fabrication Process. *JOM*, *68*(7), 1978–1982. doi:10.100711837-016-1883-4

Park, H. S., & Nguyen, T. T. (2014). Optimization of injection molding process for car fender in consideration of energy efficiency and product quality. *Journal of Computational Design and Engineering.*, *1*(4), 256–265. doi:10.7315/JCDE.2014.025

Park, J., Tari, M. J., & Hahn, H. T. (2000). Characterization of the laminated object manufacturing (LOM) process. *Rapid Prototyping Journal*, *6*(1), 36–50. doi:10.1108/13552540010309868

Park, P. J., Tahara, K., & Inaba, A. (2007). Product quality-based eco-efficiency applied to digital cameras. *Journal of Environmental Management*, *83*(2), 158–170. doi:10.1016/j.jenvman.2006.02.006 PMID:16697518

Parmenter, D. (2007). *Key Performance Indicators (KPI): Developing, Implementing, and Using Winning KPIs*. Zhurnal Eksperimental'noi i Teoreticheskoi Fiziki.

Parmenter, D. (2010). *Key Performance Indicators (KPI). Key Performance Indicators (KPI): Developing, Implementing, and Using Winning KPIs* (2nd ed.). Academic Press.

Parry, G., & Turner, C. (2006, January). Application of lean visual process management tools. *Production Planning and Control*, *17*(1), 77–86. doi:10.1080/09537280500414991

Patel, S., & Sobh, T. (2014). Manipulator Performance Measures - A Comprehensive Literature Survey. *Journal of Intelligent & Robotic Systems*, *77*(3-4), 547–570. doi:10.100710846-014-0024-y

Pattnaik, S., Jha, P. K., & Karunakar, D. B. (2014). A review of rapid prototyping integrated investment casting processes. *Proceedings of the Institution of Mechanical Engineers, Part L: Journal of Materials: Design and Applications*, *228*(4), 249-77.

Paul. (1981). *Robot Manipulators: Mathematics, Programming, and Control*. MIT Press.

Paul, B. D. (2008). A history of the concept of sustainable development: Literature review. Ann Univ Oradea. *Econ Sci Ser.*, *17*(2), 576–580.

Peças, P., Ribeiro, I., Folgado, R., & Henriques, E. (2009). A Life Cycle Engineering model for technology selection: A case study on plastic injection moulds for low production volumes. *Journal of Cleaner Production*, *17*(9), 846–856. doi:10.1016/j.jclepro.2009.01.001

Peña-Fernández, A., Wyke, S., Brooke, N., & Duarte-Davidson, R. (2014). Factors influencing recovery and restoration following a chemical incident. *Environment International*, *72*, 98–108. doi:10.1016/j.envint.2014.05.001 PMID:24874002

Petrovic, V., Vicente Haro Gonzalez, J., Jorda Ferrando, O., Delgado Gordillo, J., Ramon Blasco Puchades, J., & Portoles Grinan, L. (2011). Additive layered manufacturing: Sectors of industrial application shown through case studies. *International Journal of Production Research*, *49*(4), 1061–1079. doi:10.1080/00207540903479786

Petrovoic, Gonzale, Ferrando, Gordillo, Purchades, & Grinan. (2011). Additive Layer Manufacturing: Sectors of Industrial Applications shown through case studies. *International Journal of Production Research, 49*(4), 1061-1079.

Pham, D. T., Dimov, S. S., & Gault, R. S. (1999). Part orientation in stereolithography. *International Journal of Advanced Manufacturing Technology*, *15*(9), 674–682. doi:10.1007001700050118

Pilipović, A., Raos, P., & Šercer, M. (2011). Experimental testing of quality of polymer parts produced by laminated object manufacturing–LOM. *Tehnicki Vjesnik (Strojarski Fakultet)*, *18*, 253–260.

Pirso, J., Viljus, M., Juhani, K., & Kuningas, M. (2010). Three-body abrasive wear of TiC–NiMo cermets. *Tribology International*, *43*(1-2), 340–346. doi:10.1016/j.triboint.2009.06.014

Polifke, M. (1998). *Lost Core Technology On The Way To New Application.* Spe-Antec.

Polinder, H., Mueller, M., Scuotto, M., & Goden de Sousa Prado, M. (2007). Linear generator systems for wave energy conversion. *Proceedings of the 7th European Wave and Tidal Energy Conference.*

Porte-Durrieu, M. C., Guillemot, F., Pallu, S., Labrugère, C., Brouillaud, B., Bareille, R., ... Baquey, C. (2004). Cyclo-(DfKRG) peptide grafting onto Ti-6Al-4V: Physical characterization and interest towards human osteoprogenitor cells adhesion. *Biomaterials*, *25*(19), 4837–4846. doi:10.1016/j.biomaterials.2003.11.037 PMID:15120531

Prakash, C., & Uddin, M.S. (2017). Surface modification of β-phase Ti implant by hydroaxyapatite mixed electric discharge machining to enhance the corrosion resistance and in-vitro bioactivity. *Surface and Coatings Technology, 236*(Part A), 134-145.

Prakash, C. (2016). Powder Mixed Electric Discharge Machining an Innovative Surface Modification Technique to Enhance Fatigue Performance and Bioactivity of β-Ti Implant for Orthopaedics Application. *Journal of Computing and Information Science in Engineering*, *14*, 1–9.

Prakash, C., & .... (2016). Electric discharge machining- a potential choice for surface modification of metallic implants for orthopedics applications: A review. Proceedings of the Institution of Mechanical Engineers, Part B. *Journal of Engineering*, *230*, 231–253.

Prakash, C., Kansal, H. K., Pabla, B. S., & Puri, S. (2015). Processing and characterization of novel biomimetic nanoporous bioceramic surface on β-Ti implant by powder mixed electric discharge machining. *Journal of Materials Engineering and Performance*, *24*(9), 3622–3633. doi:10.100711665-015-1619-6

Prakash, C., Kansal, H. K., Pabla, B. S., & Puri, S. (2016). Multi-objective optimization of powder mixed electric discharge machining parameters for fabrication of biocompatible layer on β-Ti alloy using NSGA-II coupled with Taguchi based response surface methodology. *Journal of Mechanical Science and Technology.*, *30*(9), 4195–4204. doi:10.100712206-016-0831-0

Prakash, C., Kansal, H. K., Pabla, B. S., & Puri, S. (2017). Experimental Investigations in Powder Mixed Electrical Discharge Machining of Ti-35Nb-7Ta-5Zr β-Ti Alloy. *Materials and Manufacturing Processes*, *32*(3), 274–285. doi:10.1080/10426914.2016.1198018

Prakash, C., Kansal, H. K., Pabla, B. S., Puri, S., & Aggarwal, A. (2016). Electric discharge machining- a potential choice for surface modification of metallic implants for orthopedics applications: A review. *Proceedings of the Institution of Mechanical Engineers. Part B, Journal of Engineering Manufacture*, *230*(2), 231–253. doi:10.1177/0954405415579113

Prakash, C., Singh, S., Pabla, B. S., & Uddin, M. S. (2018). Synthesis, characterization, corrosion and bioactivity investigation of nano-HA coating deposited on biodegradable Mg-Zn-Mn alloy. *Surface and Coatings Technology*, *346*, 9–18. doi:10.1016/j.surfcoat.2018.04.035

Price, R. L., Waid, M. C., Haberstroh, K. M., & Webster, T. J. (2003). Selective bone cell adhesion on formulations containing carbon nanofibers. *Biomaterials*, *24*(11), 1877–1887. doi:10.1016/S0142-9612(02)00609-9 PMID:12615478

Protzman, C., Whiton, F., Kerpchar, J., Lewandowski, C., Stenberg, S., & Grounds, P. (2016). *The Lean Practitioner's Field Book: Proven, Practical, Profitable and Powerful Techniques for Making Lean Really Work*. CRC Press.

Pun, K.-F., Hui, I.-K., Lewis, W. G., & Lau, H. C. W. (2003). A multiple-criteria environmental impact assessment for the plastic injection molding process: A methodology. *Journal of Cleaner Production*, *11*(1), 41–49. doi:10.1016/S0959-6526(02)00019-7

Rajurkar, K. P., Zhu, D., McGeough, J. A., Kozak, J., & De Silva, A. (1999). New developments in electro-chemical machining. *CIRP Annals-Manufacturing Technology*, *48*(2), 567–579. doi:10.1016/S0007-8506(07)63235-1

Ramakrishna, S., Mayer, J. E., Wintermantel, E., & Leong, K. W. (2001). Wintermantel and K.W. Leong, "Biomedical applications of polymer composite materials: A review. *Composites Science and Technology*, *61*(9), 1189–1224. doi:10.1016/S0266-3538(00)00241-4

Ramezani, M., Bashiri, M., & Moghaddam, R. (2013). A new multi-objective stochastic model for a forward/reverse logistic network design with responsiveness and quality level. *Applied Mathematical Modelling*, *37*(1-2), 328–344. doi:10.1016/j.apm.2012.02.032

Rathore, P., Kota, S., & Chakrabarti, A. (2011). Sustainability through remanufacturing in India: A case study on mobile handsets. *Journal of Cleaner Production*, *19*(15), 1709–1722. doi:10.1016/j.jclepro.2011.06.016

Ratner, B. D., & Hoffman, A. S. (2004). *Biomaterials science*. Amsterdam: Elsevier.

Reed, D. N., Smith, R. P., Strattan, J. K., & Swift, R. A. (1983). A Comparison of Short Transverse Tension Test Methods. In R. J. Glodowski (Ed.), *Through-Thickness Tension Testing of Steel ASTM STP 794* (pp. 25–39). American Society for Testing and Materials. doi:10.1520/STP10004S

Rehiara. (2011). Kinematics of Adept Three Robot Arm. In Robot Arms (pp. 21-38). Rijeka, Croatia: InTech.

Rengier, Mehndiratta, von Tengg-Kobligk, Zechmann, Unterhinninghofen, Kauczor, & Giesel. (2010). *3D printing based on imaging data: review of medical applications*. Academic Press. doi:10.100711548-010-0476-x

Rezaei, J. (2015). A systematic review of multi-criteria decision-making applications in reverse logistics. *Transportation Research Procedia*, *10*, 766–776. doi:10.1016/j.trpro.2015.09.030

Ribeiro, I., Peças, P., & Henriques, E. (2008). Environmental Impact of Plastic Injection Moulds. In *3rd International Conference on Polymers and Moulds Innovations – PMI 2008*. Ghent: University College Ghent.

Ribeiro, I., Peças, P., & Henriques, E. (2011). Life cycle approach to support tooling design decisions. *ICED 11 - 18th International Conference on Engineering Design - Impacting Society Through Engineering Design.*

Ribeiro, I., Peças, P., & Henriques, E. (2012). Assessment of energy consumption in injection moulding process. *Leveraging Technology for a Sustainable World - Proceedings of the 19th CIRP Conference on Life Cycle Engineering.* 10.1007/978-3-642-29069-5_45

Ribeiro, I., Kaufmann, J., Schmidt, A., Peças, P., Henriques, E., & Götze, U. (2016). Fostering selection of sustainable manufacturing technologies - A case study involving product design, supply chain and life cycle performance. *Journal of Cleaner Production*, 112.

Ribeiro, I., Peças, P., & Henriques, E. (2013). A life cycle framework to support materials selection for Ecodesign: A case study on biodegradable polymers. *Materials & Design*, 51.

Ribeiro, I., Peças, P., & Henriques, E. (2013). Incorporating tool design into a comprehensive life cycle cost framework using the case of injection molding. *Journal of Cleaner Production*, 53.

Ribeiro, I., Peças, P., & Henriques, E. (2014). Life Cycle Engineering Framework for Technology and Manufacturing Processes Evaluation. In E. Henriques, P. Peças, & A. Silva (Eds.), *Technology and Manufacturing Process Selection: The Product Life Cycle Perspective* (pp. 217–237). Springer. doi:10.1007/978-1-4471-5544-7_11

Ribeiro, I., Pousa, C., Folgado, R., Peças, P., & Henriques, E. (2009). LCC and LCA Simplified Models to Foster the Design of Sustainable Plastic Injection Moulds. *16th CIRP International Conference on Life Cycle Engineering (LCE 2009)*, 99–104.

Rochon, G. L., Niyogi, D., Fall, S., Quansah, J. E., Biehl, L., Araya, B., ... Thiam, T. (2010). Best management practices for corporate, academic and governmental transfer of sustainable technologies to developing countries. *Clean Technologies and Environmental Policy*, *12*(1), 19–30. doi:10.100710098-009-0218-3

Rosato, D. V., & Rosato, M. G. (2000). *Injection molding handbook.* Kluwer Academic Publisher. doi:10.1007/978-1-4615-4597-2

Rosinski, S. T., & Corwin, W. R. (1998). ASTM Cross-Comparison Exercise on Determination of Material Properties Through Miniature Sample Testing. In Small Specimen Test Techniques. American Society for Testing and Materials. doi:10.1520/STP37979S

Rosochowski, A., & Matuszak, A. (2000). Rapid tooling: The state of the art. *Journal of Materials Processing Technology*, *106*(1-3), 191–198. doi:10.1016/S0924-0136(00)00613-0

Routa. (2012). Study on mechanical and tribo-performance of rice-husk filled glass–epoxy hybrid composites. *Materials & Design*, *41*(October), 131–141.

Sahai & Pawar. (2014). Studies on Mechanical Properties of Fly Ash Filled PPO Composite with Coupling Agent. International Journal of Chemical, *Environmental & Biological Sciences*, 2(4), 187-192

Salahshoor, M., & Guo, Y. B. (2013). Process mechanics in ball burnishing biomedical magnesium–calcium alloy. *International Journal of Advanced Manufacturing Technology*, *64*(1-4), 133–144. doi:10.100700170-012-4024-4

Salmoria, G. V., Paggi, R. A., Lago, A., & Beal, V. E. (2011). Microstructural and mechanical characterization of PA12/MWCNTs nanocomposite manufactured by selective laser sintering. *Polymer Testing*, *30*(6), 611–615. doi:10.1016/j.polymertesting.2011.04.007

Sanders, A., Elangeswaran, C., & Wulfsberg, J. (2016). Industry 4.0 implies lean manufacturing: Research activities in industry 4.0 function as enablers for lean manufacturing. *Journal of Industrial Engineering and Management*, *9*(3), 811–833. doi:10.3926/jiem.1940

Sankar, S., Athira, N., Raj, C.K., Jyothi, Warrier, K.G.K., & Padmanabhan, P.V.A. (2012). Room temperature synthesis of high temperature stable lanthanum phosphate–yttria nano composite. *Materials Research Bulletin*, 47.

Sankar, S., & Warrier, K. G. K. (2011). Aqueous sol-gel synthesis of LaPO4 nano rods starting from lanthanum chloride precursor. *Journal of Sol-Gel Technology*, *58*(1), 195–200. doi:10.100710971-010-2377-4

Sarahara, H., Tsutsumi, M., & Chino, M. (2005). Development of Layered Manufacturing System using Sheet Metal-Polymer Lamination for Mechanical Parts. *International Journal of Advanced Manufacturing Technology*, *27*(3-4), 268–273. doi:10.100700170-004-2163-y

Saravanamohan, M., & Anbumalar, V. (2016). Modelling and simulation of multi spindle drilling redundant SCARA robot using SolidWorks and MATLAB/SimMechanics. *Revista de la Facultad de Ingeniería*, *81*, 63–72.

Saro, T. P. S., & Sidhu, H. S. (2012). Characterization and in vitro corrosion investigation of thermal sprayed hydroxy-apatite and hydroxyapatite-titania coating on TI alloy. *Metallurigical and Materials Transaction A: Physical Metalurgy and Material Science, 43*, 4365-4376. 10.100711661-012-1175-8

Saroja Devi, M., Murugesan, V., Rengaraj, K., & Anand, P. (1998). Utilization of flyash as filler for unsaturated polyester resin. *Journal of Applied Polymer Science*, *69*(7), 1385–1391. doi:10.1002/(SICI)1097-4628(19980815)69:7<1385::AID-APP13>3.0.CO;2-T

Sarwar, B., Karypis, G., Konstan, J., & Riedl, J. (2001, April). Item-based collaborative filtering recommendation algorithms. In *Proceedings of the 10th international conference on World Wide Web* (pp. 285-295). ACM.

Satapathy, L. N. (2000). A study on the mechanical, abrasion and microstructural properties of zirconia-flyash material. *Ceramics International*, *26*(1), 39–45. doi:10.1016/S0272-8842(99)00017-6

Satish, H. (2013). Combine Effect of Rice Husk Ash and Fly Ash on Concrete by 30% Cement Replacement. *Procedia Engineering*, *51*, 35–44. doi:10.1016/j.proeng.2013.01.009

Savkin, A. (2016). *KPIs Misuse: Why Does It Happen and How Avoid It.* Retrieved from https://bscdesigner.com/avoiding-kpis-misuse.htm

Schakenraad, J. M. (1996). Cells: Their surfaces and interactions with materials. In B. D. Ratner (Ed.), *Biomaterial Science: An Introduction to Materials in Medicine* (pp. 141–147). San Diego, CA: Academic Press.

Schlotter, M. (2003). *Multibody System Simulation with SimMechanics*. Retrieved from http://www.imac.unavarra.es/DSM/download/trabajos/

Schmid, M., Simon, C., & Levy, G. N. (2009). Finishing of SLS-parts for rapid manufacturing (RM)–a comprehensive approach. *Proceedings SFF*, 1-10.

Schmidt, Merklein, Bourell, Dimitrov, Hausotte, Wegener, … Levy. (2017). Laser based Additive Manufacturing in industry and academia. *CIRP Annals - Manufacturing Technology*. doi:10.1016/j.cirp.2017.05.011

Seitz, M. A. (2007). A critical assessment of motives for product recovery: The case of engine remanufacturing. *Journal of Cleaner Production*, *15*(11-12), 1147–1157. doi:10.1016/j.jclepro.2006.05.029

Sekar, T., & Marappan, R. (2008). Experimental investigations into the influencing parameters of electrochemical machining of AISI 202. *Journal of Advanced Manufacturing Systems*, *7*(02), 337–343. doi:10.1142/S0219686708001486

Selvan, M. C. P., & Raju, N. M. S. (2012). Analysis of surface roughness in abrasive waterjet cutting of cast iron. *International Journal of Science Environment and Technology*, *1*, 174–182.

Semetay, C. (2007). *Laser Engineered Net Shaping (LENS) modeling using welding simulation concepts*. Lehigh University.

Şen, S., & Nugay, N. (2000). Uncured and cured state properties of fly ash filled unsaturated polyester composites. *Journal of Applied Polymer Science*, *77*(5), 1128–1136. doi:10.1002/1097-4628(20000801)77:5<1128::AID-APP21>3.0.CO;2-D

Senthil, S., Srirangacharyulu, B., & Ramesh, A. (2014). A robust hybrid multi-criteria decision making methodology for contractor evaluation and selection in third-party reverse logistics. *Expert Systems with Applications*, *41*(1), 50–58. doi:10.1016/j.eswa.2013.07.010

Shackleton, M., & Sødal, S. (2010). Harvesting and recovery decisions under uncertainty. *Journal of Economic Dynamics & Control*, 2533-2546.

Shahgholian, G., & Shafaghi, P. (2010). *State space modeling and eigenvalue analysis of the permanent magnet DC motor drive system.* Paper presented at the 2010 International Conference on Electronic Computer Technology (ICECT). 10.1109/ICECTECH.2010.5479987

Shams, B., & Haratizadeh, S. (2017a). Reliable Graph-based Collaborative Ranking. *Information Sciences*.

Shams, B., & Haratizadeh, S. (2017b). IteRank: An iterative network-oriented approach to neighbor-based collaborative ranking. *Knowledge-Based Systems*, *128*, 102–114. doi:10.1016/j.knosys.2017.05.002

Shanmugam, D. K., & Masood, S. H. (2009). An investigation on kerf characteristics in abrasive waterjet cutting of layered composites. *Journal of Materials Processing Technology*, *209*(8), 3887–3893. doi:10.1016/j.jmatprotec.2008.09.001

Shanmugam, D. K., Wang, J., & Liu, H. (2008). Minimisation of kerf tapers in abrasive waterjet machining of alumina ceramics using a compensation technique. *Journal of Machine Tools & Manufacture*, *48*(14), 1527–1534. doi:10.1016/j.ijmachtools.2008.07.001

Sharma, S., & Dwivedi, S. P. (2017). Effects of waste eggshells and SiC addition on specific strength and thermal expansion of hybrid green metal matrix composite. *Journal of Hazardous Materials*, *333*, 1–9. doi:10.1016/j.jhazmat.2017.01.002 PMID:28340384

Sharma, V., Chattopadhyaya, S., & Hloch, S. (2011). Multi response optimization of process parameters based on Taguchi-Fuzzy model for coal cutting by water jet technology. *Journal of Advance Manufacturing Technology*, *56*(9-12), 1019–1025. doi:10.100700170-011-3258-x

Sheate, W., Dagg, S., Richardson, J., Aschemann, R., Palerm, J., & Steen, U. (2003). Integrating the environment into strategic decision-making: Conceptualizing policy SEA. *European Environment*, *13*(1), 1–18. doi:10.1002/eet.305

Sheu, J.-B. (2007). A coordinated reverse logistics system for regional management of multi-source hazardous wastes. *Computers & Operations Research*, *34*(5), 1442–1462. doi:10.1016/j.cor.2005.06.009

Shin, C.-S., & Lin, S.-W. (2012). Evaluating fatigue crack propagation properties using miniature specimens. *International Journal of Fatigue*, *43*, 105–110. doi:10.1016/j.ijfatigue.2012.02.018

Shu, X., & Wang, R. (2017). Thermal residual solutions of beams, plates and shells due to laminated object manufacturing with gradient cooling. *Composite Structures*, *15*(174), 366–374. doi:10.1016/j.compstruct.2017.04.060

Siciliano, B., & Khatib, O. (2008). *Handbook of Robotics*. Heidelberg, Germany: Springer. doi:10.1007/978-3-540-30301-5

Si, J., Feng, H., Su, P., & Zhang, L. (2014). Design and analysis of tubular permanent magnet linear wave generator. *The Scientific World Journal*. PMID:25050388

Silalertruksa, T., Gheewala, S. H., & Pongpat, P. (2015). Sustainability assessment of sugarcane biorefinery and molasses ethanol production in Thailand using eco-efficiency indicator. *Applied Energy*, *160*, 603–609. doi:10.1016/j.apenergy.2015.08.087

SimMechanics Getting Started Guide. (2015, March 16). Retrieved from https://fenix.tecnico.ulisboa.pt/downloadFile/845043405443232/sl_using_r2015a.pdf

Singh, Haverinen, Dhagat, & Jabbour. (2010). Inkjet Printing – Process and Applications. *Advance Materials, 22*(6), 673–685.

Singh, R. (2011). Process capability study of polyjet printing for plastic components. *Journal of Mechanical Science and Technology, 25*(4), 1011-1015.

Singh, J., Singh, R., & Singh, H. (2017). Dimensional accuracy and surface finish of biomedical implant fabricated as rapid investment casting for small to medium quantity production. *Journal of Manufacturing Processes, 25*, 201–211. doi:10.1016/j.jmapro.2016.11.012

Singh, M., & Singh, Y. (2015). Implant biomaterials: A comprehensive review. *World Journal of Clinical Cases, 3*(1), 52–57. doi:10.12998/wjcc.v3.i1.52 PMID:25610850

Singh, R., & Singh, M. (2017). Surface roughness improvement of cast components in vacuum moulding by intermediate barrel finishing of fused deposition modelling patterns. *Proceedings of the Institution of Mechanical Engineers. Part E, Journal of Process Mechanical Engineering, 231*(2), 309–316. doi:10.1177/0954408915595576

Singh, R., & Singh, S. (2013). Effect of process parameters on surface hardness, dimensional accuracy and surface roughness of investment cast components. *Journal of Mechanical Science and Technology, 27*(1), 91–97. doi:10.100712206-012-1218-5

Singh, R., Singh, S., Singh, I. P., Fabbrocino, F., & Fraternali, F. (2017). Investigation for surface finish improvement of FDM parts by vapor smoothing process. *Composites. Part B, Engineering, 111*, 228–234. doi:10.1016/j.compositesb.2016.11.062

Singh, S., & Ramakrishna, S. (2017). *Biomedical applications of Additive Manufacturing: present and future*. Current Opinion in Biomedical Engineering. doi:10.1016/j.cobme.2017.05.006

Singh, S., & Singh, R. (2016). Effect of process parameters on micro hardness of Al–Al2O3 composite prepared using an alternative reinforced pattern in fused deposition modelling assisted investment casting. *Robotics and Computer-integrated Manufacturing, 37*, 162–169. doi:10.1016/j.rcim.2015.09.009

Singh, S., & Singh, R. (2016). Fused deposition modelling based rapid patterns for investment casting applications: A review. *Rapid Prototyping Journal, 22*(1), 123–143. doi:10.1108/RPJ-02-2014-0017

Singh, S., & Singh, R. (2016). Precision investment casting: A state of art review and future trends. *Proceedings of the Institution of Mechanical Engineers. Part B, Journal of Engineering Manufacture, 230*(12), 2143–2164. doi:10.1177/0954405415597844

Sivadasan, M., Singh, N.K. & Sood, A.K. (2012). Use of fused deposition modeling process in investment precision casting and risk of using selective laser sintering process. *International Journal of Applied Research in Mechanical Engineering*.

Smith, B. J., St Jean, P., & Duquette, M. L. (1996). A comparison of rapid prototype techniques for investment casting Be-Al. In *Proceedings of Rapid Prototyping and Manufacturing'96* (Vol. 2, pp. 1–11). Dearborn, MI: Conference.

Sneath, P. H. A. (1957). The application of computers to taxonomy. *Journal of General Microbiology, 17*(1), 201–226. PMID:13475686

Songa, M., Guana, K., Qinb, W., & Szpunar, J. A. (2012). Comparison of mechanical properties in conventional and small punch tests of fractured anisotropic A350 alloy forging flange. *Nuclear Engineering and Design, 247*, 58–65. doi:10.1016/j.nucengdes.2012.03.023

Spong, Hutchinson, & Vidyasagar. (2005). Robot Modelling and Control. New York: John Wiley & Sons.

Sproedt, A., Plehn, J., Schönsleben, P., & Herrmann, C. (2015). A simulation-based decision support for eco-efficiency improvements in production systems. *Journal of Cleaner Production*, *105*, 389–405. doi:10.1016/j.jclepro.2014.12.082

SrinivasaRao, P., & Birru, A. K. (2017). Effect of Mechanical Properties with addition of Molasses and FlyAsh in Green Sand Moulding. *Materials Today: Proceedings*, *4*(2), 1186–1192. doi:10.1016/j.matpr.2017.01.136

Srinivas, S., & Babu, N. R. (2011). Role of garnet and silicon carbide abrasives in abrasive waterjet cutting of aluminum-silicon carbide particulate metal matrix composites. *Journal of Applied Research in Mechanical Engineering*, *1*, 109–122.

Srivastava, T., Desikan, P., & Kumar, V. (2005). Web mining–concepts, applications and research directions. *Foundations and Advances in Data Mining*, 275-307.

Staiger, M. P., Pietak, A. M., Huadmai, J., & Dias, G. (2006). Magnesium and its alloys as orthopedic biomaterials: A review. *Biomaterials*, *27*(9), 1728–1734. doi:10.1016/j.biomaterials.2005.10.003 PMID:16246414

Subramoniam, R., Huisingha, D., Chinnam, R. B., & Subramoniam, S. (2011). Remanufacturing Decision-Making Framework (RDMF): Research validation using the analytical hierarchical process. *Journal of Cleaner Production*, 1–9.

Subulan, K., Baykasoğlu, A., Özsoydan, F., Taşan, A., & Selim, H. (2015). A case-oriented approach to a lead/acid battery closed-loop supply chain network design under risk and uncertainty. *Journal of Manufacturing Systems*, *37*, 340–361. doi:10.1016/j.jmsy.2014.07.013

Subulan, K., Taşan, A., & Baykasoğlu, A. (2015). A fuzzy goal programming model to strategic planning problem of a lead/acid battery closed-loop supply chain. *Journal of Manufacturing Systems*, *37*, 243–164. doi:10.1016/j.jmsy.2014.09.001

Subulan, K., Taşan, A., & Baykasoğlu, A. (2015). Designing an environmentally conscious tire closed-loop supply chain network with multiple recovery options using interactive fuzzy goal programming. *Applied Mathematical Modelling*, *39*(9), 2661–2702. doi:10.1016/j.apm.2014.11.004

Suh, S., & Huppes, G. (2005). Methods for Life Cycle Inventory of a product. *Journal of Cleaner Production*, *13*(7), 687–697. doi:10.1016/j.jclepro.2003.04.001

Sujith, S. S., Arunkumar, S. L., Mangalaraja, R. V., Mohamed, A. P., & Ananthakumar, S. (2014). Porous to dense $LaPO_4$ sintered ceramics for advanced refractories. *Ceramics International*, *40*(9), 15121–15129. doi:10.1016/j.ceramint.2014.06.125

Sundin, E., & Bras, B. (2005). Making functional sales environmentally and economically beneficial through product remanufacturing. *Journal of Cleaner Production*, *13*(9), 913–925. doi:10.1016/j.jclepro.2004.04.006

Sunil, B. R., Ganapathy, C., Kumarn, T. S. S., & Chakkingal, U. (2014). Processing and mechanical behavior of lamellar structured degradable magnesium hydroxyapatite implants. *Journal of the Mechanical Behavior of Biomedical Materials*, *40*, 178–189. doi:10.1016/j.jmbbm.2014.08.016 PMID:25241282

Sun, J., Chen, M., Cao, G., Bi, Y., Liu, D., & Wei, J. (2013). The effect of nano-hydroxyapatite on the microstructure and properties of Mg–3Zn–0.5Zr alloy. *Journal of Composite Materials*, *48*(7), 825–834. doi:10.1177/0021998313478259

Sun, Z., Zhou, Y., Wang, J., & Li, M. (2007). γ-Y2Si2O7, a Machinable Silicate Ceramic: Mechanical Properties and Machinability. *Journal of the American Ceramic Society*, *90*(8), 2535–2541. doi:10.1111/j.1551-2916.2007.01803.x

Sutherland, L. S., Shenoi, R. A., & Lewis, S. M. (1999). Size and scale effects in composites: II. Unidirectional laminates. *Composites Science and Technology*, *59*(2), 221–233. doi:10.1016/S0266-3538(98)00083-9

Sweeney, L., & Coughlan, J. (2008). Do different industries report corporate social responsibility differently? An investigation through the lens of stakeholder theory. *Journal of Marketing Communications*, *14*(2), 113–124. doi:10.1080/13527260701856657

Szabo, L., Oprea, C., Viorel, I.-A., & Biró, K. Á. (2007). *Novel permanent magnet tubular linear generator for wave energy converters.* Paper presented at the Electric Machines & Drives Conference. 10.1109/IEMDC.2007.382809

Szilvśi-Nagy, M., & Matyasi, G. Y. (2003). Analysis of STL files. *Mathematical and Computer Modelling*, *38*(7-9), 945–960. doi:10.1016/S0895-7177(03)90079-3

Talib, Swadi, Abed, Abed, & Karim. (2013). Vibration and Kinematic Analysis of SCARA Robot Structure. *Diyala Journal of Engineering Sciences*, *6*(3), 127–143.

Tammas-Williams & Todd. (2017). *Design for Additive Manufacturing with site-specific properties in metals and alloys*. Academic Press. . doi:10.1016/j.scriptamat.2016.10.030

Tan, X., & Pan, P. (2012). A contextual item-based collaborative filtering technology. *Intelligent Information Management Journal, 4*, 85-88.

Tan, H. X., Yeo, Z., Ng, R., Tjandra, T. B., & Song, B. (2015). A sustainability indicator framework for Singapore small and medium-sized manufacturing enterprises. CIRP Procedia, 132–137. doi:10.1016/j.procir.2015.01.028

Taşcıoğlu, S., & Akar, N. (2003). A novel alternative to the additives in investment casting pattern wax compositions. *Materials & Design*, *24*(8), 693–698. doi:10.1016/S0261-3069(03)00097-9

Taşcıoğlu, S., & Akar, N. (2007). Conversion of an investment casting sprue wax to a pattern wax by chemical agents. *Materials and Manufacturing Processes*, *18*(5), 753–768. doi:10.1081/AMP-120024973

Tateno & Kondoh. (2017). *Environmental load reduction by customization for reuse with Additive Manufacturing.* Academic Press. doi: . doi:10.1016/j.procir.2016.11.219

Tavana, N. R., Shoulaie, A., & Dinavahi, V. (2012). Analytical Modeling and Design Optimization of Linear Synchronous Motor With Stair-Step-Shaped Magnetic Poles for Electromagnetic Launch Applications. *Plasma Science. IEEE Transactions on*, *40*(2), 519–527.

Taylor, P. R. (1983). An illustrated history of lost wax casting. *Proceedings of the 17th Annual BICTA Conference.*

Tennant, C., & Roberts, P. (2001). Hoshin Kanri: Implementing the catchball process. *Long Range Planning*, *34*(3), 287–308. doi:10.1016/S0024-6301(01)00039-5

Teoh, S. H. (2000). Fatigue of biomaterials: A review. *International Journal of Fatigue*, *22*(10), 825–837. doi:10.1016/S0142-1123(00)00052-9

Thambiraj, S., & Ravi Shankaran, D. (2017). Preparation and physicochemical characterization of cellulose nanocrystals from industrial waste cotton. *Applied Surface Science*, *412*, 405–416. doi:10.1016/j.apsusc.2017.03.272

Thiriez, A., & Gutowski, T. (2006). An Environmental Analysis of Injection Molding. *Proceedings of the 2006 IEEE International Symposium on Electronics and the Environment*, 195–200. 10.1109/ISEE.2006.1650060

Thürer, M. (2013). Hoshin Kanri for the Lean Enterprise: Developing Competitive Capabilities and Managing Profit. *The Quality Management Journal*, *20*(3), 70. doi:10.1080/10686967.2013.11918356

Tiwary, V., Arunkumar, P., Deshpande, A. S., & Khorate, V. (2015). Studying the effect of chemical treatment and fused deposition modelling process parameters on surface roughness to make acrylonitrile butadiene styrene patterns for investment casting process. *International Journal of Rapid Manufacturing*, *5*(3-4), 276–288. doi:10.1504/IJRAPIDM.2015.074807

Tomczuk, B., Schroder, G., & Waindok, A. (2007). Finite-element analysis of the magnetic field and electromechanical parameters calculation for a slotted permanent-magnet tubular linear motor. *IEEE Transactions on Magnetics*, *43*(7), 3229–3236. doi:10.1109/TMAG.2007.894216

Träger, H., & Bührig-Polaczek, A. (2000). *Foundry technology*. Ullmann's Encyclopedia of Industrial Chemistry. doi:10.1002/14356007.a12_035

Transparency Market Research. (2016). *Global Dental Implants Market: Rising Incidence of Periodontal Diseases Stokes Growth*. Retrieved from https://www.transparencymarketresearch.com/pressrelease/dental-implants-market.htm

Tripathi, P. K., Bandyopadhyay, S., & Pal, S. K. (2007). Multi-objective particle swarm optimization with time variant inertia and acceleration coefficients. *Information Sciences*, *177*(22), 5033–5049. doi:10.1016/j.ins.2007.06.018

Trulli, E., Ferronato, N., Torretta, V., Piscitelli, M., Masi, S., & Mancini, I. (2017). *Sustainable mechanical biological treatment of solid waste in urbanized areas with low recycling rates*. Waste Management.

Tsai. (1999). *Robot Analysis: The Mechanics of Serial and Parallel Manipulators*. New York: John Wiley & Sons.

Tseng, P., Murray, C., Kim, D., & Di Carlo, D. (2004). Research highlights: Printing the future of microfabrication. *Lab on a Chip*, *14*(9), 1491–1495. doi:10.1039/c4lc90023e PMID:24671475

Turner, B., Strong, R., & Gold, S. (2014). A review of melt extrusion additive manufacturing processes: I. Process design and modeling. *Rapid Prototyping Journal*, *20*(3), 192–204. doi:10.1108/RPJ-01-2013-0012

Turner, B. N., & Gold, S. A. (2015). A review of melt extrusion additive manufacturing processes: II. Materials, dimensional accuracy, and surface roughness. *Rapid Prototyping Journal*, *21*(3), 250–261. doi:10.1108/RPJ-02-2013-0017

Udai, Rajeevlochana, & Saha (2011). Dynamic Simulation of a KUKA KR5 Industrial Robot using MATLAB SimMechanics. In *15th National Conference on Machines and Mechanisms* (pp. 1-8). Chennai, India: IIT Chennai.

Uddin, M. S., Hall, C., & Murphy, P. (2015). Surface treatments for controlling corrosion rate of biodegradable Mg and Mg-based alloy implants. *Science and Technology of Advanced Materials*, *16*(5), 053501. doi:10.1088/1468-6996/16/5/053501 PMID:27877829

Uddin, M. S., Rosman, H., Hall, C., & Murphy, P. (2017). Enhancing the corrosion resistance of biodegradable Mg-based alloy by machining-induced surface integrity: Influence of machining parameters on surface roughness and hardness. *International Journal of Advanced Manufacturing Technology*, *90*(5-8), 2095–2108. doi:10.100700170-016-9536-x

Umar & Bakarb. (2014). Study on Trajectory Motion and Computational Analysis of Robot Manipulator. *Jurnal Teknologi (Sciences & Engineering)*, *67*(1), 53-59. doi:10.11113/jt.v67.2206

UN. (1987). World Commission on Environment and Development, Report of The World Commission on Environment and Development: Our Common Future, Annex To General Assembly Document A/42/427.

Union, E. (2010). *History and Definitions of Eco-Efficiency*. Leonardo da Vinci Program.

Urrea, C., & Kern, J. (2012). Modelling, Simulation and Control of a Redundant SCARA Type Manipulator Robot. *International Journal of Advanced Robotic Systems*, *9*(58), 58–72. doi:10.5772/51701

Vaccaro, A. R., Singh, K., Haid, R., Kitchel, S., Wuisman, P., Taylor, W., ... Garfin, S. (2003). The use of bioabsorbable implants in the spine. *The Spine Journal*, *3*(3), 227–237. doi:10.1016/S1529-9430(02)00412-6 PMID:14589204

Vaezi, M., Chianrabutra, S., Mellor, B., & Yang, S. (2013). Multiple Material Additive Manufacturing Part 1: A Review. *Virtual and Physical Prototyping*, *8*(1), 19–50. doi:10.1080/17452759.2013.778175

Vahabzadeh, A., Asiaei, A., & Zailani, S. (2015). Reprint of "Green decision-making model in reverse logistics using FUZZY-VIKOR method". *Resources, Conservation and Recycling*, *104*, 334–347. doi:10.1016/j.resconrec.2015.10.028

Vahdani, B., Dehbari, S., & Beni, M. (2014). An artificial intelligence approach for fuzzy possibilistic-stochastic multi-objective logistics network design. *Neural Computing & Applications*, *25*(7-8), 1887–1902. doi:10.100700521-014-1679-9

Vahidgolpayegani, Wen, Hodgson, & Li. (2017). *Production methods and characterization of porous Mg and Mg alloys for biomedical applications*. Woodhead Publishing.

Van den Bergh, F., & Engelbrecht, A. P. (2006). A study of particle swarm optimization particle trajectories. *Information Sciences*, *176*(8), 937–971. doi:10.1016/j.ins.2005.02.003

van Noort, R. (2012). The future of dental devices is digital. *Dental Materials*, *28*(1), 3–12. doi:10.1016/j.dental.2011.10.014 PMID:22119539

Vaupotic, B., Brezocnik, M., & Balic, J. (2006). Use of PolyJet technology in manufacture of new product. *Journal of Achievements in Materials and Manufacturing Engineering*, *18*(1-2), 319–322.

Vermeulen, M., Claessens, T., Van Der Smissen, B., Van Holsbeke, C. S., De Backer, J. W., Van Ransbeeck, P., & Verdonck, P. (2013). Manufacturing of patient-specific optically accessible airway models by fused deposition modeling. *Rapid Prototyping Journal*, *19*(5), 312–318. doi:10.1108/RPJ-11-2011-0118

Vickers C. (n.d.). *An alternative route to metal components for prototype and low-volume production. Rapid prototyping casebook*. Academic Press.

Vijay, P., Danaiah, P., & Rajesh, K. V. (2011). Critical parameters effecting the rapid prototyping surface finish. *Journal of Mechanical Engineering and Automation*, *1*, 17–20. doi:10.5923/j.jmea.20110101.03

Vikas, S., Shashikant, Roy, A. K., & Kumar, K. (2014). Effect and Optimization of Machine Process Parameters on MRR for EN19 and EN41 materials using Taguchi. *Procedia Technology*, *14*, 204–210. doi:10.1016/j.protcy.2014.08.027

Villalba, G., Segarra, M., Chimenos, J. M., & Espiell, F. (2004). Using the recyclability index of materials as a tool for design for disassembly. *Ecological Economics*, *50*(3-4), 195–200. doi:10.1016/j.ecolecon.2004.03.026

Vinodh, S. (2010). *Assessment of sustainability using multi-grade fuzzy approach*. Clean Tech Environ Policy. doi:10.100710098-010-0333-1

Viswanathan, R., Rameshbabu, N., Kennedy, S., Sreekanth, D., Venkateswarlu, K., Sandhya, R. M., & Muthupandi, V. (2013). Plasma electrolytic oxidation and characterization of spark plasma sintered magnesium/hydroxyapatite composites. *Materials Science Forum*, *765*, 827–831. doi:10.4028/www.scientific.net/MSF.765.827

Volak, J., Novak, M., Kaiser, J., & Mentl, V. (2012, December). Fatigue testing by means of miniature test specimens. *Journal of Achievement in Materials and Manufacturing Engineering*, *55*(2), 386–389.

Vosburgh, R. M. (2007). Special edition articles. Human Resource Planning, 30(1), 7–8.

Wadhwa, S., Madaan, J., & Chan, F. (2009). Flexible decision modeling of reverse logistics system: A value adding MCDM approach for alternative selection. *Robotics and Computer-integrated Manufacturing*, *25*(2), 460–469. doi:10.1016/j.rcim.2008.01.006

Waerhaug, J. & Zander, H.A. (1956). Implantation of acrylic roots in tooth sockets. *Oral Surgery, Oral Medicine, Oral Pathology, Oral Radiology, and Endodontology*, *9*, 46–54.

Wang, Jiang, Zho, Gou, & Hui. (2017). *3D printing of polymer matrix composites: A review and prospective*. Academic Press. doi:10.1016/j.compositesb.2016.11.034

Wang, J. (2004). Techniques for enhancing the cutting performance of abrasive waterjets. *Key Engineering Materials, 257/258*, 521–526. doi:10.4028/www.scientific.net/KEM.257-258.521

Wang, J., Howe, D., & Lin, Z. (2008). Analysis of a short-stroke, single-phase, quasi-Halbach magnetised tubular permanent magnet motor for linear compressor applications. *IET Electric Power Applications, 2*(3), 193–200. doi:10.1049/iet-epa:20070281

Wang, J., Ibrahim, T., & Howe, D. (2010). Prediction and measurement of iron loss in a short-stroke, single-phase, tubular permanent magnet machine. *IEEE Transactions on Magnetics, 46*(6), 1315–1318. doi:10.1109/TMAG.2010.2042685

Wang, N., & Yang, Y. (2009). A fuzzy modeling method via Enhanced Objective Cluster Analysis for designing TSK model. *Expert Systems with Applications, 36*(10), 12375–12382. doi:10.1016/j.eswa.2009.04.048

Wang, S., Miranda, A. G., & Shih, C. (2010). A study of investment casting with plastic patterns. *Materials and Manufacturing Processes, 25*(12), 1482–1488. doi:10.1080/10426914.2010.529585

Wang, Z.-X., Shi, H.-J., Lu, J., Shi, P., & Ma, X.-F. (2008). Small punch testing for assessing the fracture properties of the reactor vessel steel with different thicknesses. *Nuclear Engineering and Design, 238*(12), 3186–3193. doi:10.1016/j.nucengdes.2008.07.013

Warner, M. C. (1993). Rapid prototyping methods to manufacture functional metal and plastic parts. Rapid Prototyping System: Fast Track to Product Realisation, 137-44.

WBCSD. (2000a). *Eco-efficiency. Creating more Value with less Impact*. World Bus Counc Sustain Dev.

WBCSD. (2000b). *Measuring ecoefficiency: a guide to reporting company performance*. Geneva, Switzerland. World Business Council for Sustainable Development.

Weidema, B., Wenzel, H., Petersen, C., & Hansen, K. (2004). The product, functional unit and reference flows in LCA. *Environ News, 70*, 46. Available from: http://gfc.force.dk/resources/777.pdf

Weidema, B.P., Ekvall, T., & Heijungs, R. (2009). *Guidelines for application of deepened and broadened LCA*. Guidel Appl Deep broadened LCA Deliv D18 Work Packag 5 CALCAS Proj. 2009, (037075), 49.

Weisenel, L., Travitzky, N., Seiber, H., & Grell, P. (2004).. *Laminated Object Manufacturing of SiSiC Composite*. Academic Press.

Wendel, B., Rietzel, D., Kühnlein, F., Feulner, R., Hülder, G., & Schmachtenberg, E. (2008). Additive processing of polymers. *Macromolecular Materials and Engineering, 293*(10), 799–809. doi:10.1002/mame.200800121

Wennerberg, A., & Albrektsson, T. (2010). On implant surfaces: A review of current knowledge and opinions. *International Journal of Oral & Maxillofacial Implants, 25*, 63–74. PMID:20209188

White, C. D., Masanet, E., & Meisner, C. (2003). Product recovery with some byte: An overview of management challenges and environmental consequences in reverse manufacturing for the computer industry. *Journal of Cleaner Production, 11*(4), 445–458. doi:10.1016/S0959-6526(02)00066-5

Whitmore. (2015). Additive Manufacturing as an enabling technology for "green" hybrid spacecraft propulsion. *Recent Advances in Space Technologies (RAST), 2015 7th International Conference*. 10.1109/RAST.2015.7208305

Wiemes, L., Pawlowsky, U., & Mymrin, V. (2017). Incorporation of industrial wastes as raw materials in brick's formulation. *Journal of Cleaner Production, 142*, 69–77. doi:10.1016/j.jclepro.2016.06.174

Wintermantel, E., & Suk-Woo, H. (2002). *Medizintechnik mit biokompatiblen Werkstoffen und Verfahren*. Berlin, Germany: Springer.

Witcher, B. J., & Chau, V. S. (1993). Balanced scorecard and hoshin kanri : Dynamic capabilities for managing strategic fit. *Management Decision*, *45*(3), 518–538. doi:10.1108/00251740710745115

Witcher, B., & Butterworth, R. (1999). Hoshin Kanri: How Xerox manages. *Long Range Planning*, *32*(3), 323–332. doi:10.1016/S0024-6301(99)00036-9

Witte, F., Feyerabend, F., Maier, P., Fischer, J., Stormer, M., Blawert, C., ... Hort, N. (2007). Biodegradable magnesium–hydroxyapatite metal matrix composites. *Biomaterials*, *28*(13), 2163–2174. doi:10.1016/j.biomaterials.2006.12.027 PMID:17276507

Wohlers, T. (2010). *Wohlers Report 2010*. Wholers Associates.

Wong & Hernandez. (2012). *International Scholarly Research Network ISRN Mechanical Engineering Volume*. doi:10.5402/2012/208760

Wood & Kennedy. (2015, April 24). *Simulating Mechanical Systems in Simulink with SimMechanics The MathWorks*. Retrieved from http://cn.mathworks.comtagteam/12634_SimMechanics.pdf

Wood, I. (2016). *Development of a layerless additive manufacturing stereolithography machine to improve surface quality and dimensional accuracy* (Doctoral dissertation). University of Ontario Institute of Technology.

World Commission on Environment and Development. (1987). *From one earth to one world: an over view*. Retrieved from http://mom.gov.af/Content/files/Bruntland_Report.pdf

Wu, H., Li, D., & Guo, N. (2009). Fabrication of integral ceramic mold for investment casting of hollow turbine blade based on stereolithography. *Rapid Prototyping Journal*, *15*(4), 232–237. doi:10.1108/13552540910979749

Wursthorn, S., Poganietz, W. R., & Schebek, L. (2011). Economic-environmental monitoring indicators for European countries: A disaggregated sector-based approach for monitoring eco-efficiency. *Ecological Economics*, *70*(3), 487–496. doi:10.1016/j.ecolecon.2010.09.033

Xiang, Q., Zhang, H., Jiang, Z., Zhu, S., & Yan, W. (2016). A Decision-making Method for Active Remanufacturing Time Based on Environmental and Economic Indicators. *International Journal of Online Engineering*, *12*(12).

Xiao, R., Cai, Z., & Zhang, X. (2012). An optimization approach to risk decision-making of closed-loop logistics based on SCOR model. *Optimization*, *61*(10), 1221–1251. doi:10.1080/02331934.2012.688827

Xiaoyong, R., Peng, Z., Hu, Y., Wang, C., Fu, Z., Yue, W., ... Hezhuo, M. (2013). Abrasive wear behavior of TiCN cermets under water-based slurries with different abrasives. *Tribology International*, *66*, 35–43. doi:10.1016/j.triboint.2013.04.002

Xu, H. H. K., Wei, L., & Jahanmir, S. (1995). Grinding force and micro crack density in abrasive machining of silicon nitride. *Journal of Materials Research*, *10*(12), 3204–3209. doi:10.1557/JMR.1995.3204

Xu, L. P., Yu, G. N., Zhang, E. L., Pan, F., & Yang, K. (2007). In vivo corrosion behavior of Mg–Mn–Zn alloy for bone implant application. *Journal of Biomedical Materials Research*, *83A*(3), 703–711. doi:10.1002/jbm.a.31273 PMID:17549695

Xu, L. P., Yu, G. N., Zhang, E. L., Pan, F., & Yang, K. (2008). In vitro corrosion behavior of Mg alloys in a phosphate buffered solution for bone implant application. *Journal of Materials Science. Materials in Medicine*, *19*(3), 1017–1025. doi:10.100710856-007-3219-y PMID:17665099

Xu, Y., Ning, G., Zhang, C., Yu, Q., & Xu, Y. (2000). Application of the miniature specimen technique to material irradiation tests and surveillance for reactor components. *International Journal of Pressure Vessels and Piping*, *77*(12), 715–721. doi:10.1016/S0308-0161(00)00066-1

Yan, F., Liu, G., Tao, N. R., & Lu, K. (2012). Strength and ductility of 316L austenitic stainless steel strengthened by nano-scale twin bundles. *Acta Materialia*, *60*(3), 1059–1071. doi:10.1016/j.actamat.2011.11.009

Yang, J., Shi, Y., Shen, Q., & Yan, C. (2009). Selective laser sintering of HIPS and investment casting technology. *Journal of Materials Processing Technology*, *209*(4), 1901–1908. doi:10.1016/j.jmatprotec.2008.04.056

Yang, Q., Lu, Z., Zhou, J., Miao, K., & Li, D. (2017). A novel method for improving surface finish of stereolithography apparatus. *International Journal of Advanced Manufacturing Technology*, *393*(5-8), 1537–1544. doi:10.100700170-017-0529-1

Yang, Z., Zhou, X., & Xu, L. (2014). Eco-efficiency optimization for municipal solid waste management. *Journal of Cleaner Production*, *104*, 242–249. doi:10.1016/j.jclepro.2014.09.091

Ye, X., Chen, M., Yang, M., Wei, J., & Liu, D. (2010). In vitro corrosion resistance and cytocompatibility of nano-hydroxyapatite reinforced Mg–Zn–Zr composites. *Journal of Materials Science. Materials in Medicine*, *21*(4), 1321–1328. doi:10.100710856-009-3954-3 PMID:20012772

Yi, S., Liu, F., Zhang, J., & Xiong, S. (2004). Study of the key technologies of LOM for functional metal parts. *Journal of Materials Processing Technology*, *150*(1), 175–181. doi:10.1016/j.jmatprotec.2004.01.035

Zeng, Q., Xu, Z., Tian, Y., & Qin, Y. (2016). Advancement in Additive Manufacturing & numerical modelling considerations of direct energy deposition process. In *Proceeding of the 14th International Conference on Manufacturing Research*. IOS Press.

Zeng, S. X., Liu, H. C., Tam, C. M., & Shao, Y. K. (2008). Cluster analysis for studying industrial sustainability: An empirical study in Shanghai. *Journal of Cleaner Production*, *16*(10), 1090–1097. doi:10.1016/j.jclepro.2007.06.004

Zhai, Y., Lados, D. A., & LaGoy, J. L. (2014). Additive Manufacturing: Making Imagination the Major Limitation. *JOM*, *66*(5), 808–816. doi:10.100711837-014-0886-2

Zhang, Liu, & To. (2017). *Role of anisotropic properties on topology optimization of additive manufactured load bearing structures*. Academic Press. . doi:10.1016/j.scriptamat.2016.10.021

Zhang, B., Bi, J., Fan, Z. Y., Yuan, Z. W., & Ge, J. J. (2008). Eco-efficiency analysis of industrial system in China: A data envelopment analysis approach. *Ecological Economics*, *68*(1-2), 306–316. doi:10.1016/j.ecolecon.2008.03.009

Zhang, E. L., He, W. W., Du, H., & Yang, K. (2008). Microstructure, mechanical properties and corrosion properties of Mg–Zn–Y alloys with low Zn content. *Materials Science and Engineering A*, *488*(1-2), 102–111. doi:10.1016/j.msea.2007.10.056

Zhang, E. L., Yin, D. S., Xu, L. P., Yang, L., & Yang, K. (2009). Microstructure, mechanical and corrosion properties and biocompatibility of Mg–Zn–Mn alloys for biomedical application. *Materials Science and Engineering C*, *29*(3), 987–993. doi:10.1016/j.msec.2008.08.024

Zhang, E., & Yang, L. (2008). Microstructure, mechanical properties and bio-corrosion properties of Mg–Zn–Mn–Ca alloy for biomedical application. *Materials Science and Engineering A*, *497*(1-2), 111–118. doi:10.1016/j.msea.2008.06.019

Zhang, E., Yang, L., Xu, J., & Chen, H. (2010). Microstructure, mechanical properties and bio-corrosion properties of Mg–Si(–Ca, Zn) alloy for biomedical application. *Acta Biomaterialia*, *6*(5), 1756–1762. doi:10.1016/j.actbio.2009.11.024 PMID:19941979

Zhang, S., Wu, Y., & Wang, Y. A. (2011). A review on abrasive waterjet and wire electrical discharge machining – high speeds. *The Open Mechanical Engineering Journal*, *5*(1), 178–185. doi:10.2174/1874155X01105010178

Zhang, W., Leu, M. C., Ji, Z., & Yan, Y. (1999). Rapid freezing prototyping with water. *Materials & Design*, *20*(2), 139–145. doi:10.1016/S0261-3069(99)00020-5

Zhao, J., Yu, Z., Yu, K., & Chen, L. (2011). Biodegradable behaviors of Mg–6%Zn–5%Hydroxyapatite biomaterial. *Advanced Materials Research*, *239–242*, 1287–1291. doi:10.4028/www.scientific.net/AMR.239-242.1287

Zhao, W., & Guo, C. (2014). Topography and microstructure of the cutting surface machined with abrasive waterjet. *Journal of Advance Manufacturing Technology*, *73*(5-8), 941–947. doi:10.100700170-014-5869-5

Zhao, Z. L., Wang, C. D., & Lai, J. H. (2016). AUI&GIV: Recommendation with asymmetric user influence and global importance value. *PLoS One*, *11*(2), e0147944. doi:10.1371/journal.pone.0147944 PMID:26828803

Zheng, B., Ertorer, O., Li, Y., Zhou, Y., Mathaudhu, S. N., Tsao, C. Y. A., & Laverni, J. E. (2011). High strength, nano-structured Mg–Al–Zn alloy. *Materials Science and Engineering A*, *528*(4-5), 2180–2191. doi:10.1016/j.msea.2010.11.073

Zheng, Z.-Q., Huang, P., Gao, D.-X., & Chang, Z.-Y. (2015). Analysis of Electromagnetic Force of the Linear Generator in Point Absorber Wave Energy Converters. *Journal of Marine Science and Technology*, *23*(4), 475–480.

Zi, B., Cao, J., & Zhu, Z. (2011). Dynamic Simulation of Hybrid-driven Planar Five-bar Parallel Mechanism Based on SimMechanics and Tracking Control. *International Journal of Advanced Robotic Systems*, *8*(4), 28–33. doi:10.5772/45683

Ziout, A., Azab, A., & Atwan, M. (2014). A holistic approach for decision on selection of end-of-life products recovery options. *Journal of Cleaner Production*, *65*, 497–516. doi:10.1016/j.jclepro.2013.10.001

Zlajpah, L. (2008). Simulation in robotics. *Mathematics and Computers in Simulation*, *79*(4), 879–897. doi:10.1016/j.matcom.2008.02.017

Related References

To continue our tradition of advancing research on topics in the field of engineering, we have compiled a list of recommended IGI Global readings. These references will provide additional information and guidance to further enrich your knowledge and assist you with your own research and future publications.

Abawajy, J. H., Pathan, M., Rahman, M., Pathan, A., & Deris, M. M. (2013). *Network and traffic engineering in emerging distributed computing applications*. Hershey, PA: IGI Global. doi:10.4018/978-1-4666-1888-6

Abu-Faraj, Z. O. (2012). Bioengineering/biomedical engineering education. In Z. Abu-Faraj (Ed.), *Handbook of research on biomedical engineering education and advanced bioengineering learning: Interdisciplinary concepts* (pp. 1–59). Hershey, PA: IGI Global. doi:10.4018/978-1-4666-0122-2.ch001

Abu-Nimeh, S., & Mead, N. R. (2012). Combining security and privacy in requirements engineering. In T. Chou (Ed.), *Information assurance and security technologies for risk assessment and threat management: Advances* (pp. 273–290). Hershey, PA: IGI Global. doi:10.4018/978-1-61350-507-6.ch011

Abu-Taieh, E., El Sheikh, A., & Jafari, M. (2012). *Technology engineering and management in aviation: Advancements and discoveries*. Hershey, PA: IGI Global. doi:10.4018/978-1-60960-887-3

Achumba, I. E., Azzi, D., & Stocker, J. (2010). Low-cost virtual laboratory workbench for electronic engineering. *International Journal of Virtual and Personal Learning Environments*, *1*(4), 1–17. doi:10.4018/jvple.2010100101

Achumba, I. E., Azzi, D., & Stocker, J. (2012). Low-cost virtual laboratory workbench for electronic engineering. In M. Thomas (Ed.), *Design, implementation, and evaluation of virtual learning environments* (pp. 201–217). Hershey, PA: IGI Global. doi:10.4018/978-1-4666-1770-4.ch014

Addo-Tenkorang, R., & Eyob, E. (2013). Engineer-to-order: A maturity concurrent engineering best practice in improving supply chains. In Industrial engineering: Concepts, methodologies, tools, and applications (pp. 1780-1796). Hershey, PA: IGI Global. doi:10.4018/978-1-4666-1945-6.ch095

Aguilera, A., & Davim, J. (2014). *Research developments in wood engineering and technology*. Hershey, PA: IGI Global. doi:10.4018/978-1-4666-4554-7

Aharoni, A., & Reinhartz-Berger, I. (2013). Semi-automatic composition of situational methods. In K. Siau (Ed.), *Innovations in database design, web applications, and information systems management* (pp. 335–364). Hershey, PA: IGI Global. doi:10.4018/978-1-4666-2044-5.ch013

Ahmad, M., Jung, L. T., & Zaman, N. (2014). A comparative analysis of software engineering approaches for sequence analysis. In *Software design and development: Concepts, methodologies, tools, and applications* (pp. 1093–1102). Hershey, PA: IGI Global. doi:10.4018/978-1-4666-4301-7.ch053

Ahrens, A., Bassus, O., & Zaščerinska, J. (2014). Enterprise 2.0 in engineering curriculum. In M. Cruz-Cunha, F. Moreira, & J. Varajão (Eds.), *Handbook of research on enterprise 2.0: Technological, social, and organizational dimensions* (pp. 599–617). Hershey, PA: IGI Global. doi:10.4018/978-1-4666-4373-4.ch031

Akbar, D. (2012). Community engagement in engineering education: Needs and learning outcomes. In M. Rasul (Ed.), *Developments in engineering education standards: Advanced curriculum innovations* (pp. 301–317). Hershey, PA: IGI Global. doi:10.4018/978-1-4666-0951-8.ch017

Alam, F., Subic, A., Plumb, G., Shortis, M., & Chandra, R. P. (2012). An innovative offshore delivery of an undergraduate mechanical engineering program. In M. Rasul (Ed.), *Developments in engineering education standards: Advanced curriculum innovations* (pp. 233–245). Hershey, PA: IGI Global. doi:10.4018/978-1-4666-0951-8.ch013

Ali, D. F., Patil, A., & Nordin, M. S. (2012). Visualization skills in engineering education: Issues, developments, and enhancement. In A. Patil, H. Eijkman, & E. Bhattacharyya (Eds.), *New media communication skills for engineers and IT professionals: Trans-national and trans-cultural demands* (pp. 175–203). Hershey, PA: IGI Global. doi:10.4018/978-1-4666-0243-4.ch011

Aljawarneh, S. (2013). Cloud security engineering: Avoiding security threats the right way. In S. Aljawarneh (Ed.), *Cloud computing advancements in design, implementation, and technologies* (pp. 147–153). Hershey, PA: IGI Global. doi:10.4018/978-1-4666-1879-4.ch010

Alkhatib, G. (2012). *Models for capitalizing on web engineering advancements: Trends and discoveries*. Hershey, PA: IGI Global. doi:10.4018/978-1-4666-0023-2

Allee, T., Handorf, A., & Li, W. (2010). Electrospinning: Development and biomedical applications. In A. Shukla & R. Tiwari (Eds.), *Intelligent medical technologies and biomedical engineering: Tools and applications* (pp. 48–78). Hershey, PA: IGI Global. doi:10.4018/978-1-61520-977-4.ch003

Alsmadi, I. (2014). Website performance measurement: Process and product metrics. In *Software design and development: Concepts, methodologies, tools, and applications* (pp. 1801–1827). Hershey, PA: IGI Global. doi:10.4018/978-1-4666-4301-7.ch086

Altarawneh, H., Alamaro, S., & El Sheikh, A. (2012). Web engineering and business intelligence: Agile web engineering development and practice. In A. Rahman El Sheikh & M. Alnoukari (Eds.), *Business intelligence and agile methodologies for knowledge-based organizations: Cross-disciplinary applications* (pp. 313–344). Hershey, PA: IGI Global. doi:10.4018/978-1-61350-050-7.ch015

Altarawneh, H., & El-Shiekh, A. (2010). Web engineering in small Jordanian web development firms: An XP based process model. In A. Tatnall (Ed.), *Web technologies: Concepts, methodologies, tools, and applications* (pp. 1696–1707). Hershey, PA: IGI Global. doi:10.4018/978-1-60566-982-3.ch091

Alzoabi, Z. (2014). Agile software: Body of knowledge. In *Software design and development: Concepts, methodologies, tools, and applications* (pp. 96–116). Hershey, PA: IGI Global. doi:10.4018/978-1-4666-4301-7.ch006

Andrade-Campos, A. (2013). Development of an optimization framework for parameter identification and shape optimization problems in engineering. In J. Davim (Ed.), *Dynamic methods and process advancements in mechanical, manufacturing, and materials engineering* (pp. 1–24). Hershey, PA: IGI Global. doi:10.4018/978-1-4666-1867-1.ch001

Andreatos, A. (2012). Educating the 21st century's engineers and IT professionals. In A. Patil, H. Eijkman, & E. Bhattacharyya (Eds.), *New media communication skills for engineers and IT professionals: Trans-national and trans-cultural demands* (pp. 132–159). Hershey, PA: IGI Global. doi:10.4018/978-1-4666-0243-4.ch009

Annamalai, C., & Ramayah, T. (2013). Reengineering for enterprise resource planning (ERP) systems implementation: An empirical analysis of assessing critical success factors (CSFs) of manufacturing organizations. In Industrial engineering: Concepts, methodologies, tools, and applications (pp. 791-806). Hershey, PA: IGI Global. doi:10.4018/978-1-4666-1945-6.ch044

Antchev, M. (2010). Other applications of converters and systems of converters. In *Technologies for electrical power conversion, efficiency, and distribution: Methods and processes* (pp. 270–299). Hershey, PA: IGI Global. doi:10.4018/978-1-61520-647-6.ch011

Anzelotti, G., & Valizadeh, M. (2010). Purpose-oriented small software: A case study for some engineering subjects. In R. Luppicini & A. Haghi (Eds.), *Cases on digital technologies in higher education: Issues and challenges* (pp. 164–178). Hershey, PA: IGI Global. doi:10.4018/978-1-61520-869-2.ch012

Anzelotti, G., & Valizadeh, M. (2010). Sights inside the virtual engineering education. In D. Russell & A. Haghi (Eds.), *Web-based engineering education: Critical design and effective tools* (pp. 160–174). Hershey, PA: IGI Global. doi:10.4018/978-1-61520-659-9.ch012

Asadi, M., Mohabbati, B., Gaševic, D., Bagheri, E., & Hatala, M. (2012). Developing semantically-enabled families of method-oriented architectures. *International Journal of Information System Modeling and Design, 3*(4), 1–26. doi:10.4018/jismd.2012100101

Augusti, G., & Feyo de Azevedo, S. (2011). Qualification frameworks and field-specific approaches to quality assurance: Initiatives in engineering and technical education. *International Journal of Quality Assurance in Engineering and Technology Education, 1*(1), 44–57. doi:10.4018/ijqaete.2011010104

Aung, Z., & Nyunt, K. K. (2014). Constructive knowledge management model and information retrieval methods for software engineering. In *Software design and development: Concepts, methodologies, tools, and applications* (pp. 253–269). Hershey, PA: IGI Global. doi:10.4018/978-1-4666-4301-7.ch014

Azad, A. K., Auer, M. E., & Harward, V. (2012). *Internet accessible remote laboratories: Scalable e-learning tools for engineering and science disciplines*. Hershey, PA: IGI Global. doi:10.4018/978-1-61350-186-3

Azar, A. T. (2013). Overview of biomedical engineering. In Bioinformatics: Concepts, methodologies, tools, and applications (pp. 1-28). Hershey, PA: IGI Global. doi:10.4018/978-1-4666-3604-0.ch001

Badr, K. B., Badr, A. B., & Ahmad, M. N. (2013). Phases in ontology building methodologies: A recent review. In M. Nazir Ahmad, R. Colomb, & M. Abdullah (Eds.), *Ontology-based applications for enterprise systems and knowledge management* (pp. 100–123). Hershey, PA: IGI Global. doi:10.4018/978-1-4666-1993-7.ch006

Baer, W., & Renfro, C. (2013). Information sources and collection planning for engineering. In S. Holder (Ed.), *Library collection development for professional programs: Trends and best practices* (pp. 128–144). Hershey, PA: IGI Global. doi:10.4018/978-1-4666-1897-8.ch008

Baghdadi, Y., & Kraiem, N. (2014). Business process modeling with services: reverse engineering databases. In R. Perez-Castillo & M. Piattini (Eds.), *Uncovering essential software artifacts through business process archeology* (pp. 177–200). Hershey, PA: IGI Global. doi:10.4018/978-1-4666-4667-4.ch007

Bakopoulou, A., Leyhausen, G., Geurtsen, W., & Koidis, P. (2013). Dental tissue engineering research and translational approaches towards clinical application. In A. Daskalaki (Ed.), *Medical advancements in aging and regenerative technologies: Clinical tools and applications* (pp. 279–312). Hershey, PA: IGI Global. doi:10.4018/978-1-4666-2506-8.ch013

Baporikar, N. (2012). Developing right graduate attributes through project-based teaching. In M. Rasul (Ed.), *Developments in engineering education standards: Advanced curriculum innovations* (pp. 64–79). Hershey, PA: IGI Global. doi:10.4018/978-1-4666-0951-8.ch004

Baraldi, E., & Nadin, G. (2012). "Network process re-engineering" in a home textile network: The importance of business relationships and actor bonds. In T. Choi (Ed.), *Fashion supply chain management: Industry and business analysis* (pp. 212–234). Hershey, PA: IGI Global. doi:10.4018/978-1-60960-756-2.ch012

Barbu, M. C., Hasener, J., & Bernardy, G. (2014). Modern testing of wood-based panels, process control, and modeling. In A. Aguilera & J. Davim (Eds.), *Research developments in wood engineering and technology* (pp. 90–130). Hershey, PA: IGI Global. doi:10.4018/978-1-4666-4554-7.ch003

Bas, T. G. (2013). Dual market(ing) in "bio-engineering high technology" new products: The risk of uncertainty and failure. *International Journal of Measurement Technologies and Instrumentation Engineering*, *3*(2), 63–74. doi:10.4018/ijmtie.2013040104

Bedi, P., Gandotra, V., & Singhal, A. (2014). Innovative strategies for secure software development. In *Software design and development: Concepts, methodologies, tools, and applications* (pp. 2099–2119). Hershey, PA: IGI Global. doi:10.4018/978-1-4666-4301-7.ch097

Bellatreche, L. (2010). *Data warehousing design and advanced engineering applications: Methods for complex construction*. Hershey, PA: IGI Global. doi:10.4018/978-1-60566-756-0

Bhattacharyya, S., & Dutta, P. (2013). *Handbook of research on computational intelligence for engineering, science, and business* (Vols. 1–2). Hershey, PA: IGI Global. doi:10.4018/978-1-4666-2518-1

Blicblau, A. S., & Richards, D. (2012). Development of "real world" project skills for engineering students. *International Journal of Quality Assurance in Engineering and Technology Education, 2*(1), 1–13. doi:10.4018/ijqaete.2012010101

Boci, E. S., Sarkani, S., & Mazzuchi, T. A. (2013). Development of a complex geospatial/RF design model in support of service volume engineering design. In M. Bartolacci & S. Powell (Eds.), *Advancements and innovations in wireless communications and network technologies* (pp. 56–67). Hershey, PA: IGI Global. doi:10.4018/978-1-4666-2154-1.ch005

Boudreaux, A., & Primeaux, B. (2014). Modular game engine design. In *Software design and development: Concepts, methodologies, tools, and applications* (pp. 1179–1199). Hershey, PA: IGI Global. doi:10.4018/978-1-4666-4301-7.ch058

Boudriga, N., & Hamdi, M. (2014). *Security engineering techniques and solutions for information systems: Management and implementation*. Hershey, PA: IGI Global. doi:10.4018/978-1-61520-803-6

Brad, S. (2010). Competitive design of web-based courses in engineering education. In D. Russell & A. Haghi (Eds.), *Web-based engineering education: Critical design and effective tools* (pp. 119–148). Hershey, PA: IGI Global. doi:10.4018/978-1-61520-659-9.ch010

Brad, S. (2010). Designing effective web-based courses in engineering. In R. Luppicini & A. Haghi (Eds.), *Cases on digital technologies in higher education: Issues and challenges* (pp. 217–240). Hershey, PA: IGI Global. doi:10.4018/978-1-61520-869-2.ch016

Bradford, M., Gingras, R., & Hornby, J. (2010). Business process reengineering and ERP: Weapons for the global organization. In K. St.Amant (Ed.), *IT outsourcing: Concepts, methodologies, tools, and applications* (pp. 211–228). Hershey, PA: IGI Global. doi:10.4018/978-1-60566-770-6.ch012

Brennan, R. W., Hugo, R., & Rosehart, W. D. (2012). CDIO as an enabler for graduate attributes assessment: A Canadian case study. *International Journal of Quality Assurance in Engineering and Technology Education, 2*(2), 45–54. doi:10.4018/ijqaete.2012040105

Burnett, M. (2012). End-user software engineering and why it matters. In A. Dwivedi & S. Clarke (Eds.), *End-user computing, development, and software engineering: New challenges* (pp. 185–201). Hershey, PA: IGI Global. doi:10.4018/978-1-4666-0140-6.ch009

Burns, G. U., & Chisohlm, C. (2011). Engineering professional development related to sustainability of quality. *International Journal of Quality Assurance in Engineering and Technology Education, 1*(1), 15–29. doi:10.4018/ijqaete.2011010102

Byrne, D., Kelly, L., & Jones, G. J. (2014). Multiple multimodal mobile devices: Lessons learned from engineering lifelong solutions. In *Software design and development: Concepts, methodologies, tools, and applications* (pp. 2014–2032). Hershey, PA: IGI Global. doi:10.4018/978-1-4666-4301-7.ch093

Cervera, M., Albert, M., Torres, V., & Pelechano, V. (2012). A model-driven approach for the design and implementation of software development methods. *International Journal of Information System Modeling and Design, 3*(4), 86–103. doi:10.4018/jismd.2012100105

Chang, L., Levy, M., & Powell, P. (2011). Small firm process re-engineering success. In M. Tavana (Ed.), *Managing adaptability, intervention, and people in enterprise information systems* (pp. 138–155). Hershey, PA: IGI Global. doi:10.4018/978-1-60960-529-2.ch007

Chinyemba, F. (2011). Mobility of engineering and technology professionals and its impact on the quality of engineering and technology education: The case of Chinhoyi University of Technology, Zimbabwe. *International Journal of Quality Assurance in Engineering and Technology Education, 1*(2), 35–49. doi:10.4018/ijqaete.2011070104

Chiong, R. (2010). *Nature-inspired informatics for intelligent applications and knowledge discovery: Implications in business, science, and engineering*. Hershey, PA: IGI Global. doi:10.4018/978-1-60566-705-8

Chiprianov, V., Kermarrec, Y., & Rouvrais, S. (2014). Integrating DSLs into a software engineering process: Application to collaborative construction of telecom services. In *Software design and development: Concepts, methodologies, tools, and applications* (pp. 570–595). Hershey, PA: IGI Global. doi:10.4018/978-1-4666-4301-7.ch028

Chis, M. (2010). Introduction: A survey of the evolutionary computation techniques for software engineering. In M. Chis (Ed.), *Evolutionary computation and optimization algorithms in software engineering: Applications and techniques* (pp. 1–12). Hershey, PA: IGI Global. doi:10.4018/978-1-61520-809-8.ch001

Chiu, D. K. (2013). *Mobile and web innovations in systems and service-oriented engineering*. Hershey, PA: IGI Global. doi:10.4018/978-1-4666-2470-2

Chu, P. K., & Wu, S. (2012). Biomaterials. In Z. Abu-Faraj (Ed.), *Handbook of research on biomedical engineering education and advanced bioengineering learning: Interdisciplinary concepts* (pp. 238–283). Hershey, PA: IGI Global. doi:10.4018/978-1-4666-0122-2.ch006

Cimellaro, G. P. (2013). Optimal placement of controller for seismic structures. In N. Lagaros, V. Plevris, & C. Mitropoulou (Eds.), *Design optimization of active and passive structural control systems* (pp. 1–33). Hershey, PA: IGI Global. doi:10.4018/978-1-4666-2029-2.ch001

Clark, R., & Andrews, J. (2012). Engineering the future. In M. Rasul (Ed.), *Developments in engineering education standards: Advanced curriculum innovations* (pp. 143–155). Hershey, PA: IGI Global. doi:10.4018/978-1-4666-0951-8.ch008

Clark, T., & Willans, J. (2013). Software language engineering with XMF and XModeler. In M. Mernik (Ed.), *Formal and practical aspects of domain-specific languages: Recent developments* (pp. 311–340). Hershey, PA: IGI Global. doi:10.4018/978-1-4666-2092-6.ch011

Coll, R. K., & Zegwaard, K. E. (2012). Enculturation into engineering professional practice: Using legitimate peripheral participation to develop communication skills in engineering students. In A. Patil, H. Eijkman, & E. Bhattacharyya (Eds.), *New media communication skills for engineers and IT professionals: Trans-national and trans-cultural demands* (pp. 22–33). Hershey, PA: IGI Global. doi:10.4018/978-1-4666-0243-4.ch003

Coll, R. K., & Zegwaard, K. E. (2012). Enculturation into engineering professional practice: Using legitimate peripheral participation to develop communication skills in engineering students. In A. Patil, H. Eijkman, & E. Bhattacharyya (Eds.), *New media communication skills for engineers and IT professionals: Trans-national and trans-cultural demands* (pp. 22–33). Hershey, PA: IGI Global. doi:10.4018/978-1-4666-0243-4.ch003

Colomb, R. M. (2013). Representation of action is a primary requirement in ontologies for interoperating information systems. In M. Nazir Ahmad, R. Colomb, & M. Abdullah (Eds.), *Ontology-based applications for enterprise systems and knowledge management* (pp. 68–76). Hershey, PA: IGI Global. doi:10.4018/978-1-4666-1993-7.ch004

Cooklev, T. (2013). The role of standards in engineering education. In K. Jakobs (Ed.), *Innovations in organizational IT specification and standards development* (pp. 129–137). Hershey, PA: IGI Global. doi:10.4018/978-1-4666-2160-2.ch007

Costa, L., Loughran, N., & Grønmo, R. (2014). Model-driven engineering, services and interactive real-time applications. In *Software design and development: Concepts, methodologies, tools, and applications* (pp. 178–202). Hershey, PA: IGI Global. doi:10.4018/978-1-4666-4301-7.ch010

Dai, Y., Chakraborty, B., & Shi, M. (2011). *Kansei engineering and soft computing: Theory and practice*. Hershey, PA: IGI Global. doi:10.4018/978-1-61692-797-4

Daud, M. F., Taib, J. M., & Shariffudin, R. S. (2012). Assessing mechanical engineering undergraduates' conceptual knowledge in three dimensional computer aided design (3D CAD). In K. Yusof, N. Azli, A. Kosnin, S. Yusof, & Y. Yusof (Eds.), *Outcome-based science, technology, engineering, and mathematics education: Innovative practices* (pp. 350–363). Hershey, PA: IGI Global. doi:10.4018/978-1-4666-1809-1.ch017

Daugherty, A., Hires, W. E., & Braunstein, S. G. (2013). Collection development for the college of engineering at Louisiana State University libraries: Liaison responsibilities and duties. In S. Holder (Ed.), *Library collection development for professional programs: Trends and best practices* (pp. 291–305). Hershey, PA: IGI Global. doi:10.4018/978-1-4666-1897-8.ch017

Davim, J. (2013). *Dynamic methods and process advancements in mechanical, manufacturing, and materials engineering*. Hershey, PA: IGI Global. doi:10.4018/978-1-4666-1867-1

de Vere, I., & Melles, G. (2013). Integrating 'designerly' ways with engineering science: A catalyst for change within product design and development. In Industrial engineering: Concepts, methodologies, tools, and applications (pp. 56-78). Hershey, PA: IGI Global. doi:10.4018/978-1-4666-1945-6.ch005

del Sagrado Martinez, J., & del Aguila Cano, I. M. (2010). A Bayesian network for predicting the need for a requirements review. In F. Meziane & S. Vadera (Eds.), *Artificial intelligence applications for improved software engineering development: New prospects* (pp. 106–128). Hershey, PA: IGI Global. doi:10.4018/978-1-60566-758-4.ch006

Dhar-Bhattacharjee, S., & Takruri-Rizk, H. (2012). An Indo-British comparison. In C. Romm Livermore (Ed.), *Gender and social computing: Interactions, differences and relationships* (pp. 50–71). Hershey, PA: IGI Global. doi:10.4018/978-1-60960-759-3.ch004

Díaz, V. G., Lovelle, J. M., García-Bustelo, B. C., & Martinez, O. S. (2014). *Advances and applications in model-driven engineering*. Hershey, PA: IGI Global. doi:10.4018/978-1-4666-4494-6

Dogru, A., Senkul, P., & Kaya, O. (2012). Modern approaches to software engineering in the compositional era. In *Machine learning: Concepts, methodologies, tools and applications* (pp. 1903–1923). Hershey, PA: IGI Global. doi:10.4018/978-1-60960-818-7.ch803

Dogru, A. H., & Biçer, V. (2011). *Modern software engineering concepts and practices: Advanced approaches*. Hershey, PA: IGI Global. doi:10.4018/978-1-60960-215-4

Doll, W. J., & Deng, X. (2013). Antecedents of improvisation in IT-enabled engineering work. In A. Dwivedi & S. Clarke (Eds.), *Innovative strategies and approaches for end-user computing advancements* (pp. 242–264). Hershey, PA: IGI Global. doi:10.4018/978-1-4666-2059-9.ch013

Dominguez, U., & Magdaleno, J. (2011). Industrial training in engineering education in Spain. In P. Keleher, A. Patil, & R. Harreveld (Eds.), *Work-integrated learning in engineering, built environment and technology: Diversity of practice in practice* (pp. 72–84). Hershey, PA: IGI Global. doi:10.4018/978-1-60960-547-6.ch004

Dormido, S., Vargas, H., & Sánchez, J. (2013). AutomatL@bs consortium: A Spanish network of web-based labs for control engineering education. In *Industrial engineering: Concepts, methodologies, tools, and applications* (pp. 679–699). Hershey, PA: IGI Global. doi:10.4018/978-1-4666-1945-6.ch039

Dyro, J. F. (2012). Clinical engineering. In Z. Abu-Faraj (Ed.), *Handbook of research on biomedical engineering education and advanced bioengineering learning: Interdisciplinary concepts* (pp. 521–576). Hershey, PA: IGI Global. doi:10.4018/978-1-4666-0122-2.ch012

Easton, J. M., Davies, J. R., & Roberts, C. (2011). Ontology engineering the "what's", "why's", and "how's" of data exchange. *International Journal of Decision Support System Technology*, *3*(1), 40–53. doi:10.4018/jdsst.2011010103

Eijkman, H., & Kayali, O. (2011). Addressing the politics of accreditation in engineering education: The benefits of soft systems thinking. *International Journal of Quality Assurance in Engineering and Technology Education*, *1*(2), 1–10. doi:10.4018/ijqaete.2011070101

El-Khalili, N. H. (2013). Teaching agile software engineering using problem-based learning. *International Journal of Information and Communication Technology Education*, *9*(3), 1–12. doi:10.4018/jicte.2013070101

Favre, L. M. (2010). MDA-based object-oriented reverse engineering. In L. Favre (Ed.), *Model driven architecture for reverse engineering technologies: Strategic directions and system evolution* (pp. 199–229). Hershey, PA: IGI Global. doi:10.4018/978-1-61520-649-0.ch010

Favre, L. M. (2010). Reverse engineering and MDA: An introduction. In L. Favre (Ed.), *Model driven architecture for reverse engineering technologies: Strategic directions and system evolution* (pp. 1–14). Hershey, PA: IGI Global. doi:10.4018/978-1-61520-649-0.ch001

Fenton, N., Hearty, P., Neil, M., & Radlinski, L. (2010). Software project and quality modelling using Bayesian networks. In F. Meziane & S. Vadera (Eds.), *Artificial intelligence applications for improved software engineering development: New prospects* (pp. 1–25). Hershey, PA: IGI Global. doi:10.4018/978-1-60566-758-4.ch001

Ferreira, R., Brisolara, L., Mattos, J. C., Spech, E., & Cota, E. (2010). Engineering embedded software: From application modeling to software synthesis. In L. Gomes & J. Fernandes (Eds.), *Behavioral modeling for embedded systems and technologies: Applications for design and implementation* (pp. 245–270). Hershey, PA: IGI Global. doi:10.4018/978-1-60566-750-8.ch010

Ferreira da Silva Oliveira, M., Wehrmeister, M. A., Assis do Nascimento, F., & Pereira, C. E. (2010). High-level design space exploration of embedded systems using the model-driven engineering and aspect-oriented design approaches. In L. Gomes & J. Fernandes (Eds.), *Behavioral modeling for embedded systems and technologies: Applications for design and implementation* (pp. 114–146). Hershey, PA: IGI Global. doi:10.4018/978-1-60566-750-8.ch005

Ferris, T. L. (2013). Engineering design as research. In *Industrial engineering: Concepts, methodologies, tools, and applications* (pp. 1766–1779). Hershey, PA: IGI Global. doi:10.4018/978-1-4666-1945-6.ch094

Frank, M. (2010). Capacity for engineering systems thinking (CEST): Literature review, principles for assessing and the reliability and validity of an assessing tool. In M. Hunter (Ed.), *Strategic information systems: Concepts, methodologies, tools, and applications* (pp. 1171–1183). Hershey, PA: IGI Global. doi:10.4018/978-1-60566-677-8.ch076

Fraser, D. (2012). A case study of curriculum development in engineering: Insights gained over two decades. In K. Yusof, N. Azli, A. Kosnin, S. Yusof, & Y. Yusof (Eds.), *Outcome-based science, technology, engineering, and mathematics education: Innovative practices* (pp. 27–49). Hershey, PA: IGI Global. doi:10.4018/978-1-4666-1809-1.ch002

Fraser, D. (2012). Curriculum initiatives to help engineering students learn and develop. In K. Yusof, N. Azli, A. Kosnin, S. Yusof, & Y. Yusof (Eds.), *Outcome-based science, technology, engineering, and mathematics education: Innovative practices* (pp. 85–106). Hershey, PA: IGI Global. doi:10.4018/978-1-4666-1809-1.ch005

Fries, T. P. (2014). Reengineering structured legacy system documentation to UML object-oriented artifacts. In *Software design and development: Concepts, methodologies, tools, and applications* (pp. 749–771). Hershey, PA: IGI Global. doi:10.4018/978-1-4666-4301-7.ch036

Funabashi, M. (2014). Transdisciplinary science and technology and service systems. In M. Kosaka & K. Shirahada (Eds.), *Progressive trends in knowledge and system-based science for service innovation* (pp. 101–127). Hershey, PA: IGI Global. doi:10.4018/978-1-4666-4663-6.ch006

Gaetano, L., Puppato, D., & Balestra, G. (2012). Modeling clinical engineering activities to support healthcare technology management. In A. Kolker & P. Story (Eds.), *Management engineering for effective healthcare delivery: Principles and applications* (pp. 113–131). Hershey, PA: IGI Global. doi:10.4018/978-1-60960-872-9.ch005

Galvão, T. A., Neto, F., & Bonates, M. F. (2013). eRiskGame: A persistent browser-based game for supporting project-based learning in the risk management context. In Industrial engineering: Concepts, methodologies, tools, and applications (pp. 1243-1259). Hershey, PA: IGI Global. doi:10.4018/978-1-4666-1945-6.ch067

Garousi, V., Shahnewaz, S., & Krishnamurthy, D. (2013). UML-driven software performance engineering: A systematic mapping and trend analysis. In V. Díaz, J. Lovelle, B. García-Bustelo, & O. Martínez (Eds.), *Progressions and innovations in model-driven software engineering* (pp. 18–64). Hershey, PA: IGI Global. doi:10.4018/978-1-4666-4217-1.ch002

Ge, W., Yang, N., Wang, W., & Li, J. (2011). Interfacial interactions: Drag. In S. Pannala, M. Syamlal, & T. O'Brien (Eds.), *Computational gas-solids flows and reacting systems: Theory, methods and practice* (pp. 128–177). Hershey, PA: IGI Global. doi:10.4018/978-1-61520-651-3.ch004

Génova, G., Llorens, J., & Morato, J. (2014). Software engineering research: The need to strengthen and broaden the classical scientific method. In *Software design and development: Concepts, methodologies, tools, and applications* (pp. 1639–1658). Hershey, PA: IGI Global. doi:10.4018/978-1-4666-4301-7.ch079

Genvigir, E. C., & Vijaykumar, N. L. (2010). Requirements traceability. In M. Ramachandran & R. de Carvalho (Eds.), *Handbook of research on software engineering and productivity technologies: Implications of globalization* (pp. 102–120). Hershey, PA: IGI Global. doi:10.4018/978-1-60566-731-7.ch008

Ghosh, S. (2010). Online automated essay grading system as a web based learning (WBL) tool in engineering education. In D. Russell & A. Haghi (Eds.), *Web-based engineering education: Critical design and effective tools* (pp. 53–62). Hershey, PA: IGI Global. doi:10.4018/978-1-61520-659-9.ch005

Gill, A. Q., & Bunker, D. (2013). SaaS requirements engineering for agile development. In X. Wang, N. Ali, I. Ramos, & R. Vidgen (Eds.), *Agile and lean service-oriented development: Foundations, theory, and practice* (pp. 64–93). Hershey, PA: IGI Global. doi:10.4018/978-1-4666-2503-7.ch004

Giraldo, F. D., Villegas, M. L., & Collazos, C. A. (2014). The use of HCI approaches into distributed CSCL activities applied to software engineering courses. In *Software design and development: Concepts, methodologies, tools, and applications* (pp. 2033–2050). Hershey, PA: IGI Global. doi:10.4018/978-1-4666-4301-7.ch094

Girardi, R., & Leite, A. (2011). Knowledge engineering support for agent-oriented software reuse. In M. Ramachandran (Ed.), *Knowledge engineering for software development life cycles: Support technologies and applications* (pp. 177–195). Hershey, PA: IGI Global. doi:10.4018/978-1-60960-509-4.ch010

Goodhew, P. (2013). Why get your engineering programme accredited? In *Industrial engineering: Concepts, methodologies, tools, and applications* (pp. 18–20). Hershey, PA: IGI Global. doi:10.4018/978-1-4666-1945-6.ch002

Goossenaerts, J., Possel-Dölken, F., & Popplewell, K. (2011). Vision, trends, gaps and a broad roadmap for future engineering. In I. Management Association (Ed.), Global business: Concepts, methodologies, tools and applications (pp. 2229-2243). Hershey, PA: IGI Global. doi:10.4018/978-1-60960-587-2.ch802

Goyal, S. B., & Prakash, N. (2013). Functional method engineering. *International Journal of Information System Modeling and Design*, *4*(1), 79–103. doi:10.4018/jismd.2013010104

Gunasekaran, A., & Shea, T. (2010). Requirements management for ERP projects. In *Organizational advancements through enterprise information systems: Emerging applications and developments* (pp. 29–45). Hershey, PA: IGI Global. doi:10.4018/978-1-60566-968-7.ch003

Gustavsson, I., Claesson, L., Nilsson, K., Zackrisson, J., Zubia, J. G., & Jayo, U. H. ... Claesson, I. (2012). The VISIR open lab platform. In A. Azad, M. Auer, & V. Harward (Eds.) Internet accessible remote laboratories: Scalable e-learning tools for engineering and science disciplines (pp. 294-317). Hershey, PA: IGI Global. doi:10.4018/978-1-61350-186-3.ch015

Habib, M. K., & Davim, J. (2013). *Engineering creative design in robotics and mechatronics*. Hershey, PA: IGI Global. doi:10.4018/978-1-4666-4225-6

Haghi, A. K. (2013). *Methodologies and applications for chemoinformatics and chemical engineering*. Hershey, PA: IGI Global. doi:10.4018/978-1-4666-4010-8

Haghi, A. K., & Noroozi, B. (2010). Adapting engineering education to the new century. In D. Russell & A. Haghi (Eds.), *Web-based engineering education: Critical design and effective tools* (pp. 30–41). Hershey, PA: IGI Global. doi:10.4018/978-1-61520-659-9.ch003

Haghpanahi, M., Nikkhoo, M., & Peirovi, H. A. (2010). Computer aided tissue engineering from modeling to manufacturing. In A. Lazakidou (Ed.), *Biocomputation and biomedical informatics: Case studies and applications* (pp. 75–88). Hershey, PA: IGI Global. doi:10.4018/978-1-60566-768-3.ch004

Hepsø, I. L., Rindal, A., & Waldal, K. (2013). The introduction of a hand-held platform in an engineering and fabrication company. In T. Rosendahl & V. Hepsø (Eds.), *Integrated operations in the oil and gas industry: Sustainability and capability development* (pp. 246–260). Hershey, PA: IGI Global. doi:10.4018/978-1-4666-2002-5.ch015

Hernández-López, A., Colomo-Palacios, R., García-Crespo, Á., & Cabezas-Isla, F. (2013). Software engineering productivity: Concepts, issues and challenges. In J. Wang (Ed.), *Perspectives and techniques for improving information technology project management* (pp. 69–79). Hershey, PA: IGI Global. doi:10.4018/978-1-4666-2800-7.ch006

Herrmann, A., & Morali, A. (2012). Interplay of security requirements engineering and reverse engineering in the maintenance of undocumented software. In J. Rech & C. Bunse (Eds.), *Emerging technologies for the evolution and maintenance of software models* (pp. 57–91). Hershey, PA: IGI Global. doi:10.4018/978-1-61350-438-3.ch003

Hochrainer, M., & Ziegler, F. (2013). Tuned liquid column gas damper in structural control: The salient features of a general purpose damping device and its application in buildings, bridges, and dams. In N. Lagaros, V. Plevris, & C. Mitropoulou (Eds.), *Design optimization of active and passive structural control systems* (pp. 150–179). Hershey, PA: IGI Global. doi:10.4018/978-1-4666-2029-2.ch007

Hochstein, L., Schott, B., & Graybill, R. B. (2013). Computational engineering in the cloud: Benefits and challenges. In A. Dwivedi & S. Clarke (Eds.), *Innovative strategies and approaches for end-user computing advancements* (pp. 314–332). Hershey, PA: IGI Global. doi:10.4018/978-1-4666-2059-9.ch017

Höhn, S., Lowis, L., Jürjens, J., & Accorsi, R. (2010). Identification of vulnerabilities in web services using model-based security. In C. Gutiérrez, E. Fernández-Medina, & M. Piattini (Eds.), *Web services security development and architecture: Theoretical and practical issues* (pp. 1–32). Hershey, PA: IGI Global. doi:10.4018/978-1-60566-950-2.ch001

Holton, D. L., & Verma, A. (2010). Designing animated simulations and web-based assessments to improve electrical engineering education. In D. Russell & A. Haghi (Eds.), *Web-based engineering education: Critical design and effective tools* (pp. 77–95). Hershey, PA: IGI Global. doi:10.4018/978-1-61520-659-9.ch007

Honnutagi, A. R., Sonar, R., & Babu, S. (2012). Quality accreditation system for Indian engineering education using knowledge management and system dynamics. *International Journal of Quality Assurance in Engineering and Technology Education*, *2*(3), 47–61. doi:10.4018/ijqaete.2012070105

Hua, G. B. (2013). Business process re-engineering. In *Implementing IT business strategy in the construction industry* (pp. 118–140). Hershey, PA: IGI Global. doi:10.4018/978-1-4666-4185-3.ch006

Hussein, B., Hage-Diab, A., Hammoud, M., Kawtharani, A., El-Hage, H., & Haj-Ali, A. (2014). Management response to improve the educational performance of engineering students: The case of the lebanese international university. In G. Khoury & M. Khoury (Eds.), *Cases on management and organizational behavior in an Arab context* (pp. 91–110). Hershey, PA: IGI Global. doi:10.4018/978-1-4666-5067-1.ch006

Hussey, M., Wu, B., & Xu, X. (2011). Co-operation models for industries and software education institutions. In *Software industry-oriented education practices and curriculum development: Experiences and lessons* (pp. 39–56). Hershey, PA: IGI Global. doi:10.4018/978-1-60960-797-5.ch003

Hussey, M., Wu, B., & Xu, X. (2011). Curriculum issues in industry oriented software engineering education. In *Software industry-oriented education practices and curriculum development: Experiences and lessons* (pp. 153–165). Hershey, PA: IGI Global. doi:10.4018/978-1-60960-797-5.ch010

Hussey, M., Wu, B., & Xu, X. (2011). Industry oriented curriculum and syllabus creation for software engineering series courses in the school of software. In *Software industry-oriented education practices and curriculum development: Experiences and lessons* (pp. 98–109). Hershey, PA: IGI Global. doi:10.4018/978-1-60960-797-5.ch006

Ilyas, Q. M. (2013). Ontology augmented software engineering. In K. Buragga & N. Zaman (Eds.), *Software development techniques for constructive information systems design* (pp. 406–413). Hershey, PA: IGI Global. doi:10.4018/978-1-4666-3679-8.ch023

Islam, S., Mouratidis, H., Kalloniatis, C., Hudic, A., & Zechner, L. (2013). Model based process to support security and privacy requirements engineering. *International Journal of Secure Software Engineering*, *3*(3), 1–22. doi:10.4018/jsse.2012070101

Jahan, K., Everett, J. W., Tang, G., Farrell, S., Zhang, H., Wenger, A., & Noori, M. (2010). Use of living systems to teach basic engineering concepts. In D. Russell & A. Haghi (Eds.), *Web-based engineering education: Critical design and effective tools* (pp. 96–107). Hershey, PA: IGI Global. doi:10.4018/978-1-61520-659-9.ch008

Jakovljevic, M. (2013). A conceptual model of creativity, invention, and innovation (MCII) for entrepreneurial engineers. In S. Buckley & M. Jakovljevic (Eds.), *Knowledge management innovations for interdisciplinary education: Organizational applications* (pp. 66–87). Hershey, PA: IGI Global. doi:10.4018/978-1-4666-1969-2.ch004

Jaroucheh, Z., Liu, X., & Smith, S. (2014). A software engineering framework for context-aware service-based processes in pervasive environments. In *Software design and development: Concepts, methodologies, tools, and applications* (pp. 71–95). Hershey, PA: IGI Global. doi:10.4018/978-1-4666-4301-7.ch005

Joiner, R., Iacovides, I., Darling, J., Diament, A., Drew, B., Duddley, J., ... Gavin, C. (2013). Racing academy: A case study of a digital game for supporting students learning of physics and engineering. In Y. Baek & N. Whitton (Eds.), *Cases on digital game-based learning: Methods, models, and strategies* (pp. 509–523). Hershey, PA: IGI Global. doi:10.4018/978-1-4666-2848-9.ch026

Jordan, M. E. (2014). Interweaving the digital and physical worlds in collaborative project-based learning experiences. In D. Loveless, B. Griffith, M. Bérci, E. Ortlieb, & P. Sullivan (Eds.), *Academic knowledge construction and multimodal curriculum development* (pp. 265–282). Hershey, PA: IGI Global. doi:10.4018/978-1-4666-4797-8.ch017

Juang, Y. (2010). WIRE: A highly interactive blended learning for engineering education. In D. Russell & A. Haghi (Eds.), *Web-based engineering education: Critical design and effective tools* (pp. 149–159). Hershey, PA: IGI Global. doi:10.4018/978-1-61520-659-9.ch011

Kadry, S. (2013). *Diagnostics and prognostics of engineering systems: Methods and techniques*. Hershey, PA: IGI Global. doi:10.4018/978-1-4666-2095-7

Kamboj, A., Kumar, S., & Singh, H. (2012). Design and development of hybrid stir casting process. *International Journal of Applied Industrial Engineering*, *1*(2), 1–6. doi:10.4018/ijaie.2012070101

Kamthan, P. (2010). A methodology for integrating the social web environment in software engineering education. In S. Dasgupta (Ed.), *Social computing: Concepts, methodologies, tools, and applications* (pp. 457–471). Hershey, PA: IGI Global. doi:10.4018/978-1-60566-984-7.ch031

Kamthan, P. (2010). A social web perspective of software engineering education. In S. Murugesan (Ed.), *Handbook of research on web 2.0, 3.0, and X.0: Technologies, business, and social applications* (pp. 472–495). Hershey, PA: IGI Global. doi:10.4018/978-1-60566-384-5.ch026

Kamthan, P. (2010). On the prospects and concerns of pattern-oriented web engineering. In G. Alkhatib & D. Rine (Eds.), *Web engineering advancements and trends: Building new dimensions of information technology* (pp. 97–128). Hershey, PA: IGI Global. doi:10.4018/978-1-60566-719-5.ch006

Kamthan, P. (2011). Using the social web environment for software engineering education. In L. Tomei (Ed.), *Online courses and ICT in education: Emerging practices and applications* (pp. 23–45). Hershey, PA: IGI Global. doi:10.4018/978-1-60960-150-8.ch003

Kamthan, P. (2013). An exploration of the social web environment for collaborative software engineering education. In N. Karacapilidis, M. Raisinghani, & E. Ng (Eds.), *Web-based and blended educational tools and innovations* (pp. 1–23). Hershey, PA: IGI Global. doi:10.4018/978-1-4666-2023-0.ch001

Kapos, G., Dalakas, V., Nikolaidou, M., & Anagnostopoulos, D. (2014). An integrated framework to simulate SysML models using DEVS simulators. In P. Fonseca i Casas (Ed.), *Formal languages for computer simulation: Transdisciplinary models and applications* (pp. 305–332). Hershey, PA: IGI Global. doi:10.4018/978-1-4666-4369-7.ch010

Karlsson, F., & Ågerfalk, P. J. (2011). Towards structured flexibility in information systems development: Devising a method for method configuration. In K. Siau (Ed.), *Theoretical and practical advances in information systems development: Emerging trends and approaches* (pp. 214–238). Hershey, PA: IGI Global. doi:10.4018/978-1-60960-521-6.ch010

Karpati, P., Sindre, G., & Matulevicius, R. (2012). Comparing misuse case and mal-activity diagrams for modelling social engineering attacks. *International Journal of Secure Software Engineering, 3*(2), 54–73. doi:10.4018/jsse.2012040103

Keleher, P., & Patil, A. (2012). Conducting an effective residential school for an undergraduate materials science and engineering course. *International Journal of Quality Assurance in Engineering and Technology Education, 2*(3), 41–46. doi:10.4018/ijqaete.2012070104

Ker, H. W. (2012). Engineering education and attitudes toward mathematics: A comparative study. *International Journal of Quality Assurance in Engineering and Technology Education, 2*(1), 63–76. doi:10.4018/ijqaete.2012010105

Kinsner, W. (2012). Challenges in the design of adoptive, intelligent and cognitive systems. In Y. Wang (Ed.), *Software and intelligent sciences: New transdisciplinary findings* (pp. 47–67). Hershey, PA: IGI Global. doi:10.4018/978-1-4666-0261-8.ch004

Kirikova, M. (2011). Domain modeling approaches in IS engineering. In J. Osis & E. Asnina (Eds.), *Model-driven domain analysis and software development: Architectures and functions* (pp. 388–406). Hershey, PA: IGI Global. doi:10.4018/978-1-61692-874-2.ch018

Kljajic, M., & Farr, J. V. (2010). Importance of systems engineering in the development of information systems. In D. Paradice (Ed.), *Emerging systems approaches in information technologies: Concepts, theories, and applications* (pp. 51–66). Hershey, PA: IGI Global. doi:10.4018/978-1-60566-976-2.ch004

Kljajic, M., & Farr, J. V. (2010). The role of systems engineering in the development of information systems. In M. Hunter (Ed.), *Strategic information systems: Concepts, methodologies, tools, and applications* (pp. 369–381). Hershey, PA: IGI Global. doi:10.4018/978-1-60566-677-8.ch026

Knoell, H. D. (2010). User participation in the quality assurance of requirements specifications: An evaluation of traditional models and animated systems engineering techniques. In M. Hunter (Ed.), *Strategic information systems: Concepts, methodologies, tools, and applications* (pp. 1623–1638). Hershey, PA: IGI Global. doi:10.4018/978-1-60566-677-8.ch106

Kof, L. (2010). From textual scenarios to message sequence charts. In F. Meziane & S. Vadera (Eds.), *Artificial intelligence applications for improved software engineering development: New prospects* (pp. 83–105). Hershey, PA: IGI Global. doi:10.4018/978-1-60566-758-4.ch005

Köhler, B., Gluchow, M., & Brügge, B. (2014). Teaching basic software engineering to senior high school students. In *K-12 education: Concepts, methodologies, tools, and applications* (pp. 1634–1649). Hershey, PA: IGI Global. doi:10.4018/978-1-4666-4502-8.ch094

Kolker, A. (2012). Efficient managerial decision-making in healthcare settings: Examples and fundamental principles. In A. Kolker & P. Story (Eds.), *Management engineering for effective healthcare delivery: Principles and applications* (pp. 1–45). Hershey, PA: IGI Global. doi:10.4018/978-1-60960-872-9.ch001

Kolker, A., & Story, P. (2012). *Management engineering for effective healthcare delivery: Principles and applications*. Hershey, PA: IGI Global. doi:10.4018/978-1-60960-872-9

Kornyshova, E., Deneckère, R., & Claudepierre, B. (2013). Towards method component contextualization. In J. Krogstie (Ed.), *Frameworks for developing efficient information systems: Models, theory, and practice* (pp. 337–368). Hershey, PA: IGI Global. doi:10.4018/978-1-4666-4161-7.ch015

Krishnan, S. (2012). Problem-based learning curricula in engineering. In M. Rasul (Ed.), *Developments in engineering education standards: Advanced curriculum innovations* (pp. 23–40). Hershey, PA: IGI Global. doi:10.4018/978-1-4666-0951-8.ch002

Ku, H., & Thonglek, S. (2011). Running a successful practice school: Challenges and lessons learned. In P. Keleher, A. Patil, & R. Harreveld (Eds.), *Work-integrated learning in engineering, built environment and technology: Diversity of practice in practice* (pp. 131–163). Hershey, PA: IGI Global. doi:10.4018/978-1-60960-547-6.ch007

Lacuesta, R., Fernández-Sanz, L., & Romay, M. D. (2014). Requirements specification as basis for mobile software quality assurance. In *Software design and development: Concepts, methodologies, tools, and applications* (pp. 719–732). Hershey, PA: IGI Global. doi:10.4018/978-1-4666-4301-7.ch034

Lai, A., Zhang, C., & Busovaca, S. (2013). 2-SQUARE: A web-based enhancement of SQUARE privacy and security requirements engineering. *International Journal of Software Innovation*, *1*(1), 41–53. doi:10.4018/ijsi.2013010104

Lane, J. A., Petkov, D., & Mora, M. (2010). Software engineering and the systems approach: A conversation with Barry Boehm. In M. Hunter (Ed.), *Strategic information systems: Concepts, methodologies, tools, and applications* (pp. 333–337). Hershey, PA: IGI Global. doi:10.4018/978-1-60566-677-8.ch024

Lansiquot, R. D. (2013). *Cases on interdisciplinary research trends in science, technology, engineering, and mathematics: Studies on urban classrooms*. Hershey, PA: IGI Global. doi:10.4018/978-1-4666-2214-2

Laribi, S., Le Bris, A., Huang, L. M., Olsson, P., & Guillemoles, J. F. (2013). Phononic engineering for hot carrier solar cells. In L. Fara & M. Yamaguchi (Eds.), *Advanced solar cell materials, technology, modeling, and simulation* (pp. 214–242). Hershey, PA: IGI Global. doi:10.4018/978-1-4666-1927-2.ch012

Lee, C. K., & Sidhu, M. S. (2013). Computer-aided engineering education: New learning approaches and technologies. In V. Wang (Ed.), *Handbook of research on teaching and learning in K-20 education* (pp. 317–340). Hershey, PA: IGI Global. doi:10.4018/978-1-4666-4249-2.ch019

Lee, J., Ma, S., Lee, S., Wu, C., & Lee, C. L. (2014). Towards a high-availability-driven service composition framework. In *Software design and development: Concepts, methodologies, tools, and applications* (pp. 1498–1520). Hershey, PA: IGI Global. doi:10.4018/978-1-4666-4301-7.ch073

Leng, J., Rhyne, T., & Sharrock, W. (2012). Visualization: Future technology and practices for computational science and engineering. In J. Leng & W. Sharrock (Eds.), *Handbook of research on computational science and engineering: Theory and practice* (pp. 381–413). Hershey, PA: IGI Global. doi:10.4018/978-1-61350-116-0.ch016

Leng, J., & Sharrock, W. (2012). *Handbook of research on computational science and engineering: Theory and practice* (Vols. 1–2). Hershey, PA: IGI Global. doi:10.4018/978-1-61350-116-0

Li, Z. J. (2013). Prototyping of robotic systems in surgical procedures and automated manufacturing processes. In *Industrial engineering: Concepts, methodologies, tools, and applications* (pp. 1969–1987). Hershey, PA: IGI Global. doi:10.4018/978-1-4666-1945-6.ch106

Lladó, C. M., Bonet, P., & Smith, C. U. (2014). Towards a multi-formalism multi-solution framework for model-driven performance engineering. In M. Gribaudo & M. Iacono (Eds.), *Theory and application of multi-formalism modeling* (pp. 34–55). Hershey, PA: IGI Global. doi:10.4018/978-1-4666-4659-9.ch003

Loo, A. (2013). *Distributed computing innovations for business, engineering, and science*. Hershey, PA: IGI Global. doi:10.4018/978-1-4666-2533-4

Ma, Z. M. (2011). Engineering design knowledge management. In D. Schwartz & D. Te'eni (Eds.), *Encyclopedia of knowledge management* (2nd ed.; pp. 263–269). Hershey, PA: IGI Global. doi:10.4018/978-1-59904-931-1.ch025

Maiti, C. K., & Maiti, A. (2013). Teaching technology computer aided design (TCAD) online. In Industrial engineering: Concepts, methodologies, tools, and applications (pp. 1043-1063). Hershey, PA: IGI Global. doi:10.4018/978-1-4666-1945-6.ch057

Male, S. A. (2012). Generic engineering competencies required by engineers graduating in Australia: The competencies of engineering graduates (CEG) project. In M. Rasul (Ed.), *Developments in engineering education standards: Advanced curriculum innovations* (pp. 41–63). Hershey, PA: IGI Global. doi:10.4018/978-1-4666-0951-8.ch003

Malmqvist, J. (2012). A comparison of the CDIO and EUR-ACE quality assurance systems. *International Journal of Quality Assurance in Engineering and Technology Education*, *2*(2), 9–22. doi:10.4018/ijqaete.2012040102

Maree, D. J., & Maree, M. (2010). Factors contributing to the success of women working in science, engineering and technology (SET) careers. In A. Cater-Steel & E. Cater (Eds.), *Women in engineering, science and technology: Education and career challenges* (pp. 183–210). Hershey, PA: IGI Global. doi:10.4018/978-1-61520-657-5.ch009

Marichal, G. N., & González, E. J. (2012). Intelligent MAS in system engineering and robotics. In *Machine learning: Concepts, methodologies, tools and applications* (pp. 175–182). Hershey, PA: IGI Global. doi:10.4018/978-1-60960-818-7.ch204

Martinez, L., Favre, L., & Pereira, C. (2013). Architecture-driven modernization for software reverse engineering technologies. In V. Díaz, J. Lovelle, B. García-Bustelo, & O. Martínez (Eds.), *Progressions and innovations in model-driven software engineering* (pp. 288–307). Hershey, PA: IGI Global. doi:10.4018/978-1-4666-4217-1.ch012

Matanovic, D. (2014). The macondo 252 disaster: Causes and consequences. In D. Matanovic, N. Gaurina-Medjimurec, & K. Simon (Eds.), *Risk analysis for prevention of hazardous situations in petroleum and natural gas engineering* (pp. 115–131). Hershey, PA: IGI Global. doi:10.4018/978-1-4666-4777-0.ch006

Matanovic, D., Gaurina-Medjimurec, N., & Simon, K. (2014). *Risk analysis for prevention of hazardous situations in petroleum and natural gas engineering*. Hershey, PA: IGI Global. doi:10.4018/978-1-4666-4777-0

Matsuo, T., & Fujimoto, T. (2012). Analogical thinking based instruction method in IT professional education. In R. Colomo-Palacios (Ed.), *Professional advancements and management trends in the IT sector* (pp. 95–108). Hershey, PA: IGI Global. doi:10.4018/978-1-4666-0924-2.ch007

Mazo, R., Salinesi, C., Diaz, D., Djebbi, O., & Lora-Michiels, A. (2012). Constraints: The heart of domain and application engineering in the product lines engineering strategy. *International Journal of Information System Modeling and Design, 3*(2), 33–68. doi:10.4018/jismd.2012040102

Mead, N. R. (2010). Benefits and challenges in the use of case studies for security requirements engineering methods. *International Journal of Secure Software Engineering, 1*(1), 74–91. doi:10.4018/jsse.2010102005

Méausoone, P., & Aguilera, A. (2014). Inventory of experimental works on cutting tools' life for the wood industry. In A. Aguilera & J. Davim (Eds.), *Research developments in wood engineering and technology* (pp. 320–342). Hershey, PA: IGI Global. doi:10.4018/978-1-4666-4554-7.ch009

Medlin, B. D., & Cazier, J. (2011). Obtaining patient's information from hospital employees through social engineering techniques: An investigative study. In H. Nemati (Ed.), *Pervasive information security and privacy developments: Trends and advancements* (pp. 77–89). Hershey, PA: IGI Global. doi:10.4018/978-1-61692-000-5.ch006

Meziane, F., & Vadera, S. (2010). *Artificial intelligence applications for improved software engineering development: new prospects*. Hershey, PA: IGI Global. doi:10.4018/978-1-60566-758-4

Meziane, F., & Vadera, S. (2012). Artificial intelligence in software engineering: Current developments and future prospects. In *Machine learning: Concepts, methodologies, tools and applications* (pp. 1215–1236). Hershey, PA: IGI Global. doi:10.4018/978-1-60960-818-7.ch504

Milanovic, N. (2011). *Engineering reliable service oriented architecture: Managing complexity and service level agreements*. Hershey, PA: IGI Global. doi:10.4018/978-1-60960-493-6

Miller, W. S., & Summers, J. D. (2013). Tool and information centric design process modeling: Three case studies. In *Industrial engineering: Concepts, methodologies, tools, and applications* (pp. 1613–1637). Hershey, PA: IGI Global. doi:10.4018/978-1-4666-1945-6.ch086

Mishra, B., & Shukla, K. K. (2014). Data mining techniques for software quality prediction. In *Software design and development: Concepts, methodologies, tools, and applications* (pp. 401–428). Hershey, PA: IGI Global. doi:10.4018/978-1-4666-4301-7.ch021

Mishra, S. (2011). Social and ethical concerns of biomedical engineering research and practice. In A. Shukla & R. Tiwari (Eds.), *Biomedical engineering and information systems: Technologies, tools and applications* (pp. 54–80). Hershey, PA: IGI Global. doi:10.4018/978-1-61692-004-3.ch003

Moeller, D. P., & Sitzmann, D. (2012). Online computer engineering: combining blended e-learning in engineering with lifelong learning. In M. Rasul (Ed.), *Developments in engineering education standards: Advanced curriculum innovations* (pp. 194–215). Hershey, PA: IGI Global. doi:10.4018/978-1-4666-0951-8.ch011

Moeller, D. P., & Vakilzadian, H. (2012). Technology-enhanced learning standard through integration of modeling and simulation into engineering study programs. In M. Rasul (Ed.), *Developments in engineering education standards: Advanced curriculum innovations* (pp. 157–177). Hershey, PA: IGI Global. doi:10.4018/978-1-4666-0951-8.ch009

Mohan Baral, L., Kifor, C. V., Bondrea, I., & Oprean, C. (2012). Introducing problem based learning (PBL) in textile engineering education and assessing its influence on six sigma project implementation. *International Journal of Quality Assurance in Engineering and Technology Education*, *2*(4), 38–48. doi:10.4018/ijqaete.2012100104

Moore, S., May, D., & Wold, K. (2012). Developing cultural competency in engineering through transnational distance learning. In R. Hogan (Ed.), *Transnational distance learning and building new markets for universities* (pp. 210–228). Hershey, PA: IGI Global. doi:10.4018/978-1-4666-0206-9.ch013

Mora, M., Gelman, O., Frank, M., Paradice, D. B., & Cervantes, F. (2010). Toward an interdisciplinary engineering and management of complex IT-intensive organizational systems: A systems view. In D. Paradice (Ed.), *Emerging systems approaches in information technologies: Concepts, theories, and applications* (pp. 1–24). Hershey, PA: IGI Global. doi:10.4018/978-1-60566-976-2.ch001

Mora, M., O'Connor, R., Raisinghani, M. S., Macías-Luévano, J., & Gelman, O. (2013). An IT service engineering and management framework (ITS-EMF). In P. Ordóñez de Pablos & R. Tennyson (Eds.), *Best practices and new perspectives in service science and management* (pp. 76–91). Hershey, PA: IGI Global. doi:10.4018/978-1-4666-3894-5.ch005

Moser, T., Biffl, S., Sunindyo, W. D., & Winkler, D. (2013). Integrating production automation expert knowledge across engineering domains. In N. Bessis (Ed.), *Development of distributed systems from design to application and maintenance* (pp. 152–167). Hershey, PA: IGI Global. doi:10.4018/978-1-4666-2647-8.ch009

Mouratidis, H. (2011). *Software engineering for secure systems: Industrial and research perspectives.* Hershey, PA: IGI Global. doi:10.4018/978-1-61520-837-1

Mourtos, N. J. (2012). Defining, teaching, and assessing engineering design skills. *International Journal of Quality Assurance in Engineering and Technology Education*, *2*(1), 14–30. doi:10.4018/ijqaete.2012010102

Moustafa, A. (2012). Damage assessment of inelastic structures under simulated critical earthquakes. In V. Plevris, C. Mitropoulou, & N. Lagaros (Eds.), *Structural seismic design optimization and earthquake engineering: Formulations and applications* (pp. 128–151). Hershey, PA: IGI Global. doi:10.4018/978-1-4666-1640-0.ch006

Nasir, M. H., Alias, N. A., Fauzi, S. S., & Massatu, M. H. (2012). Implementation of the personal software process in academic settings and current support tools. In S. Fauzi, M. Nasir, N. Ramli, & S. Sahibuddin (Eds.), *Software process improvement and management: Approaches and tools for practical development* (pp. 117–148). Hershey, PA: IGI Global. doi:10.4018/978-1-61350-141-2.ch007

Nhlabatsi, A., Nuseibeh, B., & Yu, Y. (2012). Security requirements engineering for evolving software systems: A survey. In K. Khan (Ed.), *Security-aware systems applications and software development methods* (pp. 108–128). Hershey, PA: IGI Global. doi:10.4018/978-1-4666-1580-9.ch007

Noroozi, B., Valizadeh, M., & Sorial, G. A. (2010). Designing of e-learning for engineering education in developing countries: Key issues and success factors. In D. Russell & A. Haghi (Eds.), *Web-based engineering education: Critical design and effective tools* (pp. 1–19). Hershey, PA: IGI Global. doi:10.4018/978-1-61520-659-9.ch001

Oh Navarro, E. (2011). On the role of learning theories in furthering software engineering education. In *Instructional design: Concepts, methodologies, tools and applications* (pp. 1645–1666). Hershey, PA: IGI Global. doi:10.4018/978-1-60960-503-2.ch709

Onwubiko, C. (2014). Modelling situation awareness information and system requirements for the mission using goal-oriented task analysis approach. In *Software design and development: Concepts, methodologies, tools, and applications* (pp. 460–478). Hershey, PA: IGI Global. doi:10.4018/978-1-4666-4301-7.ch023

Osis, J., & Asnina, E. (2011). Is modeling a treatment for the weakness of software engineering? In J. Osis & E. Asnina (Eds.), *Model-driven domain analysis and software development: Architectures and functions* (pp. 1–14). Hershey, PA: IGI Global. doi:10.4018/978-1-61692-874-2.ch001

Othman, R., & Awang, Z. (2012). Using multiple methods in assessing oral communication skills in the final year project design course of an undergraduate engineering program. In K. Yusof, N. Azli, A. Kosnin, S. Yusof, & Y. Yusof (Eds.), *Outcome-based science, technology, engineering, and mathematics education: Innovative practices* (pp. 263–287). Hershey, PA: IGI Global. doi:10.4018/978-1-4666-1809-1.ch013

Ozcelik, Y. (2010). IT-enabled reengineering: Productivity impacts. In K. St.Amant (Ed.), *IT outsourcing: Concepts, methodologies, tools, and applications* (pp. 371–376). Hershey, PA: IGI Global. doi:10.4018/978-1-60566-770-6.ch022

Ožvoldová, M., & Schauer, F. (2012). Remote experiments in freshman engineering education by integrated e-learning. In A. Azad, M. Auer, & V. Harward (Eds.), *Internet accessible remote laboratories: Scalable e-learning tools for engineering and science disciplines* (pp. 60–83). Hershey, PA: IGI Global. doi:10.4018/978-1-61350-186-3.ch004

Paay, J., Pedell, S., Sterling, L., Vetere, F., & Howard, S. (2011). The benefit of ambiguity in understanding goals in requirements modelling. *International Journal of People-Oriented Programming, 1*(2), 24–49. doi:10.4018/ijpop.2011070102

Palmer, S., & Holt, D. (2010). Online discussion in engineering education: Student responses and learning outcomes. In L. Shedletsky & J. Aitken (Eds.), *Cases on online discussion and interaction: Experiences and outcomes* (pp. 105–122). Hershey, PA: IGI Global. doi:10.4018/978-1-61520-863-0.ch005

Pandian, A., Ismail, S. A., & Abdullah, A. S. (2012). Communication framework to empower 21st century engineers and IT professionals. In A. Patil, H. Eijkman, & E. Bhattacharyya (Eds.), *New media communication skills for engineers and IT professionals: Trans-national and trans-cultural demands* (pp. 34–54). Hershey, PA: IGI Global. doi:10.4018/978-1-4666-0243-4.ch004

Pasupathy, K. S. (2010). Systems engineering and health informatics: Context, content, and implementation. In S. Kabene (Ed.), *Healthcare and the effect of technology: Developments, challenges and advancements* (pp. 123–144). Hershey, PA: IGI Global. doi:10.4018/978-1-61520-733-6.ch009

Pasupathy, K. S. (2010). Transforming healthcare: Leveraging the complementarities of health informatics and systems engineering. *International Journal of Healthcare Delivery Reform Initiatives*, *2*(2), 35–55. doi:10.4018/jhdri.2010040103

Pasupathy, K. S. (2011). Systems engineering and health informatics. In *Clinical technologies: Concepts, methodologies, tools and applications* (pp. 1684–1705). Hershey, PA: IGI Global. doi:10.4018/978-1-60960-561-2.ch606

Patel, C., & Ramachandran, M. (2010). Story card process improvement framework for agile requirements. In M. Ramachandran & R. de Carvalho (Eds.), *Handbook of research on software engineering and productivity technologies: Implications of globalization* (pp. 61–54). Hershey, PA: IGI Global. doi:10.4018/978-1-60566-731-7.ch006

Patro, C. S. (2012). A study on adaptability of total quality management in engineering education sector. *International Journal of Quality Assurance in Engineering and Technology Education*, *2*(4), 25–37. doi:10.4018/ijqaete.2012100103

Peña de Carrillo, C. I., Choquet, C., Després, C., Iksal, S., Jacoboni, P., Lekira, A., … Thi-Ngoc, D. P. (2014). Engineering and reengineering of technology enhanced learning scenarios using context awareness processes. In Software design and development: Concepts, methodologies, tools, and applications (pp. 1289-1313). Hershey, PA: IGI Global. doi:10.4018/978-1-4666-4301-7.ch063

Peng, F., & Li, H. (2011). A steganalysis method for 2D engineering graphics based on the statistic of geometric features. *International Journal of Digital Crime and Forensics*, *3*(2), 35–40. doi:10.4018/jdcf.2011040103

Pérez, J. L., Rabuñal, J., & Abella, F. M. (2012). Soft computing techniques in civil engineering: Time series prediction. In *Computer engineering: Concepts, methodologies, tools and applications* (pp. 1982–1997). Hershey, PA: IGI Global. doi:10.4018/978-1-61350-456-7.ch811

Pérez-Castillo, R., Rodríguez de Guzmán, I. G., & Piattini, M. (2011). Architecture-driven modernization. In A. Dogru & V. Biçer (Eds.), *Modern software engineering concepts and practices: Advanced approaches* (pp. 75–103). Hershey, PA: IGI Global. doi:10.4018/978-1-60960-215-4.ch004

Pérez-Castillo, R., Rodríguez de Guzmán, I. G., & Piattini, M. (2012). Model-driven reengineering. In J. Rech & C. Bunse (Eds.), *Emerging technologies for the evolution and maintenance of software models* (pp. 200–229). Hershey, PA: IGI Global. doi:10.4018/978-1-61350-438-3.ch008

Petkov, D., Edgar-Nevill, D., Madachy, R., & O'Connor, R. (2010). Information systems, software engineering, and systems thinking: Challenges and opportunities. In M. Hunter (Ed.), *Strategic information systems: Concepts, methodologies, tools, and applications* (pp. 315–332). Hershey, PA: IGI Global. doi:10.4018/978-1-60566-677-8.ch023

Petkov, D., Edgar-Nevill, D., Madachy, R., & O'Connor, R. (2012). Towards a wider application of the systems approach in information systems and software engineering. In *Computer engineering: Concepts, methodologies, tools and applications* (pp. 1627–1645). Hershey, PA: IGI Global. doi:10.4018/978-1-61350-456-7.ch701

Petre, L., Sere, K., & Troubitsyna, E. (2012). *Dependability and computer engineering: Concepts for software-intensive systems*. Hershey, PA: IGI Global. doi:10.4018/978-1-60960-747-0

Poels, G., Decreus, K., Roelens, B., & Snoeck, M. (2013). Investigating goal-oriented requirements engineering for business processes. *Journal of Database Management, 24*(2), 35–71. doi:10.4018/jdm.2013040103

Polgar, J., & Adamson, G. (2011). *New generation of portal software and engineering: Emerging technologies*. Hershey, PA: IGI Global. doi:10.4018/978-1-60960-571-1

Power, C., Freire, A. P., & Petrie, H. (2011). Integrating accessibility evaluation into web engineering processes. In G. Alkhatib (Ed.), *Web engineered applications for evolving organizations: Emerging knowledge* (pp. 315–339). Hershey, PA: IGI Global. doi:10.4018/978-1-60960-523-0.ch018

Praeg, C. (2011). Framework for IT service value engineering: Managing value and IT service quality. In C. Praeg & D. Spath (Eds.), *Quality management for IT services: Perspectives on business and process performance* (pp. 274–297). Hershey, PA: IGI Global. doi:10.4018/978-1-61692-889-6.ch016

Prescott, J., & Bogg, J. (2013). *Gendered occupational differences in science, engineering, and technology careers*. Hershey, PA: IGI Global. doi:10.4018/978-1-4666-2107-7

Prpic, J. K., & Moore, G. (2012). E-portfolios as a quantitative and qualitative means of demonstrating learning outcomes and competencies in engineering. In K. Yusof, N. Azli, A. Kosnin, S. Yusof, & Y. Yusof (Eds.), *Outcome-based science, technology, engineering, and mathematics education: Innovative practices* (pp. 124–154). Hershey, PA: IGI Global. doi:10.4018/978-1-4666-1809-1.ch007

Puteh, M. M., & Ismail, K. (2012). Quality assurance through innovation policy: The pedagogical implications on engineering education. In *Human resources management: Concepts, methodologies, tools, and applications* (pp. 40–49). Hershey, PA: IGI Global. doi:10.4018/978-1-4666-1601-1.ch004

Puthanpurayil, A. M., Dhakal, R. P., & Carr, A. J. (2013). Optimal passive damper positioning techniques: State-of-the-art. In N. Lagaros, V. Plevris, & C. Mitropoulou (Eds.), *Design optimization of active and passive structural control systems* (pp. 85–111). Hershey, PA: IGI Global. doi:10.4018/978-1-4666-2029-2.ch004

Ramachandran, M. (2010). Knowledge engineering support for software requirements, architectures and components. In F. Meziane & S. Vadera (Eds.), *Artificial intelligence applications for improved software engineering development: New prospects* (pp. 129–145). Hershey, PA: IGI Global. doi:10.4018/978-1-60566-758-4.ch007

Ramachandran, M. (2011). *Knowledge engineering for software development life cycles: Support technologies and applications*. Hershey, PA: IGI Global. doi:10.4018/978-1-60960-509-4

Ramachandran, M., & de Carvalho, R. (2010). *Handbook of research on software engineering and productivity technologies: Implications of globalization*. Hershey, PA: IGI Global. doi:10.4018/978-1-60566-731-7

Rasul, M., Nouwens, F., Swift, R., Martin, F., & Greensill, C. V. (2012). Assessment of final year engineering projects: A pilot investigation on issues and best practice. In M. Rasul (Ed.), *Developments in engineering education standards: Advanced curriculum innovations* (pp. 80–104). Hershey, PA: IGI Global. doi:10.4018/978-1-4666-0951-8.ch005

Razali, Z. B., & Trevelyan, J. (2012). An evaluation of students' practical intelligence and ability to diagnose equipment faults. In K. Yusof, N. Azli, A. Kosnin, S. Yusof, & Y. Yusof (Eds.), *Outcome-based science, technology, engineering, and mathematics education: Innovative practices* (pp. 328–349). Hershey, PA: IGI Global. doi:10.4018/978-1-4666-1809-1.ch016

Reimann, D. (2011). Shaping interactive media with the sewing machine: Smart textile as an artistic context to engage girls in technology and engineering education. *International Journal of Art, Culture and Design Technologies, 1*(1), 12–21. doi:10.4018/ijacdt.2011010102

Rikkilä, J. (2014). Agile, lean, and service-oriented development, continuum, or chasm. In *Software design and development: concepts, methodologies, tools, and applications* (pp. 132–163). Hershey, PA: IGI Global. doi:10.4018/978-1-4666-4301-7.ch008

Rocci, L. (2010). Engineering and environmental technoethics. In R. Luppicini (Ed.), *Technoethics and the evolving knowledge society: Ethical issues in technological design, research, development, and innovation* (pp. 146–162). Hershey, PA: IGI Global. doi:10.4018/978-1-60566-952-6.ch008

Rodrigues, D., Estrella, J. C., Monaco, F. J., Branco, K. R., Antunes, N., & Vieira, M. (2012). Engineering secure web services. In V. Cardellini, E. Casalicchio, K. Castelo Branco, J. Estrella, & F. Monaco (Eds.), *Performance and dependability in service computing: Concepts, techniques and research directions* (pp. 360–380). Hershey, PA: IGI Global. doi:10.4018/978-1-60960-794-4.ch016

Rodríguez, J., Fernández-Medina, E., Piattini, M., & Mellado, D. (2011). A security requirements engineering tool for domain engineering in software product lines. In N. Milanovic (Ed.), *Non-functional properties in service oriented architecture: Requirements, models and methods* (pp. 73–92). Hershey, PA: IGI Global. doi:10.4018/978-1-60566-794-2.ch004

Rojko, A., Zürcher, T., Hercog, D., & Stebler, R. (2012). Implementation of remote laboratories for industrial education. In A. Azad, M. Auer, & V. Harward (Eds.), *Internet accessible remote laboratories: Scalable e-learning tools for engineering and science disciplines* (pp. 84–107). Hershey, PA: IGI Global. doi:10.4018/978-1-61350-186-3.ch005

Rosado, D. G., Mellado, D., Fernandez-Medina, E., & Piattini, M. G. (2013). *Security engineering for cloud computing: Approaches and tools*. Hershey, PA: IGI Global. doi:10.4018/978-1-4666-2125-1

Rose, S., Lauder, M., Schlereth, M., & Schürr, A. (2011). A multidimensional approach for concurrent model-driven automation engineering. In J. Osis & E. Asnina (Eds.), *Model-driven domain analysis and software development: Architectures and functions* (pp. 90–113). Hershey, PA: IGI Global. doi:10.4018/978-1-61692-874-2.ch005

Russell, C. (2012). Conceptual mapping, visualisation, and systems thinking in engineering. In A. Patil, H. Eijkman, & E. Bhattacharyya (Eds.), *New media communication skills for engineers and IT professionals: Trans-national and trans-cultural demands* (pp. 72–93). Hershey, PA: IGI Global. doi:10.4018/978-1-4666-0243-4.ch006

Russell, D., & Haghi, A. (2010). *Web-based engineering education: Critical design and effective tools*. Hershey, PA: IGI Global. doi:10.4018/978-1-61520-659-9

Russo, B., Scotto, M., Sillitti, A., & Succi, G. (2010). Improving agile methods. In B. Russo, M. Scotto, A. Sillitti, & G. Succi (Eds.), *Agile technologies in open source development* (pp. 189–231). Hershey, PA: IGI Global. doi:10.4018/978-1-59904-681-5.ch012

Russo, B., Scotto, M., Sillitti, A., & Succi, G. (2010). Requirements management. In B. Russo, M. Scotto, A. Sillitti, & G. Succi (Eds.), *Agile technologies in open source development* (pp. 268–286). Hershey, PA: IGI Global. doi:10.4018/978-1-59904-681-5.ch015

Russo, M. T. (2012). Humanities in engineering education. In M. Rasul (Ed.), *Developments in engineering education standards: Advanced curriculum innovations* (pp. 285–300). Hershey, PA: IGI Global. doi:10.4018/978-1-4666-0951-8.ch016

Saidane, A., & Guelfi, N. (2014). Towards test-driven and architecture model-based security and resilience engineering. In *Software design and development: Concepts, methodologies, tools, and applications* (pp. 2072–2098). Hershey, PA: IGI Global. doi:10.4018/978-1-4666-4301-7.ch096

Sala, N. (2012). Virtual reality in architecture, in engineering and beyond. In E. Abu-Taieh, A. El Sheikh, & M. Jafari (Eds.), *Technology engineering and management in aviation: Advancements and discoveries* (pp. 336–345). Hershey, PA: IGI Global. doi:10.4018/978-1-60960-887-3.ch020

Salzmann, C., Gillet, D., Esquembre, F., Vargas, H., Sánchez, J., & Dormido, S. (2012). Web 2.0 open remote and virtual laboratories in engineering education. In A. Okada, T. Connolly, & P. Scott (Eds.), *Collaborative learning 2.0: Open educational resources* (pp. 369–390). Hershey, PA: IGI Global. doi:10.4018/978-1-4666-0300-4.ch020

Samaras, G. (2012). Human-centered systems engineering: Managing stakeholder dissonance in healthcare delivery. In A. Kolker & P. Story (Eds.), *Management engineering for effective healthcare delivery: Principles and applications* (pp. 148–171). Hershey, PA: IGI Global. doi:10.4018/978-1-60960-872-9.ch007

Sampaio, A. Z., Henriques, P. G., Cruz, C. O., & Martins, O. P. (2011). Interactive models based on virtual reality technology used in civil engineering education. In G. Vincenti & J. Braman (Eds.), *Teaching through multi-user virtual environments: Applying dynamic elements to the modern classroom* (pp. 387–413). Hershey, PA: IGI Global. doi:10.4018/978-1-61692-822-3.ch021

Sanabria, S. J., Furrer, R., Neuenschwander, J., Niemz, P., & Sennhauser, U. (2014). Bonding defect imaging in glulam with novel air-coupled ultrasound testing. In A. Aguilera & J. Davim (Eds.), *Research developments in wood engineering and technology* (pp. 221–246). Hershey, PA: IGI Global. doi:10.4018/978-1-4666-4554-7.ch006

Sánchez de la Rosa, J. L., Miranda, S. A., & González, C. S. (2013). Evaluation of transversal competences of the engineering students and their relation to the enterprise requirements. In K. Patel & S. Vij (Eds.), *Enterprise resource planning models for the education sector: Applications and methodologies* (pp. 1–17). Hershey, PA: IGI Global. doi:10.4018/978-1-4666-2193-0.ch001

Sarirete, A., & Chikh, A. (2012). A knowledge management process in communities of practice of engineering based on the SECI model for knowledge. In E. Ng, N. Karacapilidis, & M. Raisinghani (Eds.), *Evaluating the impact of technology on learning, teaching, and designing curriculum: Emerging trends* (pp. 134–149). Hershey, PA: IGI Global. doi:10.4018/978-1-4666-0032-4.ch009

Sass, L. (2013). Direct building manufacturing of homes with digital fabrication. In *Industrial engineering: Concepts, methodologies, tools, and applications* (pp. 1231–1242). Hershey, PA: IGI Global. doi:10.4018/978-1-4666-1945-6.ch066

Sauvet, B., Boukhicha, M., Balan, A., Hwang, G., Taverna, D., Shukla, A., & Régnier, S. (2012). Selective pick-and-place of thin film by robotic micromanipulation. *International Journal of Intelligent Mechatronics and Robotics*, *2*(3), 24–37. doi:10.4018/ijimr.2012070103

Schnabel, M. A. (2013). Learning parametric designing. In *Industrial engineering: Concepts, methodologies, tools, and applications* (pp. 197–210). Hershey, PA: IGI Global. doi:10.4018/978-1-4666-1945-6.ch013

Schrödl, H., & Wind, S. (2013). Requirements engineering for cloud application development. In A. Bento & A. Aggarwal (Eds.), *Cloud computing service and deployment models: Layers and management* (pp. 137–150). Hershey, PA: IGI Global. doi:10.4018/978-1-4666-2187-9.ch007

Schulz, T., Radlinski, L., Gorges, T., & Rosenstiel, W. (2011). Software process model using dynamic bayesian networks. In M. Ramachandran (Ed.), *Knowledge engineering for software development life cycles: Support technologies and applications* (pp. 289–310). Hershey, PA: IGI Global. doi:10.4018/978-1-60960-509-4.ch016

Sevillian, D. B. (2013). Aircraft development and design: Enhancing product safety through effective human factors engineering design solutions. In *Industrial engineering: Concepts, methodologies, tools, and applications* (pp. 858–886). Hershey, PA: IGI Global. doi:10.4018/978-1-4666-1945-6.ch048

Shan, T. C., & Hua, W. W. (2010). Strategic technology engineering planning. In M. Hunter (Ed.), *Strategic information systems: Concepts, methodologies, tools, and applications* (pp. 414–434). Hershey, PA: IGI Global. doi:10.4018/978-1-60566-677-8.ch029

Sharma, A., & Maurer, F. (2014). A roadmap for software engineering for the cloud: Results of a systematic review. In *Software design and development: Concepts, methodologies, tools, and applications* (pp. 1–16). Hershey, PA: IGI Global. doi:10.4018/978-1-4666-4301-7.ch001

Sharma, N., Singh, K., & Goyal, D. (2014). Software engineering, process improvement, and experience management: Is the nexus productive? Clues from the Indian giants. In *Software design and development: Concepts, methodologies, tools, and applications* (pp. 1401–1414). Hershey, PA: IGI Global. doi:10.4018/978-1-4666-4301-7.ch068

Shirk, S., Arreola, V., Wobig, C., & Russell, K. (2012). Girls' e-mentoring in science, engineering, and technology based at the University of Illinois at Chicago women in science and engineering (WISE) program. In Computer engineering: Concepts, Methodologies, tools and applications (pp. 1144-1163). Hershey, PA: IGI Global. doi:10.4018/978-1-61350-456-7.ch505

Shukla, A., & Tiwari, R. (2011). *Biomedical engineering and information systems: Technologies, tools and applications*. Hershey, PA: IGI Global. doi:10.4018/978-1-61692-004-3

Sidhu, M. S. (2010). Challenges and trends of TAPS packages in enhancing engineering education. In M. Sidhu (Ed.), *Technology-assisted problem solving for engineering education: Interactive multimedia applications* (pp. 158–166). Hershey, PA: IGI Global. doi:10.4018/978-1-60566-764-5.ch011

Sidhu, M. S. (2010). Evaluation of TAPS packages. In M. Sidhu (Ed.), *Technology-assisted problem solving for engineering education: Interactive multimedia applications* (pp. 128–147). Hershey, PA: IGI Global. doi:10.4018/978-1-60566-764-5.ch009

Sidhu, M. S. (2010). Technology assisted problem solving packages: A new approach to learning, visualizing, and problem solving in engineering. In M. Sidhu (Ed.), *Technology-assisted problem solving for engineering education: Interactive multimedia applications* (pp. 69–90). Hershey, PA: IGI Global. doi:10.4018/978-1-60566-764-5.ch006

Sidhu, M. S. (2010). *Technology-assisted problem solving for engineering education: Interactive multimedia applications*. Hershey, PA: IGI Global. doi:10.4018/978-1-60566-764-5

Sidhu, M. S., & Kang, L. C. (2012). Emerging trends and technologies for enhancing engineering education: An overview. In L. Tomei (Ed.), *Advancing education with information communication technologies: Facilitating new trends* (pp. 320–330). Hershey, PA: IGI Global. doi:10.4018/978-1-61350-468-0.ch026

Sidhu, M. S., & Kang, L. C. (2012). New trends and futuristic information communication technologies for engineering education. In S. Chhabra (Ed.), *ICTs for advancing rural communities and human development: Addressing the digital divide* (pp. 251–262). Hershey, PA: IGI Global. doi:10.4018/978-1-4666-0047-8.ch017

Simonette, M. J., & Spina, E. (2014). Enabling IT innovation through soft systems engineering. In M. Pańkowska (Ed.), *Frameworks of IT prosumption for business development* (pp. 64–72). Hershey, PA: IGI Global. doi:10.4018/978-1-4666-4313-0.ch005

Singh, T., Verma, M. K., & Singh, R. (2014). Role of emotional intelligence in academic achievement: An empirical study on engineering students. In P. Ordóñez de Pablos & R. Tennyson (Eds.), *Strategic approaches for human capital management and development in a turbulent economy* (pp. 255–263). Hershey, PA: IGI Global. doi:10.4018/978-1-4666-4530-1.ch016

Smolnik, S., Teuteberg, F., & Thomas, O. (2012). *Semantic technologies for business and information systems engineering: Concepts and applications*. Hershey, PA: IGI Global. doi:10.4018/978-1-60960-126-3

Solemon, B., Sahibuddin, S., & Ghani, A. A. (2014). Requirements engineering process improvement and related models. In *Software design and development: Concepts, methodologies, tools, and applications* (pp. 203–218). Hershey, PA: IGI Global. doi:10.4018/978-1-4666-4301-7.ch011

Sorial, G. A., & Noroozi, B. (2012). Improvement of engineering students education by e-learning. In *Virtual learning environments: Concepts, methodologies, tools and applications* (pp. 870–883). Hershey, PA: IGI Global. doi:10.4018/978-1-4666-0011-9.ch414

Srinivasan, A., López-Ribot, J. L., & Ramasubramanian, A. K. (2011). Microfluidic applications in vascular bioengineering. In A. Shukla & R. Tiwari (Eds.), *Biomedical engineering and information systems: Technologies, tools and applications* (pp. 1–30). Hershey, PA: IGI Global. doi:10.4018/978-1-61692-004-3.ch001

Steinbuch, R. (2013). Evolutionary optimization of passive compensators to improve earthquake resistance. In N. Lagaros, V. Plevris, & C. Mitropoulou (Eds.), *Design optimization of active and passive structural control systems* (pp. 250–273). Hershey, PA: IGI Global. doi:10.4018/978-1-4666-2029-2.ch011

Stephenson, S. V., & Sage, A. P. (2010). Information and knowledge perspectives in systems engineering and management for innovation and productivity through enterprise resource planning. In M. Hunter (Ed.), *Strategic information systems: Concepts, methodologies, tools, and applications* (pp. 338–368). Hershey, PA: IGI Global. doi:10.4018/978-1-60566-677-8.ch025

Strasser, T., Zoitl, A., & Rooker, M. (2013). Zero-downtime reconfiguration of distributed control logic in industrial automation and control. In *Industrial engineering: Concepts, methodologies, tools, and applications* (pp. 2024–2051). Hershey, PA: IGI Global. doi:10.4018/978-1-4666-1945-6.ch109

Sudeikat, J., & Renz, W. (2010). Building complex adaptive systems: On engineering self-organizing multi-agent systems. In M. Hunter (Ed.), *Strategic information systems: Concepts, methodologies, tools, and applications* (pp. 767–787). Hershey, PA: IGI Global. doi:10.4018/978-1-60566-677-8.ch050

Sugiyama, S., & Burgess, L. (2012). Principle for engineering service based system by swirl computing. In X. Liu & Y. Li (Eds.), *Advanced design approaches to emerging software systems: Principles, methodologies and tools* (pp. 48–60). Hershey, PA: IGI Global. doi:10.4018/978-1-60960-735-7.ch003

Sun, Z., Han, J., Dong, D., & Zhao, S. (2010). Engineering of experience based trust for e-commerce. In M. Wang & Z. Sun (Eds.), *Handbook of research on complex dynamic process management: Techniques for adaptability in turbulent environments* (pp. 342–367). Hershey, PA: IGI Global. doi:10.4018/978-1-60566-669-3.ch014

Sunindyo, W. D., Moser, T., Winkler, D., Mordinyi, R., & Biffl, S. (2013). Workflow validation framework in collaborative engineering environments. In A. Loo (Ed.), *Distributed computing innovations for business, engineering, and science* (pp. 285–299). Hershey, PA: IGI Global. doi:10.4018/978-1-4666-2533-4.ch015

Szeto, A. Y. (2014). Assistive technology and rehabilitation engineering. In *Assistive technologies: Concepts, methodologies, tools, and applications* (pp. 277–331). Hershey, PA: IGI Global. doi:10.4018/978-1-4666-4422-9.ch015

Takruri-Rizk, H., Sappleton, N., & Dhar-Bhattacharjee, S. (2010). Progression of UK women engineers: Aids and hurdles. In A. Cater-Steel & E. Cater (Eds.), *Women in engineering, science and technology: Education and career challenges* (pp. 280–300). Hershey, PA: IGI Global. doi:10.4018/978-1-61520-657-5.ch013

Tekinerdogan, B., & Aksit, M. (2011). A comparative analysis of software engineering with mature engineering disciplines using a problem-solving perspective. In A. Dogru & V. Biçer (Eds.), *Modern software engineering concepts and practices: Advanced approaches* (pp. 1–18). Hershey, PA: IGI Global. doi:10.4018/978-1-60960-215-4.ch001

Tellioglu, H. (2012). About representational artifacts and their role in engineering. In G. Viscusi, G. Campagnolo, & Y. Curzi (Eds.), *Phenomenology, organizational politics, and IT design: The social study of information systems* (pp. 111–130). Hershey, PA: IGI Global. doi:10.4018/978-1-4666-0303-5.ch007

Terawaki, Y., Takahashi, Y., Kodama, Y., & Yana, K. (2013). The development of educational environment suited to the Japan-specific educational service using requirements engineering techniques: Case study of running Sakai with PostgreSQL. In V. Kumar & F. Lin (Eds.), *System and technology advancements in distance learning* (pp. 261–270). Hershey, PA: IGI Global. doi:10.4018/978-1-4666-2032-2.ch019

Thomas, K. D., & Muga, H. E. (2012). Sustainability: The new 21st century general education requirement for engineers. In M. Rasul (Ed.), *Developments in engineering education standards: Advanced curriculum innovations* (pp. 263–284). Hershey, PA: IGI Global. doi:10.4018/978-1-4666-0951-8.ch015

Trigo, B., Olguin, G., & Matai, P. (2010). The use of applets in an engineering chemistry course: Advantages and new ideas. In D. Russell & A. Haghi (Eds.), *Web-based engineering education: Critical design and effective tools* (pp. 108–118). Hershey, PA: IGI Global. doi:10.4018/978-1-61520-659-9.ch009

Tsadimas, A., Nikolaidou, M., & Anagnostopoulos, D. (2014). Model-based system design using SysML: The role of the evaluation diagram. In P. Fonseca i Casas (Ed.), *Formal languages for computer simulation: Transdisciplinary models and applications* (pp. 236–266). Hershey, PA: IGI Global. doi:10.4018/978-1-4666-4369-7.ch008

Unhelkar, B., Ghanbary, A., & Younessi, H. (2010). *Collaborative business process engineering and global organizations: Frameworks for service integration*. Hershey, PA: IGI Global. doi:10.4018/978-1-60566-689-1

Uziak, J., Oladiran, M. T., & Kommula, V. P. (2012). Integrating general education courses into engineering curriculum: Students' perspective. In M. Rasul (Ed.), *Developments in engineering education standards: Advanced curriculum innovations* (pp. 247–262). Hershey, PA: IGI Global. doi:10.4018/978-1-4666-0951-8.ch014

Valizadeh, M., Anzelotti, G., & Salehi, A. S. (2010). Web-based approaches in engineering education. In R. Luppicini & A. Haghi (Eds.), *Cases on digital technologies in higher education: Issues and challenges* (pp. 241–256). Hershey, PA: IGI Global. doi:10.4018/978-1-61520-869-2.ch017

Valizadeh, M., Anzelotti, G., & Salehi, S. (2010). Web-based training: An applicable tool for engineering education. In D. Russell & A. Haghi (Eds.), *Web-based engineering education: Critical design and effective tools* (pp. 186–198). Hershey, PA: IGI Global. doi:10.4018/978-1-61520-659-9.ch014

Valverde, R., Toleman, M., & Cater-Steel, A. (2010). Design science: A case study in information systems re-engineering. In M. Hunter (Ed.), *Strategic information systems: Concepts, methodologies, tools, and applications* (pp. 490–503). Hershey, PA: IGI Global. doi:10.4018/978-1-60566-677-8.ch033

Vaquero, L. M., Rodero-Merino, L., Cáceres, J., Chapman, C., Lindner, M., & Galán, F. (2012). Principles, methodology and tools for engineering cloud computing systems. In X. Liu & Y. Li (Eds.), *Advanced design approaches to emerging software systems: Principles, methodologies and tools* (pp. 250–273). Hershey, PA: IGI Global. doi:10.4018/978-1-60960-735-7.ch012

Vargas, E. P. (2014). Quality, improvement and measurements in high risk software. In *Software design and development: Concepts, methodologies, tools, and applications* (pp. 733–748). Hershey, PA: IGI Global. doi:10.4018/978-1-4666-4301-7.ch035

Vasant, P. (2013). *Meta-heuristics optimization algorithms in engineering, business, economics, and finance*. Hershey, PA: IGI Global. doi:10.4018/978-1-4666-2086-5

Venkatraman, S. (2013). Software engineering research gaps in the cloud. *Journal of Information Technology Research*, *6*(1), 1–19. doi:10.4018/jitr.2013010101

Villazón-Terrazas, B. C., Suárez-Figueroa, M., & Gómez-Pérez, A. (2012). A pattern-based method for re-engineering non-ontological resources into ontologies. In A. Sheth (Ed.), *Semantic-enabled advancements on the web: Applications across industries* (pp. 17–54). Hershey, PA: IGI Global. doi:10.4018/978-1-4666-0185-7.ch002

Vinod, J. S. (2010). Dem simulations in geotechnical earthquake engineering education. *International Journal of Geotechnical Earthquake Engineering*, *1*(1), 61–69. doi:10.4018/jgee.2010090804

Virdi, J. (2014). Business risk analysis: Obsolescence management in requirements engineering. In *Software design and development: Concepts, methodologies, tools, and applications* (pp. 1736–1763). Hershey, PA: IGI Global. doi:10.4018/978-1-4666-4301-7.ch083

Walk, S., Pöschko, J., Strohmaier, M., Andrews, K., Tudorache, T., Noy, N. F., ... Musen, M. A. (2013). PragmatiX: An interactive tool for visualizing the creation process behind collaboratively engineered ontologies. *International Journal on Semantic Web and Information Systems*, *9*(1), 45–78. doi:10.4018/jswis.2013010103 PMID:24465189

Wang, J. (2013). Architects and engineers. In *Challenging ICT applications in architecture, engineering, and industrial design education* (pp. 48–64). Hershey, PA: IGI Global. doi:10.4018/978-1-4666-1999-9.ch003

Wang, J. (2013). *Challenging ICT applications in architecture, engineering, and industrial design education*. Hershey, PA: IGI Global. doi:10.4018/978-1-4666-1999-9

Wang, J. (2013). Pedagogy and curriculum in architecture and engineering. In *Challenging ICT applications in architecture, engineering, and industrial design education* (pp. 65–92). Hershey, PA: IGI Global. doi:10.4018/978-1-4666-1999-9.ch004

Wang, J. (2013). Professionalism in architecture and engineering. In *Challenging ICT applications in architecture, engineering, and industrial design education* (pp. 137–156). Hershey, PA: IGI Global. doi:10.4018/978-1-4666-1999-9.ch007

Wang, J. (2013). Reviewing engineers and introducing industrial designers. In *Challenging ICT applications in architecture, engineering, and industrial design education* (pp. 111–136). Hershey, PA: IGI Global. doi:10.4018/978-1-4666-1999-9.ch006

Wang, Y. (2010). On cognitive properties of human factors and error models in engineering and socialization. In Y. Wang (Ed.), *Discoveries and breakthroughs in cognitive informatics and natural intelligence* (pp. 93–109). Hershey, PA: IGI Global. doi:10.4018/978-1-60566-902-1.ch006

Wang, Y. (2012). Convergence of software science and computational intelligence: A new transdisciplinary research field. In Y. Wang (Ed.), *Software and intelligent sciences: New transdisciplinary findings* (pp. 1–13). Hershey, PA: IGI Global. doi:10.4018/978-1-4666-0261-8.ch001

Wang, Y., & Patel, S. (2012). Exploring the cognitive foundations of software engineering. In Y. Wang (Ed.), *Software and intelligent sciences: New transdisciplinary findings* (pp. 232–251). Hershey, PA: IGI Global. doi:10.4018/978-1-4666-0261-8.ch014

Wei, J., Chen, J., & Zhu, Q. (2010). Service science, management and engineering education: A unified model for university. *International Journal of Service Science, Management, Engineering, and Technology, 1*(2), 51–69. doi:10.4018/jssmet.2010040104

Weigang, L., Barros, A. D., & Romani de Oliveira, I. (2010). *Computational models, software engineering, and advanced technologies in air transportation: Next generation applications*. Hershey, PA: IGI Global. doi:10.4018/978-1-60566-800-0

Westh Nicolajsen, H. (2010). Limitations and perspectives on use of e-services in engineering consulting. In *Electronic services: Concepts, methodologies, tools and applications* (pp. 1280–1295). Hershey, PA: IGI Global. doi:10.4018/978-1-61520-967-5.ch078

Wilson, R., & Younis, H. (2013). *Business strategies for electrical infrastructure engineering: Capital project implementation*. Hershey, PA: IGI Global. doi:10.4018/978-1-4666-2839-7

Winter, A. (2010). The smart women – smart state strategy: A policy on women's participation in science, engineering and technology in Queensland, Australia. In A. Cater-Steel & E. Cater (Eds.), *Women in engineering, science and technology: Education and career challenges* (pp. 1–20). Hershey, PA: IGI Global. doi:10.4018/978-1-61520-657-5.ch001

Winter, R., & van Beijnum, I. (2014). Inter-domain traffic engineering using the origin preference attribute. In M. Boucadair & D. Binet (Eds.), *Solutions for sustaining scalability in internet growth* (pp. 18–38). Hershey, PA: IGI Global. doi:10.4018/978-1-4666-4305-5.ch002

Worth, D., Greenough, C., & Chin, S. (2014). Pragmatic software engineering for computational science. In *Software design and development: Concepts, methodologies, tools, and applications* (pp. 663–694). Hershey, PA: IGI Global. doi:10.4018/978-1-4666-4301-7.ch032

Wu, J. (2013). *Biomedical engineering and cognitive neuroscience for healthcare: Interdisciplinary applications*. Hershey, PA: IGI Global. doi:10.4018/978-1-4666-2113-8

Wu, Y., & Koszalka, T. A. (2013). Instructional design of an advanced interactive discovery environment: Exploring team communication and technology use in virtual collaborative engineering problem solving. In *Industrial engineering: Concepts, methodologies, tools, and applications* (pp. 117–136). Hershey, PA: IGI Global. doi:10.4018/978-1-4666-1945-6.ch009

Yadav, V. (2011). Research review: Globally distributed requirements engineering and agility. *International Journal of Innovation in the Digital Economy*, *2*(1), 1–11. doi:10.4018/jide.2011010101

Yang, J. J., Liu, J. F., Kurokawa, T., Kitamura, N., Yasuda, K., & Gong, J. P. (2013). Tough double-network hydrogels as scaffolds for tissue engineering: Cell behavior in vitro and in vivo test. In J. Wu (Ed.), *Technological advancements in biomedicine for healthcare applications* (pp. 213–222). Hershey, PA: IGI Global. doi:10.4018/978-1-4666-2196-1.ch023

Yeoh, W., Gao, J., & Koronios, A. (2010). Empirical investigation of critical success factors for implementing business intelligence systems in multiple engineering asset management organisations. In M. Hunter (Ed.), *Strategic information systems: Concepts, methodologies, tools, and applications* (pp. 2039–2063). Hershey, PA: IGI Global. doi:10.4018/978-1-60566-677-8.ch129

Yerrick, R., Lund, C., & Lee, Y. (2013). Online simulator use in the preparing chemical engineers. *International Journal of Online Pedagogy and Course Design*, *3*(2), 1–24. doi:10.4018/ijopcd.2013040101

Yilmaz, A. E., & Yilmaz, I. B. (2012). Natural language processing techniques in requirements engineering. In *Computer engineering: Concepts, methodologies, tools and applications* (pp. 533–545). Hershey, PA: IGI Global. doi:10.4018/978-1-61350-456-7.ch303

Yong, E. (2012). Literature review skills for undergraduate engineering students in large classes. In K. Yusof, N. Azli, A. Kosnin, S. Yusof, & Y. Yusof (Eds.), *Outcome-based science, technology, engineering, and mathematics education: Innovative practices* (pp. 240–261). Hershey, PA: IGI Global. doi:10.4018/978-1-4666-1809-1.ch012

Yun, H., Xu, J., Xiong, J., & Wei, M. (2013). A knowledge engineering approach to develop domain ontology. In V. Kumar & F. Lin (Eds.), *System and technology advancements in distance learning* (pp. 55–70). Hershey, PA: IGI Global. doi:10.4018/978-1-4666-2032-2.ch004

Zardari, S., Faniyi, F., & Bahsoon, R. (2013). Using obstacles for systematically modeling, analysing, and mitigating risks in cloud adoption. In I. Mistrik, A. Tang, R. Bahsoon, & J. Stafford (Eds.), *Aligning enterprise, system, and software architectures* (pp. 275–296). Hershey, PA: IGI Global. doi:10.4018/978-1-4666-2199-2.ch014

Zašcerinska, J., Ahrens, A., & Bassus, O. (2012). Enterprise 2.0 and 3.0 in education: Engineering and business students' view. In M. Cruz-Cunha, P. Gonçalves, N. Lopes, E. Miranda, & G. Putnik (Eds.), *Handbook of research on business social networking: Organizational, managerial, and technological dimensions* (pp. 472–494). Hershey, PA: IGI Global. doi:10.4018/978-1-61350-168-9.ch025

Zašcerinska, J., Ahrens, A., & Bassus, O. (2012). Enterprise 2.0 and 3.0 in education: Engineering and business students' view. In M. Cruz-Cunha, P. Gonçalves, N. Lopes, E. Miranda, & G. Putnik (Eds.), *Handbook of research on business social networking: Organizational, managerial, and technological dimensions* (pp. 472–494). Hershey, PA: IGI Global. doi:10.4018/978-1-61350-168-9.ch025

Zhang, D. (2012). Machine learning and value-based software engineering. In Y. Wang (Ed.), *Software and intelligent sciences: New transdisciplinary findings* (pp. 287–301). Hershey, PA: IGI Global. doi:10.4018/978-1-4666-0261-8.ch017

Zhang, H., Xu, C., Su, W., & Luo, H. (2014). Routing optimization for inter-domain traffic engineering under identifier network. In M. Boucadair & D. Binet (Eds.), *Solutions for sustaining scalability in internet growth* (pp. 127–147). Hershey, PA: IGI Global. doi:10.4018/978-1-4666-4305-5.ch007

Zhong, Y. (2013). Processing of 3D unstructured measurement data for reverse engineering. In S. Sirouspour (Ed.), *Advanced engineering and computational methodologies for intelligent mechatronics and robotics* (pp. 118–127). Hershey, PA: IGI Global. doi:10.4018/978-1-4666-3634-7.ch008

Zhou, Z., Wang, H., & Lou, P. (2010). *Manufacturing intelligence for industrial engineering: Methods for system self-organization, learning, and adaptation*. Hershey, PA: IGI Global. doi:10.4018/978-1-60566-864-2

Zhu, L., Jayaram, U., & Kim, O. (2011). Online semantic knowledge management for product design based on product engineering ontologies. *International Journal on Semantic Web and Information Systems, 7*(4), 36–61. doi:10.4018/jswis.2011100102

About the Contributors

M. Uthayakumar has obtained Master of Engineering in Production Engineering from Thiagarajar College of Engineering (Autonomous), Madurai, India. He did his research on bimetallic piston machining studies in the Department of Production Engineering, National Institute of Technology, Tiruchirappalli and received Ph.D from Anna University Chennai. He was awarded as young scientist by Tamil Nadu state council for Science and Technology and undergone the post doctoral fellowship on Tribology in the Department of Mechanical Engineering, Indian Institute of Technology, New Delhi. He has published currently 100 papers in the referred international journal. He has visited Cracow University of Technology Poland and Yeungnam University, Korea as visiting professor. He organized the second international conference on Advanced Manufacturing and Automation (INCAMA 2013) successfully and produced special issues in reputed journals. Currently he is working as Director (International relations) of Kalasalingam University, India.

S. Aravind Raj is currently working as an Assistant professor - Senior in the School of Mechanical Engineering at the Vellore Institute of Technology, Vellore, India. He received his doctorate degree in Industrial Engineering from the National Institute of Technology, Tiruchirappalli in 2014. His research area is focused on Agile manufacturing process and their applications in manufacturing industries. Recent work focuses on Industrial and Safety Engineering projects. He has published 16 International journals papers in leading SCI/ SCOPUS Indexed journals and 31 papers in International conferences; he is co-author for 3 book chapters. He is reviewer of reputed international journals. He has received several awards including Doctoral Scholarship and Best Paper Award.

Tae Jo Ko is a Professor of Mechanical Engineering at Yeungnam University, Korea. He received bachelor and master's degrees from the Pusan National University, Korea. He got Ph.D in mechanical engineering from POSTECH, S. Korea. He has worked for Doosan Infracore Co. Ltd. (formerly Daewoo) from 1985 to 1995. Also, he was responsible for the Gyoungbuk Hybrid Technology Institute that is regional research innovation center and initiated the idea for hybrid manufacturing. He was dean of School of Mechanical Engineering in Yeungnam University until 2016. His research interests include machine tools, metal cuttings as well as non-traditional machining. Recently, he launched the research of surface texturing using metal cutting and grinding. Also he is very much interested in CFRP machining, especially in drilling.

S. Thirumalai Kumaran is currently working as an Associate Professor in the School of Automotive and Mechanical Engineering, Kalasalingam academy of Research and Education (KARE). After completing Ph.D., he worked as an Assistant Professor at School of Mechanical Engineering, Yeungnam University, South Korea for a period of one year. Currently at KARE, he is one among the Research Head in the "Centre for Advanced Machining". Dr. S. Thirumalai Kumaran has successfully completed a research project received from the Department of Atomic Energy-Board of Studies in Nuclear Science (DAE-BRNS) worth Rs.20,56,500/-. Currently, he has one on-going project from CVRDE, Chennai. He has published more than 40 publications as the main author and co-author in SCI indexed International Journals (as of 2018). Additionally, he has published more than 30 conference papers in different countries including India, Japan and South Korea. He is acting as reviewer for several International Journals.

J. Paulo Davim received his Ph.D. degree in Mechanical Engineering in 1997, M.Sc. degree in Mechanical Engineering (materials and manufacturing processes) in 1991, Mechanical Engineering degree (5 years) in 1986, from the University of Porto (FEUP), the Aggregate title (Full Habilitation) from the University of Coimbra in 2005 and the D.Sc. from London Metropolitan University in 2013. He is Eur Ing by FEANI-Brussels and Senior Chartered Engineer by the Portuguese Institution of Engineers with an MBA and Specialist title in Engineering and Industrial Management. Currently, he is Professor at the Department of Mechanical Engineering of the University of Aveiro, Portugal. He has more than 30 years of teaching and research experience in Manufacturing, Materials, Mechanical and Industrial Engineering, with special emphasis in Machining & Tribology. He has also interest in Management, Engineering Education and Higher Education for Sustainability. He has guided large numbers of postdoc, Ph.D. and master's students as well as coordinated & participated in several financed research projects. He has received several scientific awards. He has worked as evaluator of projects for international research agencies as well as examiner of Ph.D. thesis for many universities. He is the Editor in Chief of several international journals, Guest Editor of journals, books Editor, book Series Editor and Scientific Advisory for many international journals and conferences. Presently, he is an Editorial Board member of 25 international journals and acts as reviewer for more than 80 prestigious Web of Science journals. In addition, he has also published as editor (and co-editor) more than 100 books and as author (and co-author) more than 10 books, 80 book chapters and 400 articles in journals and conferences (more than 200 articles in journals indexed in Web of Science core collection/h-index 45+/6000+ citations and SCOPUS/h-index 53+/8500+ citations).

* * *

Izzeldin Idris Abdalla was born in Kordofan Province, Sudan, in 1982. He received his B.Sc. (Hons.) degree and M.Sc. in Electrical and Electronic Engineering, specializing in Power and Energy, from Juba University, Sudan in 2005 and Universiti Teknologi PETRONAS, Malaysia in 2011, respectively. He was in the Faculty of Engineering, Department of Electrical and Electronic Engineering, Juba University from 2005 to 2008. He joined the Department of Electrical and Electronic Engineering, Universiti Teknologi PETRONAS, Malaysia in Jan. 2012 as a Ph.D. student. His areas of interest are in power electronics, power systems and power quality. His current research interests are in the analysis and design of electrical machines.

Ahmad Majdi Abdul-Rani is an Associate Professor in the Institute of Health Analytics / Mechanical Engineering Department, Universiti Teknologi PETRONAS (UTP), Malaysia. He holds a Bachelor degree in Manufacturing Engineering and Master degree in Industrial Engineering from Northern Illinois University, USA. He obtained his PhD degree in Mechanical and Manufacturing Engineering from Loughborough University, UK. His research interests are in Advanced Manufacturing system; Fabrication of biomedical implants; Rapid Prototyping and Rapid Tooling; Reverse Engineering; CIM; FMS; CAD/CAM; CNC programming and STEP-NC. A number of PhD and MSc candidates have graduated successfully under his supervision. He is currently the principal investigator for several national and international grants.

Saood Ali graduated in the field of Mechanical Engineering and completed his post-graduationin Material Science & Engineering. He is currently working as research scholar in the area of machining and surface texturing.

V. Arumugaprabu works as associate professor in the Department of Mechanical Engineering, School of Automotive and Mechanical Engineering, Kalasalingam University, Krishnankoil, Tamilnadu, India. He graduated B.E in Mechanical Engineering at AKCE in the year 2005, Graduated M.E in CAD/CAM at Mepco Schlenk Engg., college, Graduated Ph.D in Composite Materials in the year 2014 at Kalasalingam University, India. Also worked as PostDoc Research Associate in Precision Machining Lab, School of Mechanical Engineering at Yeungnam University, South Korea in the year 2017. He has very vast research experience in the field of composite materials with nearly 25 publications in various reputed SCI Journals and has nearly 40 publications in various International and National conferences. He had also completed one Ph.D thesis as joint supervisor and guided around 20 UG projects and 3 PG projects work. Having 10 years of teaching experience and 7 years of research experience. Also acting as a Co-Investigator in project titled "Machining and Erosion studies of Red mud an Industrial waste based Polymer Matrix Composite" funded by the Ministry of Environment, Forest and Climate Change (MOEF), New Delhi, India. Also having active collaboration with eminent scientists and researchers. As an output, he has made some joint publications and planned for joint proposals. He also proved himself to be a capable mentor and teacher. Also he served in the administrative work as Deputy Director (IQAC). At present he is supervising 3 scholars in the area of composite materials and machining studies. He is also acting as editorial board member for more than 10 International journals. He organized many workshops/seminar as well as National conference.

Sílvia Esteves has got her Industrial Engineering and Management degree at Faculdade de Engenharia da Universidade do Porto-Portugal, in 2002. She has joined INEGI in 2002, assuming responsibilities as project manager and senior researcher in the Ecodesign and Product Development area. The main research topics of his activity involve the study of product development and management tools, Ecodesign, Life-cycle assessment, Life-cycle cost methodologies associated both to the product development cycle regarding eco-efficiency issues.

Nadeem Faisal, BE (Mechanical Engineering, ITM University, Gwalior, India), M.E. Pursuing (Design of Mechanical Equipment, Birla Institute of Technology, Mesra, India). He has over 1 year of Industrial experience. His areas of interests are Optimization, Material Science, Product and Process Design, CAD/CAM/CAE and Rapid Prototyping. He has 2 books, 2 Book Chapter, 1 SCI Indexed international journal to his credit.

Jyoteesh Gutha is an undergraduate student at School of Mechanical Engineering, VIT University, Vellore. His areas of interest include sustainable manufacturing, costing for manufacturing.

Taib Ibrahim was born in Kedah, Malaysia in 1972. He received the B.Eng. (Hons) in electrical and electronics engineering, M.Sc. in electrical power engineering and Ph.D. in electrical machine design from Coventry University, UK in 1996, University of Strathclyde, UK in 2000 and University of Sheffield, UK in 2009, respectively. His employment experience includes Airod (M) SdnBhd and Universiti Teknologi PETRONAS (UTP). His research concentrates on designing linear and rotary electrical machines and their associated drives.

R. Deepak Joel Johnson perused Bachelors degree in Mechanical Engineering and Master of Technology in Mechanical Engineering with Manufacturing Engineering as Specialization from Karunya University, India. With two years of experience as Assistant Professor in M.Kumarasamy College of Engineering, Karur and currently doing his Full time Research in the field of Polymer Composite at Kalasalingam University, India. He has more than 10 International Publication in reputed Journals. His area of interest are Optimization techniques, Cutting Fluid Application, Machining and composite Material.

Vimal K. E. K. is an Assistant Professor, National Institute of Technology, Patna, India. He completed his Ph.D. and M.Tech from Production Engineering Department, National Institute of Technology, Tiruchirappalli, Tamil Nadu. He received his Bachelor's degree in Production Engineering (Sandwich Programme) from PSG College of Technology, Coimbatore, India. He has published 19 papers in International Journals and 30 papers in International Conferences. His areas of research interests include lean manufacturing, sustainable manufacturing, neural network and fuzzy logic.

Jayakrishna Kandasamy is currently working as an Associate professor in the Department of Mechanical Engineering, NIT Patna, India. He received his doctorate degree in production engineering from the National Institute of Technology, Tiruchirappalli in 2014, master's degree in Production engineering from P.S.G College of Technology, Coimbatore in 2009 and a bachelor's degree in mechanical engineering from Adhiyamman College of Engineering, in 2006. His research is focused on sustainable manufacturing processes and their applications in automotive industries. Jayakrishna's current work at VIT University investigates on imbibing sustainable manufacturing practices into ERP systems for Industry 4.0. He has published 21 International journals papers in leading SCI/ SCOPUS Indexed journals and 48 papers in International conferences; he has also edited one book, co-authored three book chapters. He is editorial board member and reviewer of reputed international journals. He has received several awards including MHRD Postgraduate and Doctoral Scholarships and Best Paper Awards.

Balamurugan Karnan graduated in the field of Mechanical Engineering and completed his post graduation in Production Engineering. He did his Ph.D research work on fabrication, characterization and advance machining studies of rare earth phosphate materials. He has currently published 11 papers in the referred international journal. Now, he is doing his extensive research in the area of Advance Machining and Machinable properties of MMC's, CMC'S and PMC'S. His other areas of interest include characterization and optimization by developing a suitable mathematical tool. Currently he is working as Associate Professor, Dept. of Mechanical Engineering in VFSTR (Deemed to be University), Guntur, Andhra Pradesh, India.

Sevan Katrancıoğlu is a PhD student at Marmara University Department of Industrial Engineering.

Huseyin Selcuk Kilic is currently an instructor in the Department of Industrial Engineering in Marmara University. He received the BSc, MSc and PhD degrees in Industrial Engineering from Istanbul Technical University. He studied in-plant logistics design for his PhD dissertation. His main research areas are plant logistics, reverse logistics, lean production, decision making and ergonomics. He has research papers in journals that include Applied Mathematical Modelling, International Journal of Advanced Manufacturing Technology, Computers & Industrial Engineering, Assembly Automation and Resources, Conservation and Recycling.

Kaushik Kumar, B.Tech (Mechanical Engineering, REC (Now NIT), Warangal), MBA (Marketing, IGNOU) and Ph.D (Engineering, Jadavpur University), is presently an Associate Professor in the Department of Mechanical Engineering, Birla Institute of Technology, Mesra, Ranchi, India. He has 14 years of Teaching & Research and over 11 years of industrial experience in a manufacturing unit of Global repute. His areas of teaching and research interest are Quality Management Systems, Optimization, Non-conventional machining, CAD / CAM, Rapid Prototyping and Composites. He has 9 Patents,14 Book, 6 Edited Book, 30 Book Chapters, 120 international Journal publications, 18 International and 8 National Conference publications to his credit. He is on the editorial board and review panel of 7 International and 1 National Journals of repute. He has been felicitated with many awards and honours.

Manjeet Kumar is Ph.D Scholar in the Department of Mechanical Engineering, UIET, Panjab University, Chandigarh.

Rajesh Kumar is working as Assistant Professor in the Department of Mechanical Engineering, UIET, Panjab University, Chandigarh. His area of research is bio-materials and Advanced manufacturing.

Babu Lal graduated from KCET at Manapparai with a Bachelor of Mechanical Engineering in 2006. He completed his Master of Manufacturing Engineering at A.C.Tech, Karaikudi. He started his career as Lecturer in a private engineering college in Tamillnadu, India and followed by he continued his teaching professional in different institutions and universities both in India and abroad. He is currently working as Lecturer in Manufacturing Engineering Chair under the school of Mechanical and Industrial Engineering, DDIT, Dire Dawa University, Ethiopia since 2015

Emanuel João Menezes Lourenço holds a degree in Environmental Engineering (2010) and a Master's degree in Environmental Management (2012) from the Escola Superior Agrária de Coimbra. He joined INEGI in 2012, where he develops his professional activity in projects related with efficiency and eco-efficiency assessment of production systems. He has participated in several, national and European, projects in the field of Efficiency assessment of production systems where tools and approaches have been developed to evaluate the performance of production systems and characterize overall environmental performance.

Saravanamohan M. is a Robotics doctorate from Anna University. Earned academic qualifications, degrees B.E., M.E., and SAP-ERP degrees in 1997, 2001, 2006 from renowned institutions Arulmigu Kalasalingam College of Engineering, Madurai, MIT Chennai and Pricol Academy Coimbatore respectively. Teaching and administrative experience runs more than 14years. Specialized in Mechatronics and an active designer in automation, involved in DRDO and AICTE sponsored projects on underwater robotics as a Co-investigator and investigator. At present, works at the Department of Mechatronics Engineering as Associate Professor in Kumaraguru college of Technology, Coimbatore, India.

Sultan M. T. H. is a Professional Engineer (PEng) registered under the Board of Engineers Malaysia (BEM), Charted Engineer registered with the board of the Institution of Mechanical Engineers United Kingdom, currently attached with Universiti Putra Malaysia as a Director of Aerospace Manufacturing Research Centre (AMRC), Faculty of Engineering, UPM Serdang, Selangor, Malaysia. Being a Director of AMRC, he is also appointed as an Independent Scientific Advisor to Aerospace Malaysia Innovation Centre (AMIC) based in Cyberjaya, Selangor, Malaysia. He received his Ph.D. from University of Sheffield, United Kingdom. He has about 10 years of experience in teaching as well as in research. His area of research interests includes Hybrid Composites, Advance Materials, Structural Health Monitoring and Impact Studies. So far he has published more than 100 International journal papers. He is also the Honorary Secretary of Malaysian Society of Structural Health Monitoring (MSSHM) based in UPM Serdang, Selangor, Malaysia. Currently, he is also appointed as an Associate Professor in the Department of Aerospace Engineering, Universiti Putra Malaysia.

Nursyarizal Mohd Nor obtained his Ph.D. in Electrical Engineering from Universiti Teknologi PETRONAS (UTP), Malaysia in 2009. In 2001, he obtained his M.Sc. in Electrical Power Engineering from the University of Manchester Institute of Science and Technology (UMIST), UK. His areas of specialization are power systems state estimation, power systems analysis, electrical machines and renewable energy. He has several publications at his credit.

V. Murari obtained his doctoral degree from Indian Institute of Technology Kanpur in 2011 and currently working as an assistant professor in Motilal Nehru National Institute of Technology Allahabad, India. His area of research include solid mechanics, composite materials, continuum damage modelling, computations and experiments in structures.

Sekhar N., after completing Bachelor of Engineering in Mechanical Stream in 1969 with first class, joined MOTOR Industries Company Limited, Bangalore (presently BOSCH) and worked for 31 years and voluntarily retired as a Divisional Manager (Planning & Manufacturing) later completed Master of Technology in Material Science and Engineering with distinction in 2008 and completed doctoral research in 2016 in the field of Material Science while working as Professor of Automobile Engineering at Dayananda Sagar College of Engineering, Bangalore, India. Area of Interest advanced material science, HVOF Thermal Spraying. Member of professional societies like MSAE, MIWS, MIE and MISTE.

Perumal Nallagownden obtained his B.E (Hons) in Electrical and Electronics Engineering from Portsmouth Polytechnic, U.K. and M.Sc. from University of Wales, U.K. He is an Associate professor at Department of Electrical and Electronics Engineering, Universiti Teknologi PETRONAS, Malaysia. His special area of research interest is electrical power systems. He is a member of the Institution of Engineers Malaysia and is a Professional Engineer registered with the Board of Engineers Malaysia.

Luis Miguel Oliveira obtained his PhD in Engineering Sciences at the Vrije Universiteit Brussels. His thesis dealt with Environmental Aspects of Lithium-Ion Energy Storage Systems Using a Multi-Stage Hybrid Approach Combining the Benefits of Life Cycle Assessment and Input-Output Models. He conducted research on several projects related to Environmental Impact Assessment and Energy Storage Systems. He completed Bachelor studies in the Superior Engineering Institute of Porto (ISEP) in the area of Mechanical Engineering, and obtained his Master degree in Electromechanical Engineering at Karel de Grote Hogeschool in Antwerp. He did research at the Lung Toxicology Unit of KUL on the health effects of inhaled nano-particles created in the combustion of diesel based fuels. Nowadays, Luis focuses on The sustainability aspects of advanced manufacturing systems. Additive manufacturing technologies are now the focus of his research, specifically, the combination of high performance polymers and composite materials such as continuous carbon fibers. He also focus his research on the Design-for-AM and topological optimization topics, relevant to future developments in the additive manufacturing field.

Paulo Peças has a graduation (1991) and a doctorate (2004) in Mechanical Engineering at Instituto Superior Técnico (Universidade de Lisboa). Since 1998, he is an assistant professor at Instituto Superior Técnico in the Manufacturing Technologies and Industrial Management scientific area, being responsible for programmes in Production Management and Industrial Management. During the last ten years has participated and integrated coordination teams of R&DT projects in collaboration with the industry, at both national and European levels. The research activity has focused the development of models to support decision-making, comparative analysis and best-practices implementation in the scope of Life Cycle Engineering and lean & agile manufacturing. These models foster the build-up and implementation of solutions for productivity and competitiveness increasing in several industrial sectors. He has published and edited scientific books, published 22 research papers in International Journals and 6 chapters in scientific books, 60 publications in International Conferences Proceedings and 23 dissemination articles in national magazines.

João Paulo Pereira has got his degree in Mechanical Engineering at the Faculdade de Engenharia da Universidade do Porto (FEUP), Portugal, in 1988. In 1998, he finished a MBA at the School of Management, Porto. Actually is engaged in the Doctoral Program of Mechanical Engineering at FEUP. Since 2005, he is the head of the Product and Systems Development Division of INEGI (Institute of Science and Innovation in Mechanical and Industrial Engineering, Porto), having led or collaborated on many projects with industrial companies. His research interests are in the field of methodologies and tools for product and systems design, with special focus on Sustainable Design and Life Cycle Assessment of complex systems (e.g. Machine-tools). More recently he is engaged in a new research field related with

Additive Manufacturing leading a group of researchers developing activity in the fields of new Additive Manufacturing technologies and methodologies, in particular the design of products and parts specifically designed for AM (DfAM – Design for Additive Manufacturing), through different approaches, using topology optimization, freeform design and lattice structures and, additionally, the design of systems for Additive Manufacturing, using innovative motion control systems and processes dedicated to high performance thermoplastics (reinforced or not) processing and metallic materials processing using DMD technology (Direct Metal Deposition).

Chander Prakash is Associate Professor in the School of Mechanical Engineering, Lovely Professional University, Jalandhar, India. He has received Ph.D in Mechanical Engineering from Panjab University, Chandigarh, India. His area of research is biomaterials and bio-manufacturing. He has more than 11 years of teaching and research experience. He has contributed more than 65 research papers extensively to the world in the Titanium and Magnesium based implant literature with publications appearing in Surface and Coating Technology, Materials and Manufacturing Processes, Journal of Materials Engineering and Performance, Journal of Mechanical Science and Technology, Nanoscience and Nanotechnology Letters, Proceedings of the Institution of Mechanical Engineers, Part B: Journal of Engineering Manufacture. He authored 7 book chapters. He is also editor of 2 Book: Current Trends in Bio-manufacturing" will be published in published in Springer Series in Advanced Manufacturing, Springer International Publishing AG, Dec., 2018 and Emerging Applications of 3-D Printing in Biomedical" will be published in Springer Series in "Design Science and Innovation", Springer International Publishing AG, published in July, 2019. He is also Guest Editor of 2 Journals: Guest Editor of Special Issue of "Functional Materials and Advanced Manufacturing", Facta Universitatis, Series: Mechanical Engineering (Scopus Index) along with Dr. S. S. Sidhu and Dr. S. Singh and Guest Editor of Special Issue on "Metrology in Materials and Advanced Manufacturing", Measurement and Control (Sage, SCI Index, Impact Factor: 0.8) with Dr. S. S. Sidhu and Dr. S. Singh. He is working with joint collaboration with Prof. M. S. Uddin, South University of Australia and Prof. AM Abdul-Rani, Universiti Teknologi PETRONAS, Malaysia.

Keshavamurthy R. received his Bachelor of Engineering in Mechanical Engineering, post-graduation in Manufacturing Science and Engineering and PhD in Metal Matrix Composites from Visvesvaraya Technological University, Belgaum, India in the year 2005, 2007 and 2011 respectively. Dr. Keshavamurthy started his career as Research Associate under DST Sponsored Research Project in 2006. Later in 2010, he joined as Assistant Professor in the Department of Mechanical Engineering, PES Institute of Technology, Bangalore. In 2012 he has been appointed as Scientist in Rapid Prototyping Division, at Central Manufacturing Technology Institute (CMTI), Bangalore, India. Since August 2013, He is with Dayananda Sagar College of Engineering, Bangalore. His areas of interest are Metal Matrix Composites, Thermal Spray Coatings, Additive Manufacturing and Non-Traditional Machining. He has published over 90 research papers in international journals and conferences and has filed one Indian patent in the area of Thermal spray coatings.

Sharath Kumar Reddy is an Undergraduate Student at VIT University, Vellore, working on sustainable manufacturing and advanced industrial engineering.

M. S. Shama is currently working as Assistant Professor, Department of Mechanical Engineering, Trinity College of Engineering, Trivandrum, Kerala. Her area of interest include sustainable product development, remanufacturing and costing.

Sunpreet Singh has contributed about 60 articles in the area of Additive Manufacturing, Composites, Material Development, Optimization and many other emerging research areas. Currently, he is working a Assistant Professor in Lovely Professional University of Phagwara.

Adam Słota is an assistant professor at the Faculty of Mechanical Engineering at Cracow University of Technology. He received PhD for the thesis "Application of Petri Nets in Modelling of Automated Manufacturing Systems". His main research interests focus on modelling, simulation and control of discrete event manufacturing and assembly systems as well as coordination problems in multi robot systems. His main teaching subjects are: basis of automation and control theory, design with the use of CAD systems, virtual manufacturing. He visited gave lectures at Technical Universities in European countries within the framework of CEEPUS and ERASMUS programmes.

Vijay Tambrallimath completed meritoriously Master of Technology in Energy System Engineering with silver medal, joined as a Teaching Assistant for a year in Hubli, India and later joined as an Assistant Professor in Goa, India. Presently working as an Assistant Professor in Dayananda Sagar Engineering College, Bangalore, India. A novice researcher in the field of Additive Manufacturing, enrolled for PhD in 2016 under the able guidance of Dr. Keshavamurthy. Keen interest to enhance the perspective of understanding of Additive Manufacturing technology through research with green and sustainable mark on the planet.

M. Uddin is a program director and lecturer at the School of Engineering in the University of South Sustralia. His research is focused on biomaterials, surface engineering of biomaterials, innovative manufacturing processes and optimization, design and manufacture of biomedical devices/sensors, computational and experimental analysis of medical devices, e.g. implants.

Anbumalar V. is a professor in the Department of Mechanical Engineering, Velammal College of Engineering and Technology, Madurai, India. He received his Ph.D. from Anna University Chennai, India, in 2009 and Masters from Madurai Kamaraj University. He has produced five Ph.D. and guiding 10 Ph.D. scholars. He has co-authored more than 35 international publications which include refereed journals and conferences. His research interest includes Cellular Manufacturing System, Optimization Techniques, Composite Materials, Friction Stir Welding, Biomass Briquette Manufacturing, Robotics, etc.

Jerzy Zajac is a Professor at the faculty of Mechanical Engineering, Cracow University of Technology, Poland. He currently serves as the Vice-Rector for Education of the University. He received his MSc, PhD and DSc from Cracow University of Technology. His main research activities are split into two categories: (1) manufacturing and production systems; and (2) underwater robotics. He has focused on modelling, simulation, optimization and control of manufacturing systems; information integration in production systems; and underwater bio-inspired robots.

Divya Zindani, (BE, Mechanical Engineering, Rajasthan Technical University, Kota), M.E. (Design of Mechanical Equipment, BIT Mesra), presently pursuing PhD (National Institute of Technology, Silchar). He has over 2 years of Industrial experience. His areas of interests are Optimization, Product and Process Design, CAD/CAM/CAE, Rapid prototyping and Material Selection and Manufacturing processes. He has 1 Patent, 6 Books, 1 Edited book, 10 Book Chapters, 7 Journal publications and 5 International Conference publications to his credit. He has been awarded educational excellence award by Indus Foundation.

Index

A

B

C

D

E

F

G

H

I

K

L

M

N

O

P

Q

www.ingramcontent.com/pod-product-compliance
Lightning Source LLC
LaVergne TN
LVHW080534300926
849162LV00007B/1603